Teacher's Wraparound Edition

Technology
Shaping Our World

John B. Gradwell
McGill University

Malcolm Welch
Queen's University

Eugene Martin
Southwest Texas State

Publisher
The Goodheart-Willcox Company, Inc.
Tinley Park, Illinois
www.g-w.com

The Goodheart-Willcox Company, Inc. Brand Disclaimer: Brand names, company names, and illustrations for products and services included in this text are provided for educational purposes only and do not represent or imply endorsement or recommendation by the author or the publisher.

The Goodheart-Willcox Company, Inc. Safety Notice: The reader is expressly advised to carefully read, understand, and apply all safety precautions and warnings described in this book or that might also be indicated in undertaking the activities and exercises described herein to minimize risk of personal injury or injury to others. Common sense and good judgment should also be exercised and applied to help avoid all potential hazards. The reader should always refer to the appropriate manufacturer's technical information, directions, and recommendations; then proceed with care to follow specific equipment operating instructions. The reader should understand these notices and cautions are not exhaustive.

The publisher makes no warranty or representation whatsoever, either expressed or implied, including but not limited to equipment, procedures, and applications described or referred to herein, their quality, performance, merchantability, or fitness for a particular purpose. The publisher assumes no responsibility for any changes, errors, or omissions in this book. The publisher specifically disclaims any liability whatsoever, including any direct, indirect, incidental, consequential, special, or exemplary damages resulting, in whole or in part, from the reader's use or reliance upon the information, instructions, procedures, warnings, cautions, applications, or other matter contained in this book. The publisher assumes no responsibility for the activities of the reader.

Teacher's Wraparound Edition Contents

Teacher's Wraparound Edition Contents

Technology: Shaping Our World
Teacher's Wraparound Edition

This comprehensive technology education text helps students learn how technology affects people and the world in which we live. Review the pages that follow to see what is included in this learning package.

A variety of instructional elements appear in the margins of each page of this *Teacher's Wraparound Edition*. These elements are grounded in the nature of contemporary technology education, as well as the approach to the subject advocated in the ITEA *Standards for Technological Literacy*. While the comments that follow are intended for you, the teacher, you may find it helpful to provide students with a copy and discuss them in class.

There are three major focal points for the instructional notes: the made world; technology and society; and careers.

The made world. Notes emphasize to students (and teachers) that they live in a made environment. This made environment is comprised of products and services that are designed before they are made. Therefore, a comprehensive technology education engages students in both designing and making products and services. The textbook identifies various facets of this made world and provides propositional knowledge that students will use when designing and making products. The edge and bottom notes encourage students to think about the designed world in a critical but informed way.

Technology and society. Products and services are intended to do a particular task for a particular group of people at a particular time and place. Products and services do not exist without people. These notes engage students in thoughtful discussion about issues that are important to themselves and society. They are grounded in the fact that making design decisions involves making value judgments. Personal, social, and cultural priorities impact the value judgments students make. Notes engage students in examining the values and value decisions associated with technologies.

Careers. People create and maintain products and services for other people. Therefore, students need to be made aware of the many careers and jobs associated with design, technology, and the made environment.

Features of the *Student Edition*

The student edition of *Technology: Shaping Our World* emphasizes the integration of Technology, Science, and Math in an attractive and interactive format. Features of the *Student Edition* include:

- Correlation with ITEA's national Standards for Technological Literacy.

- Covers the scope of technology, resources and technology, creating technology, technological contexts, and technology and society.

- Each chapter features Modular Connections that tie the chapter content to modular activities.

- Illustrations are referenced within the text to help students associate visual images with written material.

Key Terms. Listing of the new vocabulary covered in the chapter, enhancing student recognition of important concepts.

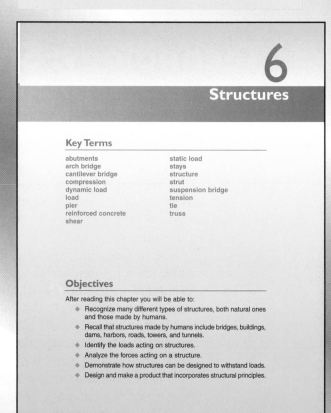

6
Structures

Key Terms

abutments	static load
arch bridge	stays
cantilever bridge	structure
compression	strut
dynamic load	suspension bridge
load	tension
pier	tie
reinforced concrete	truss
shear	

Objectives

After reading this chapter you will be able to:

- Recognize many different types of structures, both natural ones and those made by humans.
- Recall that structures made by humans include bridges, buildings, dams, harbors, roads, towers, and tunnels.
- Identify the loads acting on structures.
- Analyze the forces acting on a structure.
- Demonstrate how structures can be designed to withstand loads.
- Design and make a product that incorporates structural principles.

Objectives. Identifies the topics covered and goals to be achieved by students.

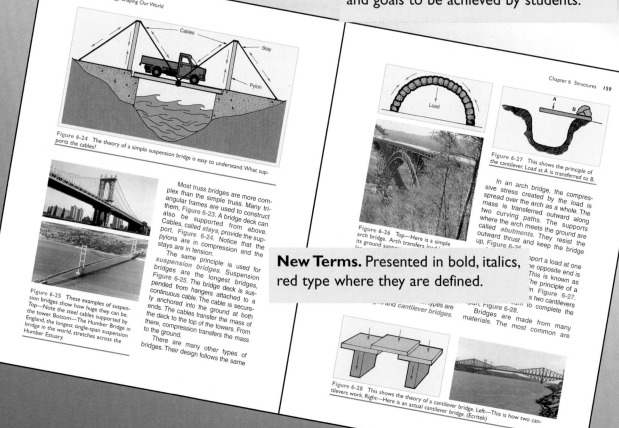

158 Technology: Shaping Our World

Figure 6-24 The theory of a simple suspension bridge is easy to understand. What supports the cables?

Figure 6-25 These examples of suspension bridges show how huge they can be. Top—Note the steel cables supported by the tower. Bottom—The Humber Bridge in England, the longest single-span suspension bridge in the world, stretches across the Humber Estuary.

Most truss bridges are more complex than the simple truss. Many triangular frames are used to construct them, Figure 6-23. A bridge deck can also be supported from above. Cables, called *stays*, provide the support, Figure 6-24. Notice that the pylons are in compression and the stays are in tension.

The same principle is used for *suspension bridges*. Suspension bridges are the longest bridges, Figure 6-25. The bridge deck is suspended from hangers attached to a continuous cable. The cable is securely anchored into the ground at both ends. The cables transfer the mass of the deck to the top of the towers. From there, compression transfers the mass to the ground.

There are many other types of bridges. Their design follows the same

Chapter 6 Structures 159

Figure 6-27 This shows the principle of the cantilever. Load at A is transferred to B.

Figure 6-26 Top—Here is a simple arch bridge. Arch transfers load to its ground support.

In an arch bridge, the compressive stress created by the load is spread over the arch as a whole. The mass is transferred outward along two curving paths. The supports where the arch meets the ground are called *abutments*. They resist the outward thrust and keep the bridge up, Figure 6-26.

...port a load at one ...e opposite end is ...This is known as ...he principle of a ...in Figure 6-27. ...s two cantilevers ...am to complete the ...Bridges are made from many materials. The most common are

...es and *cantilever bridges.*

New Terms. Presented in bold, italics, red type where they are defined.

Figure 6-28 This shows the theory of a cantilever bridge. Left—This is how two cantilevers work. Right—Here is an actual cantilever bridge. (Ecritek)

Chapter 6 Review
Structures

Summary

All structures comprise a number of connected parts. These parts provide support and withstand a load without collapsing. There are two types of loads: static and dynamic. These loads create the forces of compression, tension, and shear. Individual members of a structure must be designed to minimize the effects of these forces. The members are then connected together in such a way as to minimize bending.

Bridges provide an example of how structures are designed to resist forces. A truss bridge uses the rigidity of the triangle to resist the forces of compression and tension. Cables and pylons in a suspension bridge resist these same forces. There are many other types of bridges. As with all structures, they are designed to withstand loads and minimize bending.

Modular Connections

The information in this chapter provides the required foundation for the following types of modular activities:
- Structural Engineering
- Applied Physics
- Bridge Design
- Tower Design
- Truss Design

Summary. Provides a review of major concepts covered in the chapter for the students.

Modular Connections. These sections list several typical modular technology topics related to the chapter.

Test Your Knowledge

Write your answers to these review questions on a separate sheet of paper.

1. Name three natural structures and three structures made by humans.
2. Which of the following is NOT a natural structure?
   A. Spider's web.
   B. Bridge.
   C. Tree.

5. What for... beam loaded from abo...
6. To strengthen a beam loaded on the top su... reinforced at the:
   A. top surface only
   B. bottom surface only
   C. center
   D. top and bottom surfaces
7. Which geometric shape gives the greatest rigidity to a structure?
   A. Square.
   B. Circle.
   C. Rectangle.
   D. Triangle.
8. A beam in compression is called a:
   A. strut
   B. tie
   C. post
   D. stay

Test Your Knowledge. Questions to help students review the topics and the material covered in the chapter.

---

9. A beam in tension is called a:
   A. strut.
   B. tie.
   C. post.
   C. stay.
10. A bridge that uses a series of triangular frames is called a _____ bridge.
11. The world's longest bridges are _____ bridges.
12. Using notes and diagrams, explain how an arch bridge resists loads.
13. Using notes and diagrams, explain the principle of a cantilever bridge.
14. What are the most common materials from which bridges are built?
15. Concrete is weak in tension. How is this problem overcome?

Apply Your Knowledge

1. Look at the natural structures in the illustrations. Next look at the structures made by humans. For each of the structures made by humans, name the natural structure it most closely resembles.
2. Look at the structures in Figures 6-4 and 6-5. Write the location or address of a structure in your town that most closely resembles each one.
3. Name five different structures. For each structure, list the loads to which it is subjected. State whether each load is static or dynamic.
4. Draw a diagram of a plank bridge with a load on it. Label your diagram to show the forces of tension and compression.
5. Using only one sheet of newspaper and 4″ (10 cm) of clear tape, construct the tallest freestanding tower possible.
6. Using drinking straws and pins, construct a bridge to span a gap of 20″ (508 mm) and support the largest mass ... midpoint.

Apply Your Knowledge. Activities at the end of each chapter that encourage students to apply concepts to real-life situations and develop skills related to chapter content.

Features of the *Teacher's Wraparound Edition*

The additional elements contained in the *Teacher's Wraparound Edition* are intended to guide you in reviewing and reinforcing the chapter content. Many of these elements relate the chapter topics to the student's daily life and experiences, in addition to those that reinforce the propositional knowledge acquired from the readings. Practical applications and examples keep the students engaged in both classroom discussion and the learning process.

Links to Other Subjects. Suggestions of activities and assignments that connect the content to other subject areas.

Standards Correlations. Correlates the text material to the current Standards for Technological Literacy.

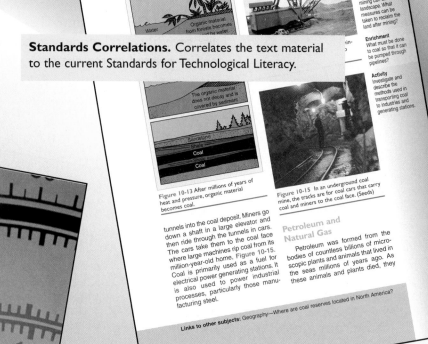

Chapter Discussion Starters. Questions and activities that relate the topics of the chapter and encourage discussion.

Career Connection. Activities involving identification of careers related to the chapter content.

Student Activity Manual Resources. Lists the Design and Make Activities (DMAs) that pertain to the chapter. Also lists the Support Tasks (STs) required to successfully complete a DMA.

Chapter Outline. Provides you with a quick overview of the main points covered in the chapter.

place ready for machining.

♦ Assembling parts together to make a product.

♦ Welding car bodies using arc and spot welding techniques.

♦ Spray painting finishes onto all kinds of manufactured goods.

♦ Inspecting finished products in quality control departments.

Robots used in nuclear power plants must be rugged, capable of doing inspection, detection, and decontamination. They must not break down, because on-the-job humans might be at great risk if they were to try to rescue them. Robots used for bomb disposal must be very strong, capable of

Standards for Technological Literacy

2

Designing and making
How can robots make the lives of handicapped people easier, safer, and more enjoyable?

Discussion
Discuss what robots can do well, such as repetitive tasks, and what they are incapable of doing, such as feeling emotions.

Reflection
Would you like to have a robot that can express when it is anxious or experiencing pleasure?

Enrichment
Have students investigate the life and work of Carl Capek.

Activity
What practical uses can you think of for the robots in Figure 13-12?

Resource
Reproducible Master, Computer Devices

Figure 13-12 A—This mobile robot was equipped with two cameras and other hardware devices to give the user better control. (Electrical and Computer Engineering Department at the University of Maine) B—Some mobile robots move around on legs. (Centre for Intelligent Machines)

Both types have three main parts:

♦ Mechanical parts that move.
♦ Computer electronics hardware that controls movement.
♦ Computer software that provides instruction.

As you read this chapter, think of the thousands of robots that are right now doing the type of work humans find boring or dangerous. They work in:

♦ Material handling, transferring parts between areas of a factory, and stacking and storing.
♦ "Pick and place" situations, picking up parts and putting them in

Figure 13-13 Pivoting on an axis, a stationary robotic welder performs repetitive welding procedures quickly and efficiently. (The Lincoln Electric Company)

Useful Web site: Have students browse www.androidworld.com to review descriptions of robots that look like humans. Students should print information they find interesting and share it with the class.

Designing and Making. Questions that increase students' awareness of the made world.

Reflect. Pose questions of a more personal nature for students to consider, not for general class discussion.

Resource. Suggests related material available in the *Student Activity Manual* or *Teacher's Resources.*

Standards for Technological Literacy

2 18

240 Technology: Shaping Our World

Discussion
Cars are usually owned or leased for a period of years, but are rarely used for more than a few hours a day. For the rest of the time, they sit in a driveway or parking lot. What kind of system could be designed so that cars can be borrowed for a short time to go shopping or to visit a friend?

Enrichment
All new vehicles in Delhi, India that transport paying pass...

Vocabulary
Figure 9-18 shows a car that uses "drive by wire" technology. What does that mean?

Safety
List the parts of a car that make up its safety system.

Resource
Transparency 9-9, Electric Motor

Resource
Activity 9-4, Balloon-Powered Hovercraft

Figure 9-18 Notice the unique design of this car. How does it differ from Figure 9-17! (General Motors)

Figure 9-19 Hybrid electric cars often employ a regenerative braking process, which can provide power to the battery or directly to the electric motors. (DaimlerChrysler)

electric motors usually take over when the car is coasting or going a set speed, Figure 9-19.

Other alternatives are the CNG, which uses compressed natural gas as fuel, and liquefied petroleum gas, which burns a mixture of mostly propane with some butane, Figure 9-20.

Transportation Systems

Modern vehicles, such as cars and buses, are complex machines composed of a number of subsystems. Automobiles have systems that con-

trol the emissions, lights, speed, and many other functions.

Systems in Vehicles

The modern car is composed of the following subsystems: electrical, emission control, computer, fuel, axles and drive train, steering, suspension, brake, heating and air conditioning, and engine. If one of these systems is not working, the car's performance and efficiency will suffer. For example, if a car has a leak in its vacuum lines, it could affect the transmission, electrical, and power brake systems. Power brakes use the vacuum to help you push the pedal. The timing of the ignition system is also affected negatively. This example shows how one small problem can eventually cause a number of other troubles.

Systems in Society

Transportation systems are not those systems confined to the physical

Technology and society: How are the priorities of car companies different from the priorities of consumer protection agencies?

Safety. Emphasize safe working practices, including hazard identification, risk assessment, and risk management.

Enrichment. Challenge students to go beyond the information presented in the chapter, including debates and additional reading.

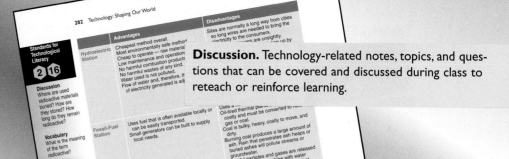

Discussion. Technology-related notes, topics, and questions that can be covered and discussed during class to reteach or reinforce learning.

Technology and Society. Topics focus on the relationship between technology and society, including benefits and disadvantages. Emphasis is placed on the criteria needed when making informed decisions about technology and assessing the impacts of those decisions on people and the environment.

Vocabulary. Emphasize important terms or present technology-related terminology.

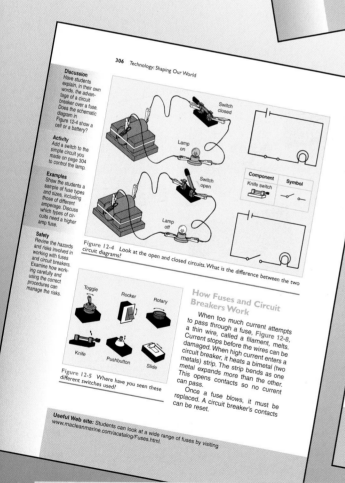

Example. Used to illustrate an important point in chapter material.

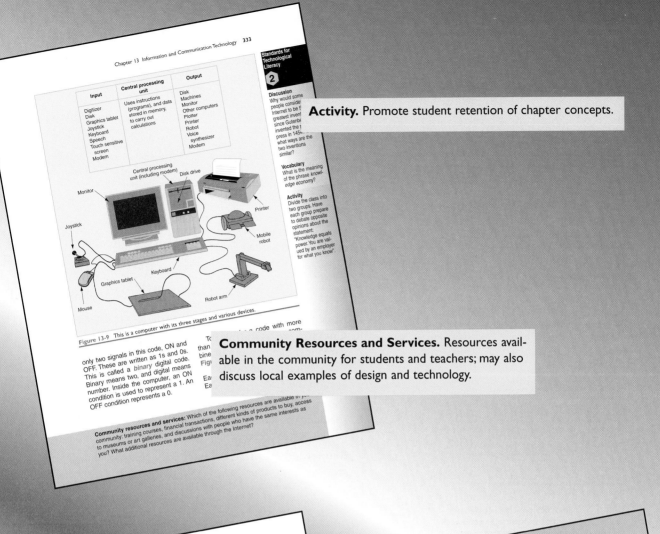

Input	Central processing unit	Output
Digitizer Disk Graphics tablet Joystick Keyboard Speech Touch sensitive screen Modem	Uses instructions (programs), and data stored in memory, to carry out calculations	Disk Machines Monitor Other computers Plotter Printer Robot Voice synthesizer Modem

Monitor
Central processing unit (including modem)
Disk drive
Printer
Joystick
Mobile robot
Graphics tablet
Keyboard
Mouse
Robot arm

Figure 13-9 This is a computer with its three stages and various devices.

only two signals in this code, ON and OFF. These are written as 1s and 0s. This is called a *binary* digital code. Binary means two, and digital means number. Inside the computer, an ON condition is used to represent a 1. An OFF condition represents a 0.

Discussion
Why would some people consider Internet to be... greatest invention since Gutenberg invented the printing press in 1454... what ways are the two inventions similar?

Vocabulary
What is the meaning of the phrase *knowledge economy?*

Activity
Divide the class into two groups. Have each group prepare to debate opposite opinions about the statement: "Knowledge equals power. You are valued by an employer for what you know".

Community resources and services: Which of the following resources are available in your community: training courses, financial transactions, different kinds of products to buy, access to museums or art galleries, and discussions with people who have the same interests as you? What additional resources are available through the Internet?

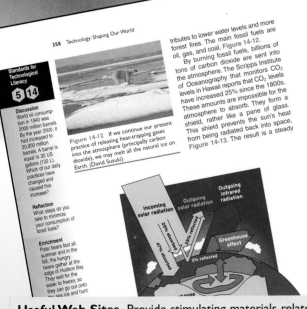

Discussion
World oil consumption in 1940 was 2000 million barrels. By the year 2000, it had increased to 20,000 million barrels. A barrel is equal to 35 US gallons (132 L). Which of our daily practices have changed and caused this increase?

Reflection
What steps do you take to minimize your consumption of fossil fuels?

Enrichment
Polar bears fast all summer and in the fall, the hungry bears gather at the edge of Hudson Bay. They wait for the water to freeze, so they can go out onto the sea ice and hunt

tributes to lower water levels and more forest fires. The main fossil fuels are oil, gas, and coal, Figure 14-12.

By burning fossil fuels, billions of tons of carbon dioxide are sent into the atmosphere. The Scripps Institute of Oceanography that monitors CO_2 levels in Hawaii reports that CO_2 levels have increased 25% since the 1800s. These amounts are impossible for the atmosphere to absorb. They form a shield, rather like a pane of glass. This shield prevents the sun's heat from being radiated back into space, Figure 14-13. The result is a steady

Figure 14-12 If we continue our present practice of releasing heat-trapping gases into the atmosphere (principally carbon dioxide), we may melt all the natural ice on Earth. (David Suzuki)

Incoming solar radiation
Outgoing solar radiation
Outgoing infrared radiation
25% absorbed
25% reflected
47% absorbed
3% reflected
Greenhouse effect

Figure 14-13 The greenhouse effect allows solar heat to pass through but does not allow it to escape from the atmosphere. (David Suzuki)

Useful Web site: Have students check predictions concerning future world oil demand. A helpful place to start searching is www.savvymotoring.com/st_010617.htm.
Technology and society: What could society be doing to ensure adequate supplies of oil and gas for future generations?

Test Your Knowledge

Write your answers to these review questions on a separate sheet of paper.
1. A collection of words or figures that have meaning is defined as ____.
2. The technologies used in storing, processing, and communicating information are, together, referred to as ____.
3. What are the differences between the old and the new ... technologies?

Answers to Test Your Knowledge Questions
1. information
2. information technology
3. The old information technology used equipment that was primarily mechanical...

to an engaged ...
Send a copy of a picture to Australia as rapidly as possible

5. Give three examples of computerized systems that affect our daily lives.
6. What are the three major technologies that make up information technology?
7. The common term for a microprocessor is ____.
8. What is the major task of a telecommunications system?
9. Telecommunications technology uses three different systems to transmit data. Describe the devices used by each system.

Transportation

9

Students will be able to:

- State the advantages and disadvantages of various modes of transportation.
- Explain the principles of various types of engines and motors.
- Recall what industries are affected by transportation systems.
- Identify what processes are involved in order for the entire transportation system to work.
- Explain how transportation technology influences everyday life.
- Discuss the environmental impact of transportation systems.

Instructional Materials

Text: Pages 226–249

Key Terms, Test Your Knowledge, Apply Your Knowledge

Activity Manual:

DMA-15, Moving Toys
ST-64, Making a Cardboard Chassis

Teacher's Resource:

Chapter Test
Supplementary Activities, *Transportation Words*
Supplementary Activities, *Elastic-Powered Vehicles*
Supplementary Activities, *Model Electric Car*
Supplementary Activities, *Balloon-Powered Hovercraft*
Supplementary Activities, *A Monorail Vehicle*
Reproducible Master, *Modes of Transportation Comparison Chart*
Transparency 9-1, *How the Four-Stroke Gasoline Engine Works* (Binder/CD only)
Transparency 9-2, *Types of Engines* (Binder/CD only)
Transparency 9-3, *Four-Stroke Gasoline Engine* (Binder/CD only)
Transparency 9-4, *Four-Stroke Diesel Engine* (Binder/CD only)
Transparency 9-5, *Steam Turbine* (Binder/CD only)
Transparency 9-6, *Turbofan Engine* (Binder/CD only)
Transparency 9-7, *Turboshaft Engine* (Binder/CD only)
Transparency 9-8, *Rocket Engine* (Binder/CD only)
Transparency 9-9, *Electric Motor* (Binder/CD only)

Instructional Strategies and Student Learning Experiences

1. Have students list as many modes of transportation as possible.
2. Have students clip pictures of various modes of transportation and make a collage.

Chapter Objectives identify the topics covered and goals to be achieved by students.

Instructional Materials are listed by chapter to assist in planning daily lessons.

Instructional Strategies and Student Learning Experiences suggestions are listed to assist you in teaching each chapter.

Standards for Technological Literacy Correlation Chart

Standards for Technological Literacy Correlation Chart	Chapter 1: pages 12–18 Chapter 2: page 22 Chapter 15: pages 386, 394
Standard 1. Students will develop an understanding of the characteristics and scope of technology.	Chapter 1: page 12
Benchmark Topics:	
New products and systems can be developed to solve problems or to help do things that could not be done without the help of technology.	Chapter 1: page 12
The development of technology is a human activity and is the result of individual or collective needs and the ability to be creative.	Chapter 2: page 22
Technology is closely linked to creativity, which has resulted in innovation.	Chapter 15: pages 386, 394
Corporations can often create demand for a product by bringing it onto the market and advertising it.	Chapter 2: pages 28–33 Chapter 3: pages 56–57 Chapter 7: pages 181–185 Chapter 9: pages 230–244 Chapter 10: pages 258–272 Chapter 11: pages 279–284 Chapter 12: pages 332–341 Chapter 15: pages 387–388
Standard 2. Students will develop an understanding of the core concepts of technology.	
Benchmark Topics:	Chapter 3: pages 56–57 Chapter 7: pages 181–182
Technological systems include input, processes, output, and, at times, feedback.	Chapter 9: pages 240–242
Systems thinking involves considering how every part relates to others.	Chapter 7: page 181
An open-loop system has no feedback path and requires human intervention, while a closed-loop system uses feedback.	Chapter 7: pages 181–185 Chapter 9: pages 239–242
Technological systems can be connected to one another.	Chapter 9: pages 240–242
Malfunctions of any part of a system may affect the function and quality of the system.	Chapter 2: pages 28–33
Requirements are the parameters placed on the development of a product or system.	Chapter 2: pages 28–33 Chapter 15: pages 387–388
...de-off is a decision process recognizing the need for careful compromises ...	Chapter 7: pages 181–185 Chapter 9: pages 230–244 Chapter 10: pages 258–272 Chapter 11: pages 279–284

Standards for Technological Literacy Correlation Chart allows you to design your class around the standards identified by the International Technology Education Association by indicating where relevant topics are covered in the text.

Basic Skills Chart identifies activities in the text and supplements that encourage the development of reading, writing, verbal, math, science, and analytical skills.

Scope-and-Sequence Chart provides assistance in outlining your courses.

Answers to Activity Questions

1. 3 seconds
2. A. Very brightly.
3.

Type of Wire	Number of Strands of Wire	Brightness of Glowing Wire	Duration of Glowing (Seconds)
Iron	1	very bright	
Iron	2		3
Iron	3	bright	3
Copper	4	dim	6
Copper	1	did not glow	20

Activity 10-7 Shooting the Ball

Time Required: 4 to 6 class periods
Materials:

- Student Activity Sheet
- Table tennis (Ping Pong®) ball
- Other materials will vary according to individual solutions

Note:

It is important to insist that each student complete each of the steps in the design process as described in their Activity Manual.

Answers to Chapter Test

1. B. A compressed spring.
2. C. Nuclear.
3. A. Heat.
4. C. Oil.
5. D. All of the above.
6. False.
7. True.
8. True.
9. True.
10. False.
11. False.
12. False.
13. True.
14. True.
15.

Answer Keys are provided for *Test Your Knowledge* questions in the text and Supplementary Activities and Chapter Tests in the *Teacher's Resources*.

The Design Process

Manufacture

Test & Evaluate

Problem

Model/ Prototype

Design Brief

Solution

Investigation

Alternative Solutions

Reproducible Masters are available for easy conversion to overhead transparencies or classroom handouts.

Features of the *Teacher's Resources*

The *Teacher's Manual, Teacher's Resource Binder,* and *Teacher's Resource CD* provide all the resources needed to instruct your technology education class. All materials are grouped together by chapter for easy reference.

The *Teacher's Resource Binder* includes all the resources found in the *Teacher's Manual,* plus Color Transparencies. The *Teacher's Resource CD* contains all the resources available in the *Teacher's Resource Binder,* plus the **Exam***View*® *Assessment Suite* and Internet Resources.

Activity 6-3
Designing and Building a Bridge

Name_____ Date _____

Objective: To build a model bridge and test it for strength.

The triangle is one of the most rigid shapes. Using this knowledge, design a bridge to span a gap of 12″ (300 mm). The bridge is to have a road or deck made from thin card. No single piece of wood is to exceed 5 1/2″ (135 mm) in length and only the amount of wood supplied may be used. The total mass of the bridge must not exceed 1 oz. (30 g). The bridge must be designed to withstand the greatest possible load. It will be tested to determine its rigidity.

Materials:
Strip balsa or pine 1/2″ × 1/4″ (12 mm × 6 mm)
Sheet balsa or cardboard (for gusset plates)
White glue
Thin card 12″ × 3″ (300 mm × 75 mm)

Procedure:

1. Draw your ideas on graph paper before commencing to build your model.
2. Cut the wood strips to length.
3. Join corners using white glue (use glue sparingly).
4. Gusset plates may be used for reinforcement where parts of the structure meet (see diagram below).

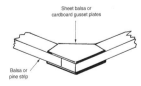

Sheet balsa or cardboard gusset plates

Balsa or pine strip

Supplementary Activities allow for greater flexibility in designing your classroom program.

Chapter 11 Test
Electricity and Magnetism

Name _____ Period _____ Score _____

Date _____

☐ Matching: Match the following terms with their definitions.

_____ 1. A device that produces an electric current as it turns.
_____ 2. Increases or decreases voltage and amperage.
_____ 3. A flow of electrons.
_____ 4. The ability of a material to attract pieces of iron or steel.
_____ 5. Electron flow that reverses direction on a regular basis.
_____ 6. The path along which electricity flows.
_____ 7. Current that does not change direction.

A. Transformer.
B. Alternating current.
C. Direct current.
D. Generator.
E. Electric current.
F. Magnetism.
G. Circuit.

...is true, circle *T*. If a statement is false, circle *F*.

...m to drive turbines.

Chapter Tests are provided to assist you in evaluating students' comprehension of content.

T   F   11. Unlike...
T   F   12. The stronger the magnetic...
T   F   13. Both a generator and an electric motor apply the...
T   F   14. A primary cell can be recharged.

☐ Essay Question: Respond to this question in the space provided below, on the back of this page, or on a separate sheet of paper.

15. List the three types of electricity generating stations and an advantage and disadvantage of each. If a generating station had to be built near your home, which would you choose? Explain why.

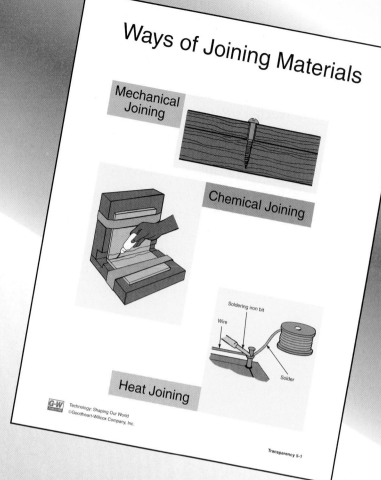

Ways of Joining Materials

Mechanical Joining

Chemical Joining

Soldering iron bit

Wire

Solder

Heat Joining

Technology: Shaping Our World
©Goodheart-Willcox Company, Inc.

Transparency 5-1

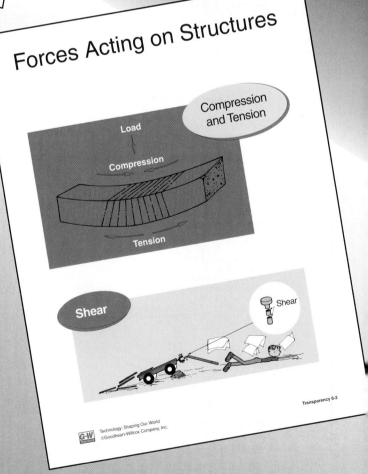

Forces Acting on Structures

Compression and Tension

Load

Compression

Tension

Shear

Shear

Technology: Shaping Our World
©Goodheart-Willcox Company, Inc.

Transparency 6-2

Color Transparencies coinciding with the textbook reinforce important technological concepts. (Available in the *Teacher's Resource Binder* and on the *Teacher's Resource CD* only.)

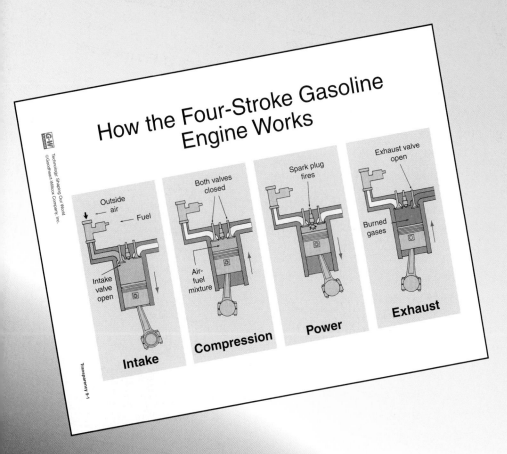

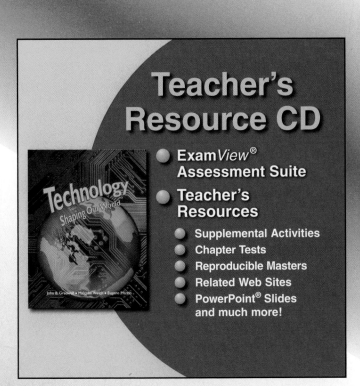

Exam*View*® **Assessment Suite** computerized test preparation allows quick and easy creation of testing materials. It contains approximately 500 questions correlated to the chapters of the textbook. (Available on the *Teacher's Resource CD* only.)

Internet Resources list Web sites related to the content covered in the textbook. (Available on the *Teacher's Resource CD* only.)

Features of the *Student Activity Manual*

The *Student Activity Manual* goes hand-in-hand with the *Technology: Shaping Our World* textbook. It includes a set of Design and Make Activities and a range of Support Tasks to provide the knowledge, skills, and understanding required for students to become technologically literate.

Support Tasks (STs) are short, highly structured, and focused activities through which the student learns the designing skills, making skills, technical knowledge, and evaluation skills needed to successfully complete a Design and Make Activity.

ST-32
Making a 3-D Framework

Name: _____ Date: _____

What You Will Learn
◆ To mark, cut, and join wood strip to make a 3-D framework.

Materials
◆ ST-32 worksheet
◆ 3/8" × 3/8" (10 mm × 10 mm) wood strip
◆ Thin cardboard
◆ PVA glue

Introduction
To make a 3D framework, you must be able to mark, cut, and join wood strip accurately. You can learn to do this by making a cube.

Procedure
1. The first step is to make a sheet of double gusset plates. To make a set of gusset plates, draw a rectangle 7 1/2" × 3 3/4" (180 mm × 90 mm) on a sheet of thin cardboard. Draw a diagonal on each square as shown below. Carefully cut off strips of triangles from the card. Cut out the individual double gussets. When folded along the center line, these gussets allow you to build up from the corners of a flat frame to make a 3-D framework.

Double Triangles

Fold Line

DMA-17
Robots

Related Chapter: Chapter 11, Electricity and Magnetism

The Context
Robots pervade all aspects of our life. Manufacturers use "pick-and-place" robots to perform complex, precise, repetitive tasks. Surgeons use robots to perform delicate surgery. Some robots are designed to work in radioactive environments, in ocean waters too deep for humans, or in deep space. Security robots can detect an intruder and set off an alarm. Toy robots amuse and intrigue people of all ages.

The Design Brief
Design and make a prototype pick-and-place robot that can move a small object. A human operator or computer can control the robot.

The Design Decisions
You can decide on the following:

The Robot Load
◆ What object is to be moved?
◆ How far the object is to be moved?
◆ In what direction the object is to be moved?

The Power Source
◆ Powered by an electric motor.
◆ Controlled by a computer.
◆ Controlled by human using a controller.

The Structure
◆ What type of structure is needed to support the manipulator?

Design and Make Activities (DMAs) require the students to design and make a product, system, or environment that is useful to them or other people.

Introduction

The *Technology: Shaping Our World* teaching package is written as an introduction to the study of technology. The authors' primary objective was to write material that students not only *can* read, but also *want* to read, and that encourages them to engage in the process of technology. It is our belief that it is just as important for all students to acquire a basic level of technological literacy as it is for them to have a basic level of language and mathematical literacy.

The *Technology: Shaping Our World* teaching package includes four products: the text, the *Teacher's Wraparound Edition,* the *Student Activity Manual*, and the *Teacher's Resources.* Using these products can help you develop an effective technology program tailored to your students' unique needs.

Throughout the *Technology: Shaping Our World* materials, the emphasis is on ideas rather than on rote learning. A variety of activities are presented to enable students to learn and apply basic concepts and principles of technology. These activities are presented in a form that encourages creativity and originality. Above all, we have tried to demonstrate to students that technology is a process to satisfy their needs, while also considering the effect that changes may bring to their environment. It is a creative process that draws on a variety of knowledge and resources.

In societies such as ours, technology, from its simplest to its most sophisticated, plays an integral role in everyone's life. Students must live with the impact of technology on the environment and on society. Because today's students are tomorrow's citizens, they should not only have an understanding of various technologies, but should be capable of making informed decisions. Such decisions may include the advisability of building a highway or a nuclear reactor close to their home, the ethics of cloning, or the benefits and costs of more industrial development resulting in increased air pollution and acid rain.

It is important that technology be understood and monitored by all members of society; therefore, a principal role of technology education is to clear away the mystery surrounding technology and to equip students with sufficient knowledge to judge technological developments wisely.

Educators have begun to realize that courses in technology should be made available to all students enrolled in secondary schools. Many new programs have already been written or are currently being developed. While there are differences between these programs, a number of fundamental points of agreement exist.

The first is that all programs have, as a starting point, a definition of technology. Definitions range from technology as technique, as a series of systems, or as a body of knowledge. For the most part, recognition is given to the idea that technology may be defined as a process. This process uses resources, such as materials, energy, and knowledge, to achieve human purpose. This definition implies that technology consists of practical tasks and knowledge of particular types as well as people, systems, and social organizations.

The second point is there is a connection between technology and science, but the two subjects are fundamentally different. It is easy to fall into the trap of referring to technology as applied science. Science is concerned with how things are. Emphasis is placed on the discovery and analysis of the laws and principles of nature. Technology is concerned with how things can be. Products and systems are designed and built to solve problems and meet human needs.

The third point is to remind us that one fundamental strength of technology education is its creativity component. Most people are thrilled when they successfully design and make something. Even simple models and prototypes are exciting to build, especially when they work.

However, a strong caution is in order. The type of practical activity in technology education should differ considerably from that used in past Industrial Arts programs. The focus must not be primarily on increasing the level of skillful performance with tools. Rather, it should be to increase the creativity and problem-solving abilities of students through structured situations. It is the difference between making a footstool from plans versus designing, making, and testing a flatpack coffee table that appeals to a young person setting up his or her first home.

The improvement of skills remains as one important part of technology education. However, the skill set to be learned is broadened to include those of problem identification, researching, designing, problem solving, and making. In addition, the breadth of knowledge sought by students is also increased to include engineering, scientific, and technological knowledge. Finally, consideration is given to the formation of values though consideration of moral and ethical decisions.

The fourth point builds on the first point. It is the recognition that technology is fundamentally a process. This process involves the following steps:

- Identifying and stating a problem.
- Analyzing the problem (assessing the situation).
- Writing a design brief.
- Proposing and communicating ideas (thinking creatively).
- Making a decision (selecting/rejecting ideas).
- Developing the idea (making a model/mock-up).
- Finalizing the solution (realization in a production form).
- Evaluating the solution (comparison with originally stated problem).

A major strength of defining technology as a process is that the process can be applied to any and all specific technologies. In other words, the design process may be used to solve problems in robotics, communications, transportation, and so on. In addition, the process is unchanging over time. Stone Age humans used basically the same process to problem-solve that we use today.

The Materials for Learning and Teaching

Students using *Technology: Shaping Our World* learn that they can participate as agents of change in the made world. They not only learn that all the products and services around them have been designed, but also that they can make a contribution that improves those products and services. They learn that the quality of the designed world profoundly affects quality of life.

Engaging students in Design and Make Activities provides skills required in a knowledge-based economy. These skills include thinking creatively to resolve ill-defined (open-ended) problems, communicating design ideas, working collaboratively, managing information, self-organization, managing risk, and the ability to be reflective.

◆ **Textbook.** Materials to support designing and making, including design process skills, making skills, and technical information.

◆ **Student Activity Manual.** A set of Design and Make Activities and a range of Support Tasks to provide the knowledge, skills, and understanding students require to become technologically literate.

◆ **Teacher's Manual.** Guidance on how to use the *Student Activity Manual* and textbook to design a scheme of work appropriate for your students, as they learn to design and make products and services. The *Teacher's Manual* also contains Supplementary Activities, answers to textbook questions, reproducible masters, and chapter quizzes.

◆ **Teacher's Resource Binder.** Provides the content of the *Teacher's Manual* and 48 transparencies in a convenient three-ring binder.

◆ **Teacher's Resource CD.** Provides the content of the *Teacher's Resource Binder* in electronic format and the **Exam***View* software program.

Teaching Students to Design and Make

The approach to learning and teaching in technology education described in the *Technology: Shaping Our World* material has, at its center, a **Design and Make Activity** (DMA). A Design and Make Activity is a significant activity in which students must use the knowledge, understanding, and skills they have been taught in an integrated and holistic way. It forms a focal point in a teaching sequence and enables students to reveal what they have learned through what they can do. A DMA requires students to intervene in and make improvements to the made world by designing something they themselves can make, then making the product they have designed. Both the product and the processes by which it is conceived, developed, and realized are significant in this activity.

Designing and making products and services presents students with a wide variety of learning opportunities. When designing, they need to use a variety of strategies, including problem-solving, creative thinking, sketching and drawing, visual imagery, critical thinking, and evaluating. When making a product, they need to measure, mark, cut, shape, and join materials.

For students to be successful in a DMA, they need particular and appropriate knowledge, skill, and understanding. These are taught through a series of **Support Tasks** (ST): short, highly structured, and focused activities. The effectiveness of this teaching and learning is evidenced through the quality of response to the Design and Make Activity.

For example, in DMA-8 Desk Lamp, students are given a design brief in which they are to design and make a desk lamp with a light source that is focused, fully adjustable for height and angle, and has an on/off switch. To be successful with this DMA, students must be able to generate, develop, and communicate ideas; mark, cut, and shape sheet materials to a high degree of accuracy; and use electrical components to build a circuit. Once a lamp has been designed and made, the student must evaluate the product and their own learning. Hence, students complete the following Support Tasks *before* tackling the design brief:

ST-3   Using a Collage to Generate Ideas
ST-5   Writing a Design Specification
ST-8   Exploring Existing Products
ST-74 What Is a Circuit?
ST-76 Building a Series Circuit
ST-77 Building a Parallel Circuit
EV-1   Evaluating the Final Product
EV-2   Unit Review

Making design decisions involves making value judgments. Making these judgments is an ongoing process that permeates each Design and Make Activity. Teachers should engage students in thoughtful discussion that helps them make decisions that are important to them as individuals and eventually to society. By balancing technical, economic, aesthetic, environmental, and moral criteria, students are able to make decisions that influence design, production, and marketing. Personal, social, and cultural priorities impact the value judgments students are making.

Types of Support Tasks

Support Tasks are short, highly structured, and focused activities through which students learn the knowledge, skills, and understanding needed to successfully complete a Design and Make Activity. Support Tasks are active. They require students to engage with design skills, technical understanding, and making skills. While a single Support Task usually addresses a quite narrow topic, a sequence of Support Tasks leads to the acquisition of a wide repertoire of design, technical, constructional, and aesthetic knowledge and skills.

Designing

These Support Tasks teach students the skills they need to generate and develop design ideas. They illustrate the importance of identifying and carefully specifying what is to be designed, to take into account the results of their research and human factors when designing. Examples include:

ST-3   Using a Collage to Generate Ideas
ST-4   Identifying User Needs and Interests
ST-13 Writing a Design Brief

Communicating Ideas

These Support Tasks teach students the skills they need in order to develop and communicate design ideas to themselves, their teacher, and other students. Students work with a range of media, including paper, card, and foam board. Examples include:

ST-19 Sketching on Isometric Grid Paper
ST-22 Showing Surface Texture
ST-23 3-D Modeling with Cardboard

Materials

These Support Tasks teach students the skills they need to mark, cut, shape, and join a variety of materials. Students learn to work with increasing accuracy to produce high-quality products. Examples include:

ST-33 Laminating Thin Plywood
ST-34 Bending Sheet Metal
ST-35 Bending Acrylic

Structures

These Support Tasks teach students about structural principles important in the design of buildings and other structures. Examples include:
ST-52 How Trusses Work
ST-55 Natural and Human-Made Structures
ST-58 Strengthening Beams

Machines and Transportation

These Support Tasks teach students the principles of simple machines and mechanisms, pneumatics and hydraulics, and ways to control movement. Examples include:
ST-59 Using Gears
ST-67 Investigating Pneumatics and Hydraulics
ST-71 How Arm Robots Work

Electricity and Electronics

These Support Tasks teach students about electric circuits and components, so they can use them in products they design and make. Examples include:
ST-76 Building a Series Circuit
ST-78 Investigating Resistance
ST-82 Drawing Circuit Diagrams

Information and Communications Technology

These Support Tasks teach students how to use information and communication technology to conduct research using the World Wide Web. They also learn to produce high-quality graphics for use with products and services they have designed and made. This group includes:
ST-86 Using Desktop Publishing to Produce Instruction Sheets
ST-87 Using the World Wide Web to Gather Information
ST-88 Using Desktop Publishing to Produce Graphics

Biotechnology

These Support Tasks provide students the knowledge they need to design healthful products. This group includes:
ST-89 Using MyPyramid
ST-90 Analyzing Food Labels

World of Work

These Support Tasks introduce students to career opportunities and how to prepare themselves for job applications. Examples include:
ST-91 Choosing a Career
ST-92 Investigating Résumés
ST-94 Writing a Cover Letter

Learning from the Past, Predicting the Future

These two Support Tasks increase students' awareness of the impacts of technology on their lives:
ST-95 Technology in the News
ST-96 Noise Pollution

Evaluation

The first of these two Support Tasks (Evaluating the Final Product) is used by students to evaluate the product or service they have designed and how well it meets their design specification. The second Support Task (Unit Review) is used when students evaluate their own learning and identify ways in which they can improve their capability for designing and making products and services. Both Support Tasks are used as the culminating activities for each Design and Make Activity.

EV-1   Evaluating the Final Product
EV-2   Unit Review

Using the Support Tasks

The *Technology: Shaping Our World Student Activity Manual* helps students use concepts presented in the text to design and make products and services. It also helps students apply what they have learned to real-life situations.

The Support Tasks in the *Student Activity Manual* are divided into sections that correspond to the chapters in the text. By reading the appropriate pages in the textbook, your students acquire much of the background information they need to complete the activities. The pages of the *Student Activity Manual* are perforated, so students can easily turn in completed activities to the teacher for evaluation. They are also drilled so students can keep completed activities in a file for easy reference.

Each Support Task is presented to the student as a worksheet and has the same structure, as the following shows:

Support Tasks   Designing

ST-12
Decorating Cardboard

Name:_____   Date:_____

What You Will Learn
To decorate cardboard using a variety of materials.

Materials
ST-12 worksheet
Pencil
Markers
Thin cardboard
PVA glue
Fabric of assorted colors and textures
Colored paper
Sand
Small glass beads

Introduction
Cardboard is used for a variety of purposes by technologists:
Making simple structures to test rigidity
Experimenting with color and texture
Making small models to explore shape, size, and form

Procedure
In this activity, you will make a refrigerator magnet and use a variety of techniques to decorate a cardboard shape.

1. Draw a natural or geometric shape on a sheet of thin cardboard. This will be used as a template. You may want to complete Support Task ST-9 (Exploring Natural Shapes) and Support Task ST-10 (Exploring Geometric Shapes) before completing this Support Task.
2. Using the template, mark and cut four pieces of thick cardboard. Each will be decorated using a different technique. Following are the four techniques:
   Use felt tip markers to decorate the first shape.
   Use PVA glue to stick pieces of textured fabric to the second shape.
   Use a glue stick to fasten colored paper to the third shape.
   Spread a thin layer of PVA glue onto the fourth shape and sprinkle sand, small glass beads, glitter, or some other fine textured material onto the shape.
3. Allow the glue to dry. To complete the magnet, use epoxy resin to fasten a small magnet to the back of one of your card shapes.

Annotations (margin labels):
- Support Task category
- Reference number
- Title
- Statement describing what the student will learn by completing the Support Task
- List of consumable materials and tools required
- Instructions to guide the student

The following table correlates each support task with the DMAs it supports and the related textbook chapters.

Overview of Support Tasks

ST-#	ST Title	Related DMAs	Relevant Chapter #
ST-1	You Are a Designer	DMA-1 Tallest Tower	2
ST-2	Brainstorming	DMA-1 Tallest Tower DMA-23 Healthy Eating	2, 3
ST-3	Using a Collage to Generate Ideas	DMA-8 Desk Lamp DMA-10 On the Move DMA-25 Fundraising	2, 3
ST-4	Identifying User Needs and Interests	DMA-2 New Uses for Everyday Products DMA-4 My Personal Prism DMA-6 Games Galore DMA-12 Temporary Shelter DMA-14 Mechanical Toys DMA-19 Electronic Jewelry DMA-22 Party Plans DMA-24 Prepacked Lunches DMA-25 Fundraising DMA-27 New Uses for Old Things	2, 3
ST-5	Writing a Design Specification	DMA-4 My Personal Prism DMA-8 Desk Lamp DMA-11 Pet House DMA-14 Mechanical Toys DMA-16 Pneumatic Ergonome DMA-21 Pop-Up Cards DMA-22 Party Plans DMA-23 Healthy Eating DMA-24 Prepacked Lunches DMA-27 New Uses for Old Things	2,3
ST-6	Making and Using Nets	DMA-5 Storage DMA-10 On the Move DMA-15 Moving Toys DMA-22 Party Plans DMA-23 Healthy Eating DMA-24 Prepacked Lunches	2
ST-7	Using a Gantt Chart to Plan	DMA-5 Storage DMA-25 Fundraising	2, 3

(continued)

Overview of Support Tasks *(continued)*

ST-#	ST Title	Related DMAs	Relevant Chapter #
ST-8	Exploring Existing Products	DMA-6 Games Galore DMA-8 Desk Lamp DMA-15 Moving Toys DMA-19 Electronic Jewelry DMA-20 Game of Chance DMA-21 Pop-Up Cards	2, 3
ST-9	Exploring Natural Shapes	DMA-7 Hanging Fun	2
ST-10	Exploring Geometric Shapes	DMA-7 Hanging Fun	2
ST-11	Investigating Pets	DMA-11 Pet House	2, 3
ST-12	Decorating Cardboard	DMA-13 Puppets DMA-15 Moving Toys DMA-21 Pop-Up Cards DMA-22 Party Plans	2
ST-13	Writing a Design Brief	DMA-23 Healthy Eating DMA-24 Prepacked Lunches DMA-27 New Uses for Old Things	2, 3
ST-14	Designing for People— Anthropometric Data	DMA-16 Pneumatic Ergonome	2
ST-15	Making an Ergonome	DMA-16 Pneumatic Ergonome	2
ST-16	Modeling Ideas with Cardboard, Wood Strip, and Dowel	DMA-16 Pneumatic Ergonome	2, 5
ST-17	Designing a Questionnaire	DMA-21 Pop-Up Cards	2, 3
ST-18	Freehand Sketching Techniques	DMA-2 New Uses for Everyday Products DMA-3 My "Dream" Furniture	3
ST-19	Sketching on Isometric Grid Paper	DMA-2 New Uses for Everyday Products DMA-3 My "Dream" Furniture DMA-10 On the Move DMA-14 Mechanical Toys	3
ST-20	Isometric Sketching on Plain Paper	DMA-3 My "Dream" Furniture DMA-10 On the Move DMA-14 Mechanical Toys	3
ST-21	Sketching in Proportion	DMA-3 My "Dream" Furniture DMA-14 Mechanical Toys	3
ST-22	Showing Surface Texture	DMA-3 My "Dream" Furniture	3
ST-23	3-D Modeling with Cardboard	DMA-10 On the Move	3, 5
ST-24	Drawing to Scale	DMA-11 Pet House	3

(continued)

Overview of Support Tasks *(continued)*

ST-#	ST Title	Related DMAs	Relevant Chapter #
ST-25	Measuring Scale Drawings—Conventional	DMA-11 Pet House	3
ST-26	Measuring Scale Drawings—Metric	DMA-11 Pet House	3
ST-27	Reading Scale Drawings—Conventional	DMA-11 Pet House	3
ST-28	Reading Scale Drawings—Metric	DMA-11 Pet House	3
ST-29	Modeling with Foam Board	DMA-11 Pet House DMA-12 Temporary Shelter	3, 4, 5
ST-30	Scaling Up—Drawing Things Bigger	DMA-13 Puppets DMA-16 Pneumatic Ergonome	3
ST-31	Using Wood Strip to Make a Flat Frame	DMA-4 My Personal Prism DMA-6 Games Galore DMA-15 Moving Toys DMA-17 Robots DMA-20 Games of Chance	4, 5
ST-32	Making a 3-D Framework	DMA-4 My Personal Prism DMA-6 Games Galore DMA-9 Showcase DMA-12 Temporary Shelter DMA-14 Mechanical Toys DMA-17 Robots	4, 5
ST-33	Laminating Thin Plywood	DMA-5 Storage	4, 5
ST-34	Bending Sheet Metal	DMA-5 Storage	4, 5
ST-35	Bending Acrylic	DMA-5 Storage	4, 5
ST-36	Cutting and Shaping Thin Plywood	DMA-7 Hanging Fun DMA-19 Electronic Jewelry DMA-23 Healthy Eating DMA-25 Fundraising	4, 5
ST-37	Cutting and Shaping Thin Aluminum	DMA-7 Hanging Fun DMA-11 Pet House DMA-25 Fundraising	4, 5
ST-38	Cutting and Shaping Acrylic	DMA-7 Hanging Fun DMA-11 Pet House DMA-19 Electronic Jewelry DMA-23 Healthy Eating DMA-25 Fundraising	4, 5

(continued)

Overview of Support Tasks *(continued)*

ST-#	ST Title	Related DMAs	Relevant Chapter #
ST-39	Exploring Manufactured Boards	DMA-10 On the Move	4
ST-40	Exploring Knockdown Fittings	DMA-10 On the Move	5
ST-41	Making a Wooden Candleholder	DMA-11 Pet House	4, 5
ST-42	Tools and Materials	DMA-25 Fundraising	5
ST-43	Paper Engineering—Making a Box Fold	DMA-21 Pop-Up Cards	5
ST-44	Paper Engineering—Making a Mouth Fold	DMA-21 Pop-Up Cards	5
ST-45	Paper Engineering—Making a Slider	DMA-21 Pop-Up Cards	5
ST-46	Paper Engineering—Making a Lift-Up Flap	DMA-21 Pop-Up Cards	5
ST-47	Paper Engineering—Making a Rotator	DMA-21 Pop-Up Cards	5
ST-48	Recycling Materials to Make New Products	DMA-27 New Uses for Old Things	4, 5, 14
ST-49	Materials—Cradle to Grave	DMA-27 New Uses for Old Things	4, 5
ST-50	New Uses for Old Things	DMA-27 New Uses for Old Things	4, 14
ST-51	Safety—Managing Risk	DMA-1 Tallest Tower DMA-3 My "Dream" Furniture DMA-4 My Personal Prism DMA-5 Storage DMA-6 Games Galore DMA-7 Hanging Fun DMA-8 Desk Lamp DMA-9 Showcase DMA-10 On the Move DMA-11 Pet House DMA-12 Temporary Shelter DMA-13 Puppets DMA-14 Mechanical Toys DMA-15 Moving Toys DMA-16 Pneumatic Ergonome DMA-17 Robots DMA-18 Keep Out! DMA-19 Electronic Jewelry	5, 15, 17

(continued)

Overview of Support Tasks *(continued)*

ST-#	ST Title	Related DMAs	Relevant Chapter #
		DMA-20 Games of Chance DMA-21 Pop-Up Cards DMA-22 Party Plans DMA-23 Healthy Eating DMA-24 Prepacked Lunches DMA-25 Fundraising DMA-27 New Uses for Old Things	
ST-52	How Trusses Work	DMA-9 Showcase DMA-11 Pet House	6
ST-53	Making Structures Rigid	DMA-9 Showcase DMA-11 Pet House DMA-12 Temporary Shelter	6
ST-54	Struts and Ties	DMA-9 Showcase DMA-12 Temporary Shelter	6, 7
ST-55	Natural and Human-Made Structures	DMA-12 Temporary Shelter	6, 7
ST-56	Forces Acting on Structures	DMA-9 Showcase DMA-12 Temporary Shelter	6, 7
ST-57	Using Beams in Structures	DMA-12 Temporary Shelter	6
ST-58	Strengthening Beams	DMA-12 Temporary Shelter	8
ST-59	Using Gears	DMA-9 Showcase DMA-14 Mechanical Toys	8
ST-60	Exploring Animal Movements	DMA-13 Puppets	8
ST-61	Levers All Around Us	DMA-13 Puppets DMA-17 Robots	8
ST-62	Exploring Linkages	DMA-13 Puppets	8, 9
ST-63	Exploring Wheels and Axles	DMA-15 Moving Toys	8, 9
ST-64	Making a Cardboard Chassis	DMA-15 Moving Toys	8
ST-65	Moments and Levers—Conventional Measurements	DMA-14 Mechanical Toys	8
ST-66	Moments and Levers—Metric Measurements	DMA-14 Mechanical Toys	8
ST-67	Investigating Pneumatics and Hydraulics	DMA-16 Pneumatic Ergonome DMA-17 Robots	7, 9

(continued)

Overview of Support Tasks *(continued)*

ST-#	ST Title	Related DMAs	Relevant Chapter #
ST-68	Understanding Systems	DMA-16 Pneumatic Ergonome DMA-17 Robots	7, 9
ST-69	Systems and Subsystems	DMA-17 Robots	7, 8, 9
ST-70	Pneumatics—One-Way Valves	DMA-17 Robots	8, 10
ST-71	How Arm Robots Work	DMA-17 Robots	8
ST-72	Using Pulleys	DMA-17 Robots	8
ST-73	Using Cams	DMA-14 Mechanical Toys	8
ST-74	What Is a Circuit?	DMA-8 Desk Lamp DMA-9 Showcase DMA-15 Moving Toys DMA-19 Electronic Jewelry	11, 12
ST-75	Electric Quiz Board	DMA-9 Showcase	12
ST-76	Building a Series Circuit	DMA-8 Desk Lamp DMA-9 Showcase DMA-15 Moving Toys DMA-19 Electronic Jewelry	12
ST-77	Building a Parallel Circuit	DMA-8 Desk Lamp DMA-9 Showcase DMA-19 Electronic Jewelry	12
ST-78	Investigating Resistance	DMA-9 Showcase	12
ST-79	Reading Resistors	DMA-9 Showcase	12
ST-80	Making a Reversing Switch	DMA-15 Moving Toys	12
ST-81	Making a Membrane Panel Switch	DMA-18 Keep Out! DMA-19 Electronic Jewelry	12
ST-82	Drawing Circuit Diagrams	DMA-18 Keep Out!	12
ST-83	Using Light Emitting Diodes	DMA-18 Keep Out! DMA-20 Game of Chance	12
ST-84	Using Diodes	DMA-20 Game of Chance	12
ST-85	Using Push-to-Make and Push-to-Break Switches	DMA-20 Game of Chance	12
ST-86	Using Desktop Publishing to Produce Instruction Sheets	DMA-20 Game of Chance DMA-24 Prepacked Lunches	13
ST-87	Using the World Wide Web to Gather Information	DMA-13 Puppets	13

(continued)

Overview of Support Tasks *(continued)*

ST-#	ST Title	Related DMAs	Relevant Chapter #
ST-88	Using Desktop Publishing to Produce Graphics	DMA-18 Keep Out! DMA-21 Pop-Up Cards DMA-22 Party Plans DMA-23 Healthy Eating DMA-24 Prepacked Lunches DMA-28 Seeing Both Sides	13
ST-89	Using MyPyramid	DMA-23 Healthy Eating DMA-24 Prepacked Lunches	14
ST-90	Analyzing Food Labels	DMA-23 Healthy Eating DMA-24 Prepacked Lunches	14
ST-91	Choosing a Career	DMA-26 Getting a Job	15
ST-92	Investigating Résumés	DMA-26 Getting a Job	15
ST-93	Using the World Wide Web to Investigate Career Opportunities	DMA-26 Getting a Job	15
ST-94	Writing a Cover Letter	DMA-26 Getting a Job	15
ST-95	Technology in the News	DMA-28 Seeing Both Sides	16
ST-96	Noise Pollution	DMA-28 Seeing Both Sides	16
EV-1	Evaluating the Final Product	Used after each DMA	
EV-2	Unit Review	Used after each DMA	

Using the Design and Make Activities

People have always designed technological products—meals, clothes, shelter, and transportation. The purpose of designing is to create change in the made world. In the *Standards for Technological Literacy*, technology is described as the process by which "humans modify the world around them to meet their needs and wants or to solve practical problems" (ITEA, 2000, p. 22). This desire and ability to constantly improve the made world is, according to Bronowski (1973), a unique feature of humans—"the ability to visualize the future…and to represent it to ourselves as images that we project and move about inside our head" (p. 56). Kimbell and Perry (2001) refer to this imaging as "designing…The process that lies at the heart of technology education" (p. 3). Central to designing is generating, developing, and communicating design ideas.

Therefore, technology education is much more than providing students with technological understanding—the ability to use and work with existing technologies. Technology education enables the development of technological *capability*, "that combination of ability and motivation that transcends understanding and enables creative development" (Kimbell, Stables & Green, 2002, p. 19).

According to the National Curriculum in the UK, technology education:

Prepares pupils to participate in tomorrow's rapidly changing technologies. They learn to think and intervene creatively to improve quality of life. The subject calls for pupils to become autonomous and creative problem solvers, as individuals and members of a team. They must look for needs, wants, and opportunities and respond to them by developing a range of ideas and making products and systems. They combine practical skills with an understanding of aesthetics, social and environmental issues, function, and industrial practices. As they do so, they reflect on and evaluate present and past design and technology, its uses and effects. (Department for Education and Employment, 1999, p. 15)

The *Technology: Shaping Our World Student Activity Manual* provides a series of Design and Make Activities through which students can acquire the ability to purposefully pursue "a task to some form of resolution that results in improvement (for someone) in the made world" (Rutland, 2002, p. 50). Each DMA requires the student to use the knowledge, skills, and understanding acquired through the completion of a series of Support Tasks. A DMA typically engages the student in the following:

◆ Writing a design specification based on an open design brief supplied by the teacher.

◆ Conducting appropriate research.

◆ Generating ideas.

◆ Developing ideas.

◆ Communicating ideas through 2-D and 3-D models.

◆ Working with resistant materials to produce a high-quality product.

◆ Evaluating the product.

◆ Evaluating their learning.

◆ Identifying ways in which to improve their capability.

The DMAs are presented as worksheets of the same structure, as shown below:

DMA category

Design and Make Activities   Communicating Ideas

Reference number

DMA-3
My "Dream" Furniture

Title

Related textbook chapter

Related Chapter: Chapter 3, Communicating Ideas

The Context

Everyone enjoys imagining a piece of furniture they would like to have in their "dream" room. It may be a bed in the shape of a space vehicle, a special table on which to play video games, or a chair in the shape of a cartoon character. The furniture may be decorated and colored to illustrate a particular mood or image.

Describing this furniture using only words is very difficult, and often impossible. A sketch, however, lets you share your ideas much more easily. Sketching allows and encourages a designer to "play" with ideas, an essential part in creative idea development. Sketches aid in evaluation of a design proposal and help to identify and restate the problems. Sketching provides a means of testing concepts that will, in turn, encourage the further generation of ideas. Annotations can be added to explain anything that cannot be shown by the sketch.

The context for the DMA

The Design Brief

Use sketches to communicate your ideas for a piece of "dream" furniture for your room. It can be a piece you have seen in a store that you adapt to meet a particular need or image, or it can be something entirely original you have created. Add notes to your sketches to explain the details.

The design brief

The Design Decisions

You can decide on the following:

The design decisions the student will make

The Product
◆ Which piece of furniture is to be modeled?
◆ What size is the furniture?
◆ What special features are required?

The Appearance
◆ What shape is the piece of furniture?
◆ What color is the piece of furniture?
◆ What materials are used to build the furniture?

The Sketches
◆ How many sketches are required?
◆ What annotation is required?
◆ What coloring is required?

Design and Make Activities   Materials

Support
Tasks to be
completed
before the
DMA

The Support Tasks
ST-9   Exploring Natural Shapes
ST-10   Exploring Geometric Shapes
ST-36   Cutting and Shaping Thin
        Plywood
ST-37   Cutting and Shaping Thin
        Aluminum
ST-38   Cutting and Shaping Acrylic
ST-51   Safety—Managing Risk
ST-5   Writing a Design Specification
EV-1   Evaluating the Final Product
EV-2   Unit Review

What You Will Learn
What the
student
will learn
◆ To identify user needs.
◆ To write a design specification.
◆ About geometric and natural shapes.
◆ About balance and equilibrium in
  structures.
◆ To mark, measure, and shape thin
  sheet materials to a high degree of
  accuracy.
◆ About the properties of wood, alu-
  minum, and acrylic.
◆ To evaluate a product against a design
  specification.
◆ To evaluate your designing and making.

What You Will Need

Stimulus Materials
◆ Photographs of mobiles by Alexander
  Calder

Tools
◆ Junior hacksaw
◆ Scissors
◆ Wire cutters
◆ Files
◆ Snips
◆ Coping saw
◆ Abrasive paper
◆ Emery cloth
◆ Hand drill
◆ Drill bits
◆ Computer with Internet access
◆ Books with pictures of mobiles

Consumables
◆ String
◆ Wire
◆ Thin sticks
◆ Colored paper
◆ 1/8" (3 mm) ply or wood
◆ 1/8" (3 mm) acrylic
◆ 1/8" (3 mm) aluminum
◆ Selection of wooden dowels
◆ Wire coat hangers

Stimulus
materials,
tools, and
consumable
materials needed
to begin the
Design and
Make Activity

The following table lists the Design and Make Activities and corresponding approximate time required, design brief, Support Tasks, and the relevant chapter in the textbook.

Overview of the Design and Make Activities

DMA-# and Title	Time (hours)	The Design Brief	Associated Support Tasks	Relevant Textbook Chapter
DMA-1 Tallest Tower	3	Working in small groups, design and make the tallest freestanding tower possible using one 8 1/2″ × 11″ sheet of thin cardboard and 4″ of clear tape.	ST-1 You Are a Designer ST-2 Brainstorming ST-51 Safety—Managing Risk EV-1 Evaluating the Final Product EV-2 Unit Review	2
DMA-2 New Uses for Everyday Products	5	Using one element of the CAMPERS strategy at a time, suggest new uses for a common household product. Make a sketch to show the original product, adding notes to indicate the materials used in its manufacture. Describe the use for which it was originally designed. Record your work using sketches and note the changes you make to the product.	ST-4 Identifying User Needs and Interests ST-18 Freehand Sketching Techniques ST-19 Sketching on Isometric Grid Paper EV-1 Evaluating the Final Product EV-2 Unit Review	2
DMA-3 My "Dream" Furniture	10	Use sketches to communicate your ideas for a piece of "dream" furniture for your room. It can be a piece you have seen in a store that you adapt to meet a particular need or image, or it can be something entirely original you have created. Add notes to your sketches to explain details.	ST-18 Freehand Sketching Techniques ST-19 Sketching on Isometric Grid Paper ST-20 Isometric Sketching on Plain Paper ST-21 Sketching in Proportion ST-22 Showing Surface Texture EV-1 Evaluating the Final Product EV-2 Unit Review	3
DMA-4 My Personal Prism	15	Design and make a "personal prism", using 3/8″ × 3/8″ (10 mm x 10 mm) wood strip. The prism can be used as a container, a stand for a small item (such as a photograph cube), or a showcase. One or more faces of the prism can be decorated or all the faces can be left open. The prism can be designed to be freestanding or suspended.	ST-4 Identifying User Needs and Interests ST-5 Writing a Design Specification ST-31 Using Wood Strip to Make a Flat Frame ST-32 Making a 3-D Framework ST-51 Safety—Managing Risk EV-1 Evaluating the Final Product EV-2 Unit Review	3

(continued)

Overview of the Design and Make Activities *(continued)*

DMA-# and Title	Time (hours)	The Design Brief	Associated Support Tasks	Relevant Textbook Chapter
DMA-5 Storage	20	Design and make a storage unit, using only one sheet of material, to hold and display a set of flat items (e.g., CDs, DVDs, coasters). The unit should be designed either for you or for a friend.	ST-6 Making and Using Nets ST-7 Using a Gantt Chart to Plan ST-33 Laminating Thin Plywood ST-34 Bending Sheet Metal ST-35 Bending Acrylic ST-51 Safety—Managing Risk EV-1 Evaluating the Final Product EV-2 Unit Review	4
DMA-6 Games Galore	20	Design and make a high-quality, durable toy or game that will amuse and intrigue a bedridden patient and that can be played with on a bed tray. The game should include a set of rules or instructions and be appropriately packaged. The toy or game should not require batteries or household power.	ST-4 Identifying User Needs and Interests ST-8 Exploring Existing Products ST-31 Using Wood Strip to Make a Flat Frame ST-32 Making a 3-D Framework ST-51 Safety—Managing Risk EV-1 Evaluating the Final Product EV-2 Unit Review	4
DMA-7 Hanging Fun	25	Design and make a mobile with asymmetric balance. The mobile should be based on a particular theme, for example, animals, fish, geometric shapes, transportation, or science fiction. The mobile should contain a minimum of six hanging shapes.	ST-9 Exploring Natural Shapes ST-10 Exploring Geometric Shapes ST-36 Cutting and Shaping Thin Plywood ST-37 Cutting and Shaping Thin Aluminum ST-38 Cutting and Shaping Acrylic ST-51 Safety—Managing Risk EV-1 Evaluating the Final Product EV-2 Unit Review	5
DMA-8 Desk Lamp	25	Design and make a desk lamp in which the light source is focused and fully adjustable for height and angle. The lamp must contain an on/off switch.	ST-3 Using a Collage to Generate Ideas ST-5 Writing a Design Specification ST-8 Exploring Existing Products ST-51 Safety—Managing Risk ST-74 What Is a Circuit? ST-76 Building a Series Circuit ST-77 Building a Parallel Circuit EV-1 Evaluating the Final Product EV-2 Unit Review	5
DMA-9 Showcase	30	Design and make a display system for a collection of items belonging to you or a friend. The display may be static or may revolve and it may incorporate lighting.	ST-32 Making a 3-D Framework ST-51 Safety—Managing Risk ST-52 How Trusses Work ST-53 Making Structures Rigid ST-54 Struts and Ties ST-56 Forces Acting on Structures ST-59 Using Gears ST-74 What Is a Circuit? ST-75 Electric Quiz Board ST-76 Building a Series Circuit ST-77 Building a Parallel Circuit ST-78 Investigating Resistance ST-79 Reading Resistors EV-1 Evaluating the Final Product EV-2 Unit Review	6

(continued)

Overview of the Design and Make Activities *(continued)*

DMA-# and Title	Time (hours)	The Design Brief	Associated Support Tasks	Relevant Textbook Chapter
DMA-10 On the Move	30	Design and make a flat-pack coffee table that will appeal to a young person setting up his or her first home. The pack must contain all of the components required to assemble the table. Assembly should involve only common household tools. The packaging must not only protect the table during shipment but also advertise the company.	ST-3 Using a Collage to Generate Ideas ST-6 Making and Using Nets ST-19 Sketching on Isometric Grid Paper ST-20 Isometric Sketching on Plain Paper ST-23 3-D Modeling with Cardboard ST-39 Exploring Manufactured Boards ST-40 Exploring Knockdown Fittings ST-51 Safety—Managing Risk EV-1 Evaluating the Final Product EV-2 Unit Review	6
DMA-11 Pet House	30	Design and make a home for a household pet. The home must be suitable for transporting the pet and easy to clean.	ST-5 Writing a Design Specification ST-11 Investigating Pets ST-24 Drawing to Scale ST-25 Measuring Scale Drawings— Conventional ST-27 Reading Scale Drawings— Conventional ST-29 Modeling with Foam Board ST-37 Cutting and Shaping Thin Aluminum ST-38 Cutting and Shaping Acrylic ST-41 Making a Wooden Candle Holder ST-51 Safety—Managing Risk ST-52 How Trusses Work ST-53 Making Structures Rigid EV-1 Evaluating the Final Product EV-2 Unit Review	7
DMA-12 Temporary Shelter	25	Design and make a scale model of a rapid-assembly shelter. The shelter must be designed so its sections can be easily fitted together, dismantled, stored, and transported. It must also protect occupants from the weather.	ST-4 Identifying User Needs and Interests ST-29 Modeling with Foam Board ST-32 Making a 3-D Framework ST-51 Safety—Managing Risk ST-53 Making Structures Rigid ST-54 Struts and Ties ST-55 Natural and Human-Made Structures ST-56 Forces Acting on Structures ST-57 Using Beams in Structures ST-58 Strengthening Beams EV-1 Evaluating the Final Product EV-2 Unit Review	7
DMA-13 Puppets	10	Design and make a lever-operated puppet to illustrate the appearance and movements of an animal on the World Wildlife Fund list of endangered species. The puppets will be sold during a special evening at the school to raise money to send to the World Wildlife Fund.	ST-12 Decorating Cardboard ST-30 Scaling Up—Drawing Things Bigger ST-51 Safety—Managing Risk ST-60 Exploring Animal Movements ST-61 Levers All Around Us ST-62 Exploring Linkages ST-87 Using the World Wide Web to Gather Information EV-1 Evaluating the Final Product EV-2 Unit Review	8

(continued)

Overview of the Design and Make Activities *(continued)*

DMA-# and Title	Time (hours)	The Design Brief	Associated Support Tasks	Relevant Textbook Chapter
DMA-14 Mechanical Toys	20	Design and make an automaton that will amuse and intrigue. The automaton may be for an individual of your choice. Alternatively, it may be designed to attract the attention of a group of people in order to encourage them to join a club.	ST-4 Identifying User Needs and Interests ST-5 Writing a Design Specification ST-19 Sketching on Isometric Grid Paper ST-20 Isometric Sketching on Plain Paper ST-21 Sketching in Proportion ST-32 Making a 3-D Framework ST-51 Safety—Managing Risk ST-59 Using Gears ST-65 Moments and Levers—Conventional Measurements EV-1 Evaluating the Final Product EV-2 Unit Review	8
DMA-15 Moving Toys	15	Design and make a controllable, battery-powered model vehicle for yourself or a special friend. The vehicle can be based on a real vehicle, a vehicle from a book or a movie, or can be a fantasy vehicle developed from your own imagination. The appearance of the vehicle should appeal to the individual who will use it.	ST-6 Making and Using Nets ST-8 Exploring Existing Products ST-12 Decorating Cardboard ST-31 Using Wood Strip to Make a Flat Frame ST-51 Safety—Managing Risk ST-63 Exploring Wheels and Axles ST-64 Making a Cardboard Chassis ST-74 What Is a Circuit? ST-76 Building a Series Circuit ST-80 Making a Reversing Switch EV-1 Evaluating the Final Product EV-2 Unit Review	9
DMA-16 Pneumatic Ergonome	20	Design and make a pneumatic ergonome that will assist with the design of a consumer product.	ST-5 Writing a Design Specification ST-14 Designing for People—Anthropometric Data ST-15 Making an Ergonome ST-16 Modeling Ideas with Cardboard, Wood Strip, and Dowel ST-30 Scaling Up—Drawing Things Bigger ST-51 Safety—Managing Risk ST-67 Investigating Pneumatics and Hydraulics ST-68 Understanding Systems EV-1 Evaluating the Final Product EV-2 Unit Review	10
DMA-17 Robots	20	Design and make a prototype pick-and-place robot that can move a small object. A human operator or computer can control the robot.	ST-31 Using Wood Strip to Make a Flat Frame ST-32 Making a 3-D Framework ST-51 Safety—Managing Risk ST-61 Levers All Around Us ST-67 Investigating Pneumatics and Hydraulics ST-68 Understanding Systems ST-69 Systems and Subsystems ST-70 Pneumatics—One-Way Valves ST-71 How Arm Robots Work EV-1 Evaluating the Final Product EV-2 Unit Review	11

(continued)

Overview of the Design and Make Activities *(continued)*

DMA-# and Title	Time (hours)	The Design Brief	Associated Support Tasks	Relevant Textbook Chapter
DMA-18 Keep Out!	18	Design and make an electrical circuit that allows someone to warn you they want to enter your room. A membrane panel switch located on your door should control the circuit. The switch should be easy to use and visually attractive.	ST-51 Safety—Managing Risk ST-81 Making a Membrane Panel Switch ST-82 Drawing Circuit Diagrams ST-88 Using Desktop Publishing to Produce Graphics EV-1 Evaluating the Final Product EV-2 Unit Review	11
DMA-19 Electronic Jewelry	25	Design and make a piece of jewelry for a particular person that will attract attention through the use of LEDs. The LEDs should be controlled by a switch that can be hidden in a pocket.	ST-4 Identifying User Needs and Interests ST-8 Exploring Existing Products ST-36 Cutting and Shaping Thin Plywood ST-38 Cutting and Shaping Acrylic ST-51 Safety—Managing Risk ST-74 What Is a Circuit? ST-76 Building a Series Circuit ST-77 Building a Parallel Circuit ST-81 Making a Membrane Panel Switch ST-83 Using Light Emitting Diodes EV-1 Evaluating the Final Product EV-2 Unit Review	12
DMA-20 Game of Chance	20	Design and make a game of chance for two players that uses diodes and LEDs in an electronic circuit. The game should also include a set of rules and scoring system.	ST-8 Exploring Existing Products ST-31 Using Wood Strip to Make a Flat Frame ST-51 Safety—Managing Risk ST-83 Using Light Emitting Diodes ST-84 Using Diodes ST-85 Using Push-to-Make and Push-to-Break Switches ST-86 Using Desktop Publishing to Produce Instruction Sheets EV-1 Evaluating the Final Product EV-2 Unit Review	12
DMA-21 Pop-Up Cards	18	Working in small groups, design and make a set of high-quality pop-up cards to be sold in local stores to raise money for a charity of your choice. The illustration on the cards and the mechanisms can amuse and intrigue, illustrate the work of the charity, or have a theme of your choice. The cards are to be packaged in an attractive format that advertises the work of the charity. Each card should be approximately 5″ × 7″ (125 mm × 175 mm) and should weigh less than 1 ounce (30 grams), so that it can be sent by regular mail.	ST-5 Writing a Design Specification ST-8 Exploring Existing Products ST-12 Decorating Cardboard ST-17 Designing a Questionnaire ST-43 Paper Engineering—Making a Box Fold ST-44 Paper Engineering—Making a Mouth Fold ST-45 Paper Engineering—Making a Slider ST-46 Paper Engineering—Making a Lift-Up Flap ST-47 Paper Engineering—Making a Rotator ST-51 Safety—Managing Risk ST-88 Using Desktop Publishing to Produce Graphics EV-1 Evaluating the Final Product EV-2 Unit Review	13

(continued)

Overview of the Design and Make Activities *(continued)*

DMA-# and Title	Time (hours)	The Design Brief	Associated Support Tasks	Relevant Textbook Chapter
DMA-22 Party Plans	18	Design and make a combined pack and counter-top sales display for a new range of party balloons. The pack will contain 20 balloons, 5 each of 4 designs. The counter-top unit will display all four designs.	ST-4 Identifying User Needs and Interests ST-5 Writing a Design Specification ST-6 Making and Using Nets ST-12 Decorating Cardboard ST-51 Safety—Managing Risk ST-88 Using Desktop Publishing to Produce Graphics EV-1 Evaluating the Final Product EV-2 Unit Review	13
DMA-23 Healthy Eating	25	Design and make a variety of display materials, leaflets, and promotional items to promote healthy eating habits among high school students. In addition, design and make a display system for a location in your school.	ST-2 Brainstorming ST-5 Writing a Design Specification ST-6 Making and Using Nets ST-13 Writing a Design Brief ST-36 Cutting and Shaping Thin Plywood ST-38 Cutting and Shaping Acrylic ST-51 Safety—Managing Risk ST-88 Using Desktop Publishing to Produce Graphics ST-89 Using MyPyramid ST-90 Analyzing Food Labels EV-1 Evaluating the Final Product EV-2 Unit Review	14
DMA-24 Prepacked Lunches	20	Working in teams of four, design a series of menus for prepacked lunches that the cafeteria can prepare. Design and make a prototype packaging that can be used to carry the food for a particular field trip. The illustration on the packaging should reflect the nature of the field trip, as well as contain some information about good dietary habits.	ST-4 Identifying User Needs and Interests ST-5 Writing a Design Specification ST-6 Making and Using Nets ST-13 Writing a Design Brief ST-51 Safety—Managing Risk ST-88 Using Desktop Publishing to Produce Graphics ST-89 Using MyPyramid ST-90 Analyzing Food Labels EV-1 Evaluating the Final Product EV-2 Unit Review	14
DMA-25 Fundraising	30	Working in small groups, design and make a simple, attractive item that can be sold by an environmental group to raise funds and advertise its activity. The item must reflect the work of the group and be suitable for small batch production.	ST-3 Using a Collage to Generate Ideas ST-4 Identifying User Needs and Interests ST-7 Using a Gantt Chart to Plan ST-36 Cutting and Shaping Thin Plywood ST-37 Cutting and Shaping Thin Aluminum ST-38 Cutting and Shaping Acrylic ST-42 Tools and Materials ST-51 Safety—Managing Risk EV-1 Evaluating the Final Product EV-2 Unit Review	15

(continued)

Overview of the Design and Make Activities *(continued)*

DMA-# and Title	Time (hours)	The Design Brief	Associated Support Tasks	Relevant Textbook Chapter
DMA-26 Getting a Job	10	Design and make a career passport/exit portfolio that includes a résumé, a business style letter of introduction, a job application, and a portfolio of your work.	ST-91 Choosing a Career ST-92 Investigating Résumés ST-93 Using the World Wide Web to Investigate Career Opportunities ST-94 Writing a Cover Letter EV-1 Evaluating the Final Product EV-2 Unit Review	15
DMA-27 New Uses for Old Things	20	Design and make a product that can be used to store small items in the home. The product should be made mostly from materials that are normally thrown away.	ST-4 Identifying User Needs and Interests ST-5 Writing a Design Specification ST-13 Writing a Design Brief ST-48 Recycling Materials to Make New Products ST-49 Materials—Cradle to Grave ST-50 New Uses for Old Things ST-51 Safety—Managing Risk EV-1 Evaluating the Final Product EV-2 Unit Review	16
DMA-28 Seeing Both Sides	20	Design and make an information pack that presents both sides of a controversial environmental issue. The pack should be suitable for distribution to a particular audience, for example, adults in a shopping mall or students at your school. It should encourage readers to think critically and consider alternatives. It should use data from different sources and contain verifiable references.	ST-88 Using Desktop Publishing to Produce Graphics ST-95 Technology in the News ST-96 Noise Pollution EV-1 Evaluating the Final Product EV-2 Unit Review	16

Using the Textbook

The *Technology: Shaping Our World* textbook provides the knowledge to assist students in becoming capable in technology education—to design and make products and services that meet a need, function well, are made to a high standard, and are sensitive to social and environmental issues. Students should refer to the textbook as they complete both the Support Tasks and the Design and Make Activities.

The textbook contains sixteen chapters that describe the design process skills, technical information, and making skills students need to learn as they develop their capability.

Textbook Contents

1. Technology and You—An Introduction

As an introduction to the scope of technology in our lives, this chapter includes:

◆ A definition of technology.

◆ The benefits of technology.

◆ Reference to the ever-increasing rate of technological change.

◆ The relationship between science and technology.

2. Generating and Developing Ideas

This chapter provides the student with techniques for generating and developing design ideas. It includes information about:

◆ Identifying design problems.

◆ Steps in designing.

◆ Developing a design proposal.

◆ Using the elements and principles of design.

◆ Steps in making a prototype.

◆ Evaluating a prototype.

3. Communicating Ideas

This chapter provides techniques for communicating design ideas. Included is information about:

◆ Isometric and perspective sketching.

◆ Orthographic drawing.

◆ Computer-aided design.

4. Types of Materials

This chapter introduces students to:

◆ Properties of resistant materials.

◆ Characteristics of materials.

◆ Processes required to change raw materials into processed stock.

5. Processing Materials

This chapter provides the student with techniques for marking, cutting, shaping, joining, and finishing materials they use when making a product that they have designed. It includes information about:

◆ Correct tools and techniques required to shape materials.

◆ How to join materials.

◆ How to select an appropriate finish for a range of materials.

◆ How to work safely.

6. Structures

This chapter provides the knowledge and information required to design structures. The information addressed includes:

◆ Types of structures.

◆ Forces acting on structures.

◆ Design of structures.

7. Construction

This chapter introduces students to buildings and systems, including:

- The types of residential buildings.
- Component parts and function of house construction.
- Construction materials.
- Open-loop and closed-loop systems.

8. Machines

This chapter provides the knowledge and information students need to design mechanical products that incorporate:

- Any of the six simple machines.
- Gears.
- The principles of hydraulics and pneumatics.

9. Transportation

This chapter provides information about:

- Various modes of transportation.
- The principles of various types of engines and motors.
- Elements of a transportation system.
- The impact of transportation systems on everyday life and the environment.

10. Energy

This chapter provides an introduction to:

- Various forms of energy.
- Changing the form of energy.
- Renewable and nonrenewable energy sources.

11. Electricity and Magnetism

This chapter provides basic information on:

- Generating and distributing electricity.
- Electron theory.
- Laws of magnetism.
- Operation of an electric motor.
- Electricity from batteries.

12. Using Electricity and Electronics

This chapter provides the knowledge and information required to design products that incorporate an electric circuit. Information covered includes:

- Types of circuits.
- The function of common electronic components.

13. Information and Communication Technology

This chapter provides an introduction to:

- New and old information and communication technologies.
- The role of information and communication technologies in society.
- The role of robots in our daily lives.
- Use of microelectronics.

14. Agriculture, Biotechnology, Environmental Technology, and Medical Technology

This chapter introduces students to foods and medical technologies that affect their lives and health, including:

- Growing, harvesting, genetically engineering, and processing food.
- The use of biotechnology in foods and medicines.
- The environmental impact of agriculture and biotechnologies.
- Nutritional requirements of humans.
- Healthy eating.

15. The World of Work

This chapter introduces:

- The primary, secondary, and tertiary sectors of the economy.

◆ Traditional and modern manufacturing processes.

◆ Careers.

16. Learning from the Past, Predicting the Future

This chapter reviews past technological achievements and encourages students to consider future scenarios. It includes:

◆ Early technologies.

◆ A history of inventions.

◆ The impact of technology on the individual and society.

◆ Predictions for technological advances.

The *Technology: Shaping Our World* text is straightforward and easy to read. Hundreds of photographs and charts attract student interest and highlight key concepts. References to the illustrations are included in the text material to help students associate visual images with the written material. This helps reinforce learning.

Each chapter begins with a list of objectives and key terms. This helps students understand what is expected of them after studying the chapter, and how they can benefit from mastering the material. A summary, modular connections, and questions follow each chapter to promote learning and expand the material presented. A glossary at the back of the book helps students learn the meaning of terms. A complete index is provided to assist in finding information quickly and easily.

Using the *Key Terms* Section

A listing of technical words appears at the beginning of each chapter under the heading *Key Terms*. This list is designed to help students identify important terms

within the chapter. These terms are presented in **bold, *italicized type*** in the text where they are defined to provide easy recognition while reading.

Discussing these words with students helps familiarize them with the concepts being introduced. To be sure students are familiar with these important terms, you may want to ask them to:

◆ Look up the dictionary definition and explain each term in their own words.

◆ Relate each term to the topic being studied.

◆ Match terms with appropriate definitions to reinforce the students' knowledge.

◆ Find examples of how the terms are used in current newspaper and magazine articles, reference books, and other related materials.

Using the *Modular Connections* Section

Some technology education programs incorporate modular projects into the technology education curriculum. The *Modular Connections* sections lists several typical modular technology topics related to the chapter. The *Technology: Shaping Our World* textbook provides the theory and background needed to complement the modular laboratory.

Using the *Test Your Knowledge* Section

The questions at the end of each chapter, under the heading *Test Your Knowledge*, cover the basic information presented in the chapter. They consist of a variety of true/false, completion, multiple choice, and essay questions. These review questions are designed to help students recall, organize, and use the information presented in the text. Answers to these questions appear in the *Teacher's Wraparound Edition* and the *Teacher's Resources.*

Using the *Apply Your Knowledge* Section

The activities in this section give students opportunities to increase their knowledge through firsthand experiences. These activities allow students to apply many of the concepts learned in the chapter to real-life situations. Suggestions for both individual and group work are provided in varying degrees of difficulty. Therefore, you may choose and assign activities according to the students' interests and abilities.

Achieving Breadth and Balance

Technology courses throughout North America vary considerably in terms of the concepts taught and length of time each is allotted. The *Technology: Shaping Our World Student Activity*

Manual is written to provide maximum flexibility. The teacher may select topics according to the ability of a given class, time allotted, and resources available.

If students are to become capable, they must be given the opportunity to work in a range of media (wood, graphics, electronics) and to design and make products and services in a variety of contexts (home, school, business). Therefore, it is important that the teacher choose a sequence of Design and Make Activities that provides this breadth.

At the same time, it is important that any scheme of work provide balance. For example, a 7th grade student completing three Design and Make Activities focusing on electricity and electronics, would not experience a balanced program. While other areas (structures, materials) could be studied in grades 8 and 9, it is better to provide balance in each grade level. Hence, there should be balance *within* a grade and *across* grades.

The following sequences suggest programs that provide both breadth and balance. Clearly, these suggestions should be adapted to meet the needs of your students, the time allotted, and the facilities available.

DMA Sequence for Grade 7

Sequence	Title of Design and Make Activity	Focus of DMA
1	My "Dream" Furniture	Communicating Ideas
2	Storage	Materials
3	Puppets	Machines and Transportation
4	Keep Out!	Electricity and Electronics
5	Pop-Up Cards	Information and Communications Technology

DMA Sequence for Grade 8

Sequence	Title of Design and Make Activity	Focus of DMA
1	My Personal Prism	Communicating Ideas
2	Hanging Fun	Materials
3	On the Move	Structures
4	Mechanical Toys	Machines and Transportation
5	Electronic Jewelry	Electricity and Electronics

DMA Sequence for Grade 9

Sequence	Title of Design and Make Activity	Focus of DMA
1	Desk Lamp	Materials
2	Temporary Shelter	Structures
3	Robots	Electricity and Electronics
4	Healthy Eating	Biotechnology
5	Fundraising	World of Work

Progression and Differentiation

Progression

If learning in technology education is to be meaningful for the student, it must be relevant, build progressively on previous work, and be differentiated.

A student's progress can be gauged by the increased complexity of design decisions as the student moves through a sequence of Design and Make Activities, and as they move from grade to grade. There is the expectation of such progress in the nature of the DMAs. For example, designing and making a flatpack coffee table (DMA-10 On the Move) at grade 9 requires more complex design decisions than designing and making a mobile with asymmetric balance (DMA-7 Hanging Fun) at grade 8. The latter, in turn, is a more complex activity than designing and making a photograph cube (DMA-4 My Personal Prism) at grade 7.

In sequencing a series of DMAs, the teacher must ensure that new tasks offer new challenges. The increasing demands of the design decisions are the key to progression. For example, the design specification for a Design and Make Activity can be defined both in terms of the number of design decisions to be made and the range of the decisions. Students begin with few criteria and progress to working with many. Similarly, they can begin with simple functional requirements and progress to working with more complex social and environmental ones. Modeling ideas may begin with simple 3-D materials and progress through sketches to freehand orthographic drawings to using computer-aided design. Evaluating a final product may progress from an egocentric view to considering the views of other users.

Differentiation

It is possible to achieve differentiation with Design and Make Activities by negotiating the complexity of the specification that an individual student's final product has to meet. The more capable students in your class may be required to work at a more demanding specifications on the same Design and Make Activity than the less able students. In other words, a Design and Make Activity can be tackled at a number of different levels, according to the ability and previous experience of the student.

At the same time, the teacher must plan a unit of work that makes appropriate demands on the individual student. The task for the teacher is to ensure that each unit undertaken is appropriate for the learner at his or her particular stage of development.

Assessing Capability

Technology education is concerned with developing students' capability. This capability requires students to combine their designing abilities with knowledge, skill, and understanding in order to design and make products or services. In his 1997 book, *Assessing Technology: International Trends in Curriculum and Assessment*, Richard Kimbell has defined capability as "that combination of skills, knowledge, and motivation that transcends understanding and enables pupils creatively to intervene in the world and 'improve' it" (p. 46). This is quite different from students acquiring a range of separate skills and abilities as achievements in their own right. The following questions guide the assessment strategy adopted by the teacher:

◆ What is important to learn in technology education that needs to be assessed?

◆ What kind of evidence allows teachers to assess the students' work?

As described earlier, a Design and Make Activity is a significant activity in which students have to use the knowledge, understanding, and skills they have been taught in an integrated and holistic way. It forms a focal point in a teaching sequence and enables students to reveal what they have learned through what they can do. Both the product and the processes by which it is conceived, developed, and realized are significant in this activity.

What evidence allows the teacher to assess capabilities? There are two types of evidence available to any teacher: transitory evidence and permanent evidence. Transitory evidence may be collected through teacher observation of students, as well as through teacher interaction with students. Transitory evidence is often left as a gestalt impression of the student inside the teacher's mind's eye. It is only available to the teacher, but open to scrutiny if some attempt at record keeping is made. Transitory evidence should be collected while students are completing Support Tasks and allows for ongoing formative assessment of student learning.

No formal assessment occurs while students are completing Support Tasks, for they are acquiring the knowledge, skills, and understanding that enable them to demonstrate capability. The effectiveness of teaching during and learning from Support Tasks is evidenced through the quality of response to the Design and Make Activity.

Evidence from the DMA—The Designer's Portfolio

Designing is an intensely personal business. A designer's drawings, from preliminary doodles to finished renderings and accurate plans, are in some ways as intimate as an artist's sketchbook. This is particularly so for the early work where the ideas emerge and are developed into an incomplete and uncertain design. A student's portfolio provides evidence of his or her struggle to bring their ideas into the reality of a product. It provides evidence of the intellectual and practical endeavors that turn ideas into products that can be used and evaluated. The student's portfolio can tell a clear, internally consistent story of the student's decisions as they were designing and making a product. A portfolio is a powerful assessment tool in technology education. It can provide a stop-motion glimpse of the student's flow of creative activity, and is a central element in summative assessment.

A student's portfolio should include:

◆ A title page.

◆ Descriptions of the context for the designing and making.

◆ Descriptions of the problem.

◆ Design briefs.

◆ Descriptions of the user.

◆ Evidence of research that investigates existing products.

◆ Lists of specifications for each product to be designed.

◆ Evidence of the generation of ideas using 2-D and 3-D modeling techniques.

◆ Evidence of the development of ideas using 2-D and 3-D modeling techniques.

◆ Critical reflection on those ideas.

◆ Appropriate uses of communication techniques.

◆ Evidence of a plan for making the product.

◆ Descriptions of how the product will be tested.

◆ Evidence of testing.

◆ Results of testing and reflection on those results.

The aim of this structure is not to produce uniform work across the class, rather the opposite. The structure provided allows students to concentrate on fully developing their own ideas as part of a class in which there is a culture of sharing and cooperation to everyone's benefit, as opposed to isolation. The individual signature of each student is developed and revealed. Every student gains from sharing ideas and working with a partner. The worth of the work in the designer's portfolio is recognized and valued. The teacher should find the situation manageable and see students making progress. The contents of a designer's portfolio provide insights into the mind of the student.

Two Support Tasks are provided to help in assessing a student's capability. The first, entitled EV-1: *Evaluating the Final Product*, allows students to evaluate the product or service they have designed and made against the design specification they wrote. This Support Task has two parts.

In Part 1, students work in small groups to assess the work of their group members. Each group member is required to provide written feedback to his or her peer. It is important to stress to the class that feedback should be positive and constructive. The feedback should provide clear indications of how the product or service meets the design specification, ways in which it is faulty, and a clear statement outlining ways it could be improved. The Support Task contains a series of questions (Questions for the Group) to help students with this assessment.

In Part 2 of the Support Task, the designer/maker assesses his or her work. This assessment is, once again, guided by a series of questions. The designer/maker should take into account the feedback from peers. As a final step, the students and the teacher should agree on an assessment level that considers the feedback from peers, self-assessment, and teacher's assessment.

The second Support Task associated with assessment is entitled EV-2: *Unit Review*. The purpose of this Support Task is to engage students in discussion and evaluate their learning. It is important that students, both as individuals and as a class, are given the opportunity to reflect on their experiences in technology education, to think about their successes, and, particularly, to identify ways in which they can improve their capability. At the end of the discussion, the class can be asked to write a statement of improvement for their next Design and Make Activity.

Using the Exam*View*® Assessment Suite

At the end of each textbook chapter, there are questions entitled *Test Your Knowledge* and *Apply Your Knowledge*. To answer these questions, students frequently refer to relevant parts of the chapter.

Many teachers give more formal tests using a set of questions that can vary each time the test is administered. The **Exam**View® Assessment Suite provides the opportunity to select a unique set of questions whenever it is appropriate to test the acquisition of propositional knowledge. The **Exam**View® Assessment Suite is included on the *Technology: Shaping Our World Teacher's Resource CD*.

Safety

Students who work through a technology course are expected to develop confidence in practical situations and a commitment to safe procedures. The teacher's role in promoting safety is twofold. He or she must be responsible for the prevention of accidents in the technology laboratory and must also educate students to be safety conscious at all times. The latter is particularly important, as the course provides students with ample opportunities to use a variety of materials and gain experience on a number of tools and machines.

The traditional approach to teaching safe practice in school workshops and laboratories is to adopt a behaviorist approach. Provide students with a long list of procedures and cautions, demonstrate safe use of tools and equipment, and then have each student copy the teacher's practice. Often students must pass a safety test before they are allowed to use a particular tool or piece of equipment.

One major weakness of this approach is that any list of safety rules is not transferable to other instances. For example, the specific safety rules to be observed when using a jointer are not much help when a student is using a soldering iron.

A constructivist approach to learning safe working habits is more appropriate. Following a demonstration of a particular tool, machine, or process, the teacher should engage students in a discussion in which they identify the hazards inherent in the situation, assess the risks, and describe how to manage the risks.

The basic rules for safe and responsible behavior in the workshop should be particularly emphasized and reinforced in the early activities. This provides a foundation on which positive attitudes toward safe practice can be reinforced throughout the course.

To encourage students to adhere to safety rules, the teacher should ensure that:

◆ The general appearance of the technology room is clean and well organized.

◆ There is sufficient storage for tools, materials, and student work.

◆ Benches are not cluttered with unnecessary tools and waste materials.

◆ Scrap boxes are provided for various materials.

◆ Safety zones are marked around machines.

◆ Combustible materials, potentially explosive substances, and fuel sources do not pose a fire hazard within the room and are stored in fire-resistant metal cabinets.

◆ Only sufficient flammable liquids for immediate use are kept, and these are stored in approved safety containers.

◆ All combustible waste materials are placed in spring-lid metal containers.

◆ Unwanted flammable and combustible materials are disposed of daily in an approved manner.

References

Bronowski, J. (1973). *The Ascent of Man*. London: British Broadcasting Corporation.

Department for Education and Employment. (1999). *Design and Technology: The National Curriculum for England*. London: Her Majesty's Stationary Office.

International Technology Education Association. (2000). *Standards for Technological Literacy: Content for the Study of Technology*. Reston, VA: International Technology Education Association.

Kimbell, R. (1997). *Assessing Technology: International Trends in Curriculum and Assessment*. Buckingham, UK: Open University.

Kimbell, R., & Perry, D. (2001). *Design and Technology in a Knowledge Economy*. London: Engineering Council.

Kimbell, R., Stables, K., & Green, R. (2002). The nature and purpose of design and technology. In G. Owen-Jackson (Ed.), *Teaching Design and Technology in Secondary Schools* (pp. 19-30). London: RoutledgeFalmer.

Rutland, M. (2002). Links across design and technology. In G. Owen-Jackson (Ed.), *Teaching Design and Technology in Secondary Schools* (pp. 48-63). London: RoutledgeFalmer.

Using Other Resources

Allowing students to see, analyze, and work with examples can reinforce in-class learning. Providing samples of materials, pictures, demonstration samples, and articles related to technology can greatly enhance student learning. Students can use these items in many activities related to the text.

You may be able to acquire some items through local stores. Many stores may be willing to donate items. You may be able to obtain pamphlets and project ideas by contacting manufacturers.

Magazines, catalogs, and sales brochures are excellent sources of photos. Having a large quantity of photos available for clipping and mounting will prove helpful to students. Students may analyze and discuss the pictures in a variety of activities.

Current magazines and journals are also good sources of articles on various aspects of technology. Having copies in the classroom encourages students to use them for research and ideas as they study technology. Other information may be obtained through various trade and professional organizations.

The following publications may be helpful to you or your students:

Alcorn, P.A., *Social Issues in Technology: A Format for Investigation.* Englewood Cliffs, NJ: Prentice-Hall, Inc.

Anscombe, I., and Gere, C., *Arts and Crafts in Britain and America.* New York: Van Nostrand Reinhold Co.

Ardley, N. (ed.), *The Children's Encyclopedia of Science.* London: Optimum.

Ardley, N., and Manley, D. (ed.), *The Amazing World of Machines.* Leicester, England: Galley Press.

Arnold, R.B., *A First Electronics Course.* Cheltenham, UK: Stanley Thornes (Publishers) Ltd.

Ayensu, E., (ed.), *The Timetable of Technology.* New York: Hearst Books.

Beazley, M., (ed.), *The International Book of the Forest.* London: Simon and Schuster.

Belliston, L., Edwards, D., and Hanks, K., *Design Yourself.* San Diego, CA: William Kaufmann, Inc.

Bisacre, M., et al (eds.), *The Illustrated Encyclopedia of Technology.* London: Marshall Cavendish Books Ltd.

Bonnet, K.R., and Oldfield, G., *Everyone Can Build a Robot Book.* New York: Simon and Schuster.

Boyd, W.T., *Fiber Optics: Communications, Experiments, and Projects.* Indianapolis, IN: Howard W. Sams & Co.

Botting, D., Gerrard, D., and Osborne, K., *The Technology Connection: The Impact of Technology on Canada.* Vancouver: CommCept Publishing.

Bramwell, M., (ed.), *The International Book of Wood.* London: Mitchell Beazley Publishers, Ltd.

Briggs, Asa, *The Power of Steam: An Illustrated History of The World's Steam Age.* Chicago: University of Chicago Press.

Brown, A.E., and Jeffcott Jr., H.A., *Absolutely Mad Inventions.* New York: Dover Publications.

Brown, W.C. and Kicklighter, C.E., *Drafting for Industry.* Tinley Park, IL: Goodheart-Willcox Publisher.

Canada Mortgage and Housing Corporation, *Canadian Wood-Frame House Construction.* Ottawa: CMHC.

Canada Mortgage and Housing Corporation, *Residential Standards, Canada.* Ottawa: CMHC.

Carassa, F., et al., *Quest for Space.* New York: Crescent Books.

Cave, J., *Technology in School: A Handbook of Practical Approaches and Ideas.* London: Routledge Kegan Paul.

Cipolla, C., and Birdsall, D., *The Technology of Man.* New York: Holt, Rinehart and Winston.

Clark, D., (ed.), *The Encyclopedia of Inventions: The Story of Technology through the Ages.* London: Marshal Cavendish, Ltd.

Cooper, A., *Visual Science: Electricity.* New Jersey: Silver Burdett Company.

Corn, J.J. (ed.), *Imagining Tomorrow: History, Technology and the American Future.* Cambridge, MA: MIT Press.

Council of Ministers of Education, *Metric Style Guide.* Toronto: Council of Ministers of Education.

Cowan, H.J., and Wilson, F., *Structural Systems.* New York: Van Nostrand Reinhold.

Cross, A. (eds.), and McCormick, R. (eds.), *Technology in Schools.* London: Taylor & Francis, Inc.

d'Estaing, V.G., *The World Almanac Book of Inventions.* New York: World Almanac Publications.

Daintith, J., et al (eds.), *Junior Science Encyclopedia.* Feltham, England: Hamlyn Publishing Group.

Daniels, J., *The Anatomy of the Car.* London: Simon & Schuster.

Davies, H., and Wharton, M., *Inside the Chip: How it Works and What it Can Do.* London: Usborne Publishing.

Do Bono, E., *Lateral Thinking—A Textbook of Creativity.* Markham, Ontario: Penguin Books.

Du Vall, J.B., Mauhan Jr., G.R., and Berger, E.G., *Getting the Message: The Technology of Communication.* Clifton Park, NY: Delmar Learning.

Ellul, J., *The Technological Society.* New York: Vintage Books.

Forester, T., *High-Tech Society.* Cambridge, MA: MIT Press.

Fuller, Buckminster. *Critical Path.* London: Hutchinson & Co.

Gerrish, H.H., Dugger, W.E., and Roberts, R.M., *Electricity and Electronics.* Tinley Park, IL: Goodheart-Willcox Publisher.

Ghitelman, D., *The Space Telescope.* New York: Michael Friedman Publishing Group, Inc.

Gordon, J.E., *The New Science of Strong Materials: or Why Things Don't Fall Through the Floor.* New York: Penguin.

Gordon, J.E., *Structures: or Why Things Don't Fall Down.* Cambridge, MA: Da Capo Press.

Graham, I., Watts, L. (ed.), *The Usborne Guide to Computer and Video Games.* Burlington, Ontario: Hayes Publishing Ltd.

Hamilton, B., *Brainteasers and Mindbenders.* New Jersey: Prentice Hall, Inc.

Harahan, J., *Design in General Education.* London: Design Council.

Hardy, W., *A Guide to Art Nouveau Style.* London: Quintet Publishing Ltd.

Henson, H., *Robots.* New York: Warwick Press.

Hicks, G.A., Heddie, G.M., and Bridge, P.A., *Design and Technology: Metal.* Oxford: Pergamon.

Hirol, T., *Kites: Sculpting the Sky.* New York: Pantheon Books.

Hirsch, A.J., *Physics: A Practical Approach.* Toronto: John Wiley and Sons.

Hoadley, R.B., *Understanding Wood: A Craftman's Guide to Wood Technology.* Newton, CT: Tauton Press.

Jacobs, J.A., and Kilduff, T.F., *Engineering Materials Technology.* Englewood Cliffs, NJ: Prentice-Hall.

Johnson, G.K., *Environmental Tips: How You Can Save This Planet.* Calgary, Alberta: Detselig Enterprises, Ltd.

Joliands, D., (ed), *Science Universe Series: Machines, Power & Transportation.* New York: Arco Publishing, Inc.

Jones, R.E., and Robb, J.L., *Discovering Technology: Communication.* Orlando, FL: Harcourt Publisher.

Karwatka, D., and Kozak, M.R., *Discovering Technology: Energy, Power and Transportation.* Orlando, FL: Harcourt Publisher.

Kerrod, R., *Collins Guide to Modern Technology.* London: William Collins Sons.

Kicklighter, C.E., *Architecture: Residential Drawing and Design.* Tinley Park, IL: Goodheart-Willcox Publisher.

Kimbell, R., *Design Education: The Foundation Years.* London: Routledge Kegan Paul.

Kingston, J., *How Bridges Are Made.* New York: Facts on File Publications.

Klemm, F., *A History of Western Technology.* New York: Charles Scribner's Sons.

Knight, D. C., *Robotics: Past, Present, and Future.* New York: William Morrow.

Knowles, D., *Automechanics: Understanding the New Technology.* Englewood Cliffs, NJ: Prentice-Hall, Inc.

Koff, R.M., *How Does it Work?* New York: Bonanza Books.

Koshland, Jr., D.E. (ed.), *Biotechnology: The Renewable Frontier.* Washington, DC: The American Association for the Advancement of Science.

Littrell, J.J., *From School to Work: A Cooperative Education Book.* Tinley Park, IL: Goodheart-Willcox Publisher.

Macaulay, D., *Cathedral: The Story of Its Construction.* Boston: Houghton Mifflin Co.

Macaulay, D., *Pyramid.* Boston: Houghton Mifflin Co.

Marden, A., *Design and Realization.* Oxford: Oxford University Press.

Markert, L.R., and Backer P.R., *Contemporary Technology.* Tinley Park, IL: Goodheart-Willcox Publisher.

Marshall, A.R., (ed.), *School Technology in Action.* London: English Universities Press.

Mathe, J., *Leonardo's Inventions.* New York: Crescent Books.

Matt, S.R., *Electricity and Basic Electronics.* Tinley Park, IL: Goodheart-Willcox Publisher.

McEvedy, C., *The Century World History FactFinder.* London: Century Publishing.

Micklus, C.S., *Odyssey of the Mind. Problems to Develop Creativity.* New Jersey: Creative Competitions.

Micklus, C.S., *Problems! Problems! Problems!* New Jersey: Creative Competitions.

Millet, R., *Design and Technology—Plastics.* Oxford: Pergamon Press.

Millet, R., and Storey, E.W., *Design and Technology—Wood.* Oxford: Pergamon Press.

Mitchell, B., *The Illustrated Dictionary of Twentieth Century Designers.* London: Quarto Publishing.

Mitchell, J., (ed.), *The Illustrated Reference Book of Man and Machines.* London: Windward.

Mitchell, J., (ed.), *The Illustrated Reference Book of Modern Technology.* London: Windward.

Mitchell, J., (ed.), *The Illustrated Reference Book of The Ages of Discovery.* London: Windward.

Murphy, J., *Weird and Wacky Inventions.* London: Angus and Robertson.

Myring, L., and Graham, I., *Information Revolution.* London: Usborne Publishing.

Myring, L., and Kimmitt, M., *Lasers: What They Can Do and How They Work.* London: Usborne Publishing.

Naisbitt, J., *Megatrends: Ten New Directions Transforming Our Lives.* New York: Warner Books.

National Geographic Society, *How Things Are Made.* Washington, DC: National Geographic Society.

Niesewand, N., *The Complete Interior Designer.* London: Macdonald.

Nostbakken, J., and Humphrey, J., *The Canadian Inventions Book.* Toronto: Greey de Pencier Books.

Ontario Ministry of Energy, *Turn on the Sun.* Toronto: Ontario Ministry of Energy.

Orme, A.H., and Yarwood, A., *Design and Technology.* London: Hodder and Stoughton.

Pacey, A., *The Culture of Technology.* Oxford: Basil Blackwell Publisher, Ltd.

Papanek, V., *Design for the Real World—Human Ecology and Social Change.* London: Thames and Hudson.

Penrose, R., *The Emperor's New Mind.* Oxford: University Press.

Priest, J., *Energy—Principles, Problems, Alternatives.* Reading, MA: Addison Wesley.

Potter, T., and Guild, J., *Robotics: What Robots Can Do and How They Work.* London: Usborne Publishing.

Ranzi, C., *Seventy Million Years of Man.* New York: Greenwich House.

Rehg, J.A., *Introduction to Robotics: A System Approach.* New Jersey: Prentice-Hall, Inc.

Rifkin, J., and Howard, Ted, *Entropy: Into the Greenhouse World.* New York: Bantam Books.

Rifkin, J., *Time Wars: The Primary Conflict in Human History.* New York: Simon & Schuster.

Rothschild, J., *Teaching from a Feminist Perspective: A Practical Guide.* New York: Pergamon Press.

Seymour, R.D., Ritz, J.M., Cloghessy, F.A., *Exploring Communications.* Tinley Park, IL: Goodheart-Willcox Publisher.

Sharp, M., *Robot World.* London: Piper Books.

Stevenson, J., *Visual Science: Telecommunications.* New Jersey: Silver Burdett Co.

Strandh, S., *Machines: An Illustrated History.* Gothenburg, Sweden: Nordbok.

Suzuki, D., *The Sacred Balance.* Vancouver, British Columbia: Greystone Books.

Terry, C., and Thomas, P. (eds.), *Teaching and Learning with Robots.* New York: Croom Helm.

Training and Retraining, Inc., *Understanding Electricity and Electronic Circuits.* Indianapolis, IN: Howard W. Sams & Co.

Training and Retraining, Inc., *Understanding Electricity and Electronic Principles.* Indianapolis, IN: Howard W. Sams & Co.

Van Do Lemme, Arlo, *A Guide to Art Deco Style.* Rexdale, Ontario: Quintet Publishing, Ltd.

Vergara, W.C., *Science for Everyday Life.* London: Sphere Books.

Walker, J.R., and Mathis, B.D., *Exploring Drafting.* Tinley Park, IL: Goodheart-Willcox Publisher.

Wallace, J., *The Deep Sea.* New York: Michael Friedman Publishing Group, Inc.

Wanat, J.A., Pfeiffer, E.W., and VanGulik, R., *Learning for Earning.* Tinley Park, IL: Goodheart-Willcox Publisher.

Wehmeyer, L.B., *Futuristics.* New York: Franklin Watts.

Weiner, J., *The Next One Hundred Years: Shaping The Fate of Our Living Earth.* New York: Bantam.

Weitzman, D., *Windmills, Bridges, and Old Machines.* New York: Charles Scribner's Sons.

Williams, I.W., *The History of Invention.* New York: Facts on File Publications.

Williams, P., and Jinks, D., *Design and Technology 5–12.* London: Falmner Press.

Williams, P., *Teaching Craft, Design and Technology 5–13.* London: Croom Heim.

Wright, R.T., *Manufacturing and Automation Technology.* Tinley Park, IL: Goodheart-Willcox Publisher.

Wyatt, V., *Inventions: An Amazing Investigation.* Toronto: Greey de Pencier Books.

Planning Your Program

Technology: Shaping Our World is divided into sixteen chapters. The chapters are organized to present the information in a logical progression of topics concerning the principles of technology. The *Teacher's Resources* include three charts to aid in program planning:

◆ ***Standards for Technological Literacy* Correlation Chart.** This chart lists benchmark topics for grades 6–8 from the *Standards for Technological Literacy* and identifies the specific locations where each topic is addressed in the textbook.

◆ **Basic Interdisciplinary Skills Chart.** This chart identifies topics within *Technology: Shaping Our World* that specifically encourage the development of basic skills. Reading, writing, verbal, math, science, and analytical skills are identified.

◆ **Scope and Sequence Chart.** This chart identifies the major topics addressed in a technology education program and identifies the specific textbook content applicable to each topic.

Aims and Goals

The technologically literate student should be able to:

◆ Appreciate that we are living in a technological world and we must interact effectively with all aspects of technology that surround us daily.

◆ Learn and apply the steps common to the process of technology. These steps involve them in:
  ◆ Developing their creative potential.
  ◆ Visually recording and communicating design ideas.
  ◆ Working with a variety of materials in order to produce an object.

◆ Examine the specific technologies related to structures, machines, and energy. This helps students to relate the processes listed above to the technology at work they encounter in their everyday lives.

◆ Recognize the ever-changing nature of technology and attempt to forecast future developments.

General Principles of Instruction

Students learn in a variety of ways, and no two students are exactly alike. They may be visual learners who enjoy reading or viewing illustrations or auditory learners who gain the most from discussions with fellow students or knowledgeable experts. They may be tactile learners who learn best when they are able to manipulate objects and create in a three-dimensional form. One of the most exciting and pedagogically valuable characteristics of an introductory technology course is that it involves students in a variety of types of learning experiences that appeal to all three types of learners. These experiences range from the exclusively intellectual (reading, researching, and writing reports) through applied research (data collection, design, and drawing), to "hands-on" activities (model making, construction or prototypes, and project building).

An introductory technology course permits variety not only in the activities of students, but also in the way that the class can be organized. Variety in routine is extremely important in maintaining high motivation. There are a number of ways in which variety may be ensured. At times, students may be required to work individually—researching, generating ideas, or making drawings. However, it is important that students working on individual assignments have opportunities to share their ideas with the group. In this way, the class benefits from the original ideas of individual students.

At other times, students may work in small groups, completing research or group experiments. Design problem discussion and reaction are best achieved when the entire class works as a group. The class may also be convened as a group for special presentations, including demonstrations, films, and guest speakers.

The most important form of instruction in an introductory technology course is individualized instruction. Individualized instruction takes many different forms. Ideally, each student should be able to move through a course at a rate that is personally comfortable. Education becomes a more meaningful process when a student assumes responsibility for learning and becomes involved in planning a schedule of learning activities and setting goals.

Students who have never been given the freedom to work at their own pace require an adjustment period. It may take some students quite a long time to make the transition from teacher-directed learning to self-pacing. Teachers should be particularly alert to the needs of shy, insecure, or retiring students, because these students will not otherwise seek the help they require. Students who are accustomed to making high grades by memorizing material and by doing exactly what the teacher wants may also need special attention. They may feel extremely frustrated and lost when placed in the unfamiliar situation of directing their own learning.

Role of the Teacher

It has already been noted that students must accept a greater responsibility for their learning. At all times, they should be active participants in class. The role of the teacher, therefore, changes. He or she should:

◆ Avoid lecture-type lessons as much as possible. Emphasis should be placed on the discovery of laws and principles through observation, reflection, intuition, and experimentation.

◆ Plan realistic experiments, use simple and attractive models, and provide activities that take into consideration the students' backgrounds, interests, capabilities, and age.

◆ Make it clear that he or she is not a walking encyclopedia to explain everything and answer every question, but rather an animator, a guide who shares the process of discovery with his or her students.

◆ Emphasize the interdisciplinary aspects of the course, particularly the interrelationship with the natural sciences, social sciences, and language.

The Relationship between Theory and Practical Work

An emphasis on a combination of theory and practical work is crucial to the success of this program. The theoretical component involves the design stage, which requires logic and creativity as well as technological, scientific, and mathematical knowledge. The practical component centers on production, which requires imagination and initiative as well as manual skills, perception of form and detail, and a sense of organization.

Teachers who have previously taught courses with a workshop component have depended heavily on the use of project work. Building or constructing articles has proved to be a method of teaching theoretical concepts in a way that is very motivating to students. Such activity is pedagogically desirable, since it:

◆ Permits the development of a student's capabilities on an individual basis.

◆ Develops skills in problem solving.

◆ Requires the application of knowledge.

◆ Encourages cooperation with other students.

◆ Requires both divergent and convergent thinking.

◆ Promotes self-discipline and responsibility.

◆ Develops creative thought.

◆ Involves intuition and ingenuity.

In contrast to Industrial Arts courses, where the focus of assessment is the end product of the student's work, an introductory technology course concerns itself with the procedures used to get there. Students become confident only when they can apply ideas and see that they work. Therefore, technology becomes not merely the acquisition of facts but the active implementation of creative ideas to solve a real problem. From such experiences, long-term learning is achieved.

Assessment and Evaluation

The diversity of learning experiences and class organization, coupled with a wide-ranging subject matter, provides ample opportunity for the technology teacher to plan changes in the cognitive, psychomotor, and affective behavior of students in the knowledge and understanding, practical skills, and attitudes displayed by the student. As a result, a number of different kinds of assessment are needed to determine whether or not the students are attaining the objectives specified for the course.

The various methods of assessing a student's work may be classified into two forms: formative assessment and summative assessment. Formative assessment is used to inform the student and the teacher of two things:

◆ The extent to which an objective has been reached.

◆ The nature and source of any difficulties that have been experienced.

Formative assessment is concerned with the observation and gauging of individual progress to keep the student on the right track. A teacher should continually assess the work and attitudes of students in order to respond appropriately and helpfully to their needs. Without this assessment, there can be no teaching in any meaningful sense of the word.

Summative assessment, on the other hand, attempts to observe and gauge attainment (or nonattainment) of a recognized standard. It tells the student and the teacher whether or not a set of objectives has been achieved. It is carried out at the end of a set of learning activities. In the case of an introductory technology course, it may be carried out at the end of a theme, term, or year.

Interpretations of performance may be norm-referenced or criterion-referenced. A norm-referenced interpretation distinguishes between those who have reached a certain standard (usually based on what the average student is expected to achieve) and those who have not. By contrast, criterion-referenced tests are designed to reveal the difficulties that students are having in learning a particular skill or knowledge and are student-centered. The reference is to a particular student and his or her attainment without reference to other students. Their purpose is to find out what a particular student can or cannot do and, if possible, *why* he or she cannot do something. Such assessment must be based firmly on the objectives of the course. The objectives are the criteria by which a teacher judges whether or not a student is being successful in the course.

At particular times throughout the course, students may be tested to determine their knowledge of theory. The evaluation of theoretical knowledge is usually accomplished by written tests, including multiple choice, short-answer, and extended-answer questions.

Most learning experiences, however, should not only involve the acquisition of theoretical knowledge. Whenever possible, the activities should require a combination of cognitive, psychomotor, and affective learning. The student should be required to apply theoretical knowledge to solve technical problems, while considering the possible effect on society. A typical example would involve students discovering the principles of magnetism, using this knowledge to build an electric motor, and studying the nonpolluting aspect of an electric motor compared to a gasoline engine.

The quest for objectivity in the assessment of design projects is a troubling one. It is a simple matter to set up relatively objective tests of technological knowledge or even manipulative skill. These are areas in which there are right or wrong answers. However, no such answers exist to design problems. A teacher's personal likes and dislikes may influence the mark given. Design draws out and develops the individual thinking of students. That is why it is such a valuable educational exercise and such a problem for examiners.

The project evaluation form provides one possible scheme for assessing Design and Make Activities. It may be used as a reproducible master. Note that there are four main areas of assessment:

◆ Conception
◆ Design
◆ Production, testing, and evaluation
◆ Personal qualities

It should be understood that a project may not always involve the student in all four areas. For example, the teacher may require only that students conceive and draw solutions to a particular problem. Production, testing, and evaluation may be delayed until later in the course or may not be required at all. In this case, the teacher should make use of the relevant sections adjusting the maximum marks possible for each column according to need. If a project involves only conception and design, each of the items in the first 10 columns would be assigned 10 marks, for a total of 100 marks.

In addition to the assessment of cognitive and psychomotor skills, it is also important to assess changes in the attitudes of students. The affective domain includes objectives that describe changes in interest, attitudes, values; the development of appreciations and adequate adjustment; and an understanding of the effect that any product has on society. The teacher should develop a scheme to assess student's degree of behavior modification in relation to his or her:

◆ Interest in technology.
◆ Ability to cooperate with others in the effective use of shared facilities.
◆ Enjoyment in working with tools and materials.
◆ Confidence and adaptability in practical situations
◆ Respect for skilled work.
◆ Recognition of the need to plan.
◆ Appreciation of good design, sound construction, and appropriate use of materials in his or her own work and as a potential consumer.
◆ Recognition of the value of drawings as a means of communication of ideas and information.
◆ Commitment to safe procedures.

To summarize, a technology course involves a student in a wide variety of learning experiences. These range from the formality of the classroom lecture through open-ended design and drawing to experimentation with the use of tools and materials. Consequently, a number of different types of assessment are needed to determine the attainment of the objectives established for cognitive, psychomotor, and affective behavior. With some degree of confidence, it can be argued that technology teachers use methodologies that are revolutionizing teaching. An effective technology teacher is one whose teaching is:

◆ Problem-centered (rather than focused on information gathering)

◆ Idea-centered (as opposed to being fact-centered)

◆ Experience-oriented and interdisciplinary (not subject-centered)

◆ Focused on individual evaluation (rather than norm-referenced evaluation)

◆ Community-centered (not confined to the classroom)

◆ Multimedia-friendly (not using mostly "talk and chalk")

◆ Focused on the importance of intrinsic motivation (as opposed to extrinsic motivation)

Teaching Techniques

A main goal of this *Teacher's Wraparound Edition* is to help you provide students with learning experiences that enable them to comprehend the material in the text. Students should be able to retain what they learn and apply it to new and varied situations. An important aspect of this goal is to help students achieve personal success and satisfaction in their everyday living.

Before examining specific methodologies for teaching and learning there are a few general guidelines that teachers may follow in any school activity.

◆ SAY "NO." Search in your mind for your "bottom line." Students may be testing you constantly, but what they are really looking for are guidelines. Your support and concern for them can be seen in your ability to stick to your clear and well-articulated parameters.

◆ COMMUNICATE. Yes, verbal language is important. The genuine way you say hello and the way you ask how they are when you can see that something is troubling them. Verbal and nonverbal communication are of equal importance. Your smile not only shows you are relaxed but that you are approachable. Such accessibility is reinforced when you move among the students guiding them and commenting on their work.

◆ BE REAL. Hook into the technological age. Combine your ideas with the world that surrounds your students every day of their lives including their electronic equipment, music, and heroes.

◆ BE YOURSELF. Don't try to imitate the style of other talented teachers you have seen in the past. Use your own talents. Draw on previous work experience, talents, hobbies, and any experiences that show you are a real person and not just a teacher!

◆ SHARE FEELINGS. Show genuine emotions by sharing laughter, joy, and disappointments. Give students the message that success is something to be celebrated in a cooperative way and that successes can be both large and small.

◆ STRETCH AND EXTEND. Set targets for your students. Do not, however, let them assume that such goals represent the upper limits. Show them that they can achieve more, they are capable of more, and that you expect them to reach higher.

Teaching Styles

No style of teaching is the best. What works for one teacher may be ineffective for another. Furthermore, students vary along a continuum from those who are always actively involved and willing to teach themselves to those who are passive and content to sit and absorb facts to those who may simply sit! They are equally diverse in their ability to visualize concepts, varying from those who are abstract thinkers to those who must have concrete examples for all concepts.

While recognizing this diversity, there are some general principles you can use to help your students become more actively involved in the learning process.

1. Make learning pleasurable. One way to do this is to involve students in lesson planning. When possible, allow students to select the modes of learning they most enjoy. This increases their interest in the topics under discussion and motivates them to continue learning.

2. Make learning real. You can do this by relating the subject matter to issues that concern students. Students gain the most from their learning when they can apply classroom information to real-life situations.

3. Make learning varied. Try using several different techniques to teach the same concept. Students can learn by using each of their senses: seeing, hearing, touching, tasting, and smelling. The more senses students use, the easier it is for them to retain information.

4. Make learning success-oriented. Experiencing success increases students' self-esteem and helps them mature. You can help students experience success by providing for their individual learning styles. You can also encourage them to expand their ability to learn from other styles. For instance, some students work well in groups, while others work better on their own. Those who work best in groups can be encouraged to develop independent learning skills. Those who work best alone can be encouraged to interact with others. In this way, students develop skills they can continue to use long after finishing your course.

5. Make learning personal. Establishing rapport with your students helps them feel comfortable about sharing their feelings and ideas in group discussions.

A variety of teaching techniques may be grouped according to the different goals you have for your students. Keep in mind that not all techniques work equally well in all classrooms. A technique that works beautifully with one group of students may not be as successful with another group. The techniques you choose depend on the topic, your teaching goals, and the needs of your students. Nevertheless, all teachers should try to be lively and animated using a variety of gestures and variations in pitch and tone to revive those students who come to school unmotivated by the skills and concepts you desire to share with them. Motivation is increased when a teacher comes to class fully prepared so that:

◆ Interest is stimulated at the very start.

◆ There is always something to do.

◆ The lesson provides a challenge.

◆ Such points as "how long" and "how much" are evident.

◆ There is some air of mystery.

◆ Success leads to a good feeling.

◆ Interaction with the teacher is enjoyable.

Keep in mind that all human beings have the same needs. We all, for example, want to matter, have our name said out loud, be understood, become successful according to our own definition, be loved, be happy, and lead fulfilling lives. At the same time, it must be recognized that the sense of alienation and of not belonging is indeed a troublesome aspect of growing up these days. Sometimes the stress of present-day living is too much for parents, the extended family, or even community resources, for them to be able to reach out to our youth. The result is that our children are experiencing a greater sense of loss and unimportance in our social structure. They need to feel that they matter in a world where schools are increasingly more impersonal, our family life more rushed, and our neighborhoods more anonymous.

The following techniques may be used to meet various goals with your students.

Helping Students Gain Basic Information

One group of teaching techniques designed to help you meet the goal of conveying information to your students centers on reading and lecture. Using a number of variations can make these techniques seem less common and more interesting. For instance, students may enjoy taking turns to read aloud as a change of pace from silent reading. Lectures can be energized through the use of flip charts and overhead transparencies.

Other ways of presenting basic information include the use of outside resources. Guest speakers, individually or as part of a discussion panel, can present classroom material. Guest lectures can be videotaped to show again to other classes. In addition to such videotapes, students also enjoy the sensory stimulation of videos and slides to gain basic knowledge.

Helping Students Question and Evaluate

A second group of teaching techniques helps students develop analytical and judgmental skills. These techniques aid you in meeting your goal to help your students go beyond what they see on the surface. As you employ these techniques, encourage your students to think about points raised by others. Have them evaluate how new ideas relate to their attitudes about various subjects.

Questions serve many functions. They may test preparation, arouse interest, promote understanding, or lead the student to new insights. Review questions consolidate learning and a series of questions can stimulate critical thinking. Most often, however, questions test for achievement by referring to the lesson objectives. Good questions use straightforward language, challenge the student to think, and are adapted to the interests of the students and appropriate for the purpose intended, whether asking for facts or encouraging creative comments. One key point when asking a question is to be prepared to wait up to three seconds for a response.

Students normally do not ask questions. Once you have created an atmosphere that shows questions are welcomed (even the obvious and silly questions) there may still be silent students. What do you do with these students? Experience has shown that it is better to leave them alone until they naturally enter the discussion. By all means, watch them out of the corner of your eye for the moment when they are willing to participate, and then bring them into the lesson.

Discussion is an excellent technique for helping students consider an issue from a new point of view. Using discussion may seem easy. But in reality, discussions require a great deal of advance planning and preparation. The size of the group being taught and the physical arrangement should be carefully considered. Students may be less likely to contribute in a large group, causing the discussion to become more of a lecture. Having the room arranged so students can see one another enhances the discussion.

Discussion can take a number of forms. Large group discussions involve the entire class and should be reserved for smaller sized classes. Buzz groups usually consist of two to six students. They discuss an issue among themselves and then appoint a spokesperson to report back to the entire class. Debates are a type of discussion involving a two-sided issue. The class is divided into two groups on opposing sides of the issue. Each group then selects a panel of three or four students to present the points of their side.

Helping Students Participate in Discussion

Another group of teaching techniques is designed to help you meet the goal of stimulating student discussion. Discussion is a good technique only when students are willing and able to participate. Sometimes students have limited knowledge or feel awkward about discussing certain issues in personal terms. In these cases, you may need to encourage their discussion through other means.

A number of techniques can be used to foster ideas and encourage students to interact. Case studies, surveys, stories, and pictures can all be used to promote discussion. These techniques allow students to react to or evaluate situations in which they are not directly involved. Open-ended sentences can also be used to stimulate discussion. However, this technique must be used with discretion. Students should not be asked to complete sentences dealing with confidential issues in front of their classmates.

The fishbowl is another method for stimulating class discussion. In this method, a small interactive group of about five to eight students is encircled by a larger observation group. It is useful for discussing controversial issues due to the rule that observers cannot talk during the fishbowl. Positions can be reversed after an initial discussion period, allowing some of the observers to become the participants.

Helping Students Apply Learning

Techniques in a fourth group assist you in meeting the goal of helping students use what they have learned.

Simulation games and role-playing allow students to practice solving problems and making decisions under non-threatening circumstances. By playing the roles of other people, students can examine others' feelings as well as their own. They can thus determine how they might react or cope when confronted with a similar situation in real life.

Role-plays can be structured, giving actors written accounts of the situation, or be spontaneous, created in response to a classroom discussion. Students may act out a role as they themselves see the role being played, or they may attempt to act out the role as they presume a person in that position would behave. Roles are not rehearsed and lines are composed on the spot. The follow-up discussion should focus on the feelings and emotions felt by the participants and the manner in which the problem was resolved. The discussion should help students consider how they would apply the information to similar situations in their own lives.

Helping Students Develop Creativity

Many favorite teaching techniques probably came about through structured learning, by watching others in action, or based on professional research. Others were likely the result of failure—the times when we make a mistake and realize what should have been done. What is bad about failure is when we attempt to prevent it from happening within the learning process or do not learn from it. It is a necessary step, whether through trial and error or serendipitous learning. Some students have been trained to bypass failure at any cost even to the point that they diminish their own possibilities for

success in order to circumvent the less-than-positive experience failure brings us. The important point is that you ensure that students have the ability to risk and fail *as long as they also keep going*.

A few techniques can be used to meet the goal of helping your students generate new ideas. Meditation gives students the opportunity to come up with ideas of their own. In brainstorming, students use the suggestions of others to help spawn new thoughts. Students are encouraged to spontaneously express any thoughts or reactions that come to mind. No evaluation or criticism of ideas is allowed. Some activities in the *Student Activity Manual* are specifically designed to develop creativity.

Helping Students Review Information

Most students experience various degrees of anxiety over tests or exams. As a teacher, you can help minimize test anxiety by employing some common sense ideas that help your students review material and prepare for tests:

◆ Test frequently so each test does not count for such a high percentage of the term mark and, therefore, causes less anxiety.

◆ Use tests as only one of the means of evaluating your students.

◆ Inform students well in advance about upcoming tests so they may study sufficiently before a test. Do not try to teach new material just before the test is given.

◆ Have students rewrite their notes in short phrases or words a week or so before the test. This makes them put concepts in their own words.

◆ Consider allowing legalized crib sheets of a controlled size and number; for example, one 8 1/2″ × 11″ page.

◆ Discuss the weight and value of the exam. Put it in perspective to the total course mark.

◆ Be calm in discussing the test yourself. Students will take this cue from you.

◆ Tell them the type of exam they can expect (multiple choice, essay, true-false, etc.).

◆ Conduct a thorough review before the test.

In addition, there are techniques that can be used to meet the goal of helping students retain knowledge. Games can be used to drill students on vocabulary, classifications, and facts. Review games are available commercially or you and your students may want to devise your own. The use of crossword puzzles, word puzzles, and mazes can make the review of vocabulary terms more interesting for students.

Teaching Students of Varying Abilities

The students in your classroom represent a wide range of ability levels. The special needs students that are being mainstreamed require special teaching strategies in order to meet their learning requirements. On the other hand, gifted students must not be overlooked. Their needs must also be met in the same classroom setting. It is a challenge to adapt your daily lessons to meet the needs of all these students. To help you meet these needs, these strategies may be used with mainstreamed students:

◆ Before assigning a chapter in the text, ask students to look up the definitions of the words listed in the *Key Terms* section at the beginning of each chapter. These terms are defined in the glossary at the back of the text. Have your students write out the definitions. Then have them take turns reading the definitions out loud. Have students explain what they think the definitions mean in their own words. At other times, you might want to ask your students to guess what they think the words mean before they look up the definitions. You also might ask students to use the new words in sentences or to point out sentences in the text where the new terms are used.

◆ When introducing a new chapter, go over previously learned information that may be necessary for your students to use in understanding the new material. Review previously learned vocabulary terms they will encounter again.

◆ Break the chapters up into smaller parts. Have students read only one section at a time. Define the terms, answer the *Test Your Knowledge* questions, and discuss the concepts presented in that section before proceeding to the next. Also, assign activities from the *Student Activity Manual* that relate to that section to reinforce the concepts presented.

◆ Have students answer the *Test Your Knowledge* questions, given at the end of each chapter in the text. This helps students focus on the essential information contained in the chapter.

◆ Use the buddy system. Pair students who read well with students who do not read well. Use student aides or parent volunteers to give more individual help to students who need assistance.

◆ Select a variety of learning experiences to reinforce the learning of each concept. Choose those that encourage active participation and involvement. Also, select activities that help slow learners relate the information to real-life situations. Aim for over-learning.

◆ Give directions orally, as well as in writing. Explain all assignments thoroughly, but as simply as possible. Ask questions to be certain your students understand what they are to do. Encourage your students to ask for help if they need it.

◆ Use the overhead projector and the transparency masters included in the *Teacher's Resources.* A visual presentation of concepts increases your students' ability to comprehend the material. Develop your own transparencies for use in reviewing key points covered in each chapter.

If you have advanced or gifted students in your class, you should also provide opportunities that challenge them. These students should be given more assignments that involve critical thinking and problem solving. Because these students are more capable of independent work, they can use the library to research topics in greater depth.

Cooperative learning is an effective way to ensure that each student plays a role in learning a new concept, in spite of his or her learning difficulties. Through cooperative learning, students are organized into small groups. The groups are structured to ensure the participation of all group members in each activity according to their individual abilities. Group members work together by helping and encouraging one another. Each has a role to play in the group, which is normally 3 to 6 members. The teacher's role in cooperative learning is to specify clear objectives, ensure heterogeneous grouping, explain the learning activities and the importance of interdependence, monitor the collaborative efforts, intervene when necessary to improve group skills, and evaluate the effectiveness of the group and the achievement of the students.

Effective Communication Skills

Communicating information to students is one important responsibility for teachers. Effective communication involves not only sending clear messages to students, but also receiving and interpreting feedback from them. The following are some suggestions for improving communication with your students:

1. Be aware of students' nonverbal behavior. Shifts in body position and eye contact may give you a clue to how a student is feeling at a particular time.

2. Be aware of your own body language. Work for a natural, relaxed, attentive body position. Eye contact should be natural and comparatively constant.

3. Your tone of voice sends nonverbal signals to your students. Use a warm tone. Avoid being threatening or judgmental.

4. Listen carefully to your students. Follow their comments. Do not assume that you know what a student wants to say or where a

conversation should lead. Avoid changing the topic or interrupting. You may be surprised at what a student really wants to communicate.

5. Do not be afraid of silence. Students sometimes need time to think, especially when answering questions that deal with personal problems or feelings.

6. Phrase your questions in an open-ended manner. Students should not be able to answer your questions with a simple "yes" or "no." Try asking questions beginning with "Could you tell me," "What," and "How."

7. Avoid asking questions that begin with "Why." They generally have an interrogating sound and tend to cause students to be defensive.

8. Avoid asking one student multiple questions. Also avoid asking leading questions that suggest an answer you want to hear.

9. Reflect the feelings you hear from a student. This helps you understand the student. It also helps the student better understand himself or herself. An example of a reflective statement is: "It seems that you are feeling angry with me because I asked you to rework your assignment."

10. Paraphrase information a student gives you. This helps you understand exactly what the student is saying and prevent any misconstructions. Paraphrasing also helps the student organize his or her thoughts. A paraphrase might sound like this: "I understand you to say that you were not able to do your assignment because you had to study for a math test instead."

11. Try to avoid using responses such as "good" or "right." These statements cause closure on student thinking. Rather, respond with "How did you come to that idea?" or "Are there any other suggestions?"

12. Be human. You can express your own feelings and experiences if they are relevant to a situation. This also helps students relate to you better.

Help students formulate plans to solve problems. Be sure that a student's course of action is based on his or her own plan and not yours. Be sure the student has the plan well in mind by allowing him or her to restate the solution. Do not forget to follow up on a student's progress. Help the student recognize his or her own successes and overcome his or her failures. If necessary, help the student develop a new plan.

Classroom Management

Classroom management is part and parcel of your philosophy of teaching. By having a rich bank of knowledge about behavior and management, you are equipping yourself with the power to make intelligent choices about which behavior principles you adopt as your own, which management strategies best suit your personal style, and what discipline measures you are comfortable enough to employ. There is no single answer about classroom management, no panacea, and no one way to do the right thing.

Ideally, discipline problems should occur very rarely. Thorough lesson planning involving a variety of methods that are interesting, appeal to many styles of learning, and gradually reinforce a concept keeps students on task and leads to success. Lack of preparation often results in boredom and students acting out. Therefore, a vital part of preparation is your plan for what students do the minute they walk into your classroom. If you set the tone for work immediately, you avoid having to call students to order by shouting above the noise. One example is to have a short quiz each day. Each quiz counts for only a few points, but it is waiting on the students' desks as they walk into the classroom. While students are writing, the teacher is able to take attendance.

Avoidance is also an important concept. Avoid being too friendly with students, but be sure to know their names and treat each with dignity and respect. You are their teacher, not their best buddy. Avoidance includes avoiding confrontation. Do not argue with a student in front of peers. You will most likely lose, as their pride will often not let them back down. You may also be driven to ever-increasing heights of frustration when an obstreperous young person stands his or her ground. Discipline in private after the class has been dismissed.

Certain rules are necessary to create the parameters for behavior and to maintain a critical amount of order. It is important that rules be kept clear, simple, and to a minimum. The enforcement of these rules should be consistent and fair. Much of the effect of making rules comes from the modeled behavior by the teacher, the understanding that students have input, a structured learning environment, and dealing with classroom problems within the classroom, not in the principal's office.

Even when rules and consequences are clearly stated, some students still try to get away with breaking the rules. What they are testing is not the rule or the consequence, but the effectiveness of the entire process. The student wants to see if the authorities are able to systematically keep the rules/

consequences structure in place. Once this message is sent out and the structure makes sense to the students, much of the misbehavior decreases.

Consequences should offer choices to the teacher and to the student. This allows for preservation of dignity and nurturing of personal power. Consequences should be discussed and administered in private. This permits all parties to "cool off" and avoids backing anyone into a corner, thus "saving face." Consequences should occur after a reasonable amount of wait time has transpired. This enhances the chance of both parties being heard through active listening. Consequences can range from simple warnings to in-school suspensions. They may be rigid or flexible, logical or outlandish, but consequences should always be specific, consistent actions that lead to long-term learning on the part of the student, not to punishing of behavior. For example, a punishment for graffiti on a school wall could be the washing of the wall, as well as the creation of a painted mural by the offender to be hung in the school. A good consequence should lead to some type of restitution.

Promoting Your Program

Not everyone is aware of the importance of technology classes. You can make people more aware through good public relations. Good public relations can increase your enrollment, gain the support of administrators and peers, and achieve recognition in the community. It can also work to improve the image of your program and your department.

In promoting your program, remember to approach people in a positive way without pointing out weaknesses of other programs. Public relations projects should make people aware of the benefits of your program.

The following are some suggestions for promoting your program:

◆ Create opportunities for peers, administrators, and board members to become more aware of what you are offering. Talk to teachers and administrators about your program's benefits at appropriate meetings and in less formal settings. Invite them to visit your class. Consider attending faculty meetings of other departments and board of education meetings to let these groups know what you are doing. Let them know that you are interested in what they are doing as well.

◆ Provide services related to your program for other students or the community. Service projects increase public awareness while creating a spirit of mutual support. Examples of services include informational brochures, videos, panel presentations, and hosting guest speakers.

◆ Contact local media for coverage of newsworthy projects. Newsworthy projects include those that are unusual or that spotlight outstanding individuals or groups. Submit press releases for upcoming events. Have students submit articles to school newspapers.

◆ Display student work in school showcases or display areas. Include a description of your program with the display.

◆ Work with parents and other community resources. Invite them to be guest speakers for your class. Send them newsletters to let them know what is happening in your classes. Contact parents through phone calls or letters when their children have made positive progress. Likewise, ask for their support in solving problem situations.

◆ Work with other faculty members on interdisciplinary projects.

◆ Research areas of student weaknesses and work to emphasize these areas in your class. (SAT or other test scores may be analyzed to find weaknesses.) This *Teacher's Wraparound Edition* gives you ideas for activities that strengthen specific skills. Let your administrator know that you are working to improve these skills in your class. Show your administrator examples of work that indicate student improvement in these skills.

◆ Advertise your class on school bulletin boards, in flyers, and in student publications. Ask your students what they feel they have gained by taking the class. Use some of their ideas in your advertisement. (If you decide to use direct quotes, be sure to get permission from your students.)

Acknowledgments

We appreciate the contributions of the following Goodheart-Willcox authors in this introduction: *Teaching Techniques* by Ruth Bragg and *Effective Communication Skills* by Mary G. Westfall. In addition, the contributions of John Gradwell and Judi Leonard in their booklets *Planning for Teaching* are recognized.

Standards for Technological Literacy Correlation Chart

The International Technological Education Association (ITEA) and its Technology for All Americans Project developed the *Standards for Technological Literacy: Content for the Study of Technology* to identify the essential core of technological knowledge and skills for students in grades K–12. This work defined twenty separate standards, divided into five broad categories. Within each standard, benchmark topics are defined for four different grade levels:

◆ Grades K–2
◆ Grades 3–5
◆ Grades 6–8
◆ Grades 9–12

The following chart lists the standards and the benchmark topics for grades 6–8. Adjacent to each benchmark topic is the chapter and page reference identifying material in *Technology: Shaping Our World* relating to the benchmark topic.

Standards for Technological Literacy Correlation Chart	
Standard 1. Students will develop an understanding of the characteristics and scope of technology.	Chapter 1: pages 12–18 Chapter 2: page 22 Chapter 15: pages 386, 394
Benchmark Topics:	
New products and systems can be developed to solve problems or to help do things that could not be done without the help of technology.	Chapter 1: page 12
The development of technology is a human activity and is the result of individual or collective needs and the ability to be creative.	Chapter 1: page 12
Technology is closely linked to creativity, which has resulted in innovation.	Chapter 2: page 22
Corporations can often create demand for a product by bringing it onto the market and advertising it.	Chapter 15: pages 386, 394
Standard 2. Students will develop an understanding of the core concepts of technology.	Chapter 2: pages 28–33 Chapter 3: pages 56–57 Chapter 7: pages 181–185 Chapter 9: pages 230–244 Chapter 10: pages 258–272 Chapter 11: pages 279–284 Chapter 12: pages 332–341 Chapter 15: pages 387–388
Benchmark Topics:	
Technological systems include input, processes, output, and, at times, feedback.	Chapter 3: pages 56–57 Chapter 7: pages 181–182
Systems thinking involves considering how every part relates to others.	Chapter 9: pages 240–242
An open-loop system has no feedback path and requires human intervention, while a closed-loop system uses feedback.	Chapter 7: page 181
Technological systems can be connected to one another.	Chapter 7: pages 181–185 Chapter 9: pages 239–242
Malfunctions of any part of a system may affect the function and quality of the system.	Chapter 9: pages 240–242
Requirements are the parameters placed on the development of a product or system.	Chapter 2: pages 28–33
Trade-off is a decision process recognizing the need for careful compromises among competing factors.	Chapter 2: pages 28–33 Chapter 15: pages 387–388

(continued)

Standards for Technological Literacy **Correlation Chart** *(continued)*	
Different technologies involve different sets of processes.	Chapter 7: pages 181–185 Chapter 9: pages 230–244 Chapter 10: pages 258–272 Chapter 11: pages 279–284 Chapter 12: pages 332–341
Maintenance is the process of inspecting and servicing a product or system on a regular basis in order for it to continue functioning properly, to extend its life, or to upgrade its capability.	Chapter 7: pages 180–181
Controls are mechanisms or particular steps that people perform using information about the system that causes systems to change.	Chapter 7: pages 181–182
Standard 3. Students will develop an understanding of the relationships among technologies and the connections between technology and other fields of study.	Chapter 2: pages 24–34 Chapter 7: pages 181–185 Chapter 13: pages 328–331 Chapter 16: pages 415–431
Benchmark Topics:	
Technological systems often interact with one another.	Chapter 7: pages 181–185
A product, system, or environment developed for one setting may be applied to another setting.	Chapter 2: pages 24–34
Knowledge gained from other fields of study has a direct effect on the development of technological products and systems.	Chapter 13: pages 328–331 Chapter 16: pages 415–431
Standard 4. Students will develop an understanding of the cultural, social, economic, and political effects of technology.	Chapter 1: pages 12–18 Chapter 16: pages 409–416
Benchmark Topics:	
The use of technology affects humans in various ways, including their safety, comfort, choices, and attitudes about technology's development and use.	Chapter 1: pages 12–18 Chapter 16: pages 409–416
Technology, by itself, is neither good nor bad, but decisions about the use of products and systems can result in desirable or undesirable consequences.	Chapter 16: pages 409–416
The development and use of technology poses ethical issues.	Chapter 16: pages 409–416
Economic, political, and cultural issues are influenced by the development and use of technology.	Chapter 16: pages 409–416

(continued)

Standards for Technological Literacy Correlation Chart *(continued)*	
Standard 5. Students will develop an understanding of the effects of technology on the environment.	Chapter 14: pages 348–362 Chapter 16: pages 413, 422–424, 426
Benchmark Topics:	
The management of waste produced by technological systems is an important societal issue.	Chapter 14: pages 351–360
Technologies can be used to repair damage caused by natural disasters and to break down waste from the use of various products and systems.	Chapter 14: pages 351, 354–362
Decisions to develop and use technologies often put environmental and economic concerns in direct competition with one another.	Chapter 14: pages 350–357 Chapter 16: pages 413, 422–424, 426
Standard 6. Students will develop an understanding of the role of society in the development and use of technology.	Chapter 1: pages 12–18 Chapter 16: pages 408–412
Benchmark Topics:	
Throughout history, new technologies have resulted from the demands, values, and interests of individuals, businesses, industries, and societies.	Chapter 1: pages 12–18 Chapter 16: pages 408–412
The use of inventions and innovations has led to changes in society and the creation of new needs and wants.	Chapter 1: pages 12–18 Chapter 16: pages 408–412
Social and cultural priorities and values are reflected in technological devices.	Chapter 1: pages 12–18
Meeting societal expectations is the driving force behind the acceptance and use of products and systems.	Chapter 1: pages 12–18
Standard 7. Students will develop an understanding of the influence of technology on history.	Chapter 1: pages 12–18 Chapter 6: pages 154–160 Chapter 8: pages 192–193, 207–212 Chapter 16: pages 408–409
Benchmark Topics:	
Many inventions and innovations have evolved by using slow and methodical processes of tests and refinements.	Chapter 1: pages 12–18
The specialization of function has been at the heart of many technological improvements.	Chapter 8: pages 192–193, 207–212

(continued)

Standards for Technological Literacy Correlation Chart *(continued)*	
The design and construction of structures for service or convenience have evolved from the development of techniques for measurement, controlling systems, and the understanding of spatial relationships.	Chapter 6: pages 154–160
In the past, an invention or innovation was not usually developed with the knowledge of science.	Chapter 1: pages 12–18 Chapter 16: pages 408–409
Standard 8. Students will develop an understanding of the attributes of design.	Chapter 3: pages 24–49
Benchmark Topics:	
Design is a creative planning process that leads to useful products and systems.	Chapter 3: pages 34–49
There is no perfect design.	Chapter 3: pages 34–49
Requirements for a design are made up of criteria and constraints.	Chapter 3: pages 24–34
Standard 9. Students will develop an understanding of engineering design.	Chapter 3: pages 24–34 DMA-1 ST-2
Benchmark Topics:	
Design involves a set of steps, which can be performed in different sequences and repeated as needed.	Chapter 3: pages 24–34
Brainstorming is a group problem-solving process in which each person in the group presents his or her ideas in an open forum.	DMA-1 ST-2
Modeling, testing, evaluating, and modifying are used to transform ideas into practical solutions.	Chapter 3: pages 24–32
Standard 10. Students will develop an understanding of the role of troubleshooting, research and development, invention and innovation, and experimentation in problem solving.	Chapter 2: pages 24–49 Chapter 15: pages 386–391 DMAs 1–28
Benchmark Topics:	
Troubleshooting is a problem-solving method used to identify the cause of a malfunction in a technological system.	Chapter 2: pages 32–34 DMAs 1–28
Invention is a process of turning ideas and imagination into devices and systems. Innovation is the process of modifying an existing product or system to improve it.	Chapter 2: pages 24–49 Chapter 25, pages 386–391 DMAs 1–28
Some technological problems are best solved through experimentation.	Chapter 2: pages 25–27 DMAs 1–28

(continued)

Standards for Technological Literacy Correlation Chart *(continued)*	
Standard 11. Students will develop abilities to apply the design process.	DMAs 1–28 STs 6, 23, 29, 32
Benchmark Topics:	
Apply a design process to solve problems in and beyond the laboratory-classroom.	Chapter 2: pages 27–34, 52
Specify criteria and constraints for the design.	Chapter 2: pages 24–26, 52
Make two-dimensional and three-dimensional representations of the designed solution.	Chapter 2: pages 26–27, 31–32, 52
Test and evaluate the design in relation to pre-established requirements, such as criteria and constraints, and refine as needed.	Chapter 2: pages 32–34, 52
Make a product or system and document the solution.	Chapter 2: pages 27–34, 52
Standard 12. Students will develop the abilities to use and maintain technological products and systems.	DMAs 5–16, 18–19, 21–22 STs 18–73, 86–88
Benchmark Topics:	
Use information provided in manuals, protocols, or by experienced people to see and understand how things work.	Chapter 2: pages 28–30
Use tools, materials, and machines safely to diagnose, adjust, and repair systems.	Chapter 5: pages 112–113 Chapter 8: page 224 Chapter 15: pages 388–389
Use computers and calculators in various applications.	Chapter 13: pages 328–336 Chapter 16: pages 435–436
Operate and maintain systems in order to achieve a given purpose.	Chapter 7: pages 180–181
Standard 13. Students will develop the abilities to assess the impact of products and systems.	STs 17, 95–96 EVs 1–2
Benchmark Topics:	
Design and use instruments to gather data.	Chapter 13: pages 332–336
Use data collected to analyze and interpret trends in order to identify the positive or negative effects of a technology.	Chapter 1: pages 12–16 Chapter 9: pages 244–245, 249 Chapter 13: page 345 Chapter 15: pages 384–385, 404 Chapter 16: pages 409–413, 434–436

(continued)

Standards for Technological Literacy Correlation Chart (continued)	
Identify trends and monitor potential consequences of technological development.	Chapter 1: pages 12–16 Chapter 9: pages 244–245, 249 Chapter 15: pages 384–385, 404 Chapter 16: pages 409–413, 434–436
Interpret and evaluate the accuracy of the information obtained and determine if it is useful.	Chapter 13: pages 328–341
Standard 14. Students will develop an understanding of and be able to select and use medical technologies.	Chapter 1: pages 12, 15–16 Chapter 14: pages 350–367 Chapter 16: pages 411–412, 424–426
Benchmark Topics:	
Advances and innovations in medical technologies are used to improve healthcare.	Chapter 1: pages 12, 15–16 Chapter 14: pages 364–367 Chapter 16: pages 411–412, 422–426
Sanitation processes used in the disposal of medical products help to protect people from harmful organisms and disease, and shape the ethics of medical safety.	Chapter 14: pages 354–367 Chapter 16: pages 422–426
The vaccines developed for use in immunization require specialized technologies to support environments in which a sufficient amount of vaccines are produced.	Chapter 14: pages 363–367 Chapter 16: pages 422–426
Genetic engineering involves modifying the structure of DNA to produce novel genetic make-ups.	Chapter 14: pages 350–354 Chapter 16: page 424
Standard 15. Students will develop an understanding of and be able to select and use agricultural and related biotechnologies.	Chapter 14: pages 348–354
Benchmark Topics:	
Technological advances in agriculture directly affect the time and number of people required to produce food for a large population.	Chapter 14: pages 348–349
A wide range of specialized equipment and practices is used to improve the production of food, fiber, fuel, and other useful products and in the care of animals.	Chapter 14: pages 348–349

(continued)

Standards for Technological Literacy **Correlation Chart** *(continued)*	
Biotechnology applies the principles of biology to create commercial products or processes.	Chapter 14: pages 350–354
Artificial ecosystems are human-made complexes that replicate some aspects of the natural environment.	Chapter 14: page 349
The development of refrigeration, freezing, dehydration, preservation, and irradiation provide long-term storage of food and reduce the health risks caused by tainted food.	Chapter 14: pages 348–349
Standard 16. Students will develop an understanding of and be able to select and use energy and power technologies.	Chapter 8: page 217 Chapter 10: pages 252–257 Chapter 11: pages 278–283
Benchmark Topics:	
Energy is the capacity to do work.	Chapter 10: page 252
Energy can be used to do work, using many processes.	Chapter 10: pages 252–253
Power is the rate at which energy is converted from one form to another or transferred from one place to another, or the rate at which work is done.	Chapter 8: page 217
Power systems are used to drive and provide propulsion to other technological products and systems.	Chapter 10: pages 253–256 Chapter 11: pages 278–283
Much of the energy used in our environment is not used efficiently.	Chapter 10: page 257
Standard 17. Students will develop an understanding of and be able to select and use information and communication technologies.	Chapter 2: pages 56–67 Chapter 13: pages 328–331
Benchmark Topics:	
Information and communication systems allow information to be transferred from human to human, human to machine, and machine to human.	Chapter 13: pages 328–331
Communication systems are made up of a source, encoder, transmitter, receiver, decoder, and destination.	Chapter 2: pages 56–57
The design of a message is influenced by such factors as the intended audience, medium, purpose, and nature of the message.	Chapter 2: page 57
The use of symbols, measurements, and drawings promotes clear communication by providing a common language to express ideas.	Chapter 2: pages 59–67

(continued)

Standards for Technological Literacy Correlation Chart *(continued)*	
Standard 18. Students will develop an understanding of and be able to select and use transportation technologies.	Chapter 9: pages 228, 240–242
Benchmark Topics:	
Transporting people and goods involves a combination of individuals and vehicles.	Chapter 9: pages 228, 240–242
Transportation vehicles are made up of subsystems, such as structural, propulsion, suspension, guidance, control, and support, that must function together for a system to work effectively.	Chapter 9: pages 240–242
Governmental regulations often influence the design and operation of transportation systems.	Chapter 9: pages 240–242
Processes, such as receiving, holding, storing, loading, moving, unloading, delivering, evaluating, marketing, managing, communicating, and using conventions are necessary for the entire transportation system to operate efficiently.	Chapter 9: pages 228, 240–242
Standard 19. Students will develop an understanding of and be able to select and use manufacturing technologies.	Chapter 14: page 365 Chapter 15: pages 378–392
Benchmark Topics:	
Manufacturing systems use mechanical processes that change the form of materials through the processes of separating, forming, combining, and conditioning them.	Chapter 15: pages 378–392
Manufactured products may be classified as durable and non-durable.	Chapter 15: pages 378–392
The manufacturing process includes the designing, development, making, and servicing of products and systems.	Chapter 15: pages 378–392
Chemical technologies are used to modify or alter chemical substances.	Chapter 14: page 365
Materials must first be located before they can be extracted from the earth through such processes as harvesting, drilling, and mining.	Chapter 15: pages 378–392
Marketing a product involves informing the public about it as well as assisting in selling and distributing it.	Chapter 15: pages 378–392

(continued)

Standards for Technological Literacy Correlation Chart *(continued)*	
Standard 20. Students will develop an understanding of and be able to select and use construction technologies.	Chapter 7: pages 166, 172–175, 181–185 DMA-12
Benchmark Topics:	
The selection of designs for structures is based on factors such as building laws and codes, style, convenience, cost, climate, and function.	Chapter 7: pages 166, 172–174
Structures rest on a foundation.	Chapter 7: pages 174–175
Some structures are temporary, while others are permanent.	Chapter 7: page 166 DMA-12
Buildings generally contain a variety of subsystems.	Chapter 7: pages 181–185

Scope and Sequence Chart

In planning your program, you may want to use this Scope and Sequence Chart to identify the major concepts presented in each chapter of the *Technology: Shaping Our World* text. Refer to the chart to select topics that meet your curriculum needs. The chart is divided into four groups of chapters. Within these groups, bold numbers indicate the chapters in which the concepts are found. Topics and their corresponding page numbers follow the chapter numbers for easy reference to the text.

Chapters 1–4

Core Concepts of Technology

1: Understanding Technology (12)

2: Solving a Problem (24); The Design Process (27)

3: Communication System (56); Purposes of Communication (57)

4: Types of Materials (87)

Connections to Other Fields of Study

1: How Science and Technology Are Related (16)

2: Elements of Design (34); Principles of Design (39)

3: Forms of Communication (58); Computer-Aided Drafting (73)

4: Properties of Materials (82); Types of Materials (87)

Technology and Its Impacts

1: How Technology Affects Us (12); Technology: Both Old and New (16)

Design and Problem Solving in Technology

2: Solving a Problem (24); The Design Process (27); Defining the Problem (28); Determining the Design Brief (28); Investigating (28); Developing Alternative Solutions (30); Choosing a Solution (30); Making Models and Prototypes (31); Testing and Evaluating (32); Manufacturing (34)

3: Drawing Techniques (67); Alphabet of Lines (69); Dimensioning (69); Using Drawing Techniques (70)

Technological Tools and Materials

4: Properties of Materials (82); Physical Properties (83); Mechanical Properties (83); Thermal Properties (83); Chemical Properties (85); Optical Properties (85); Acoustical Properties (86); Electrical Properties (86); Magnetic Properties (87); Types of Materials (87); Woods (87); Metals (92); Plastics (Polymers) (96); Ceramics (98); Composite Materials (101); Biomaterials (103)

Technological Systems and Processes

3: Communication System (56)

Applications of Technology

1: How Technology Affects Us (12)

2: Solving a Problem (24); Defining Need (24); Asking Questions (25); Finding Solutions (26); Model Making (26)

Chapters 5–8

Core Concepts of Technology

5: Safety (112); Shaping Materials (113); Finishing Materials (137)

6: Types of Structures (151)

7: What Is a System? (181)

8: Simple Machines (194); Gears (207); Pressure (212); Friction (218)

Connections to Other Fields of Study

5: Safety (112)

6: What Structures Have in Common (148); Types of Structures (151); Forces Acting on Structures (153)

7: The Structure of a House (166); Planning for a Home (169); Heating System (182); Electrical System (183); Plumbing Systems (183); Communication System (184); New Automated Systems (185)

8: Levers (194); Gears (207); Pressure (212); Mechanism (215)

Design and Problem Solving in Technology

5: Shaping Materials (113)

6: Types of Structures (151); Forces Acting on Structures (153); Designing Structures to Withstand Loads (154)

7: Planning for a Home (169)

Technological Tools and Materials

5: Safety (112); Shaping Materials (113); Joining Materials (124); Nails (125); Screws (125); Nuts and Bolts (128); Rivets (130); Knockdown (KD) Joints (130); Chemical Joining (132); Heat Joining (136); Finishing Materials (137)

7: The Structure of a House (166)

Technological Systems and Processes

Applications of Technology

Chapters 9–12

Core Concepts of Technology

Connections to Other Fields of Study

Technology and Its Impacts

Design and Problem Solving in Technology

Technological Tools and Materials

Technological Systems and Processes

Applications of Technology

Chapters 13–16

Core Concepts of Technology

Connections to Other Fields of Study

13: Information Technology (328); Computers (332); Robots (334); Virtual Reality (336); Microelectronics (336); Telecommunications (337)

14: Agriculture (349); Genetic Engineering (350); Protecting Our Environment (354)

15: Automation (382); CNC and FMS (385); Business and Office (394); Communications (394); Health (394); Hospitality and Recreation (395); Marketing and Distribution (395); Personal Services (396); Public and Social Services (398); Transportation (398)

16: Computers and Electronics (416); Telecommunications and Networking (418); Media and Entertainment (419); Nanotechnology and Materials (421); Energy and the Environment (422); Biotechnology and Healthcare (424); Transportation and Cities (426); Privacy, Security, and Defense (429)

Technology and Its Impacts

13: Information Technology (328); Computers (332); Telecommunications (337)

14: Genetic Engineering (350); Protecting Our Environment (354); Keeping Our Air Clean (355); Regenerating the Soil (356); Purifying the Water (356); Using Appropriate Types of Energy (357); Taking Action (360); Good Health (362)

15: Nonrenewable Raw Materials (377); Mechanization and Mass Production (380); Growth in Jobs (393); Looking to the Future (393)

16: The Beginning of Technology (408); The Present: How Technology Affects Us (409); Unforeseen Accidents (413); The Future (413); The New Technologies (416)

Design and Problem Solving in Technology

14: Genetic Engineering (350)

15: The Production System (386); Designing (387); Planning (388); Controlling Production (389)

Technological Tools and Materials

15: The Primary Sector: Processing Raw Materials (377); The Secondary Sector: Manufacturing Products (376); Tooling Up (388)

Technological Systems and Processes

Applications of Technology

Technological Enterprises

Basic Interdisciplinary Skills

The *Basic Interdisciplinary Skills Chart* has been designed to identify those activities in the *Technology: Shaping Our World* text and *Student Activity Manual* that specifically encourage the development of basic skills.

Academic areas addressed in the chart include reading, writing, verbal (other than reading and writing), math, science, and analytical.

◆ **Reading** activities include assignments designed to improve comprehension of information presented in the chapter. Some are designed to improve understanding of vocabulary terms.

◆ **Writing** activities allow students to practice composition skills, such as letter writing, informative writing, and creative writing.

◆ **Verbal** activities encourage students to organize ideas, develop interpersonal and group speaking skills, and respond appropriately to verbal messages. Activities include oral reports and interviews.

◆ **Math** activities require students to use basic principles of math, as well as computation skills, to solve typical problems.

◆ **Science** activities call for students to use fundamental principles of science to solve typical problems.

◆ **Analytical** activities involve the higher-order skills needed for thinking creatively, making decisions, solving problems, visualizing information, reasoning, and knowing how to learn.

Activities are broken down by chapter, and a page number is given to locate the activity.

Basic Skills Chart

	Chapters 1–4
Reading	**Text:** 1: Understanding Technology (12); Technology: Both Old and New (16); How Science and Technology Are Related (16) 2: Solving a Problem (24); The Design Process (27); Elements of Design (34); Principles of Design (39) 3: Communication System (56); Forms of Communication (58); Drawing Techniques (67); Computer-Aided Design (73) 4: Properties of Materials (82); Types of Materials (87)
Writing	**Student Activity Manual:** ST-1 You Are a Designer ST-5 Writing a Design Specification ST-13 Writing a Design Brief ST-17 Designing a Questionnaire DMA-1 Tallest Tower DMA-6 Games Galore
Verbal	**Text:** 3: Communication System (56); Forms of Communication (58) **Student Activity Manual:** DMA-1 Tallest Tower
Math	**Text:** 3: Forms of Communication (58) **Student Activity Manual:** ST-24 Drawing to Scale ST-25 Measuring Scale Drawings—Conventional ST-26 Measuring Scale Drawings—Metric ST-27 Reading Scale Drawings—Conventional ST-28 Reading Scale Drawings—Metric ST-30 Scaling Up—Drawing Things Bigger DMA-1 Tallest Tower DMA-5 Storage

(continued)

Basic Skills Chart *(continued)*

	Chapters 1–4
Science	**Text:** 1: How Science and Technology Are Related (16) 2: Solving a Problem (24); The Design Process (27) 3: Properties of Materials (82); Types of Materials (87) **Student Activity Manual:** ST-5 Writing a Design Specification ST-8 Exploring Existing Products ST-11 Investigating Pets ST-13 Writing a Design Brief ST-50 New Uses for Old Things DMA-2 New Uses for Everyday Products DMA-6 Games Galore
Analytical	**Text:** 2: Solving a Problem (24); The Design Process (27) **Student Activity Manual:** ST-4 Identifying User Needs and Interests ST-8 Exploring Existing Products ST-14 Designing for People—Anthropometric Data ST-50 New Uses for Old Things DMA-2 New Uses for Everyday Products DMA-6 Games Galore
	Chapters 5–8
Reading	**Text:** 5: Safety (112); Shaping Materials (113); Joining Materials (124); Finishing Materials (137) 6: What Structures Have in Common (148); Types of Structures (151); Forces Acting on Structures (153); Designing Structures to Withstand Loads (154)

(continued)

Basic Skills Chart *(continued)*

	Chapters 5–8
	7: The Structure of a House (166); Planning for a Home (169); Systems in Structures (181) 8: Simple Machines (194); Gears (207); Pressure (212); Mechanism (215); Friction (218)
Writing	**Student Activity Manual:** ST-55 Natural and Human-Made Structures ST-60 Exploring Animal Movements ST-61 Levers All Around Us ST-62 Exploring Linkages
Verbal	**Student Activity Manual:** ST-60 Exploring Animal Movements
Math	**Text:** 5: Shaping Materials (113) 7: Planning for a Home (169) 8: Simple Machines (194); Gears (207); Pressure (212); Mechanism (215) **Student Activity Manual:** ST-65 Moments and Levers—Conventional Measurements ST-66 Moments and Levers—Metric Measurements DMA-10 On the Move
Science	**Text:** 5: Shaping Materials (113); Joining Materials (124); Finishing Materials (137) **Student Activity Manual:** ST-52 How Trusses Work ST-53 Making Structures Rigid ST-57 Using Beams in Structures ST-58 Strengthening Beams ST-59 Using Gears ST-61 Levers All Around Us

(continued)

Basic Skills Chart *(continued)*

	Chapters 5–8
	ST-62 Exploring Linkages
	ST-67 Investigating Pneumatics and Hydraulics
	ST-68 Understanding Systems
	ST-69 Systems and Subsystems
	ST-70 Pneumatics—One-Way Valves
	ST-72 Using Pulleys
	DMA-10 On the Move
	DMA-16 Pneumatic Ergonome
Analytical	**Text:** 5: Planning for a Home (169) **Student Activity Manual:** ST-52 How Trusses Work ST-53 Making Structures Rigid ST-54 Struts and Ties ST-56 Forces Acting on Structures ST-59 Using Gears ST-61 Levers All Around Us ST-67 Investigating Pneumatics and Hydraulics ST-68 Understanding Systems ST-69 Systems and Subsystems ST-70 Pneumatics—One-Way Valves ST-72 Using Pulleys DMA-10 On the Move DMA-16 Pneumatic Ergonome

(continued)

Basic Skills Chart *(continued)*

	Chapters 9–12
Reading	**Text:** 9: Modes of Transportation (228); Engines and Motors (230); Transportation Systems (240); Exploring Space (242); The Impact of Transportation (244) 10: Forms of Energy (253); Where Does Energy Come From? (258) 11: Generating Electricity (278); Transmission and Distribution of Electricity (281); What Is Electricity? (285); Magnets and Magnetism (287); The Generation of Electricity Using Magnetism (290); Electric Motors (294); Cells and Batteries (295) 12: Electric Circuits (304); AND and OR Gates (309); Conductors, Insulators, and Semiconductors (310); Measuring Electrical Energy (313); Electronics (316)
Writing	**Student Activity Manual:** DMA-20 Game of Chance
Verbal	
Math	**Text:** 12: Measuring Electrical Energy (313); Electronics (316) **Student Activity Manual:** ST-79 Reading Resistors DMA-17 Robots DMA-18 Keep Out! DMA-20 Game of Chance
Science	**Text:** 9: Engines and Motors (230); The Impact of Transportation (244) 10: Forms of Energy (253); Where Does Energy Come From? (258) 11: Generating Electricity (278); What Is Electricity? (285); Magnets and Magnetism (287); The Generation of Electricity Using Magnetism (290); Cells and Batteries (295) 12: Conductors, Insulators, and Semiconductors (310)

(continued)

Basic Skills Chart *(continued)*

	Chapters 9–12
	Student Activity Manual: ST-76 Building a Series Circuit ST-77 Building a Parallel Circuit ST-78 Investigating Resistance ST-79 Reading Resistors ST-82 Drawing Circuit Diagrams ST-83 Using Diodes DMA-17 Robots DMA-18 Keep Out! DMA-19 Electronic Jewelry DMA-20 Game of Chance
Analytical	**Text:** 9: Transportation Systems (240); The Impact of Transportation (244) 11: Generating Electricity (278) 12: Electric Circuits (304); Measuring Electrical Energy (313) **Student Activity Manual:** ST-76 Building a Series Circuit ST-77 Building a Parallel Circuit ST-78 Investigating Resistance ST-79 Reading Resistors ST-82 Drawing Circuit Diagrams ST-83 Using Light Emitting Diodes ST-84 Using Diodes ST-85 Using Push-to-Make and Push-to-Break Switches DMA-17 Robots DMA-18 Keep Out! DMA-19 Electronic Jewelry DMA-20 Game of Chance

(continued)

Basic Skills Chart *(continued)*

	Chapters 13–16
Reading	**Text:** 13: Information Technology (328); Computers (332); Microelectronics (336); Telecommunications (337) 14: Food Production (348); Protecting Our Environment (354); Taking Action (360); Good Health (362) 15: The Primary Sector: Processing Raw Materials (377); The Secondary Sector: Manufacturing Products (378); The Tertiary Sector: Providing Services (392) 16: The Beginning of Technology (408); The Present: How Technology Affects Us (409); The Future (413) **Student Activity Manual:** ST-89 Using MyPyramid ST-90 Analyzing Food Labels ST-91 Choosing a Career ST-95 Technology in the News
Writing	**Student Activity Manual:** ST-86 Using Desktop Publishing to Produce Instruction Sheets ST-87 Using the World Wide Web to Gather Information ST-88 Using Desktop Publishing to Produce Graphics ST-89 Using MyPyramid ST-90 Analyzing Food Labels ST-91 Choosing a Career ST-92 Investigating Résumés ST-93 Using the World Wide Web to Investigate Career Opportunities ST-94 Writing a Cover Letter ST-95 Technology in the News ST-96 Noise Pollution EV-1 Evaluating the Final Product EV-2 Unit Review DMA-21 Pop-Up Cards

(continued)

Basic Skills Chart *(continued)*

	Chapters 13–16
	DMA-22 Party Plans DMA-23 Healthy Eating DMA-24 Prepacked Lunches DMA-26 Getting a Job DMA-28 Seeing Both Sides
Verbal	**Student Activity Manual:** ST-96 Noise Pollution EV-1 Evaluating the Final Product DMA-21 Pop-Up Cards
Math	**Text:** 13: Computers (332)
Science	**Text:** 13: Information Technology (328); Computers (332) 14: Food Production (348); Protecting Our Environment (354); Taking Action (360); Good Health (362) 15: The Primary Sector: Processing Raw Materials (377); The Secondary Sector: Manufacturing Products (378) 16: The Present: How Technology Affects Us (409); The Future (413) **Student Activity Manual:** ST-86 Using Desktop Publishing to Produce Instruction Sheets ST-89 Using MyPyramid ST-90 Analyzing Food Labels ST-91 Choosing a Career ST-92 Investigating Résumés ST-93 Using the World Wide Web to Investigate Career Opportunities

(continued)

Basic Skills Chart *(continued)*

	Chapters 13–16
Analytical	**Text:** 13: Computers (332) 14: Food Production (348); Protecting Our Environment (354); Taking Action (360); Good Health (362) 15: The Primary Sector: Processing Raw Materials (377); The Secondary Sector: Manufacturing Products (378) 16: The Present: How Technology Affects Us (409); The Future (413) **Student Activity Manual:** ST-86 Using Desktop Publishing to Produce Instruction Sheets ST-89 Using MyPyramid ST-90 Analyzing Food Labels ST-91 Choosing a Career ST-95 Technology in the News ST-96 Noise Pollution EV-1 Evaluating the Final Product EV-2 Unit Review

Technology
Shaping Our World

John B. Gradwell
McGill University

Malcolm Welch
Queen's University

Eugene Martin
Southwest Texas State

Publisher
The Goodheart-Willcox Company, Inc.
Tinley Park, Illinois
www.g-w.com

The Goodheart-Willcox Company, Inc. Brand Disclaimer: Brand names, company names, and illustrations for products and services included in this text are provided for educational purposes only and do not represent or imply endorsement or recommendation by the author or the publisher.

The Goodheart-Willcox Company, Inc. Safety Notice: The reader is expressly advised to carefully read, understand, and apply all safety precautions and warnings described in this book or that might also be indicated in undertaking the activities and exercises described herein to minimize risk of personal injury or injury to others. Common sense and good judgment should also be exercised and applied to help avoid all potential hazards. The reader should always refer to the appropriate manufacturer's technical information, directions, and recommendations; then proceed with care to follow specific equipment operating instructions. The reader should understand these notices and cautions are not exhaustive.

The publisher makes no warranty or representation whatsoever, either expressed or implied, including but not limited to equipment, procedures, and applications described or referred to herein, their quality, performance, merchantability, or fitness for a particular purpose. The publisher assumes no responsibility for any changes, errors, or omissions in this book. The publisher specifically disclaims any liability whatsoever, including any direct, indirect, incidental, consequential, special, or exemplary damages resulting, in whole or in part, from the reader's use or reliance upon the information, instructions, procedures, warnings, cautions, applications, or other matter contained in this book. The publisher assumes no responsibility for the activities of the reader.

Drawings by: Nadia Graphics

Cover Photo Credit: Comstock Images

Library of Congress Cataloging-in-Publication Data

Gradwell, John B.
    Technology shaping our world / John B. Gradwell, Malcolm Welch, Eugene Martin.
      p. cm.
    Includes index.
    ISBN-10: 1-59070-706-0
    ISBN-13: 978-1-59070-706-7
    1. Technology--Textbooks. 2. Engineering--Textbooks.
I. Welch, Malcolm. II. Martin, Eugene. III. Title.

T47.G69 2006
600—dc22   2006041212

Preface

Technology: Shaping Our World is written to help you understand the technological world around you. This book introduces you to the various technologies and shows how they have used basic scientific principles.

First you will study the solving of problems using the design process. Various chapters will introduce the major technologies: Communications, Manufacturing, Construction, and Transportation. The importance of materials, structures, systems, energy, people, machines, and information to all technological activity becomes clear as you study. You will also learn about the job opportunities in technology.

The technology of our earliest ancestors was simple; still, it helped them to live better lives. You will see how the invention of simple machines marked the advance of humans struggling to control their environment. Often, new developments affected only a small group of people. Travel being limited, a new plow might be developed and used in only one village. Even so, life was improved for the village through greater crop yields.

Technology today is usually more complicated. A technological decision might affect an entire city, a whole continent, or even the entire planet. Millions of people have experienced great changes in their lives as a result of new technology. Consider the new technologies affecting your life: computers, robots, surgical techniques, medicines, composites, microelectronics, biotechnology, and telecommunications. Among these technologies there are sure to be new inventions unheard of today.

It is very important that you begin to understand technology so that you can make intelligent decisions about its use. While a new technology may improve your life, you will need to consider its side effects. With the development of new technology comes the responsibility to control that technology so that no harm is done to humans, animals, or the environment. Your understanding of technology will help you care for our Earth and its people. In this way you will help shape our world. Your educated decisions will help provide a healthier environment.

Technology is an exciting subject. It affects what you wear, how you travel, what you eat, how you communicate with others, the comfort of your home, and much more. Technology is improving at an incredibly fast pace and expanding its reach into new areas every day. It is important to understand how and why technological changes come about. This book is written to help you become more informed and involved in technology and the world that is daily being shaped by it.

John Gradwell
Malcolm Welch
Eugene Martin

Brief Contents

Contents

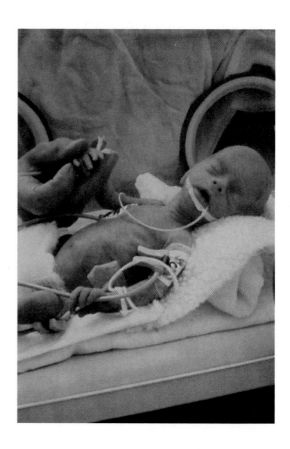

Contents

Contents

Contents

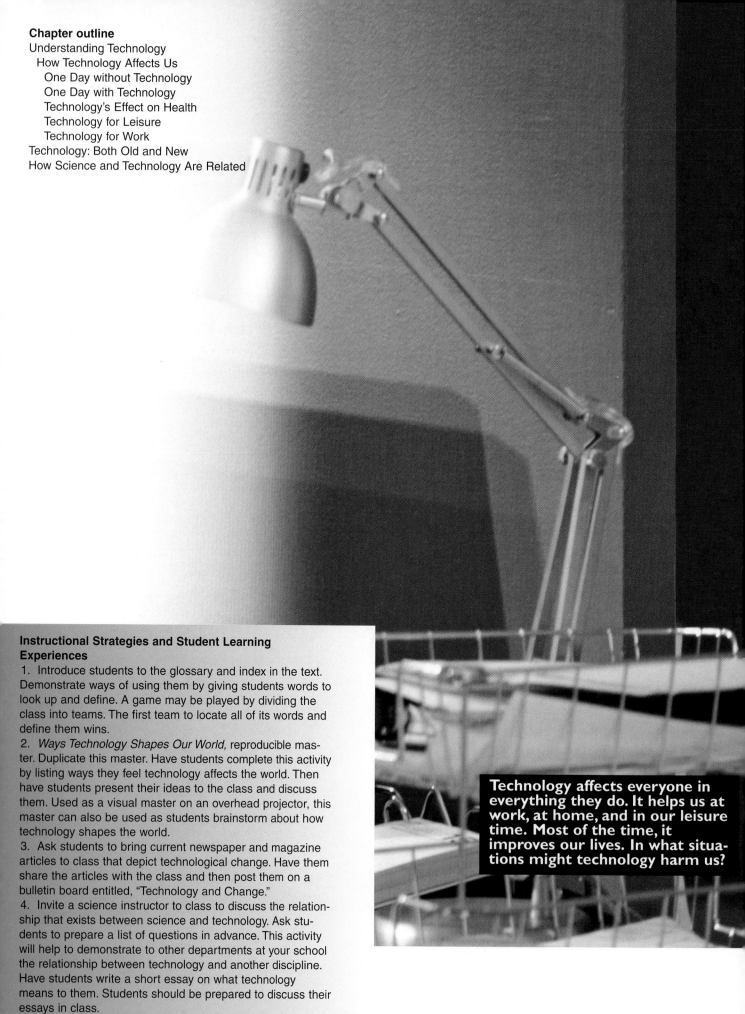

Instructional Strategies and Student Learning Experiences

1. Introduce students to the glossary and index in the text. Demonstrate ways of using them by giving students words to look up and define. A game may be played by dividing the class into teams. The first team to locate all of its words and define them wins.

2. *Ways Technology Shapes Our World,* reproducible master. Duplicate this master. Have students complete this activity by listing ways they feel technology affects the world. Then have students present their ideas to the class and discuss them. Used as a visual master on an overhead projector, this master can also be used as students brainstorm about how technology shapes the world.

3. Ask students to bring current newspaper and magazine articles to class that depict technological change. Have them share the articles with the class and then post them on a bulletin board entitled, "Technology and Change."

4. Invite a science instructor to class to discuss the relationship that exists between science and technology. Ask students to prepare a list of questions in advance. This activity will help to demonstrate to other departments at your school the relationship between technology and another discipline. Have students write a short essay on what technology means to them. Students should be prepared to discuss their essays in class.

Technology affects everyone in everything they do. It helps us at work, at home, and in our leisure time. Most of the time, it improves our lives. In what situations might technology harm us?

Technology and You— An Introduction

Discussion
Ask students what comes to mind when they hear the terms *design* and *technology*. Ask them to describe how their lives are affected by elements of both design and technology.

Career connection
Make a list of careers and describe the technological products and services involved in each.

Key Terms

designing
science
scientist
technologist
technology

Objectives

After reading this chapter you will be able to:

◆ Define "technology."

◆ Describe the benefits of technology and list some of the products it has developed.

◆ Identify the technological changes that have occurred in your lifetime.

◆ Explain the relationship that exists between technology and science.

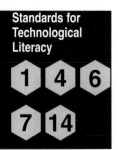

Standards for Technological Literacy

1 4 6
7 14

Designing and making
Have students describe the characteristics of a product they consider well designed and made. Are products that are well designed always well made?

Discussion
Which products in your home have a significant positive effect on your life? Which products have a significant negative effect on your life?

Vocabulary
Discuss the difference between the terms *technical*, *technique*, and *technology*.

Resource
Activity 1-1, *Picture Your World*

Space stations and robots… synthetic skin and artificial organs…supersonic aircraft… composite materials…"high-tech" sports equipment—these are all part of our lives. We hear about them every day. We use some of these items ourselves. We take them for granted. So what do they have in common? They are all products of technology. See **Figure 1-1** through **Figure 1-4**.

Understanding Technology

Designing involves generating and developing ideas for new and improved products and services that satisfy people's needs. *Technology* involves using tools and materials to make the designed products. Technological products meet your needs throughout the day and night.

Figure 1-1 The International Space Station proves that construction in space is a reality. (NASA)

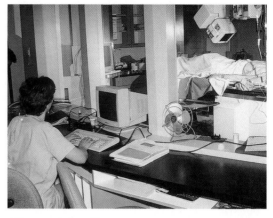

Figure 1-2 Electronic equipment is necessary in the hospital operating room. We are healthier because of computers and modern electronics. (RVH)

How Technology Affects Us

It is exciting when rockets hurl satellites into space. It is just as exciting to learn about the development of the artificial heart. Most of us are interested in the latest advancements of technology. However, our daily lives are more directly affected by less spectacular products of technology. For example, we all benefit from the use of personal computers, DVD players, digital cameras, e-mail, and the Internet.

One Day without Technology

Imagine that you had lived thousands of years ago. You wake up when the sun rises. There are no alarm clocks to ring. You would crawl out of your bed—a pile of animal

Useful Web site: Students can use the dictionaries at www.techtionary.com and www.hyperdictionary.com to find the definition of any new technical and technological terms used in this chapter.

Technology and society: Have students select one technology from this page and discuss who has benefited from its introduction and who has been disadvantaged.

Figure 1-3 Small jet aircraft not only facilitate international commerce but are also used for photo surveying and high-altitude mapping. (Bombardier)

skins spread over branches cut from trees. Your animal skin clothing hangs loosely around your body. There are no zippers, Velcro™, or even strings to fasten them.

Leaving your cave, you find that a pile of fallen rocks has partially blocked the entrance. You move the rocks by hand. There are no carts, and the wheel has not yet been invented.

What about breakfast? Want to make waffles or pour a bowl of cereal? No chance! There might be a leftover bone from yesterday's kill, but no microwave oven to reheat it.

Figure 1-4 Our leisure time is more enjoyable with sports equipment made from modern synthetics. (Laser)

If your mouth still tastes of breakfast, too bad. Toothbrushes and toothpaste do not exist. You scrape congealed sap off the bark of a tree and chew it to freshen your breath. The day is warm and sunny, and you feel like taking the day off. Unfortunately, you can't if you want to eat. Your entire family spends most of each day collecting enough food to survive.

Are you starting to appreciate the comforts and conveniences provided by modern technology? Technology plays a crucial role in our lives. Almost everything we do depends in some way on the products and services that form our technological society. These products and services are frequently the result of very sophisticated technologies. Often we take them for granted. It is only when we stop to consider what we do each day that we realize how important technology is to us.

One Day with Technology

Do you understand how important technology is and how much you depend on it? Think carefully about some of the items encountered during a typical day, **Figure 1-5.**

Your day might begin with your being awakened by a digital alarm clock/radio. The music you hear is transmitted (sent) from a station a distance away. You stumble out of bed into a hot shower. The flow and temperature of the water can be adjusted. You dry and style your hair and brush your teeth with plastic objects. Your clothes are made from a mixture of natural and synthetic fibers.

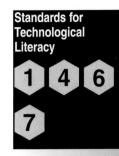

Standards for Technological Literacy

1 4 6 7

Designing and making
Have students describe the design of any modern product that makes their lives easier. Ask students to describe the materials used in any modern product that makes their lives safer.

Discussion
How would your life be changed by the sudden disappearance of a technology you normally take for granted?

Activity
Make a list of modern technologies used in your daily life. Next, list the equivalent technology used 100 years ago.

Resource
Activity 1-2, *Technology Words*

Links to other subjects: Discuss how modern technologies affect the way students study and learn other subjects.

Standards for Technological Literacy

① ④ ⑥

⑦

Designing and making
How does the design of kitchen appliances make time spent on food preparation more enjoyable?

Discussion
In what ways has technology changed the way your family eats?

Activity
Have students identify materials most suitable for food storage products.

Safety
How does the design of school buses ensure the safety of passengers?

Figure 1-5   How many examples of modern conveniences do you see on these shelves? (Ecritek)

In the kitchen, **Figure 1-6,** a variety of automatic appliances help you prepare your breakfast. Toast is ejected from the toaster. The electric kettle shuts off when the water boils. A microwave oven cooks your bacon for a preselected period of time.

Before leaving the house, you dress in clothing made from waterproof

synthetic fibers. As you leave, the timer built into the thermostat automatically lowers your home's temperature. If you live in an apartment building, you press a button to summon an elevator. Pressing a second button brings you to the ground floor. Once there, the doors will open automatically. You leave the building through a front door controlled by an electromagnetic lock.

Out on the street, vehicles use energy to move people and goods. You may use a bus or subway train to travel to school, **Figure 1-7.**

Your classroom uses electric lighting and may be heated by natural gas. The tables, desks, chairs, and cabinets have been made in a factory and transported to the school. The room is probably equipped with a computer, video projector, VCR, and a public address system.

Normally, we give little thought to these technological products and services on which we rely. Without them, your morning would have been far less comfortable and less safe.

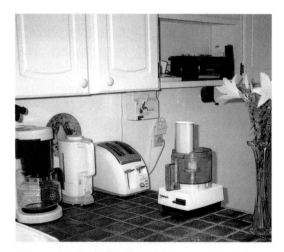

Figure 1-6   What would a cave dweller think of all the appliances in today's kitchens? (Ecritek)

Figure 1-7   How has technology changed the way students travel to school? (Ecritek)

Technology and society: Have students describe how modern technologies have made their homes safer and more comfortable.

Are you starting to recognize your dependency on the products of technology? Today, most people lead very "technological" lives. You have seen that, even during the first few hours of your day, technology is basic to your comfort and way of life. Are there any activities that do not depend on technology?

As the day progresses, you will continue to make use of many other products and services. For example, your school, is an artificial environment created by technology, **Figure 1-8**. Schools, as well as homes, are heated in the winter and cooled in the summer. They may be insulated to conserve energy.

Technology's Effect on Health

Your health depends on medical technology. Dentists use a wide variety of miniaturized tools and equipment to repair, replace, straighten, and keep your teeth in the best possible condition, **Figure 1-9**. Surgeons are able to replace a damaged heart valve with one made of metal and plastic. Diabetics can wear a tiny, computer-controlled infusion pump, **Figure 1-10**. It automatically delivers a supply of insulin to the wearer. Such advances in medical care help us to lead healthier and happier lives.

Technology for Leisure

Even in your leisure time, the products of technology surround you. Computer-designed tennis racquets use fiberglass, graphite, or ceramic to replace the conventional laminated wood. Road-racing bicycle frames,

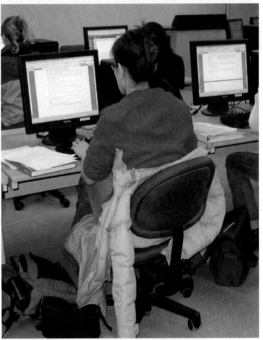

Figure 1-8 Many technology workshops and classrooms in schools use computers and data projectors to enchance teaching and learning. Internet access facilitates research for projects, word processing programs assist with writing reports, and drawing software supports designing products. (TEC)

Standards for Technological Literacy

1 4 6 7 14

Designing and making
Have students identify technological products and services that affect their health and well-being.

Discussion
Have students identify some of the materials used in modern sports equipment.

Reflection
Which of the activities you enjoy do not involve technological products?

Technology and society: Have students interview their parents and grandparents to identify how health care and leisure activities have changed in their lifetime, due to new technologies.

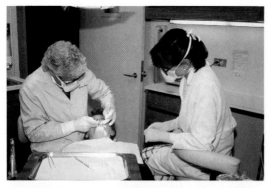

Figure 1-9 Technology has done much to ensure that we can maintain healthy teeth and gums. (Ecritek)

Figure 1-11, are now made of epoxy-glued aluminum or carbon fiber, and have a mass (weight) of less than half the average steel frame. Golf clubs use a super-hard aircraft alloy to replace hardwood. Jogging and ski suits are made of fabrics that keep out wind, rain, and snow, yet allow perspiration vapor to escape.

Technology for Work

Computer software can help you learn mathematics, create a paper, and check spelling and grammar, **Figure 1-12**. Using wireless Internet access on a pocket computer with a

Figure 1-10 This pump disperses insulin through a catheter into the body, according to information programmed into its microprocessor. (Minton)

Figure 1-11 This is a specially designed recumbent bicycle made for speed. It is low to ground and streamlined for as little wind resistance as possible. (www.uisreno.com/~photography/)

Figure 1-12 People use computers to help them do their work. A—Secretaries use computers to write reports and keep appointments and schedules. (Jack Klasey) B—CAD drafters make designs for architecture, electrical circuits, and many other things using computer programs.

miniature keyboard, one can check movie listings, verify bank funds, or buy gifts all while riding home on a bus.

Technology: Both Old and New

Technology is not really new, but new technology is developed daily. When early humans developed the bow and arrow, they were using technology. They solved a problem by designing and making something. The objects technologists make are often good solutions to the problems. Sometimes, however, the solutions are barely adequate, and others may be miserable failures.

How Science and Technology Are Related

Some people have suggested that technology and science are very similar. Many people think that technology is applied science.

While it is true that technology and science help one another, they are very different. *Science* is concerned with the laws of nature. *Scientists* (biologists, chemists, and physicists) seek to discover and understand these laws. Technology encompasses everything in the made world. *Technologists* (designers, inventors, engineers, and craftspeople) use the discoveries of scientists, along with tools and materials, to design and make products that change the made world, in order to fulfill human needs. See **Figure 1-13**.

People often assume that scientific knowledge came first—before technology. This is not true. Many technological advances came before the understanding of the principles behind them. For example, the wheel and axle was used long before humans understood the physical principles governing levers.

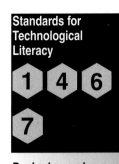

Designing and making
What products must designers streamline to reduce wind resistance?

Discussion
Have students think about one room in their home and name the various materials used to make the products contained.

Safety
Does the low profile of a recumbent bicycle make it safer to ride? What means could be taken to ensure that the rider and vehicle are more visible to other road users?

Links to other subjects: Have students identify the similarities and differences between their science studies and topics in technology classes.

Technology and society: Many technologies are designed to reduce the amount of work performed by humans. What are examples of these technologies? How are people happier and healthier as a result?

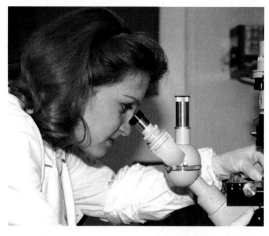

Figure 1-13 Scientists work to provide reliable knowledge of natural processes and events. Technologists use tools and materials to make products that improve people's lives.

These examples will help you to understand the relationship between technology and science:

◆ Technologists invented and built the early telescopes. Scientists used these telescopes to observe and calculate the distance from Earth to the planets. In turn, these scientific observations were used by technologists in the design of space vehicles.

◆ Scientists study the flow and formation of rivers. Technologists design and build dams across rivers.

◆ Technologists built the first steam engines. Scientists studied these engines to develop the laws of thermodynamics.

◆ Scientists study the causes and control of diabetes. Technologists design and build portable computer-controlled insulin pumps for diabetics.

◆ Technologists shape glass into tubes, bottles, and flasks. Scientists use these objects in experiments to analyze the chemical composition of substances.

◆ Scientists study the atomic theory. Technologists use the theory to build nuclear power stations.

This book explains and explores some of the knowledge and skills you will need:

◆ To understand technology.

◆ To use technology wisely.

◆ To create simple technological products for yourself and others.

Technology and society: What criteria should you consider when buying a product?

Chapter 1 Review
Technology and You— An Introduction

Modular Connections

The information in this chapter provides the required foundation for the following types of modular activities:

- Life Skills
- Practical Skills
- Introduction to Technology
- Development of Technology
- Explorations in Technology
- Technology and the Environment
- Technology Transfer
- Technological Systems

Test Your Knowledge

Write your answers to these review questions on a separate sheet of paper.

1. Identify three situations in which technological advances are affecting our daily lives.

2. Review the description in the text of a typical day in today's world. Then:
   A. List the technological objects that you use during the first few hours of the day.
   B. List five other technological objects, not included in the description, that you or a member of your family may use.

3. Technology is best defined as _____.
   A. discovering and understanding laws of nature
   B. designing machines and tools
   C. the skilled use of hand tools
   D. solving problems by designing and making objects

Answers To Test Your Knowledge Questions
1. Student response.
2. Student response.
3. D. solving problems by designing and making objects *(continued)*

19

4. Technology is an important subject to study because _____.
   A. most students will become engineers
   B. we live in a technological society
   C. it teaches you how to repair appliances
   D. it teaches you how to sketch

5. Give one technology-related word for each of the following letters of the alphabet.
   C _____
   E _____
   M _____
   S _____

Apply Your Knowledge

1. Explain the difference between science and technology.

2. Describe how technology affected your life during lunch hour today.

3. Collect pictures to illustrate different aspects of technology. Paste your pictures onto a sheet of paper to form a collage.

4. Compare old and present technologies by answering these questions.
   A. What would a doctor do to determine whether your arm was broken:
      before 1900?
      after 2000?
   B. How would the weekly laundry be cleaned:
      before 1900?
      after 2000?
   C. If you lived in a small village how would you travel to school:
      before 1900?
      after 2000?
   D. What means would you use to research and write a term paper:
      before 1900?
      after 2000?
   E. How would you communicate with family and friends who lived in another country:
      before 1900?
      after 2000?
   F. How would most homes be heated:
      before 1900?
      after 2000?

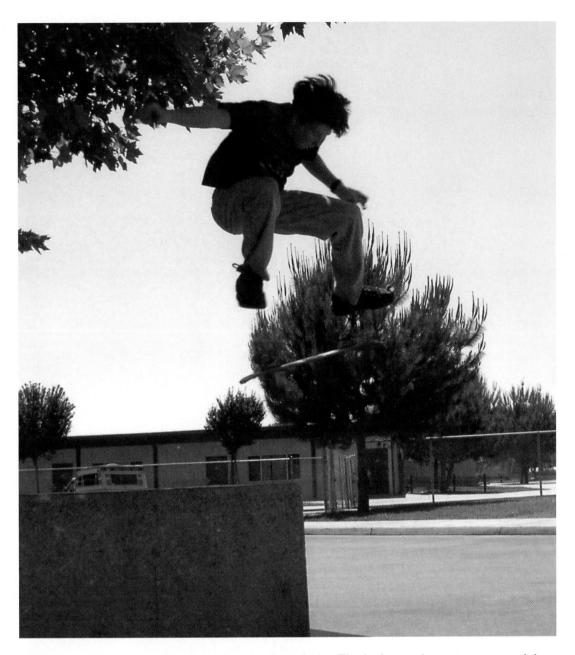

Technology is changing the way we live, work, and play. Think about what your parents did for fun when they were your age. How has technology changed the way students spend their free time today? (Ecritek)

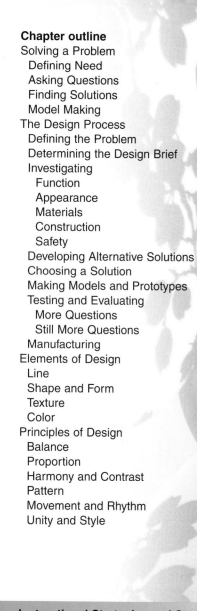

Instructional Strategies and Student Learning Experiences

1. Bring to class a variety of objects, such as a rock, a newspaper, a plant, a glass, an apple, a nail, soil, etc. Have students distinguish between objects found in nature and those made by humans.

2. *The Design Process,* reproducible master. Bring to class an item, such as a toy, kitchen gadget, or a tool. Using the master on an overhead projector, have students describe how they think the item was probably designed using the design process.

3. Have students divide into small groups. Ask each group to identify a design problem. Have them apply the steps of the design process to develop a solution to the problem and to build a prototype. Have the class evaluate the prototype.

4. Have students observe various activities in school. Ask them to list technological problems that could be solved using a technological device, such as a locker organizer, a pen and pencil holder, a lunch box that keeps cold foods cold and hot foods hot, etc. Have students select one of the problems identified and write a design brief, build a prototype, and evaluate the results.

5. Have students list the elements of design. Then have them analyze various objects and describe how each of the elements of design was used.

6. Have students list the principles of design. Ask them to bring objects to class and identify how the principles of design were applied.

Creating technology requires not only logic and intelligence but also creativity and ingenuity.

Generating and Developing Ideas

Standards for Technological Literacy

Key Terms

alternative solution
balance
contrast
design brief
designer
design process
elements of design
ergonomics
feedback
form
function
harmony
investigation

line
model
pattern
preproduction series
primary color
problem
proportion
prototype
rhythm
secondary color
shape
style
texture

Objectives

After reading this chapter you will be able to:

◆ Distinguish between objects found in nature and those made by humans.

◆ List strategies used when designing.

◆ Identify and define a design problem.

◆ Explain the steps in the design process.

◆ Use design process skills to generate and develop a solution to a problem.

◆ Recall the elements of design.

◆ Identify the principles of design.

◆ List in order the steps to build a prototype.

Discussion
Why is generating new and creative ideas an important human activity?

Career connection
Ask students to identify careers in which designing products and services is the central task. Have students browse the Web to find examples of the work completed by people working in these careers.

Student Activity Manual Resources
DMA-1, *Tallest Tower*
ST-1, *You Are a Designer*
ST-2, *Brainstorming*
ST-51, *Safety—Managing Risk*
EV-1, *Evaluating the Final Product*
EV-2, *Unit Review*

DMA-2, *New Uses for Everyday Products*
ST-18, *Freehand Sketching Techniques*
ST-19, *Sketching on Isometric Grid Paper*
ST-4, *Identifying User Needs and Interests*
EV-1, *Evaluating the Final Product*
EV-2, *Unit Review*

Standards for Technological Literacy

3 8 9

Designing and making
What problems have students solved by designing a product or service?

Discussion
Why is design an important topic to study?

Vocabulary
What words do you associate with the term *design*?

Activity
Have students create a collage illustrating the diversity of the made world.

The first person to shape a rock, making it a sharper tool, was solving a problem in technology. He or she was designing a product to meet a need.

Today every product we use has to be designed by someone, somewhere. These products may be as simple as a paper clip or as complex as a computer. In most cases, the product was designed because either the designer thought the product was needed for survival or that it would improve the quality of life.

In today's technological world, a *designer* creates and carries out plans for new products and structures. She or he will use special techniques. These help ensure a successful product or structure.

Most people, at some time in their lives, have solved a problem in technology. By so doing, they acted as designers, **Figure 2-1**. Building a sand castle or a tree house, constructing shelves for books or trophies, and decorating a birthday cake all involve designing. In each case, there was a need. In each case, the designer worked through a series of steps to arrive at a solution. This chapter will describe all the steps technologists use to solve problems. **Figure 2-2** describes a typical design problem and how the designer solved it.

Solving a Problem

Imagine you have been given a stack of small pieces of different colored paper. You have decided to use

Figure 2-1   Before this tree house could be built, someone had to solve design problems. What problems could you foresee in supporting the tree house? (Ecritek)

these to take telephone messages. Some were placed at the side of the telephone. When you came back an hour later, they were scattered over the floor. What could have happened? Perhaps, as in **Figure 2-3**, the wind blew them onto the floor when the door was opened. Someone may have bumped into the table knocking them to the floor. Obviously, there is a *problem*.

Defining Need

Let's think about exactly what is needed. You need a method of holding a stack of paper neatly and securely next to a telephone so that a message can be written on the top sheet. This

Links to other subjects: Art—How does the process of designing in technology differ from designing in art?

Useful Web site: Students can use the dictionaries at www.techtionary.com and www.hyperdictionary.com to find the definition of any new technical and technological terms used in this chapter.

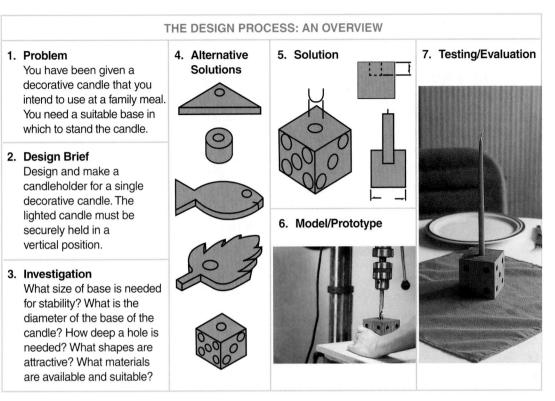

THE DESIGN PROCESS: AN OVERVIEW

1. **Problem**
You have been given a decorative candle that you intend to use at a family meal. You need a suitable base in which to stand the candle.

2. **Design Brief**
Design and make a candleholder for a single decorative candle. The lighted candle must be securely held in a vertical position.

3. **Investigation**
What size of base is needed for stability? What is the diameter of the base of the candle? How deep a hole is needed? What shapes are attractive? What materials are available and suitable?

4. **Alternative Solutions**

5. **Solution**

6. **Model/Prototype**

7. **Testing/Evaluation**

Figure 2-2 What alternative design solutions can you think of for a candleholder? (Ecritek)

Figure 2-3 Every design project begins as a problem.

statement describes clearly what is needed. Such a statement is called a *design brief*. It describes the problem.

Asking Questions

The first step in solving the problem is fun. You become a bit of a detective! You investigate. The kinds of questions you might ask include:

◆ What is the paper size?
◆ How many sheets are to be held?
◆ How much space is there beside the telephone?

Technology and society: Have students work in small groups to identify needs that must be met in designing a new bus shelter.

Standards for Technological Literacy

3 8 9

Designing and making
Have students describe different solutions to the problem of displaying indoor plants.

Discussion
What is the difference between making a model in technology education and assembling a model from a kit?

Discussion
How would you collect information to help design a new product that must meet an established need?

Vocabulary
Make a list of the different ways that the term *model* is used.

◆ Should a pen or pencil be attached?
◆ Will it be easy to write on the top sheet?
◆ Can the sheets be easily removed one at a time?
◆ What materials are available to make the product?
◆ How much material is available?
◆ Which material would be best for appearance and strength?

Finding Solutions

As you think about these questions, a number of *alternative solutions* may come to mind. Perhaps you would consider:

◆ Gluing the sheets together to make a pad.
◆ Using some kind of spring clip.
◆ Punching holes in the sheets and hanging them from a peg.
◆ Making a small box or container.

To remember all of these ideas, it is useful to sketch each one. You could also add some notes. Then, each sketch is better understood.

As a general rule, record your ideas on paper as they occur to you, **Figure 2-4**.

Suppose that you have sketched a number of alternative solutions. You can choose the one you think will work best. Let us assume that you prefer a container. The next step is to develop the idea of a container further. **Figure 2-5** shows a number of shapes and details.

Once again it is time to make a decision. What you choose will combine the best elements of these shapes and details. The container shown in **Figure 2-5C** is the preferred solution.

Model Making

At this point, you should make a full-size cardboard *model*, **Figure 2-6**. This model allows you to test and evaluate your solution. The best way to do this is to put it by the telephone and try it out, **Figure 2-7**. Does it satisfy the need stated in the design brief? Is the paper held neatly and securely? Can you easily write a message on the top sheet?

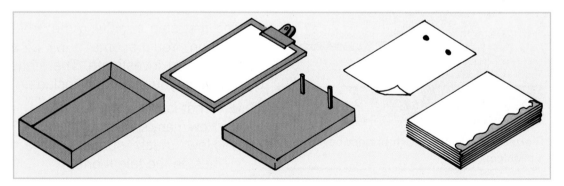

Figure 2-4 These sketches show possible solutions to the paper holder problem.

Technology and society: How can you determine whether a product is well designed?

Community resources and services: Invite an architect to talk with the class about their use of models in professional work.

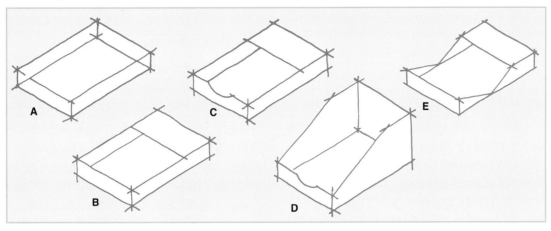

Figure 2-5 These sketches further refine the box idea in **Figure 2-4.** They show more details and more shapes.

Standards for Technological Literacy

3 **8** **9**

Designing and making
What materials have you used to make a model of an object?

Discussion
Why do you think it's important to model ideas?

Activity
List the steps you would follow to design and make a product that holds and organizes pencils in a briefcase.

If the model works, the next step is to make and test a prototype. You would use an appropriate material, such as plastic.

If you were a manufacturer, you would make a *preproduction series*. This is a small number of samples to be tested by typical consumers. These users would tell you how the product works. They would also tell if further modifications are required. Further, they would tell how much they would pay for it. Such information is called *feedback*.

The final step is to manufacture the paper holder. (Manufacturing is making products in a workshop or factory.) Manufacturing is often done by mass production.

The Design Process

Do you see that solving a problem involves working through a number of steps? It does not just happen. It is a careful and well thought-out

Figure 2-6 Making a model will show the likeness of the actual box.

Figure 2-7 Testing a prototype means trying out a working model.

Links to other subjects: What other subjects use models? What types of models are used and what do they represent?

Useful Web site: Have students browse www.biothinking.com to view alternative product solutions that are environmentally responsible. Students should print out information they find interesting to share with the class.

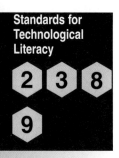

Standards for Technological Literacy

2 3 8

9

Designing and making
Why is writing a design brief an important early step in designing and making a product?

Discussion
Which of the products you use on a daily basis need design improvements?

Activity
Write a design brief for a product designed to help people retrieve small objects that have fallen behind a piece of heavy furniture.

Resource
Reproducible Master, *The Design Process*

procedure called the *design process*, **Figure 2-8**. Now, let us look at each of these steps in greater detail.

Defining the Problem

As you've just learned, the process of designing begins when there is a need. Wherever there are people, there are problems needing solutions. In some cases, the designer may have to invent a product. An example might be a game for someone with limited sight.

At other times, the designer may change an existing design. For example, if the handle of a pot becomes too hot to touch, it must be redesigned.

Designers also improve existing products. They make the product work even better. Could the chair in the waiting room of a bus or train station be altered so that waiting seems shorter?

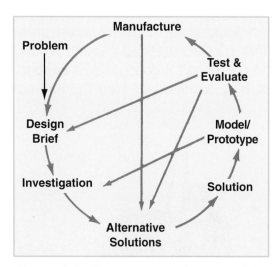

Figure 2-8 Every time we design, we follow these steps. We call these steps the design process.

Determining the Design Brief

A design brief should describe simply and clearly what is to be designed. The design brief cannot be vague. Some examples of problems and design briefs follow:

Problem: Visually impaired people cannot play many of the indoor games available to sighted people.

Design Brief: Design a game of dominoes that can be played by visually impaired.

Problem: The handle of a pot becomes too hot to hold when the pot is heated.

Design Brief: Design a handle that remains cool when the pot is heated.

Problem: Waiting time in a bus or train station seems too long. There is nothing to do.

Design Brief: Modify the seats so that a small television can be attached.

Investigating

Writing a clearly stated design brief is just one step. Now you must write down all the information you think you may need. Just write down your thoughts as they occur. Some things to consider are the following: function, appearance, materials, construction, and safety.

Links to other subjects: Describe the process of problem solving related to other subjects, such as Math, Science, and Psychology.

Technology and society: Which daily problems faced by elderly people can be solved by technology and which cannot?

Community resources and services: Invite an industrial designer, architect, or fashion designer into the class to discuss their particular design process.

Function

No matter how beautiful, an object that does not *function* well should never have been made. A functional object must solve the problem described in the design brief. The basic question to ask is: "What is the use of the article?"

Human beings like to be comfortable in the things they do. The products they use should be both easy and efficient to use. This is sometimes difficult to achieve. Humans vary in many ways. What suits one person often is not right for someone else. The study of how a person, the products used, and the environment can be best fitted together is called *ergonomics*. Ergonomics includes these considerations:

- Body sizes—can people fit the object?
- Body movement—can everything be reached easily?
- Sight—can everything be seen easily?
- Sound—can important sounds be heard and are annoying ones eliminated?
- Touch—are parts that a person touches comfortable?
- Smell—are there any unpleasant smells?
- Taste—are any materials toxic?
- Temperature—is the environment too hot, too cold, or comfortable?

Not all of these apply to every product. Look at **Figure 2-9**. In the design of a computer console, the seat, keyboard, and screen adjust in

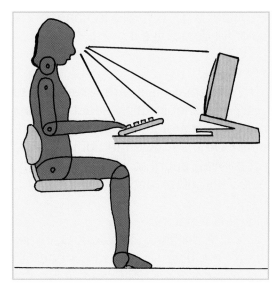

Figure 2-9 How could this computer station be designed so as to fit people of different heights?

various directions. Different sizes of people can use the same console.

Figure 2-10 shows two pairs of scissors. Those on the left have been designed to fit most people's hands. Their color makes them easily seen. The plastic is warm to the touch.

Appearance

How will the object look? The shape, color, and texture should make the object attractive. Once the project

Figure 2-10 Which of these pairs of scissors would you rather use? Why? (TEC)

Technology and society: Have students work in small groups to identify popular products that are poorly designed. Students should give reasons for their choices.

Useful Web site: Have students browse www.baddesigns.com and read one article about a product that functions poorly.

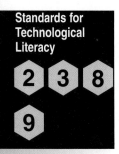

is made, will I be able to change the color if I do not like it?

Materials

What materials are available to you? You should think about the cost of these materials. Are they affordable? Do they have the right physical properties, such as strength, rigidity, color, and durability?

Construction

Will it be hard to make? Consider what methods you will need to cut, shape, form, join, and finish the material.

Safety

The object you design must be safe to use. It should not cause accidents.

Now that you know the questions to ask, you can begin looking for the answers. Where do you begin? Consider these sources:

- Existing solutions—look around you for similar articles, examine them, and collect pictures showing examples of other people's solutions.

- Libraries—search in your school or local library for magazines, books, and catalogs with relevant information and pictures.

- Experts—seek out people in industries, schools, and colleges who have this type of problem in their daily work.

Developing Alternative Solutions

Information in hand, you are ready to develop your own designs. You should produce a number of

solutions. In fact, as you were gathering your information, some ideas may have come to mind. Perhaps some were completely new ideas. Others may have been variations of existing ideas.

It is very important that you write or draw every idea on paper as it occurs to you. This will help you remember and describe them more clearly. It is also easier to discuss them with other people if you have a drawing.

These first sketches do not have to be very detailed or accurate. They should be made quickly. The important thing is to record all your ideas. Do not be critical. Try to think of lots of ideas, even some wild ones. The more ideas you have, the more likely you are to end up with a good solution.

Figure 2-11 shows a page from a designer's notebook. The design brief reads: "Design a container to hold at least four pens and three pencils. Items must be easily identified and removed." The designer thought of eight different solutions.

Choosing a Solution

You may find that you like several of the solutions. Eventually, you must choose one. Usually careful comparison with the original design brief will help you to select the best.

You must also consider:

- Your own skills.
- The materials available.
- Time needed to build each solution.
- Cost of each solution.

Deciding among the several possible solutions is not always easy. It helps to summarize the design

Technology and society: Why is it important that many different design versions of the same product be made available to the consumer?

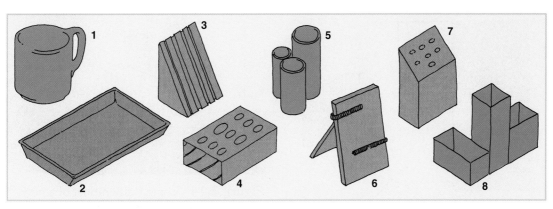

Figure 2-11 Eight possible solutions for a pencil holder design problem. Can you think of others?

requirements and solutions. Put the summary in a chart, **Figure 2-12**.

Three solutions, numbers 5, 7, and 8, satisfy all of the design requirements. Which would you choose? In cases like this, let it be the one you like best. The designer chose number 7.

In the next step, make a detailed drawing of the chosen solution. This drawing must include all of the information needed to make the pencil holder, **Figure 2-13**. It should include:

◆ The overall dimensions.
◆ Detail dimensions.

Design Requirements	Alternative Solutions							
	1	2	3	4	5	6	7	8
Holds 4 pens?	✓	✓	✓	✓	✓	✓	✓	✓
Holds 3 pencils?	✓	✓	✓	✓	✓	✓	✓	✓
Pens and pencils separated?			✓	✓	✓		✓	✓
Are pens and pencils easily removed and replaced?	✓	✓	✓	✓	✓		✓	✓
Is container stable?	✓	✓	✓	✓	✓		✓	✓
Attractive?					✓		✓	✓
Possible to make?		✓	✓	✓	✓	✓	✓	✓
Uses appropriate materials?	✓	✓	✓	✓	✓	✓	✓	✓
Tools are avalable?		✓	✓	✓	✓	✓	✓	✓
Materials are available?	✓	✓	✓	✓	✓	✓	✓	✓

Figure 2-12 This chart allows you to evaluate the solutions at a glance.

Technology and society: Why is thorough testing of a newly designed product an environmentally responsible step?

Designing and making
What questions must be asked about every product that is tested and evaluated?

Reflection
What questions do you ask yourself before buying a product?

Discussion
How would you test and evaluate a new cereal that claims to be healthier than others?

Safety
What general safety questions must a designer ask when testing and evaluating a new product?

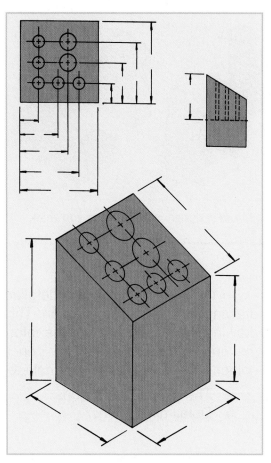

Figure 2-13 When dimensions are added, this detailed drawing will tell what size to make the holder and where to place the holes.

◆ The material to be used.
◆ How it will be made.
◆ What finish will be required.

Now you can choose what to do next. You can make a model and later a prototype, or you can go directly to making a prototype.

Making Models and Prototypes

A model is a full-size or small-scale simulation of an object, **Figure 2-14**.

Figure 2-14 Using a clay model of a new car, designers can view the car from various angles. They can then correct errors they see. (Buick)

Architects, engineers, and most designers use models.

Models are one more step in communicating an idea. It is far easier to understand an idea when seen in three-dimensional form. A scale model is used when designing objects that are very large. Buildings, ships, and planes are a few examples. In a scale model, the size is reduced. For example, a scale model could be built at one tenth of the full size.

A *prototype* is the first working version of the designer's solution. It is generally full-size and often handmade. For a simple object, such as the pencil holder, the designer probably would not make a model. He or she may go directly to a prototype, **Figure 2-15**. However, the designer would plan the steps for making the object. For example, the designer would:

◆ Select the materials.
◆ Plan the steps for cutting and shaping the material.
◆ Choose the correct tools.
◆ Cut and shape material.
◆ Apply finish.

Technology and society: Show students samples of fast food packaging. Have them evaluate the products in terms of their visual appeal, functional efficiency, and environmental impact. What ethical questions are raised by certain types of packaging?

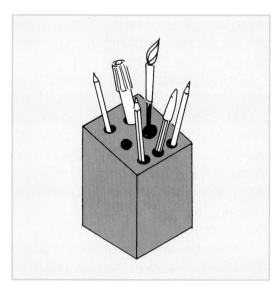

Figure 2-15 A prototype can be tested in a real-life situation.

The steps will vary depending on the object you are making. Some products have many parts. They must be assembled. The important thing is that you plan ahead.

Testing and Evaluating

Testing and evaluating answers three basic questions:

◆ Does it work?
◆ Does it meet the design brief?
◆ Will modifications improve the solution?

The question, "Does it work?" is basic to good design. It has to be answered. This same question would be asked by an engineer designing a bridge, by the designer of a subway car, or by an architect planning a new school. If you were to make a mistake in the final design of the pencil holder,

what would happen? The result might simply be unattractive. At worst, the holder would not work well. Not so if a designer makes mistakes in a car's seat belt design. Someone's life may be in danger!

More Questions

Testing and evaluating the pencil holder will provide answers to other questions:

◆ Will it hold at least four pens and three pencils?
◆ Do the pens and pencils fit the holes without being too tight or too loose?
◆ Is the container stable?
◆ Is it attractive?

Still More Questions

Other products may raise special questions. These must be answered. For example, if the product has several parts we may need to ask:

◆ How efficiently does it work?
◆ Will it last?
◆ Does it need maintenance?
◆ Will it need spare parts?
◆ Is it attractive?

All of these questions should be answered "yes." If not, the designer changes the design.

With corrections made to the prototype, it is time to make up a small number of samples. The samples are given to typical consumers. The consumers report their experiences to the manufacturer. Did it work well? How could it be improved? Is it attractive? Is it priced right? Designers use this feedback to make final changes.

Designing and making
What problems is a designer likely to encounter if he/she does not market test a new product?

Reflection
When you buy a new product, are you more interested in appearance than function?

Vocabulary
What terms describe well-designed products?

Activity
Have students work in small groups to evaluate the design of various wristwatch faces (use illustrations from magazines).

Technology and society: What questions should consumers ask before buying any product?
Community resources and services: What resources exist in your community to protect the consumer from poorly designed products and services?

Those making design changes must also remember that the product must be sold at a reasonable profit.

Manufacturing

The company is satisfied with the design. It knows that it is will sell. It must decide how many to make. Products may be mass-produced in low volume or high volume. Specialized medical equipment or airplanes are produced in the hundreds. Other products, such as nuts and bolts, are produced in large volume. Millions may be made.

The task of making the product is divided into jobs. Each worker trains to do one job. As workers complete their special jobs, the product takes shape. Mass production saves time. Since workers train to do a particular job, each becomes skilled in that job. Also, automatic equipment does such things as cut, weld, and spray.

Elements of Design

Designers are aware that people's needs and wants change with time. Products that are popular one year may not sell the next year. Anticipating these changes ensures that new products are designed and made around the same time that the old product is starting to lose customers, **Figure 2-16**. Young designers like you can be good at developing new ideas, as you don't have a financial investment in something that was made earlier. The first computers for

home use were designed and made by two young people in their garage, not by a large company.

People who design use the term *elements of design*. The term means the things you see when you look at an object. There are four elements. They are line, shape and form, texture, and color. You will find them combined in every object. The following pages will describe each of the elements.

Products are designed to satisfy needs. As we saw in the previous section, it is important that products function well. An easy chair must be comfortable. A pen must write. An airplane must fly. At the same time they must be attractive.

All objects appeal to our senses. We buy clothing that *looks* good. We enjoy the smooth *feel* of the polished wooden arm of a chair. A meal on a plate must not only look attractive but also should *taste* and *smell* good. The *sound* of musical chimes is preferable to the harsh sound of a door buzzer.

Figure 2-16 Are all products mass-produced in the same volume? Compare airplane manufacturing to making toys!

Technology and society: In the mass production process, a worker often repeats a small task thousands of times each day. What are the advantages and disadvantages to the worker?

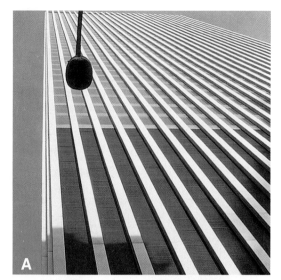

When you see something you like, ask yourself what is it you like about the product. Is it the color? Is it the shape or form? Is it the texture? Think about how each of these elements affects the appearance of the product.

Line

Lines describe the edges or contours of shapes. They show how an object will look when it has been made. Lines can also be used to create some special effects. For example, straight lines suggest strength, direction, and stability, **Figure 2-17**. What feeling do you get from curved and jagged lines, **Figure 2-18**? What do heavy or thin lines suggest, **Figure 2-19**?

A

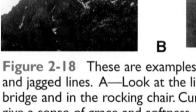

B

Figure 2-17 How we use straight lines in our designs. A—Vertical lines show strength. (Ecritek) B—Diagonal lines give a sense of movement. C—Horizontal lines give a feeling of stability (firmness). (Christopharo)

Figure 2-18 These are examples of curved and jagged lines. A—Look at the lines in the bridge and in the rocking chair. Curved lines give a sense of grace and softness. (Ecritek) B—A mountain ridge and a saw blade have jagged lines. Do they seem harsh and unfriendly to you?

Designing and making
Why is it important for designers to think about shape, form, and texture?

Reflection
What would my world be like if every building was the same shape and form?

Enrichment
Collect some natural objects (such as leaves, flower petals, and rocks) and examine their shape, form, and texture. What human-made products have similar characteristics?

Links to other subjects: In what other subjects are line, shape, form, and texture important? Are they used in the same ways?

Technology and society: How do designers use shape and form to create product identity? Find and sketch examples in your Technology notebook.

Standards for Technological Literacy

8

Designing and making
Have students identify the advantages and disadvantages of various table lamp shapes and forms.

Reflection
Do you prefer the shape and form of furniture to be primarily straight lines or curved lines? Explain your answer.

Activity
Working in small groups, have the class identify products in which texture is a very important design feature.

Shape and Form

All objects occupy space or possess volume. We say that *shape* is two-dimensional, **Figure 2-20**, and *form* is three-dimensional, **Figure 2-21**.

Shapes and forms may be geometric, organic, or stylized. Geometric shapes can be drawn using rulers, compasses, or other instruments, **Figure 2-22**. Organic and stylized shapes and forms are usually drawn freehand. **Figure 2-23** and **Figure 2-24** show both types.

Texture

Texture refers to the way a surface feels or looks. We can describe a surface as rough, smooth, hard, slippery, fuzzy, or coarse. Sandpaper feels rough. Glass feels smooth. Rock feels hard. Ice feels cold and slippery. A fur coat feels warm and fuzzy. Sand feels coarse. **Figure 2-25** and **Figure 2-26** show texture in wood and stone.

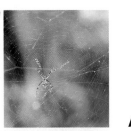

A

B

Figure 2-19 These are examples of thin lines and heavy lines. A—We find thin lines in nature and in manufactured products. We think of them as weak. (TEC) B—Heavy lines show extra strength.

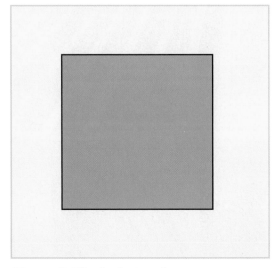

Figure 2-20 In design, shape means two dimensions—width and height.

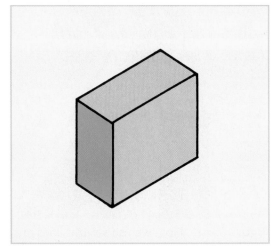

Figure 2-21 Form has three dimensions—width, height, and depth.

Technology and society: Why should designers pay special attention to the way products look and feel?

Figure 2-22 Here are a few examples of geometric shapes. A—Bricks are rectangles. (TEC) B—Pipes are circles. (Ecritek) C—A house may combine several geometric shapes. Do you see rectangles and triangles? (Ecritek)

Figure 2-23 There are many organic shapes and forms in nature. A—A blossom is a natural organic shape. (Ecritek) B—In the past, architecture has copied natural organic shapes found in leaves and flowers. (Ecritek) C—The curved, free-form shape of this chair makes it a good example of organic shape. (Christopharo)

Standards for Technological Literacy

8

Reflection
Do you prefer geometric, organic, or stylized shapes and forms?

Activity
Have students sketch a flower to show its shape. Next, have them sketch just one petal of the same flower to show its shape. Finally, have students sketch a product that includes a shape similar to the petal.

Technology and society: Have the class work in small groups to identify a product with a shape and form appealing to every member of the group. Ask them to describe why they find the shape and form of the product attractive. Does the rest of the class find the product attractive for the same reasons?

Figure 2-24 Look at these stylized shapes and forms. A—Anyone will recognize these stylized signs without the need for language. (Ecritek) B—Notice the detail on this stylized art form. (Christopharo)

A designer can choose materials according to their natural texture. She or he might also choose materials because of the way the texture can be changed.

Color

We only see color when light shines on objects. Sunlight appears to be white. In fact, it is a mixture of seven different colors. All objects react to light energy by either absorbing light energy or reflecting the various bands of light. An object that reflects all the light appears to be white. One that absorbs all the light appears to be black. Grass reflects the green bands of light and absorbs the other.

Figure 2-25 The texture of wood is pleasing. A—Here is wood in its natural state. (TEC) B—Sawed and nailed to a roof, wood shingles have a different texture from a tree. C—Sanded, stained, and polished wood has still another texture. (TEC)

When a beam of light shines through a glass prism, the path of light is bent. Each color bends at a different angle. The colors can then be seen individually. These seven colors form a spectrum, **Figure 2-27**.

The three most important colors are red, yellow, and blue. These are called *primary colors*. If you mix equal parts of two primary colors you

Figure 2-26 The texture of stone is attractive in scenery, in a wall, or used as art or jewelry. (Christopharo, TEC)

obtain a *secondary color*. Red plus yellow gives orange. Yellow plus blue gives green. Blue plus red gives violet. Mixing equal parts of a primary and a secondary color creates a *tertiary color*, **Figure 2-28**.

When you are selecting colors you may want them to harmonize or to contrast. Harmony means the colors naturally go together. You will find harmonizing colors next to one another on the color wheel, **Figure 2-29**. They are similar.

For example, if you first choose orange then the harmonizing colors are red-orange and yellow-orange.

On the other hand you may want your colors to contrast with each other. In that case you would select colors that are at the opposite side of the color wheel. For example, blue contrasts with orange. Contrasting colors are also called *complementary colors*.

Designers use colors to produce certain reactions or effects, **Figure 2-30**. Traffic signs use red to indicate danger. Yellow serves as a traffic warning. We also associate colors with objects. An apple is red. Grass is green. Charcoal is black, and milk is white.

Colors also produce emotions. Some are bright, while others are dull. Some are exciting, while others are boring. Colors are very personal. What is eventually right or wrong may depend on your own choice.

Principles of Design

You learned earlier that line, shape and form, texture, and color are the elements of design. You can think of these as building blocks that can be put together in many different ways. There are also guidelines for combining these elements. These guides are

Community resources and services: How are the elements of design used to enhance the beauty of the parks and recreation areas in your neighborhood?

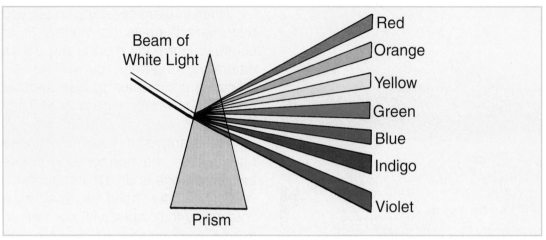

Figure 2-27 The seven colors in white light separate when shone through a glass prism. Where do you see this happen in nature?

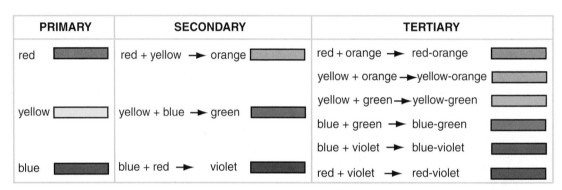

PRIMARY	SECONDARY	TERTIARY
red	red + yellow → orange	red + orange → red-orange
		yellow + orange → yellow-orange
		yellow + green → yellow-green
yellow	yellow + blue → green	blue + green → blue-green
		blue + violet → blue-violet
blue	blue + red → violet	red + violet → red-violet

Figure 2-28 This chart shows you how the mixing of primary colors can produce the secondary and tertiary colors. Which do you like best?

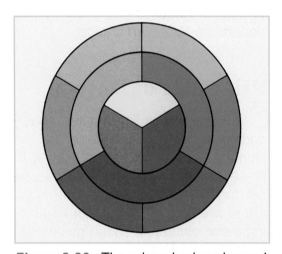

Figure 2-29 The color wheel can be used to pick colors that harmonize or contrast to each other.

called the principles of design. They include balance, proportion, harmony and contrast, pattern, movement and rhythm, and unity and style.

Balance

You can think of balance as a tightrope walker moving along a cable. She or he keeps balance using arms and a balancing pole. It is important to match or balance the mass of the body on both sides, **Figure 2-31**. *Balance* is also very important in design. It means that the mass is evenly spread over the space

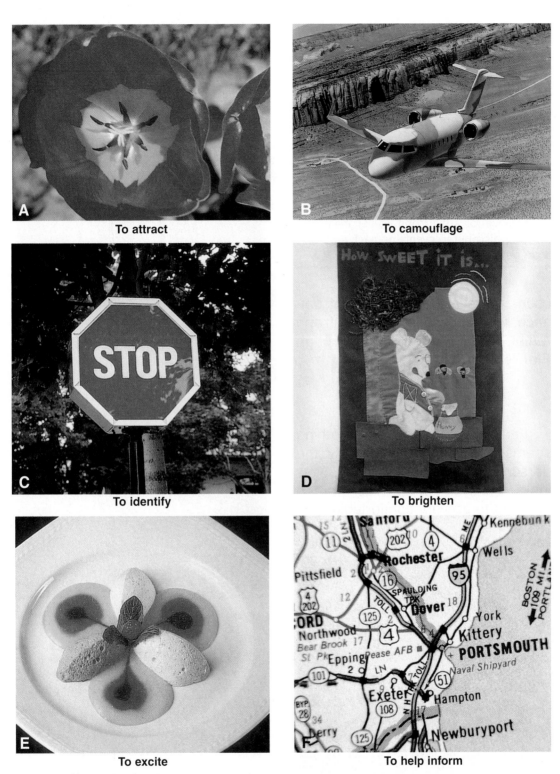

Standards for Technological Literacy

8

Discussion
How can color be used to make meals more appetizing?

Activity
Have students examine road maps to identify how color is used to convey information.

Safety
How are colors used to identify hazardous materials and situations?

Figure 2-30 Color can be used to create special effects. (Ecritek, Christopharo, TEC, Bombardier, Ecritek, TEC)

Links to other subjects: Geography—How is color used to identify features on a map?

Technology and society: Identify products and services that use color to make life easier or safer.

Figure 2-31 Think of balance in design like balance for a tightrope walker. Mass must be distributed evenly on each side of a centerline. (TEC)

used. There are three types of balance: symmetrical, asymmetrical, and radial. See **Figure 2-32** through **Figure 2-34**.

Proportion

Look at **Figure 2-35**. Something seems to be wrong. Obviously the person is too big for the chair. Now look at **Figure 2-36**. The person appears to be very comfortable. The

Figure 2-32 An object, such as this church building, is symmetrical if one half is a mirror image of the other half. (Ecritek)

relationship between the person and the chair seems to be right. The relationship between the sizes of two things is called *proportion*. In

Figure 2-33 In an asymmetrical design, the two sides are in balance visually, but they are not mirror images.

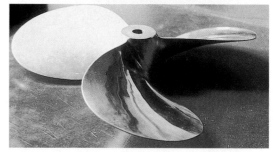

Figure 2-34 In radial balance, the mass moves outward in all directions from a point at or near the middle. Two examples are shown.

Technology and society: What health costs result from people buying and using furniture that is designed with the wrong proportions?

Useful Web site: Search the Web for home product designs that exhibit symmetrical, asymmetrical, and radial balance characteristics. Print and share interesting designs with the class. Helpful places to start include www.gomod.com, www.urbanicons.com, and www.retrospective.net.

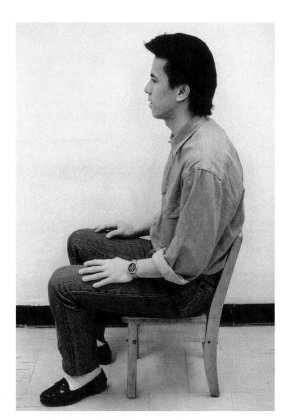

Figure 2-35 When one object is too large or too small for another object, they are out of proportion.

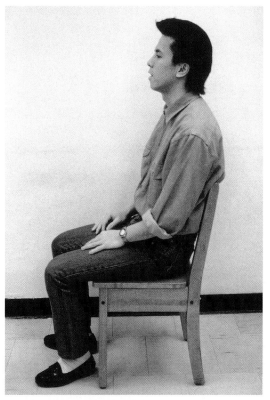

Figure 2-36 When two related objects in a design are right for each other, they are in proportion. (Christopharo)

Figure 2-35, the chair and the person are not in proportion. In Figure 2-36, they are in proportion.

Proportion may apply to the relationship between objects. It can also apply to the parts of an object. Look at the doors and drawers of the cabin in an executive jet plane, Figure 2-37. Their size is related to the overall size of the cabin. They are in proportion.

For thousands of years people have admired the proportions found in nature. The Greeks worked out a mathematical formula. It describes proportions found in nature. They called this formula the "golden mean." The golden mean has a ratio of 1:1.618. (In the case of a rectangle, the long side is a little more than 1 1/2 times longer than the shorter side.) A golden rectangle may be drawn using the following procedure:

1. Draw a base line.

2. Draw a square. The length of one side of the square is the length of the short side of the rectangle.

3. Measure halfway along the base of the square as shown in Figure 2-38. Put the point of your compass here. Now draw an arc from the top corner of the square to the base line.

Links to other subjects: History—Have students examine classical Greek architecture and identify where the "golden mean" is used in structures.

Standards for Technological Literacy

8

Designing and making
Identify and describe products in which the construction materials are in harmony with the function.

Reflection
Identify what makes you feel in harmony with your surroundings.

Activity
Collect pictures of both natural and human-made objects that illustrate the principle of harmony.

Resource
Activity 2-4, *Proportion*

Figure 2-37 Notice the cabinetry in a jet airplane. Doors and drawers must be scaled down in size so they are in proportion to the space. (Bombardier)

4. The point where the arc touches the base line is the right hand corner of the rectangle. Draw a vertical line upward from it. Then extend the top line of the square to complete the rectangle.

The golden mean also appears in the human body and many living things. In **Figure 2-39**, the lion's proportions fit the golden mean.

Mathematics is important to designers. Still, they do not rely on mathematics alone to decide the proportions of an object. They must adjust the proportions until they look right. Look at the chest of drawers, **Figure 2-40**. Notice that the drawers at the bottom are deeper than those at the top. If the drawers were of the same depth the chest of drawers would seem to be top heavy.

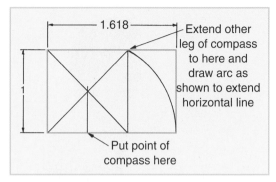

Figure 2-38 To draw a golden rectangle, start with a square.

Harmony and Contrast

Observe the best figure skaters and you will notice that their movements seem to flow with the music. We say they are in harmony with the music. *Harmony* is the condition in which two things, such as color or musical notes, naturally go together.

Designers use the idea of harmony in the objects they create. Buildings and their environment should be in harmony. The dishes in **Figure 2-41** go together well. Their shapes and colors are in harmony.

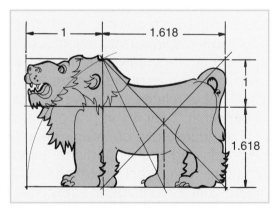

Figure 2-39 The lion's body fits the golden mean so often found in nature.

Useful Web site: Have students browse Web sites of award-winning furniture designers, such as www.thosmoser.com and www.greendesigns.com. Students should print out information they find inspirational and share it with the rest of the class.

Figure 2-40 Why should the bottom drawers be deeper than the upper ones? (Ecritek)

Figure 2-41 These objects are in harmony. Both color and shape go well together.

Sometimes designers want to surprise you. They may want to make you feel excited about what you see. They may simply want to catch your attention. To do this, designers create an obvious difference between things. This difference is called *contrast*.

You may wear bright clothes with contrasting colors, **Figure 2-42A**. The red cross on an ambulance contrasts with its white background. The jagged mountain contrasts with the calm waters of the lake. Both harmony and contrast are used to make a designer's work attractive, **Figure 2-42B**. Harmony makes you feel comfortable. Contrast adds excitement.

Pattern

What does the word *pattern* mean to you? A pattern is a design in which a shape is repeated many times. Look around you. Where can you see patterns? They are in unexpected places. As shown in **Figure 2-43**, patterns are found in nature and in objects people have designed and made. Sometimes they are used to make an uninteresting surface appear more attractive. At other times, the pattern may serve a particular function.

Movement and Rhythm

Ocean waves create a pattern. The lines on an oscilloscope have a similar pattern, **Figure 2-44**. Both suggest movement. Because they are repeating patterns, they are also said to have *rhythm*.

Standards for Technological Literacy

8

Designing and making
How do designers use contrast and pattern in the development of safety equipment?

Reflection
Do you select harmonious or contrasting clothing colors when dressing for a special occasion?

Activity
Identify household items that make use of a pattern to add to their appeal. Explain how the pattern makes the item more attractive.

Useful Web site: Have students browse Web sites of chinaware manufacturers, such as www.denbypottery.co.uk, www.lenox.com, and www.placesettings.com. Students should print out china patterns that appeal to them and share with the class.

Designing and making
Look at Figure 2-43. What other natural objects have distinctive patterns? What other products have patterns as an integral part of their design?

Enrichment
Have students bring in items from different cultures to illustrate their distinctive colors and patterns.

Vocabulary
In design education, the term *pattern* refers to a shape repeated many times. What other meanings does the term *pattern* have?

A

B

B

C

Figure 2-42 Look at two types of contrast. A—Here is contrast through use of color. B—This picture shows contrast through use of lines and shapes.

Figure 2-43 A—Do you see a pattern in this orange? Can you think of other patterns in nature? B—Sometimes a pattern has a function. Here the squares are needed to play a game of chess. C—Some patterns are created by arrangement of parts. Others are created by the material. Some patterns are applied. Which type is this? (TEC)

Technology and society: Have students research the use of movement in local architecture. Students should find examples of stationary architectural objects that give a sense of movement and share with the class.

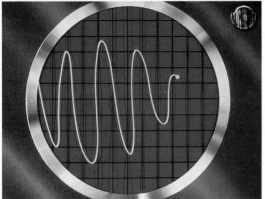

Figure 2-44 Patterns suggest movement. Both the ocean waves and the wavy pattern on the oscilloscope have rhythm because the pattern is repeated many times. (Laser)

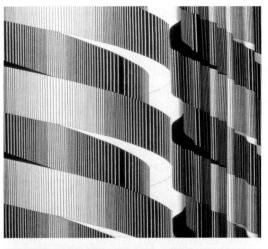

Figure 2-45 Shapes and lines give a sense of movement in objects created by humans. (Ecritek)

In **Figure 2-45**, the spiral of the printed pattern suggests a feeling of movement. The tulip bowl creates a sense of movement through its use of shape and line. Both the pattern and the bowl also have rhythm.

Unity and Style

To sum up what you have learned:
◆ The elements of visual design are line, shape and form, texture, and color.

◆ The principles of visual design are balance, proportion, harmony and contrast, pattern, and movement and rhythm.

Technology and society: Have students discuss how the use of design elements and principles make human-made environments more attractive.

Designing and making
Have students discuss the relative importance of style and function elements in products.

Reflection
Which products do you purchase because they are "in style"? Which do you purchase because they function well and are reasonably priced?

Activity
Have students sketch a structure in their neighborhood that illustrates several design elements and principles.

Resource
Activity 2-2, *Elements and Principles of Design*

All of these are used when designing an object. However, do not think of them as separate. Remember that they are all related. A well-designed product must have a sense of unity. Within the design there must be a sense of belonging or similarity among the parts, **Figure 2-46**. Remember, also, that the principles do not provide hard and fast rules. They act only as a guide. You must be the judge as to what is right or wrong. You must make the design decisions.

You can design an object any way you like. You can design according to your own style. There is no need to copy what other people have done. Each designer has his or her unique style. For example, Mies van der Rohe designed curved steel furniture. Notice how different his chair is made from the one you might ordinarily see in a home, **Figure 2-46**.

Style depends on many things:

◆ The availability and cost of materials.

◆ The tools and techniques available to shape the materials.

◆ Cultural preferences.

◆ A knowledge of the elements and principles of design, **Figure 2-47**.

Figure 2-46 Each of these structures has a sense of unity. (Ecritek)

Useful Web site: Have students browse www.fondationlecorbusier.asso.fr to examine the architecture of Le Corbusier. Students should print out information they find interesting and share it with the class.

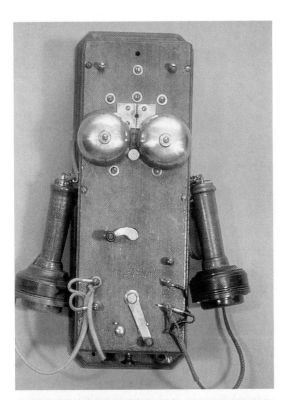

Figure 2-47 What factors do you think have influenced the design of these telephones? (TEC)

Standards for Technological Literacy

8

Designing and making
Why is it important for products to be visually appealing and functional, as well as safe to use?

Reflection
Do you consider the made world to be attractive? If not, what changes would you make?

Activity
Select one product and investigate how its design has changed over the past 100 years.

Useful Web site: Have students visit www.baddesigns.com to investigate how often poorly designed products reach the market. Students should print information they find interesting and share it with the class.

Technology and society: What strategies can you use to help generate and develop ideas for a new product or service?

Chapter 2 Review
Generating and Developing Ideas

Summary

Most people have solved problems and acted as designers. This chapter has described how problems in technology are solved by working through a series of steps. These steps first involve identifying a problem and writing a design brief. Investigating the problem further to find out all the information needed is followed by producing a number of alternative solutions. From these alternatives, one solution is chosen. Detailed drawings are then made. A model or prototype is built and tested. After modification, the product is ready for mass production.

Products must function well. They must also look and feel good. Their attractiveness is a result of their line, shape, form, texture, and color. These elements of design are combined using the principles of design. These principles are balance, proportion, harmony, contrast, pattern, movement, rhythm and style.

Modular Connections

The information in this chapter provides the required foundation for the following types of modular activities:

◆ Technical Visualization

◆ Technology Problem Solving

◆ Research and Development

◆ Research and Design

◆ Engineering Design

◆ Creative Solutions

◆ Prepared Speech

◆ Business Presentations

◆ Marketing

◆ Extemporaneous Presentation

Test Your Knowledge

Write your answers to these review questions on a separate sheet of paper.

1. Why are most new products invented?

2. List the eight steps in the design process.

3. The first step in the design of a storage container is to _____.
   A. buy the wood, glue, nails, and hinges
   B. prepare the tools you will need
   C. list and measure all the items to be stored
   D. decide its color and shape

4. The most important information for a designer planning a new seat for a bus is the _____.
   A. type of metal or plastic to be used
   B. color of the seat material
   C. time taken to manufacture each one
   D. average size of people using it

5. Given a design problem, an engineer would sketch several possible solutions because _____.
   A. there is a good range of ideas from which to choose
   B. it is difficult to decide which is the best solution
   C. similar objects can be made using different materials
   D. many people want to see the sketches

6. What is a prototype and why would one be built?

7. The study of how a person, the products used, and the environment can be best fitted together is called _____.

8. What specific questions would you ask if you were testing and evaluating a new wheelchair?

9. Think of one object that you have seen and find attractive. Describe in your own words, (a) the object and (b) how the elements of design make it attractive to you.

10. Construct a golden rectangle with a short side of 2″ (50 mm).

Answers to Test Your Knowledge Questions
1. To satisfy a need.
2. Defining the Problem. Determining the Design Brief. Investigating. Developing Alternative Solutions. Choosing a Solution. Making Models and Prototypes. Testing and Evaluating. Manufacturing.
3. C. list and measure all the items to be stored
4. D. average size of people using it
5. A. there is a good range of ideas from which to choose
6. A prototype is a full-scale working model used to identify weaknesses or errors in the design.
7. ergonomics
8. Can a handicapped person get into and out of the chair easily? Are the controls comfortable and easy to use? Is the chair maneuverable? Is there a place for storage of personal items? Is it attractive? Is it affordable?
9. Student responses will vary.
10. Refer to Figure 2-39, page 44 of the text.

Apply Your Knowledge

1. Collect or draw pictures illustrating three natural objects. Collect three more pictures to show the equivalent technical objects. For example, if you collected a picture of a bird's nest, a natural object, the equivalent technical object would be a house, a human-made object.

2. Collect four pictures to illustrate two elements of design and two principles of design.

3. Carefully observe some activities in your home. Make a list of technological problems that need to be resolved. For example, storing spices in the kitchen or making sure that your baby sister doesn't fall downstairs.
   A. From the list of problems above, identify one that you will try to resolve. Write a design brief for the problem.
   B. Make a list of the questions you will have to answer to solve the design problem.
   C. Generate a number of solutions to the problem and select the most appropriate solution.
   D. Make a list of the steps involved to build a prototype of your solution.
   E. Build, test, and evaluate your prototype. Keep a good record, in the form of notes and sketches, of all changes and modifications. Make recommendations for further improvements.

4. Design and build a device that will make one task easier in the life of a person with a special need, that is, a handicapped, sick, or elderly person.

5. Sketch as many different designs as you can showing how compact discs could be stored. Consider horizontal, vertical, and diagonal designs.

6. Which products existed before those listed below that filled a similar need? Why were the new products much better?
   A. ice cream cones (became popular in 1900s)
   B. sneakers (1910s)
   C. Band-Aids (1920s)
   D. refrigerators (1930s)
   E. nylon stockings (1940s)
   F. television set (1950s)
   G. nuclear reactors (1960s)
   H. skateboards (1970s)
   I. cellular phones (1980s)
   J. E-mail (1990s)

7. New products are generally designed to fulfill a need, but are some products designed that we really do not need? For example, when we have triple-blade razors, do we really need ones with four or five blades? Are there some features of a product in your home that you never use?

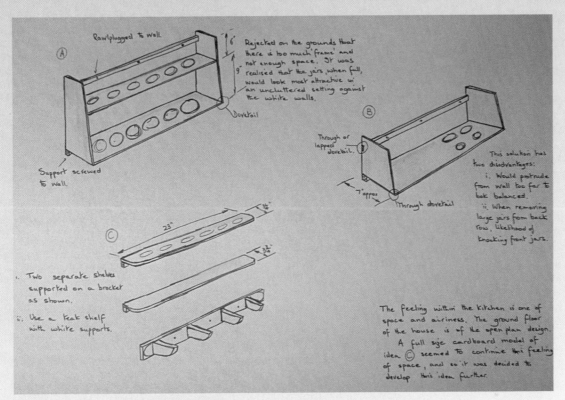

When generating designs and blueprints, always document all your ideas before you decide on the final and best solution.

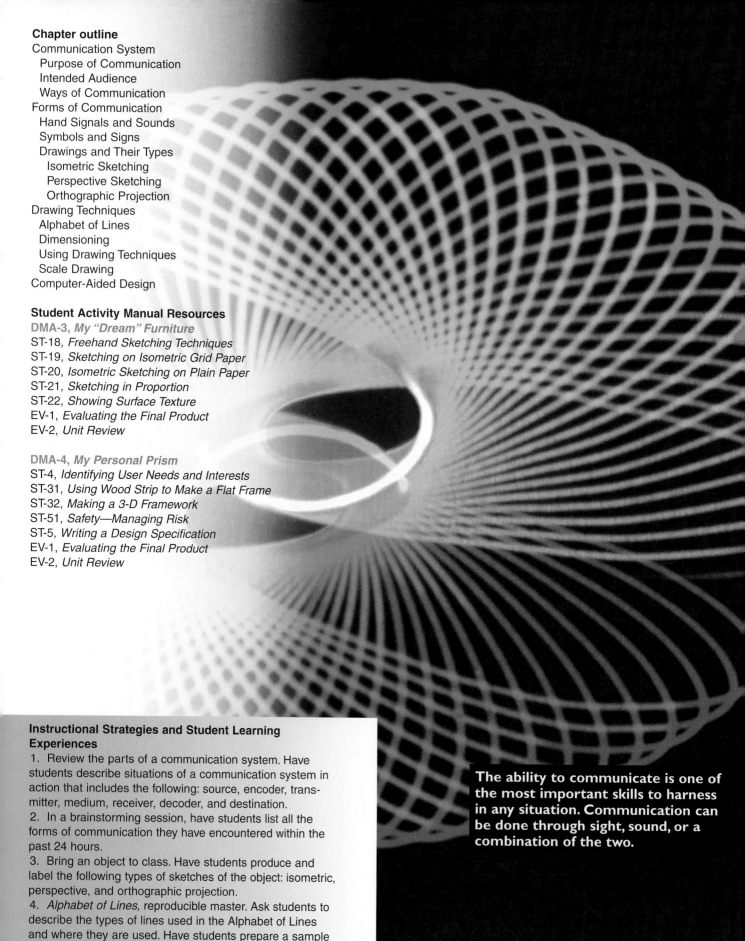

Chapter outline

Student Activity Manual Resources

DMA-3, *My "Dream" Furniture*
ST-18, *Freehand Sketching Techniques*
ST-19, *Sketching on Isometric Grid Paper*
ST-20, *Isometric Sketching on Plain Paper*
ST-21, *Sketching in Proportion*
ST-22, *Showing Surface Texture*
EV-1, *Evaluating the Final Product*
EV-2, *Unit Review*

DMA-4, *My Personal Prism*
ST-4, *Identifying User Needs and Interests*
ST-31, *Using Wood Strip to Make a Flat Frame*
ST-32, *Making a 3-D Framework*
ST-51, *Safety—Managing Risk*
ST-5, *Writing a Design Specification*
EV-1, *Evaluating the Final Product*
EV-2, *Unit Review*

Instructional Strategies and Student Learning Experiences

1. Review the parts of a communication system. Have students describe situations of a communication system in action that includes the following: source, encoder, transmitter, medium, receiver, decoder, and destination.

2. In a brainstorming session, have students list all the forms of communication they have encountered within the past 24 hours.

3. Bring an object to class. Have students produce and label the following types of sketches of the object: isometric, perspective, and orthographic projection.

4. *Alphabet of Lines,* reproducible master. Ask students to describe the types of lines used in the Alphabet of Lines and where they are used. Have students prepare a sample drawing demonstrating the use of each of the lines.

5. Invite a representative from a computer company or store to demonstrate a CAD system to the class. Encourage students to ask questions and, if possible, allow them to try various CAD techniques.

The ability to communicate is one of the most important skills to harness in any situation. Communication can be done through sight, sound, or a combination of the two.

Communicating Ideas

Discussion
In small groups, have students discuss the various ways they communicate ideas to others. Ask students to describe the advantages and disadvantages of each method they identify.

Career connection
Have students identify the careers in which communication of ideas is a central task. Have students browse the Web to find examples of work completed by people working in these careers.

Key Terms

audio communication
audiovisual communication
CAD
communication
communication technology
construction line
drafting
feedback
isometric paper
isometric sketching
line drawing
medium

object line
orthographic projection
perspective sketching
receiver
scale drawing
source
storage
symbol
view
virtual reality
visual communication

Objectives

After reading this chapter you will be able to:

◆ Define communication technology.

◆ Describe the components of a communication system.

◆ List the six basic reasons for communicating.

◆ Explain the three basic types of communication technology.

◆ List various forms of communication.

◆ Explain how drawings get across ideas more efficiently than do words.

◆ Communicate ideas by means of isometric and perspective sketches.

◆ Draw simple objects using the principles of orthographic projection.

◆ Determine which form of drawing is best for a given situation.

◆ Describe how the computer has changed drafting practices.

The exchange of information or ideas between two or more living beings is known as *communication*. This is a big word for a simple act. It came from the Latin, communo. It means to "pass along."

Communication is more than sending a message. The message must be received and understood. If this does not happen, there is no communication. *Communication technology* is the transmitting and receiving of information using technical means. Tools and equipment are used to assist in the delivery of the message. There are many kinds of communication. Their elements form a system.

Communication System

In today's advanced age of technology, information and telecommunication devices have various inputs, processes, and outputs that control the sending and receiving of information and data. One of the most common inputs is the analog voice signal. These days, our voices are digitally encoded and decoded during the transmission process. This is done to reduce noise and allow for clearer signals.

A complete communication system has several parts, **Figure 3-1**:

1. *Source*—The starting point of a message.

2. *Encoder*—Changes your message into another form for transmission.

3. *Transmitter*—Sends your encoded message toward its destination.

4. *Medium*—The wired or wireless means used to send the information.

5. *Receiver*—Accepts the encoded information and relays it to the decoder.

6. *Decoder*—Translates the encoded message into an understandable form.

7. *Destination*—Where the message goes after being sent.

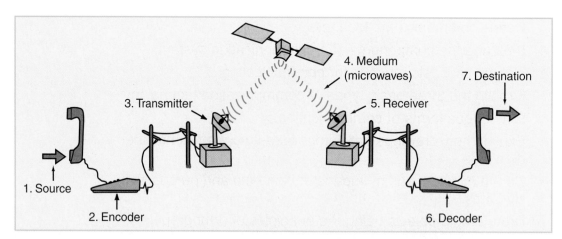

Figure 3-1   Talking to a friend is an example of using a communication system.

Links to other subjects: History—How have communication devices evolved in the last 100 years?

Useful Web site: Students can use the dictionaries at www.techtionary.com and www.hyperdictionary.com to find the definition of any new technical and technological terms used in this chapter.

These seven steps are used to transmit and receive a message. In addition, messages are often placed in *storage* so they can be retrieved at a later time.

Let's look at an example of a communication system. Suppose that you have an idea to meet in a specified chat room online with a friend from another country. You call your friend on the phone and say, "Let's meet online tonight in this chat room." Do you see the system? You are the source. The telephone encodes the message. Satellites transmit your message around the world. A satellite dish acts as the receiver of your information and relays the message to your friend's phone, which decodes the message. Finally, your friend is the destination of the message.

Suppose your friend cannot meet online tonight. "Sorry, I can't. I must study tonight." Your friend provided you with *feedback,* which is a response to the receiver's question or statement. This demonstrates the whole system over again. It also proves that your message was heard and understood. Your friend then says, "But let's remember to do it Friday night." Then both of you will store the information in your mind. When Friday comes, you will remember to meet online. That is an example of retrieval.

Purpose of Communication

There are six basic reasons for communicating. They are the following:

- Inform.
- Persuade.
- Educate.
- Entertain.
- Control.
- Manage.

Local news broadcasts daily inform the public on such topics as weather, traffic, entertainment, sports, and government. Salespersons and TV commercials try to persuade us that a certain product is important, necessary, and reasonably priced. Teachers and professors at schools, colleges, and through online classes educate students on a variety of courses and studies. Popular comedians, musicians, and movie stars entertain us on stage, at the theaters, and on television. Electrical engineers use data signals to control robots. Supervisors manage businesses and employees under their leadership.

Intended Audience

The design of a message is dependent on the audience that the speaker is trying to reach. Politicians often state certain views on issues in order to gain the favor of the majority of the voters. Movie directors and writers make scary movies to draw people that love horror films to the movie theaters. Stand-up comedians draw people that love to laugh to comedy clubs. Musicians write songs on specific issues that grab people's attention, or they play certain styles of popular music in order to get more people to buy their CDs.

Technology and society: Who has benefited from the increased ease and frequency of communication between people? Who has not experienced benefits?

Ways of Communication

There are three basic ways or types of communication. All are based on our sense of hearing and sight.

◆ *Visual communication* presents ideas in a form we can see. Thoughts are changed into words, symbols, and pictures. Do you understand that a stoplight, a street sign, a photograph, a computer, and a book are giving out visual messages?

◆ *Audio communication* is messages that can be heard. However, they cannot be seen. Examples? How about the buzzer that tells you class is over? Do you have a doorbell at your home that tells you someone is at the door? Telephones, radios, and CD players all use audio communication.

◆ Some communication can be both seen and heard. This is known as *audiovisual communication*. You are receiving audiovisual messages when you watch and listen to television, DVDs, and movies.

Forms of Communication

All forms of communication use a code or symbols. For example, *dog* and *perro* are letter symbols that communicate the idea of a certain animal. However, the same message can be given in a quite different "language." See **Figure 3-2**.

Hand Signals and Sounds

Simple movements or sounds can replace spoken and written messages. Look at **Figure 3-3**. These are signals that most people anywhere in the world would understand. Can you think of other signals that you might use? What about the signal to be quiet?

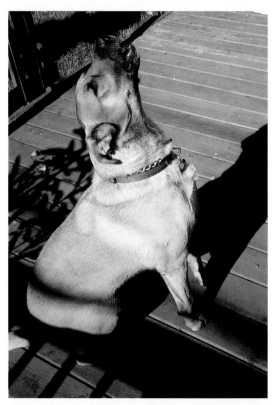

Figure 3-2 Dogs can communicate various messages, such as when a stranger approaches, when they need to go outside, or when they are hungry. (Ecritek)

Links to other subjects: How are specific ideas communicated in other subjects (for example: symbols used in math, scientific notations, and drafting symbols)?

Figure 3-3 Sounds and hand signals have meaning. What message is being sent in each of these pictures?

What about sound signals? If you know any Morse code, you will recognize the dots and dashes being sent out by the ship in **Figure 3-3** as the international distress signal. People of all languages know this signal.

Humans, however, are not the only earth dwellers that exchange messages with sounds and "body" language. Sea animals, such as dolphins and whales, have systems of sound to exchange messages among their own kind. Deer and beaver use their tails to signal danger.

Humans developed nonverbal ways of communicating long before formal language was developed. Cave dwellers drew on walls to tell their experiences. We still use such methods.

Symbols and Signs

Simple pictures and shapes are one of the most effective methods of communication. These symbols can warn, instruct, and direct without using words. They "speak" in a hundred languages all at the same time, **Figure 3-4**.

Drawings and Their Types

Objects and ideas can also be represented using lines and shapes such as the ones shown in **Figure 3-5**. These are known as *line drawings*. Designers, drafters, technicians, engineers, and architects must be able to make such line drawings. How do you suppose the designer would ever have explained how to build the parts in **Figure 3-6** without drawings? When discussing technical details, "a picture is worth a thousand words."

Drawings of the ideas to be communicated are called *drafting.* Drafting has always been known as the "language of industry." It prevents confusion about the size and shape of an object or structure. Some types of drawings look a great deal like photographs. Others show only one surface or an object or structure.

Standards for Technological Literacy

17

Designing and making
Why is it important for a designer to have good drafting skills?

Activity
Have students design a symbol to warn people not to feed animals at the zoo.

Resource
Activity 3-2, *Creating a Symbol*

Resource
Activity 3-3, *Logos*

Resource
Transparency 3-2, *Types of Drawings*

Links to other subjects: How do drawings completed in technology education classes differ from drawings completed in other subjects, especially art and science?

Useful Web site: Browse the following Web sites for additional information on signs and symbols in communication: www.symbols.com, www.get2testing.com/pictograms.htm, and www.widgit.com.

Figure 3-4 What message is each of these signs and symbols giving?

There are three types of drawings: isometric, perspective, and orthographic projection. These three types are summarized in Figure 3-7. All three may be sketched freehand or drawn using manual drafting equipment or computer-aided drafting systems.

Isometric Sketching

Sketching is the simplest type of drawing and one of the quickest ways to share an idea. While ideas are developing in your imagination, isometric sketching records them for further discussion, Figure 3-8.

Technology and society: What problems would people face if symbols were not used to communicate information?

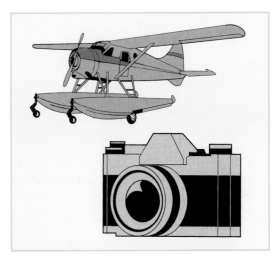

Figure 3-5 Designers use line drawings such as these to communicate designs to others.

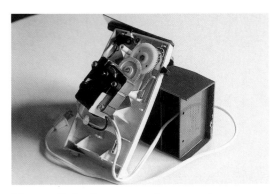

Figure 3-6 A new product requires one or more drawings to describe it well enough so others can build it. (Ecritek)

Isometric sketches can be easily drawn on *isometric paper.* This paper has a grid of vertical lines. Other lines are drawn at 30° to the horizontal. To sketch an isometric box that is six squares long, three squares wide, and four squares high:

1. First draw the front edge of the block (line 1). Draw lines 2 and 3 to show the bottom edges of the box. These three lines represent isometric axes. See **Figure 3-9**.

Isometric

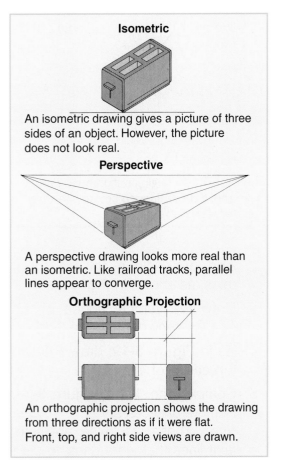

An isometric drawing gives a picture of three sides of an object. However, the picture does not look real.

Perspective

A perspective drawing looks more real than an isometric. Like railroad tracks, parallel lines appear to converge.

Orthographic Projection

An orthographic projection shows the drawing from three directions as if it were flat. Front, top, and right side views are drawn.

Figure 3-7 Which drawing looks most realistic?

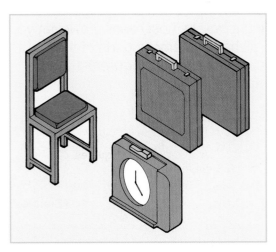

Figure 3-8 Designers often communicate their ideas using isometric sketches.

Standards for Technological Literacy

17

Designing and making
Why is isometric sketching an essential skill for the designer?

Discussion
Based on Figure 3-7, what is the difference between an isometric sketch and a perspective sketch?

Links to other subjects: Ask students to describe how the skill of isometric sketching will help in subjects other than Technology education.

Useful Web sites: Have students search the Web for examples of the use of isometric sketches in different industries and careers. Helpful places to start include www.tpub.com/engbas/5-30.htm and www.unleash.com/ronm/isometric.

Discussion
How does isometric paper help in drawing a dimensional sketch?

Activity
Have students collect examples of isometric and perspective sketches from magazines and share with the class.

Resource
Activity 3-4, *Isometric "Cheese Block"*

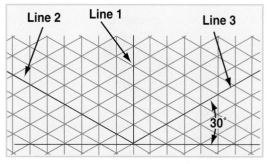

Figure 3-9 To begin an isometric sketch, make these three lines to set up the three axes (edges) of the sketch.

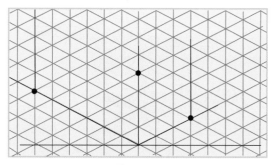

Figure 3-11 Add two more vertical lines to represent the visible vertical edges of the box.

2. Add dimensions to the three axes, as shown in **Figure 3-10**.
3. Draw the vertical edges of the box, as shown in **Figure 3-11**.
4. As in **Figure 3-12**, draw the two top edges of the box.
5. Complete the sketch, **Figure 3-13**. Remove unnecessary *construction lines*. Those would be any thin, faint lines that were used to start your drawing. Darken the outline of the object.

Whatever the shape of the object to be drawn, it is easiest to begin by drawing a box. In most cases, however, you will have to remove parts of the box to create the shape, **Figure 3-14**.

Sometimes you will want to add a piece. Could you make a sketch of a box with a small block added to one side? This method of isometric

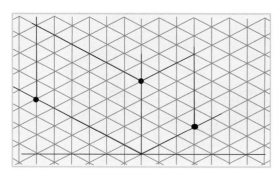

Figure 3-12 Next, begin to sketch in the top edges of the box.

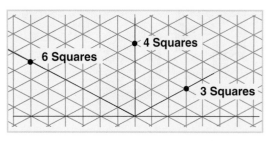

Figure 3-10 Next, establish the three dimensions of the isometric sketch.

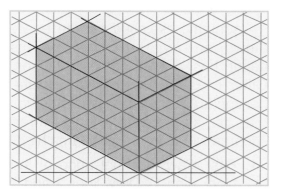

Figure 3-13 The box is completed.

Community resources and services: Invite a professional designer to the class to discuss how they use sketching in their work.

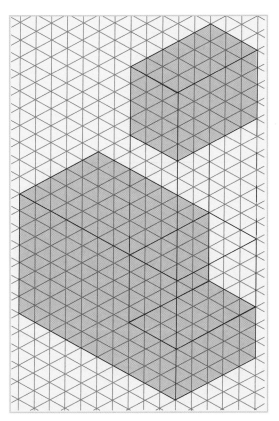

Figure 3-14 Removing part of the original box will create a new shape.

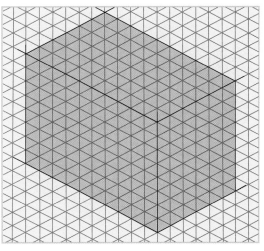

Figure 3-15 To draw a house, make a box on an isometric grid.

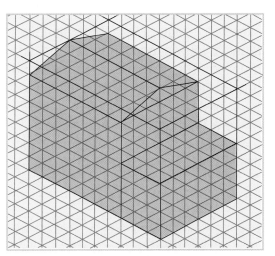

Figure 3-16 The house takes shape. See how points on the grid are used to draw in the roof lines.

sketching may be used to draw a simplified house.

1. Lightly construct an isometric box as in **Figure 3-15**. Use the dimensions eight squares long, four squares wide, and six squares high.

2. Construct the basic shape of the house by removing a corner of the box, **Figure 3-16**. Add lines for the roof.

3. Add details, including windows and doors, **Figure 3-17**.

4. Complete the line work by removing unnecessary construction lines. Darken the remaining lines to form the building's shape, **Figure 3-18**.

Community resources and services: List some careers in which the ability to sketch ideas is an important skill. Visit a career fair to find more information about the skills and qualifications necessary to succeed in the listed careers.

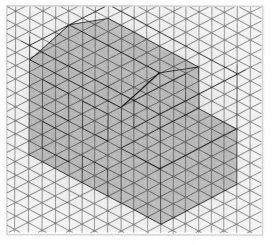

Figure 3-17 Add details such as doors and windows.

While isometric paper makes sketching easy, it has one disadvantage. It leaves grid lines on the final drawing. These could confuse someone looking at your drawing. Designers often prefer to sketch on plain paper. To make a freehand isometric sketch of a rectangular block on plain paper, use the method shown in **Figure 3-19**.

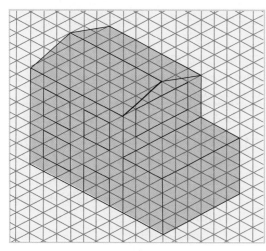

Figure 3-18 The sketch of the house is completed.

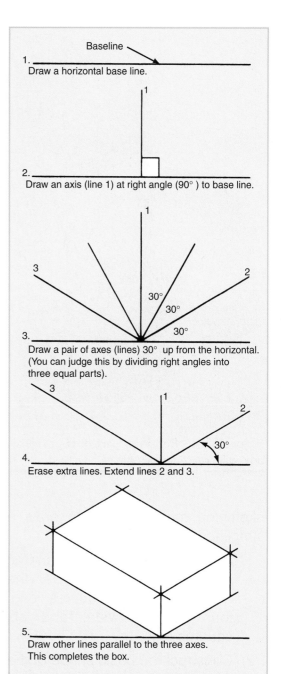

Figure 3-19 You can make a freehand isometric sketch by following these steps.

Community resources and services: Visit a local architect's office to see how sketches are used to generate and develop ideas for projects.

Figure 3-20 In real life, objects at a distance seem narrower or shorter than they are close up. (Ecritek)

Perspective Sketching

Look at the photograph of the railroad station, **Figure 3-20**. What do you notice about the parallel lines of the tracks? They appear to converge. What do you notice about the height of the lampposts? Those farther away appear shorter. What do you notice about the width of the platform? The farther away it is the narrower it appears. Of course, railroad tracks don't converge, lampposts don't get shorter, and platforms don't become narrower.

Perspective sketching provides the most realistic picture of objects. The sketches are drawn to show objects as we would actually see them. Parallel lines converge and

vertical lines become shorter as they disappear into the distance. Refer to **Figure 3-21** as you read the following steps for drawing a perspective sketch of a block:

1. Draw a faint horizontal line. Think of this as representing the horizon. Mark two points, one at each end of the line. These are vanishing points (VP).

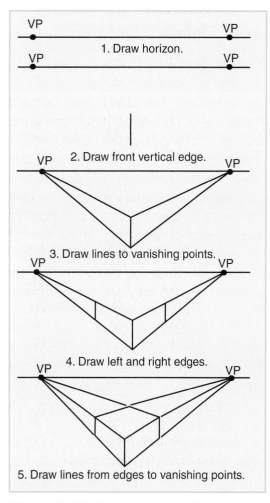

Figure 3-21 In a perspective drawing, lines become shorter as they recede into the distance.

Standards for Technological Literacy

Reflection
Where have you seen examples of parallel lines appear to converge, as in Figure 3-20?

Enrichment
Use a library or the Internet to investigate Albrecht Durer's theory of perspective sketching.

Activity
Collect photographs and pictures in which perspective is very evident. Share them with the class.

Resource
Activity 3-7, *Reading a Metric Ruler*

Useful Web site: Gather more information on the methods and uses of perspective sketches by visiting http://drawsketch.about.com/cs/perspective, http://partner.galileo.org/tips/davinci/perspective.html, and www.chasrowe.com.

2. Draw the front vertical edge of the block.

3. Draw faint lines from each end of the vertical edge to the vanishing points.

4. Draw vertical lines to represent the left and right edges of the block. The length of these vertical sides will be shorter than the real object.

5. Join the top of these vertical lines to the vanishing points. Darken the outline of the object.

Orthographic Projection

You have seen how isometric and perspective sketches are simple methods of recording your ideas and communicating them to other people. They give a general idea of the shape and features of an object. Unfortunately, there are some disadvantages to isometric and perspective sketches. For example, they do not describe the shape of an object exactly because of distortion at the corners, nor do they provide complete information for the object to be made.

Orthographic projection overcomes both these problems. This kind of drawing shows each surface of the object "square on," that is, at right angles to the surface. In this way, you see the exact shape, or view, of each surface. Complete information is usually given by drawing three *views:* front, top, and right side. To understand how a view is produced, imagine that you are the person in **Figure 3-22**. Because you are looking at the object square on, you will only see the area that is colored red. Since

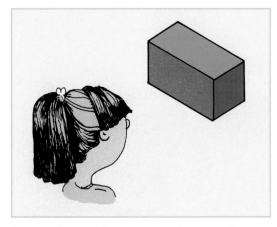

Figure 3-22 An orthographic view "sees" only one side of an object.

this is the front of the object, this view is called the front view.

To produce a top view, imagine you are above the object looking down on the top. The view you would see is shown in blue. The right-side view, shown in green, is drawn by looking at the right side square on. These three views are always arranged as shown in **Figure 3-23**. Note: Work on squared paper. Remember to keep all lines in steps 1-3 as faint as possible.

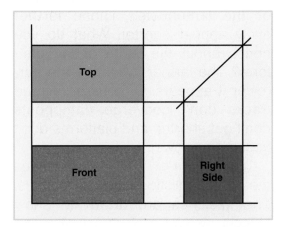

Figure 3-23 This is the proper arrangement for orthographic views of an object.

Community resources and services: Invite an engineer into the classroom to present engineering project drawings and explain how they are used in manufacturing.

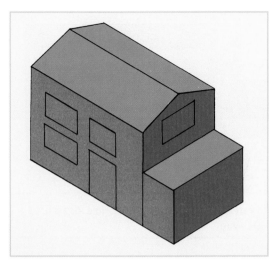

Figure 3-24 Here is an isometric view of a house. The next drawing will show it as an orthographic drawing.

To draw an orthographic projection of the house in **Figure 3-24**, complete the steps described in **Figure 3-25**.

Drawing Techniques

The orthographic views shown in **Figure 3-25** were sketched on squared (grid) paper. This is a quick method. Its disadvantage, however, is that the grid could be confused with the lines of the drawing.

An alternative to grid paper is the use of plain paper and drawing instruments. The instruments most often used are the T-squares, 45° and 30°/60° set squares (drafting triangles), compass, and scale (ruler). The following are some techniques for drawing with these instruments:

◆ As a general rule, when drawing lines with a T-square, draw in the direction the pencil is leaning, **Figure 3-26**.

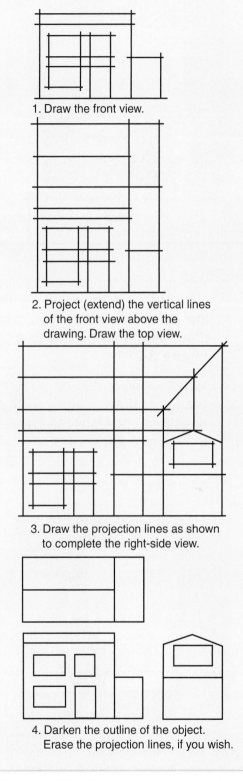

1. Draw the front view.

2. Project (extend) the vertical lines of the front view above the drawing. Draw the top view.

3. Draw the projection lines as shown to complete the right-side view.

4. Darken the outline of the object. Erase the projection lines, if you wish.

Figure 3-25 This is how orthographic drawings are developed.

Links to other subjects: History—Have students find technical drawings of early machinery (such as the steam engine, letter press, and telegraph).

Designing and making
Discuss why ortho-graphic drawing is NOT a useful tech-nique for generating and developing ideas.

Activity
Have students use instruments to make orthographic draw-ings of simple objects in the class-room.

Resource
Activity 3-10, *Measuring Using Conventional Measurement II*

Resource
Activity 3-18, *Orthographic Drawing Using Instruments I*

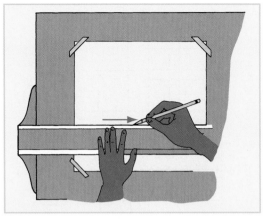

Figure 3-26 Always draw in the direction that the pencil is leaning.

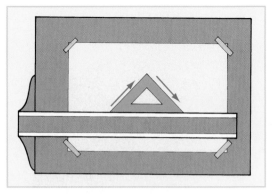

Figure 3-28 Sloping lines are drawn in this way.

◆ When drawing vertical lines with drafting triangles, lean the pencil away from you and draw lines from bottom to top, **Figure 3-27**.

◆ When using a 45° or 30°/60° triangle, draw lines in the directions shown by the arrows in **Figure 3-28**.

◆ Hold a compass between the thumb and forefinger and rotate clockwise. Lean the compass slightly in the direction of the rotation as you draw a circle, **Figure 3-29**.

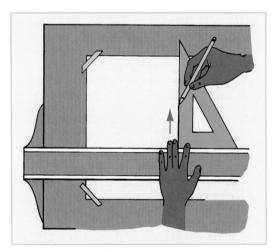

Figure 3-27 Note how the pencil is sloped when drawing vertical lines using a triangle.

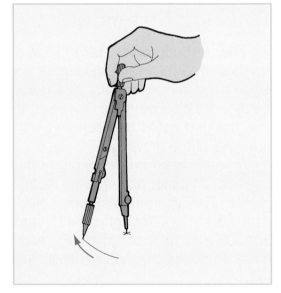

Figure 3-29 When using a compass, draw circles or arcs lightly at first. Make repeated turns to darken the line.

Links to other subjects: Have students identify the various drawing instruments used in other subjects (such as art, graphic design, and math) and how they are used.

Alphabet of Lines

There are a number of different types of lines used to produce orthographic drawings. Each line is used for a particular purpose and cannot be used for anything else. Look at the casting in **Figure 3-30**. The alphabet of lines can be used to produce orthographic drawings of this casting. **Figure 3-31** shows the rules for use of lines.

Hidden lines are short, evenly-spaced dashes. They show the hidden features (lines or shapes) of an object, **Figure 3-32**. Hidden lines almost always begin and end with a dash touching with the line where they start and end (1). This rule is not followed when the dash would continue a visible detail line (2). Dashes should join at corners (3) and (4). The dashes of parallel hidden lines that are close together should be staggered (5).

Type of line	Example
Construction line Thin, faint lines used to start a drawing.	
Object or visible line Darker, thicker lines used to show the outline of the object.	
Hidden line Short and evenly-spaced dashes used to show hidden features.	
Centerline Alternating long and short dashes to show the centers of holes.	
Extension and dimension lines Thin lines used to show the size of an object and its parts.	

Figure 3-31 The alphabet of lines explains the types of lines and where they are used in drafting.

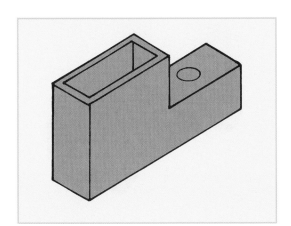

Figure 3-30 An isometric view of a metal casting.

Dimensioning

Most drawings include two types of dimensions: overall dimensions and detail dimensions. To fully describe the size and shape of the view in **Figure 3-33A**, you need two overall and two detail dimensions.

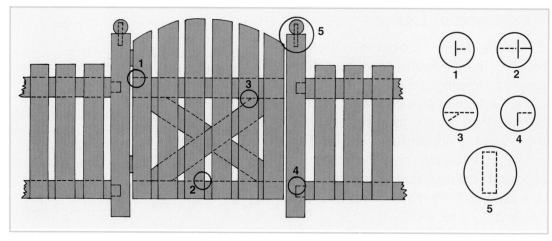

Figure 3-32 When you look at a board fence, parts of the lines are hidden by the boards. A drawing would show the hidden parts as short, evenly-spaced dashes.

If a hole is added to this view, then you must add some dimensions as in **Figure 3-33B**. The size of the hole and its location must be shown. What is important is to show the exact position of the center of the hole.

Using Drawing Techniques

The following method is used to produce a set of orthographic views using plain paper and instruments. You will also need to refer to the alphabet of lines, **Figure 3-31**. You are going to draw the truck in **Figure 3-34**.

Detail dimension

Height: overall dimension

Detail dimension

Length: overall dimension

A

Diameter dimension

B

Figure 3-33 These two examples show overall and detail dimension lines.

Figure 3-34 If you were to change this isometric drawing of a toy truck into an orthographic drawing, how many views would be needed?

Technology and society: How does the mass production of consumer products rely, in part, on the accurate dimensioning of orthographic drawings?

1. Draw the front view using construction lines, **Figure 3-35**.

2. Draw the top view directly above the front view, **Figure 3-36**. Project all vertical lines upward as shown.

3. Draw the edge of the right-side view (line a), **Figure 3-37**. Distances x and y should be the same. Project the lower edge of the top view (line b) to intersect line a. Project the lines of the front view to the right (lines c).

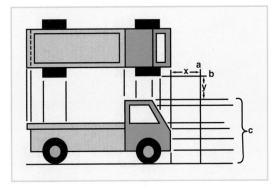

Figure 3-37   Now we are ready to draw the right-side view. Lines a, b, and c have been drawn.

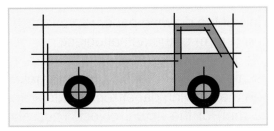

Figure 3-35   The first step is to make a front view using construction lines. Take measurements from the isometric view in Figure 3-34.

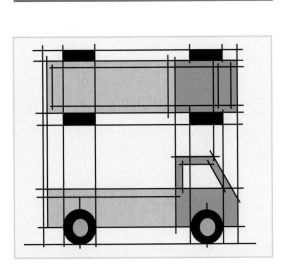

Figure 3-36   Extend the vertical lines to start the top view. Measure the drawing in Figure 3-34 to get line lengths.

4. Draw a 45° line as shown in **Figure 3-38**. Project horizontally the lines from the top view to the 45° line, then vertically to the right-side view.

5. Darken the *object lines.* These are the lines that show the outline of your object. Erase construction lines if necessary, **Figure 3-39**.

6. Add the dimension lines and dimensions, **Figure 3-40**.

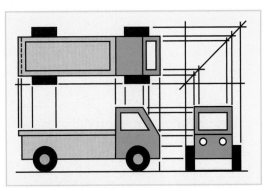

Figure 3-38   This drawing shows how to find the size of the front view.

Designing and making

Is it always necessary to draw three orthographic views of an object before making it?

Discussion

What are the essential differences between isometric sketches and orthographic drawings?

Useful Web site: Search the Web for more information on orthographic drawing. Helpful places to start include www.stemnet.nf.ca/DeptEd/g7/ortho.htm, www.knowledgetree.ca, and www.tpub.com/steelworker1/28.htm.

Figure 3-39 Orthographic drawing of the truck is almost completed.

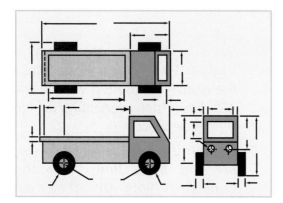

Figure 3-40 The final step adds dimension lines and dimensions.

Scale Drawing

A *scale drawing* is a drawing that is larger or smaller than the object by a fixed ratio. It is made when an object is either too large to fit onto the paper or too small to see the details. Examples of scaled drawings are an architect's drawing of a building, a cartographer's map, and an electronic engineer's schematic of a printed circuit.

If you wanted to draw a full-size front view of the tanker truck in **Figure 3-41**, you would need a piece of paper larger than the truck. Full-size is a scale of 1:1. Each inch (or millimeter) of the drawing paper represents 1″ (or 1 mm) of the actual object.

In a drawing one-half full-size (a scale of 1:2), each inch (or millimeter) on the drawing paper represents 2″ (or 2 mm) of the actual object. Thus, the actual object would be twice the size of the views on the drawing paper.

If an object to be drawn is very small, it may be necessary to prepare drawings to a scale larger than

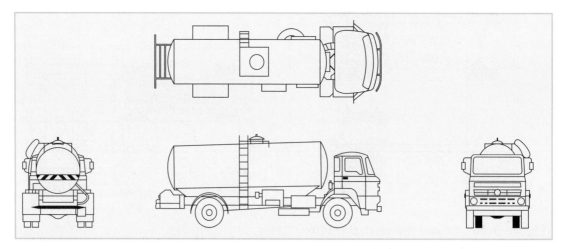

Figure 3-41 This is a scale drawing of a tanker truck.

Links to other subjects: How are scale drawings used in other subjects, like math, geography, and graphic design?

full-size. Such a scale is referred to as an enlarged scale. In a drawing that is twice full-size (a scale of 2:1), each 2″ (or 2 mm) on the drawing paper represents 1″ (or 1 mm) of the actual object. The parts of the compass shown in **Figure 3-42** are drawn twice their actual size.

Computer-Aided Design

In the past, many people worked at making drawings. They used the tools just described today, however, fewer drafters are using drafting boards. Instead, they make drawings using computers.

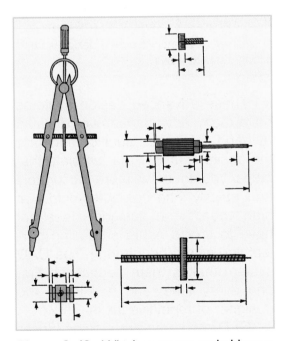

Figure 3-42 Which parts are scaled larger than their actual size?

This new form of drawing is called *CAD*. It stands for computer-aided design. A typical CAD system has three types of devices or parts:

- Input device. It gives information or instructions to the computer.
- Processor. This part carries out the instructions.
- Output device. This part actually makes the drawing.

To design using a CAD system, a mouse or digitized tablet is used to draw just as if the screen were a piece of paper, **Figure 3-43.** The designer can create the drawings, add details, and call up title blocks and other standard information. Corrections can be done quickly. Parts of a drawing that are used repeatedly can be stored in a disk file and loaded into the drawing when needed so nothing is drawn twice. Three-dimensional CAD allows the drafter to not only create but also rotate, scale, and manipulate 3-D objects. In the case of a house, it could be viewed from any angle and the future owner could actually see how the finished building would look.

In some CAD programs, including those for architecture drawing, the software will add the elevations (front and side views) automatically to a plan drawing (overhead view). It is even possible to do a virtual "walk through" to give the viewer a feeling of actually being in various rooms of the building.

In the most advanced systems, designers use goggles to view a virtual reality projection screen, called a power wall. *Virtual reality* is an artificial

Designing and making
How has CAD changed the way everyday products are designed and made?

Vocabulary
What do the acronyms *CAD* and *CAM* mean?

Technology and society: How does the introduction of CAD systems impact employees of a drawing office?

Useful Web site: Search the Internet for additional CAD information and related careers. Helpful sites include www.cadsystems.com, www.cad-portal.com, and www.cadinfo.net.

Designing and making
Will the continuing improvements in CAD systems make freehand sketches obsolete?

Reflection
How does your ability to communicate using a variety of techniques improve your overall communication skills?

Figure 3-43 Computer-aided design (CAD) systems allow a designer to view a design from any angle and to zoom in and out for close-ups and long-distance views.

environment provided by a computer that creates sights, sounds, and, occasionally, feelings through the use of a mask, speakers, and gloves or other sensory-enhancing devices. Computerized images of a proposed product can be displayed in three-dimensional form from any angle allowing an engineer to spot errors before the device is built. Detecting design flaws before building expensive prototypes is an important application of virtual reality and leads to better manufactured goods made more quickly and cheaply.

These advanced design systems have many advantages over drawing by hand. CAD is an additional tool for the designer. It is like a template that helps you to draw more accurately and quickly. The computer works at high speed. The designer does not have to spend hours producing perfect line work and lettering. The CAD system makes them perfect the first and every time, and it relieves the designer of repetitive tasks.

CAD programs are moving closer to integrate all persons involved in designing, making, and supplying

Technology and society: Discuss the ways in which society is changing as a result of the introduction of new communication systems.

materials for a project. For example, an architect designing a building will know, at any point, the cost of materials. The materials list can be sent directly to the building contractors along with drawings and contract documents. Features like these will leave more time for creative work. Other programs link CAD programs with computer-aided manufacturing (CAM). Today, most industries use some form of rapid plastic prototypes. See Figure 3-44. Design information is sent directly to a prototype machine, where a complete prototype part is made in one operation. By building plastic models, designers can highlight a product's strengths and weaknesses, eliminating costly mistakes and using prototypes in real-life field tests.

Still, CAD cannot replace the individual. It cannot think for the designer. Concepts and principles learned in drawing and design classes continue to be important. Just as you need knowledge of mathematical concepts when using a hand calculator, so the designer needs knowledge of drawing and design when using a CAD system.

Designing and making
How might rapid prototyping change the way you design and make products in the school workshop?

Reflection
Describe how learning to sketch can help you in your daily life.

Vocabulary
What is a prototype?

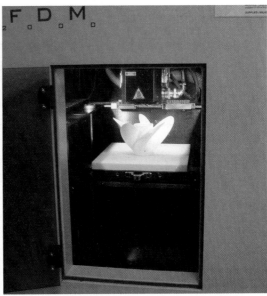

Figure 3-44 This prototype of a ship's propeller has been designed and produced using a computer-aided design (CAD) and computer-aided manufacturing (CAM) system.

Links to other subjects: Have students investigate how machines controlled by computers are changing the way people work in fields other than technology.

Useful Web site: Have students access the rapid prototyping Web site at www.cc.utah.edu/~asn8200/rapid.html, click on the "What is Rapid Prototyping?" link, and read the article that follows.

Chapter 3 Review
Communicating Ideas

Summary

Communication technology refers to the transmitting and receiving of information using technical means. Communication itself is the exchange of information or ideas between two or more living beings.

The purposes of communication are to inform, persuade, educate, entertain, control, and manage. People choose one or more of these purposes of communication when trying to reach an intended audience.

The parts of a complete communication system include the following: source, encoder, transmitter, medium, receiver, decoder, destination, storage, and retrieval.

When we tell people about our ideas, we are communicating. All forms of communication use a code or symbols. Examples include hand signals, sounds, symbols, signs, and drawings.

Of these, the most important for the designer is drawing. Designers most commonly use three types of drawing: isometric sketches, perspective sketches, and orthographic projection.

Isometric sketches are used to record ideas quickly on paper. Perspective sketches show objects as you would see them. Neither isometric nor perspective sketches describe the shape exactly. Orthographic projection shows the exact shape of each surface of an object. Orthographic views can be drawn freehand on squared paper, or by using instruments. Most drawings today are made using computer-aided design (CAD).

Sometimes an object is too large to be drawn on paper, or too small to be seen clearly. Scale drawings are made when the object is either too large to fit onto the paper or too small to see all the details.

Modular Connections

The information in this chapter provides the required foundation for the following types of modular activities:

- Graphic Communications
- Graphic Design
- Imaging Technology
- Technical Design
- Engineering Design
- Drafting
- Architectural Drafting
- Mechanical Drafting
- Computer-Aided Design
- Computer Graphics
- Technical Sketching
- 3D Modeling
- Animation
- Airbrush Design
- Audio Communications
- Music and Sound
- Digital Music
- Marketing
- Business Presentations
- Prepared Speech
- Extemporaneous Presentation
- Technical Writing
- Technical Research/Report Writing
- Portfolio Development
- Desktop Publishing

Test Your Knowledge

Write your answers to these review questions on a separate sheet of paper.

1. Name the six parts of a complete communication system in the proper sequence.

2. List seven ways in which people communicate ideas to one another.

3. State the six basic reasons of communication.

4. Name the three types of drawing. Make small sketches to show each type.

5. A sketch is a drawing made _____.
   A. using a straightedge
   B. quickly to communicate an idea
   C. with the help of drawing instruments
   D. in pencil only

6. How many sides of a rectangular block are shown in an isometric drawing?

7. Which type of drawing provides the most realistic picture of an object?
   A. Isometric sketch
   B. Plan view
   C. Perspective sketch
   D. Orthographic drawing

8. A(n) _____ drawing describes the exact shape of each surface of an object.

9. The front view of an object is usually the view which _____.
   A. shows its width
   B. best describes its shape
   C. shows its depth
   D. is the smallest

10. List the most common instruments used to draw orthographic views.

11. What technique is used by a drafter or architect to draw a large object on a small piece of paper?

12. List the advantages of computer-aided design.

Apply Your Knowledge

1. Create a symbol that can be used at a zoo to communicate the message: "Do not feed the animals!"

2. On isometric paper, draw a cube with sides measuring 12 squares. Remove pieces from the block to create a "Swiss cheese" effect.

3. Design a logo (name or symbol) that you could use on your own personalized work sheets. Letters, geometric shapes, natural shapes, and simplified pictures are most appropriate for a logo.

4. Draw an isometric sketch, a perspective sketch, and an orthographic projection of a die.

5. Make an isometric sketch of either a tool you have used in the technology lab or an object used in the kitchen.

6. Select a day of the week. List all the ways that information is communicated to you during that day.

7. International Morse Code is a means of communication invented before telephone and e-mail. Check the symbols used. Now design a series of symbols that could be used to send a message by e-mail without using letters or words.

8. Research one career related to the information you have studied in this chapter and state the following:
   A. The occupation you selected.
   B. The education requirements to enter this occupation.
   C. The possibilities of promotion to a higher level at a later date.
   D. What someone with this career does on a day-to-day basis. You might find this information on the Internet or in your library. If possible, interview a person who already works in this field to answer the four points. Finally, state why you might or might not be interested in pursuing this occupation when you finish school.

Chapter outline

Student Activity Manual Resources

Instructional Strategies and Student Learning Experiences

1. Bring to class a bag containing about 10 items. Have students state what materials were probably used in producing the items.

2. *Mechanical Properties*, reproducible master. Using this master and various materials, demonstrate the mechanical properties of those materials.

3. *Characteristics of Wood*, reproducible master. Use this master to review the various characteristics of wood.

4. Bring samples of various types of wood for students to touch and see. Prepare a display of the wood samples.

5. *Characteristics of Metals*, reproducible master. Use this master to review the various characteristics and uses of metals.

6. Ask students to prepare a display of various types of metals.

7. *Characteristics of Plastics*, reproducible master. Use this master to review various characteristics of plastics.

8. Bring a variety of plastic items to class. Ask students to determine whether the item is a thermoplastic or a thermoset plastic.

9. *Characteristics of Ceramics*, reproducible master. Use this master to review the characteristics and uses of ceramics.

10. Bring various types of ceramic materials to class. Have students identify them and make a display of them.

Every material has certain properties, such as mechanical properties and thermal properties. Materials, such as the steel used in this bicycle frame, are selected for these properties.

Materials

4

Key Terms

alloy	opaque
atom	photosynthesis
biomaterial	plasticity
ceramic	plasticizer
composite	polymer
conductor	primary material
corrosion	semiconductor
elasticity	softwood
fatigue	strength
ferrous	synthetic
hardness	thermal expansion
hardwood	thermoplastic
insulator	thermoset
magnetic	toughness
molecule	translucent
nonferrous	transparent

Objectives

After reading this chapter you will be able to:

◆ List the principal properties of materials.

◆ Identify various materials and their characteristics.

◆ Describe the major processes used to change raw materials into standard stock.

◆ Select materials to meet the needs of a particular product.

Discussion
Have students list products in which the design makes use of the special characteristics of a material. Describe the properties of the material that make it an essential part of the design.

Career connection
Have students identify careers in which the manufacture of materials is a central task. Students can browse the Web to find examples of the work completed by people in these careers.

Designing and making
In what ways do the materials designers choose for products affect your daily life?

Vocabulary
What words do you use to describe materials?

Activity
Have students list all the materials they see or use in the course of one day. Working in small groups, students should create categories for the materials and compare with the class.

Resource
Activity 4-1, *Materials You Use*

Designing and building objects always involves the use of materials, **Figure 4-1.** Concrete is used for walkways, cotton for clothing, fiberglass for insulation, copper for electrical wiring, and plastic for bottles.

In fact, materials have been so important that some of the major periods of history have been named after materials. You may have heard of the Stone Age, the Bronze Age, and the Iron Age.

If we were to name the current period or "age" by the name of a material, it might be the "Age of Composites." *Composite* materials are combinations of different materials. Water skis used to be made of laminated wood. Thin strips of wood were glued together. Today, they are made of a foam-core wrapped with layers of fiberglass and graphite reinforcement.

Designers need to know about the properties of many different materials in order to choose the most appropriate material for the object being designed. However, they must first learn how the product must function. The following are some questions they should ask:

◆ Does the product have to withstand heat?

◆ Is color important?

◆ Should it be heavy or light?

◆ How strong does it have to be?

◆ Must it withstand bad weather?

◆ Does it have to conduct electrical current?

Figure 4-1 Composite materials are combinations of materials. Concrete is a mixture of stone, gravel, and cement. (D & S Advertising, Inc.)

These types of questions will narrow the choice of materials. The next step is to determine whether the material is available and affordable. Equally important are having the tools and knowledge to work the material.

Properties of Materials

Designers and engineers judge materials by their properties. These tell how the material can be expected to perform. Properties of materials can be grouped as follows:

◆ Physical.

◆ Mechanical.

◆ Thermal.

◆ Chemical.

◆ Optical.

◆ Acoustical.

◆ Electrical.

◆ Magnetic.

Useful Web site: Students can use the dictionaries at www.techtionary.com and www.hyperdictionary.com to find the definition of any new technical and technological terms used in this chapter.

Technology and society: Why are some materials less harmful to the environment than others?

Physical Properties

Physical properties give a material its size, density, porosity, and surface texture. You can describe any material or product by physical properties. Consider an eraser, for example. It may be 2 $\frac{1}{2}$″ long, 1″ wide, and $\frac{3}{8}$″ thick (64 × 25 × 10 mm). It is not very dense. Therefore, its mass is small. Its surface is smoother than a pine board. However, it is not as smooth as glass.

Mechanical Properties

Mechanical properties are the ability of a material to withstand mechanical forces. An elastic band will stretch and return to its original shape. A diving board will spring back. The head of a hammer will withstand sharp blows.

The common mechanical properties, as shown in **Figure 4-2**, are:

◆ *Strength*.
◆ *Elasticity*.
◆ *Plasticity*.
◆ *Hardness*.
◆ *Toughness*.
◆ *Fatigue*.

Thermal Properties

Thermal properties control how a material reacts to heat or cold. Materials will generally expand when heated and shrink when cooled. They will also conduct heat.

Sometimes the expansion of metals causes problems. On a hot day, railroad tracks may expand and buckle. This characteristic of metals, called *thermal expansion*, can be useful. The flashing lights on a Christmas tree are an example. A strip of brass is joined to a strip of invar. (Invar is an alloy or mixture of iron and nickel.) The two are clamped at one end. When heated, the brass expands more than the invar. Thus, a change in temperature will cause the bimetal strip to bend. This movement can be used to open and close an electrical circuit. When the bimetal strip is cool, the lights are on. When the strip heats up, it bends, **Figure 4-3**. The lights go off.

Thermal conductivity is a measure of how easily heat flows through a material. All metals conduct heat. Some do it better than others. Copper is a good conductor of heat. The copper bottom of a frying pan quickly conducts heat from the stove element to the pan. The metal pan then conducts this heat to the food.

Thermal insulators are materials that do not conduct heat well. Nonmetallic materials are generally thermal insulators. Plastic and wood handles on saucepans prevent heat being conducted from the hot metal to your hand. A cooler used for camping may have a casing filled with polyurethane foam to keep out heat. Foam panels or fiberglass batts are used to insulate walls and ceilings in order to reduce heat losses in a home, **Figure 4-4**.

Designing and making
Which of the products you use have the mechanical properties listed on this page? Describe why the mechanical property is important to each product.

Discussion
Have students identify household products in which the thermal property of materials are a critical part of the design.

Activity
Collect small products that are made from a single material and share with the class.

Resource
Activity 4-3, *Testing Materials—Thermal Conductivity*

Resource
Transparency 4-1, *Mechanical Properties*

Useful Web site: Browse the Dow Chemical Company Web site at www.dow.com to investigate how the properties of materials is crucial to product development.

Designing and making
Which of the tasks you complete in a typical day make use of the mechanical properties of materials?

Enrichment
How are the mechanical properties strength, elasticity, and plasticity used in cooking and baking?

Resource
Activity 4-2, *Testing Materials— Hardness*

Resource
Reproducible Master, *Mechanical Properties*

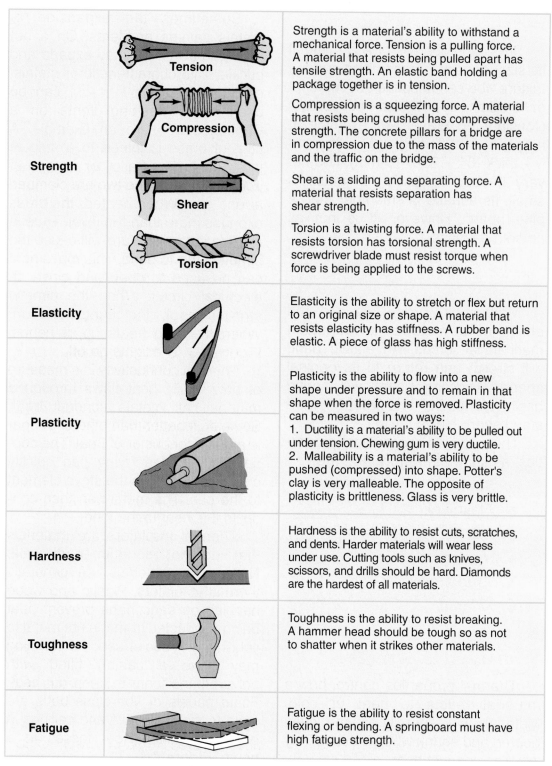

Strength	Tension Compression Shear Torsion	Strength is a material's ability to withstand a mechanical force. Tension is a pulling force. A material that resists being pulled apart has tensile strength. An elastic band holding a package together is in tension. Compression is a squeezing force. A material that resists being crushed has compressive strength. The concrete pillars for a bridge are in compression due to the mass of the materials and the traffic on the bridge. Shear is a sliding and separating force. A material that resists separation has shear strength. Torsion is a twisting force. A material that resists torsion has torsional strength. A screwdriver blade must resist torque when force is being applied to the screws.
Elasticity		Elasticity is the ability to stretch or flex but return to an original size or shape. A material that resists elasticity has stiffness. A rubber band is elastic. A piece of glass has high stiffness.
Plasticity		Plasticity is the ability to flow into a new shape under pressure and to remain in that shape when the force is removed. Plasticity can be measured in two ways: 1. Ductility is a material's ability to be pulled out under tension. Chewing gum is very ductile. 2. Malleability is a material's ability to be pushed (compressed) into shape. Potter's clay is very malleable. The opposite of plasticity is brittleness. Glass is very brittle.
Hardness		Hardness is the ability to resist cuts, scratches, and dents. Harder materials will wear less under use. Cutting tools such as knives, scissors, and drills should be hard. Diamonds are the hardest of all materials.
Toughness		Toughness is the ability to resist breaking. A hammer head should be tough so as not to shatter when it strikes other materials.
Fatigue		Fatigue is the ability to resist constant flexing or bending. A springboard must have high fatigue strength.

Figure 4-2 Some common mechanical properties of materials.

Links to other subjects: List examples, from other subjects, showing how the mechanical properties of materials play an important role.

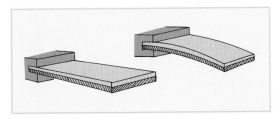

Figure 4-3 A bimetal strip can be used as a switch. Thermal expansion makes it work.

Figure 4-4 Insulating materials slow down the passage of heat or cold trying to move through walls of homes. (Ecritek)

Chemical Properties

A material's chemical properties affect how it reacts to its surroundings. Steel rusts. Glass becomes pitted. Plastics become etched and brittle. These are all the result of a chemical reaction. The environment changes these materials. The reaction is called *corrosion*.

Probably the most familiar example of corrosion is the rust on a car body. What is corrosion? When a material contacts both air and water, there is a chemical change:

Iron + Oxygen + Water =

Iron Oxide (Rust)

Sometimes the water contains dissolved chemicals, such as the salt used on roads to melt snow or the salt present in sea spray. Then rusting occurs much faster. Since water and air cause certain metals to rust, we can prevent rusting by covering the metal with paint.

For centuries, guns and tools have been coated with oil and grease. Paint, varnish, and enamel have been widely used to protect ships, trains, cars, and bridges. Nails and heating ducts are galvanized (coated with zinc). Food cans are plated with tin. Many decorative objects are electroplated. (A coating of nickel, chromium, copper, silver, or gold is applied to their surface.)

Optical Properties

Optical properties are a material's reaction to light. Materials react to light in several important ways. One has to do with how well they transmit light that strikes them. Some materials cannot transmit light at all. When a material stops light, we say it is *opaque*. A roller blind in your bedroom should be made of an opaque material. *Translucent* materials, waxed paper and stained glass, for example, allow some light to pass through. However, you cannot see clear images through them. Other

Technology and society: How do the chemical properties of materials affect the environment?

materials allow all light to pass through. These materials are *transparent*. Clear glass windows are an example.

The second optical property of a material is color. The color of a material affects its ability to absorb or reflect light. (Light is the visible part of the sun's energy.) Light, as well as other kinds of radiant energy, is reflected by shiny, smooth surfaces. It is absorbed by dark, dull surfaces. A car with black upholstery is far more uncomfortable on a hot day than one with a white interior.

The ability of a material to absorb heat can be useful. The pipes of a solar panel are painted black so the panel will absorb more heat from the sun.

Acoustical Properties

Acoustical properties in a material control how it reacts to sound waves. Sound waves are pressure waves that are carried by air, water, and other materials. They are what the ears "hear." All sound energy is produced by vibrations. Sound energy will travel through some materials. For example, a piece of string tightly stretched between two tin cans will carry a voice message over a short distance. Materials used in most musical instruments also transmit and amplify sound.

The speed of sound in a material depends on the spacing of the molecules and how easily the molecules move. (A *molecule* is a group of atoms. An *atom* is the smallest possible particle of matter.) Sound travels faster in aluminum than in pine. This is true only because the molecules in aluminum are closer together. They transmit sound energy more easily. See **Figure 4-5**.

Materials vary in their ability to absorb sound. For example, acoustical tiles or heavy carpeting both absorb sound. The sound waves become trapped in the air pockets of the material. Hard materials such as the walls of a canyon reflect sound. Call out a name and it will bounce back at you as an echo.

Electrical Properties

Some materials will conduct electricity, while others will not. This is controlled by electrical properties. Materials that will carry an electric current are called *conductors*. Those that will not conduct current are called *insulators*.

Figure 4-5 Vibrating guitar strings cause the air inside the guitar sound box to resonate.

Technology and society: To which products could sound absorbing materials be added? In what ways would our lives be improved if more products were made using sound absorbing materials?

Metals are good electrical conductors. Some are better than others. One of the best is copper, which is used in cables that supply electricity to lights, appliances, and machines.

Wires carrying electrical current must be insulated. They are covered with materials that are poor conductors. Insulators made from ceramics or plastics are used.

Between these two extremes of good conductors and good insulators is a third type of material. This material is called a *semiconductor*. It allows electricity to flow only under certain conditions. Silicon and germanium are two important semiconductors. They are used in the production of transistors. Semiconductor materials are also used in devices that detect heat or light, such as thermocouples in furnaces and photovoltaic sensors in streetlights, turning them on when it gets dark.

Magnetic Properties

A *magnetic* material will be attracted to a magnet. The most common magnetic materials are iron, nickel, cobalt, and their alloys. Most other materials, such as wood, plastic, and glass, are nonmagnetic. How a material reacts to magnetism is known as the material's magnetic properties.

Types of Materials

Many materials exist in nature. These are called natural or *primary materials*. Some examples of primary

materials include: trees, animal skins, clay, crude oil, and iron ore. Very seldom are these materials used in their natural state. Usually they are changed so that they are more useful. Some examples are the following: trees are cut into boards or planks, animal skins are treated to make leather, clay is baked into pots, crude oil is refined into gasoline, and iron ore is processed to produce steel. Other materials are created by people. These are called *synthetic*. Examples include plastic, glass, and cement.

For convenience, you can classify most solids under five groups: woods, metals, plastics, ceramics, and composites.

Woods

There are two families of wood, *softwoods* and *hardwoods*. Softwood trees are coniferous, **Figure 4-6**. Coniferous trees retain their needlelike leaves and are commonly called evergreen trees. Three examples of softwood trees are cedar, pine, and

Figure 4-6 A coniferous tree is a tree that is "cone bearing." (NFB)

spruce. Hardwood trees have broad leaves, which they usually lose in the fall, **Figure 4-7**. They are also known as deciduous trees. Examples are birch, cherry, and maple.

The terms softwood and hardwood refer to the botanical origins of woods, not their physical hardness. For example, balsa wood is botanically a hardwood, yet it is physically very soft.

Like all living things, trees need nutrients. They must have them to live and grow. Their roots absorb water and mineral salts. Carbon dioxide is absorbed through the leaves. A process known as *photosynthesis* uses energy from sunlight to convert the carbon dioxide, water, and minerals into the valuable nutrients that trees need, **Figure 4-8**.

If you look at the cross section of a tree trunk, **Figure 4-9**, you will see the various parts of the tree. Bark, the outer layer, protects the tree. Beneath is a very thin, single-cell layer called the cambium. Here new wood cells are created, adding to the sapwood.

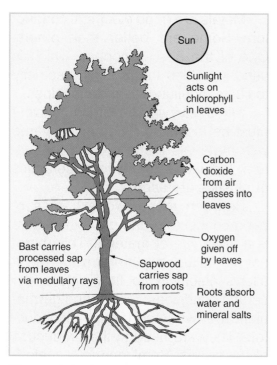

Figure 4-8 A tree uses sunlight, water, carbon dioxide, and minerals to "feed" itself.

Figure 4-7 Hardwoods come from deciduous trees. Most of these trees lose their leaves in the fall.

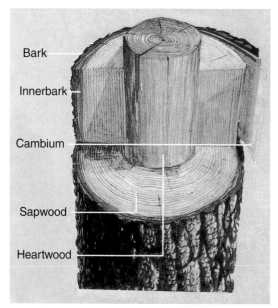

Figure 4-9 This cross section of a tree trunk shows the parts of a tree. The dark lines running outward from the center are medullary rays.

Community resources and services: Are the types of trees growing locally used commercially?

The center part of the trunk is formed of the older, dead wood. This is called the heartwood. Water and minerals move through the trunk by the medullary rays.

The internal structure of wood is cellular, **Figure 4-10**. The cells conduct sap, store food, and provide the support system for the tree. During the spring, when there is plenty of water available, larger cells are produced, which form the spring growth. During the drier summer season, the cells are smaller, heavier, and thicker-walled.

The characteristics and properties of typical softwoods and hardwoods are summarized in **Figure 4-11**. These qualities are important to know if wood is to be put to best use.

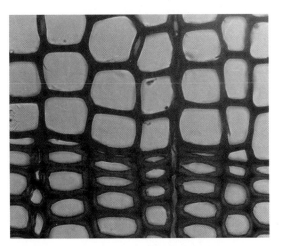

Figure 4-10 The cellular structure of trees transports food needed for growth. Growth is greatest during the spring season when more moisture is available.

To change the living tree into usable pieces of wood, the tree must be cut down, carried to the sawmill, sawn into boards, and dried in kilns. These steps are shown in **Figure 4-12**.

Wood that is used to make furniture is seasoned to reduce its water content to about the same as the air surrounding it. Traditionally, this has been done by stacking the wood outside, which allows the moisture to evaporate. The process usually takes a year or more. Nowadays, to speed up the process, the wood is dried in a special oven called a kiln. Even kiln-dried wood will still expand or contract as the moisture content in the air changes with the seasons.

Drying out causes the wood to shrink unevenly. This leads to warping as shown in **Figure 4-13**. Expansion, contraction, and warping are major problems. They are most troublesome when large pieces of solid wood are needed to make, for example, a tabletop. A way to overcome these problems is to join narrow boards edge-to-edge, **Figure 4-14**. However, this solution is time-consuming and costly.

Today, the usual solution is to use manufactured boards. The most common are plywood, blockboard, particleboard, and hardboard.

Plywood

Plywood is the strongest manufactured board and is usually the

Designing and making
Why does wood remain the most popular material used in the design and manufacture of quality furniture?

Reflection
How do you feel when you go for a walk amongst trees?

Reflection
Do your feelings change depending on the different seasons of the year?

Activity
Have students examine solid wood products in their home to check for problems caused by drying and shrinking. Where is the problem occurring and what is the problem?

Resource
Transparency 4-2, *Characteristics of Softwood and Hardwood*

Community resources and services: Have students visit a local mill to examine how logs are converted into lumber and then stored. Students should visit a local building supplier to examine how sawn lumber is stored.

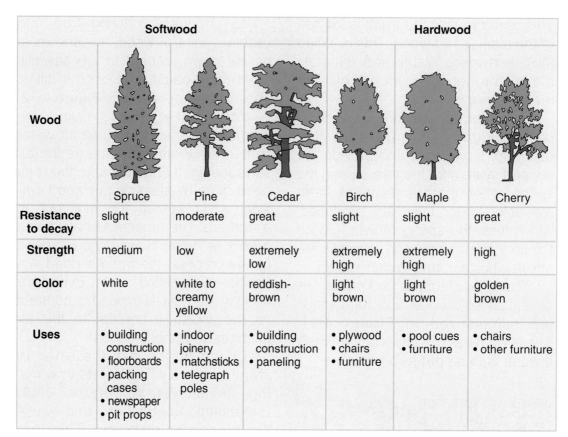

Wood	Softwood			Hardwood		
	Spruce	Pine	Cedar	Birch	Maple	Cherry
Resistance to decay	slight	moderate	great	slight	slight	great
Strength	medium	low	extremely low	extremely high	extremely high	high
Color	white	white to creamy yellow	reddish-brown	light brown	light brown	golden brown
Uses	• building construction • floorboards • packing cases • newspaper • pit props	• indoor joinery • matchsticks • telegraph poles	• building construction • paneling	• plywood • chairs • furniture	• pool cues • furniture	• chairs • other furniture

Figure 4-11 These are the characteristics and uses of common types of wood.

Figure 4-12 Woods go through these processing steps from standing trees to finished lumber.

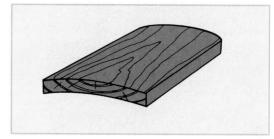

Figure 4-13 Uneven expansion and contraction causes boards to warp.

most expensive. It consists of an odd number of veneers (thin sheets of wood) glued together so the grain in one is at right angles to the grain in the layers above and below it. See **Figure 4-15**. Gluing the veneers in this

Technology and society: What is the impact of clear cutting on the environment?

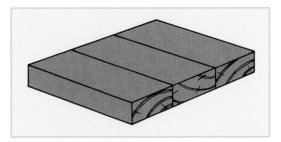

Figure 4-14 Woodworkers join narrow boards edge-to-edge to reduce warpage.

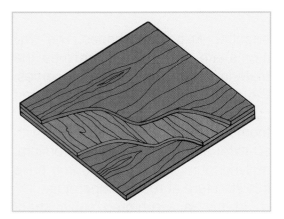

Figure 4-15 Plywood is always made with an odd number of plies (layers). Can you guess why?

way prevents the wood from twisting and warping.

The veneers are cut from a log. The log is mounted on a huge lathe-like machine, **Figure 4-16**. A large blade cuts the wood, unrolling it like a reel of paper.

Blockboard

Blockboard consists of a core of softwood strips faced with a veneer on each side, **Figure 4-17**. The grain of the face veneer is at right angles to the direction of the core strips. Blockboard is strong and durable, but expensive.

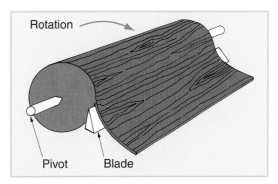

Figure 4-16 Rotary cut veneer is "peeled" from a log with a sharp blade.

Particleboard

Particleboard consists of a core of wood chips. They are bonded together with an adhesive and pressed into a flat sheet. It can be left in this form or faced with a more expensive veneer, such as mahogany or teak. Particleboard is rather brittle, difficult to join, but cheap. It is used in the manufacture of less expensive furniture, **Figure 4-18**.

Hardboard

Hardboard is made from wood fiber. These fibers are treated with chemicals. Then they are reformed

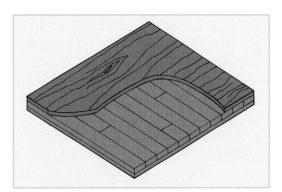

Figure 4-17 Blockboard always has a core of solid pieces of softwood.

Technology and society: How does the increased use of manufactured boards benefit the environment?

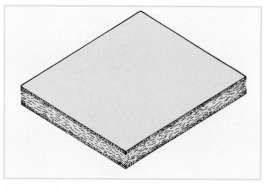

Figure 4-18 Particleboard is made from wood chips. They are glued and then pressed into a sheet. Sometimes the sheet is faced with a veneer.

Figure 4-19 Here are the three kinds of hardboard. Sometimes a decorative face is fixed to one side. (Christopharo)

into sheets using heat and pressure. Hardboard is usually smooth on one side and textured on the other.

Tempered hardboard can be made by impregnating standard hardboard with resin and heat curing. Tempering improves water resistance, hardness, and strength. Hardboard is made in thin sheets and is cheap. It is used to cover hidden parts of cheap furniture or for drawer bottoms. **Figure 4-19** shows several types.

Metals

Metals are inorganic materials. There are two families of metals: *ferrous* and *nonferrous*. The word ferrous is from the Latin word *ferrum*. It means "iron." Thus, any metal or alloy that contains iron is a ferrous metal. Metals or alloys that do not have iron as their basic component are called nonferrous metals. **Figure 4-20** shows some properties of common ferrous and nonferrous metals.

When magnified, the internal structure of metals can be seen to be crystalline. The crystals are made of atoms arranged in boxlike shapes. The way atoms are combined determines the material's structure and its properties. **Figure 4-21** shows the body-centered cubic structure. The atoms arrange themselves into a cubic structure with an extra atom at its center. Chromium, molybdenum, tungsten, and iron have this structure at room temperature.

In the face-centered cubic structure, **Figure 4-22**, each face of the cube has an additional atom at its center. Copper, silver, gold, aluminum, nickel, and lead have this structure.

Crystals, **Figure 4-23**, are the basic units of metals. As molten metal cools from the liquid state, atoms bond themselves together permanently, forming crystals. The crystals pack themselves together like the pieces of a jigsaw puzzle. Because temperature changes the way atoms are arranged, a particular metal may have different forms. Above 55°F (13°C), tin can is seen as a shiny metal. As the temperature drops, the atoms rearrange and the metallic tin begins to change into a nonmetallic gray powder. In cold cathedrals, tin organ pipes have sometimes disintegrated!

Links to other subjects: Science—Investigate the atomic structure of materials.

	Metal	Important Content	Melting Temp.	Resistance to Corrosion	Characteristics	Color	Uses
Ferrous	Cast Iron	93% iron 3% carbon	2200°F (1204°C)	poor	-hard -brittle -heavy	dark-gray	-bodies of machine tools -engine blocks -bathtubs -vises -pans
	Mild Steel	99% iron 0.25% carbon	2500°F (1371°C)	very poor	-stronger and less brittle than iron -can be easily joined by welding	gray	-girders in bridges -tubes in bicycle frames -nuts and bolts -car bodies
	High-Carbon (tool steel)	0.60-1.30% carbon	2500°F (1371°C)	very poor	-not easy to machine, weld, or forge	gray	-cutting tools -drill bits -self-tapping screws -wrenches -railroad rails
Nonferrous	Aluminum	base metal	1220°F (660°C)	high	-light, soft, and malleable -conducts heat and electricity -very difficult to solder or weld	gray	-cooking utensils -foil -siding -window fames -aircraft
	Tin	base metal	450°F (232°C)	excellent	-soft -nontoxic -shiny -most often used as an alloying agent	silver	-rolled foil -tubes and pipes -tinplate -ingredient of solder -galvanizing
	Copper	base metal	1980°F (1083°C)	high	-tough and malleable -good conductor of heat and electricity -expensive -easily joined by soldering	reddish-brown	-electrical wiring -cables -water pipes

Figure 4-20 This chart shows the characteristics and uses of ferrous and nonferrous metals.

All crude (unprocessed) metals are found buried in the ground. Most are in the form of ore. This means that they are mixed with rock and other impurities. The ore is extracted from the ground by mining or quarrying. It is then crushed, and the waste earth and rock is removed. The remaining ore is sintered (formed by heat) into pellets.

Iron ore, limestone, and coke are required for making iron. The iron ore, coke, and limestone are dumped into the top of a blast furnace. The mixture is heated to 2912°F (1600°C). The

Designing and making
How do the characteristics of metals affect the design of products?

Vocabulary
What does the term *alloy* mean?

Resource
Transparency 4-3, *Characteristics and Uses of Ferrous and Nonferrous Metals*

Resource
Reproducible Master, *Characteristics of Metals*

Technology and society: What is the environmental impact of mining for metal ores?

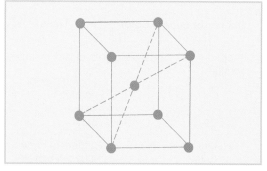

Figure 4-21 This is a body-centered cubic structure with one atom at the center. This arrangement is found in ferrous metal. (The dots represent atoms.)

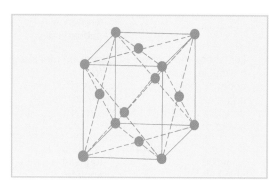

Figure 4-22 This is the structure of the atoms of nonferrous metals. Each face of the structure has an atom in its center.

Cast Iron

Aluminum

Figure 4-23 These photographs of magnified samples of metal show some of the different crystal arrangements that can happen when metals are being formed. (Alcan)

burning coke and very hot air melt the iron ore and limestone. The limestone mixes with the ashes and waste rock, forming a waste called slag. The molten iron sinks to the bottom of the furnace, and the iron and slag are tapped off separately. The iron is poured into large containers. It is ready to become steel. **Figure 4-24** shows the entire process.

Metals are rarely used in their pure state. They are often too soft and ductile. They are often mixed with other metals to produce an *alloy*, see **Figure 4-25**. Just as one liquid can dissolve in another, so can metals. Alloying can increase strength or hardness, inhibit rust, and change color and electrical or thermal conductivity. For example, mixing 10% aluminum with copper produces an alloy with approximately three times the strength of pure copper. Similarly, gold can be mixed with other metals for increased strength. The most common alloys are silver and copper. The amount of gold is given in terms of carats. Pure gold has a carat of 24. An alloy with 50% gold is called 12 carat gold. Jewelry usually varies from nine carat (37.5% gold) to 18 carat (75% gold).

Useful Web site: Have students browse www.sma-inc.com/SMAPaper.html, www.sma-inc.com/Applications.html, and www.azom.com to investigate the properties and uses of shape memory and super elastic alloys.

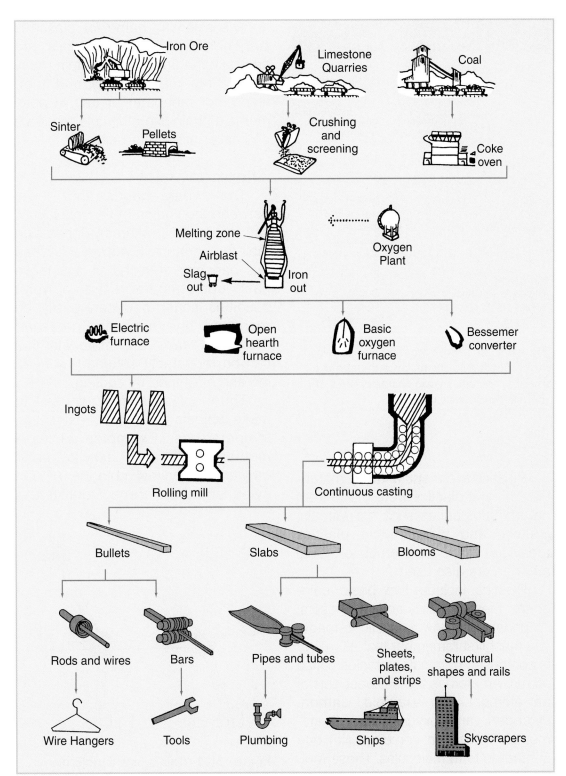

Discussion
Describe the processing of raw materials to produce steel. Identify products made from steel.

Activity
Divide the class into small groups and have each group investigate the processing of a different metal (other than iron or steel). Have each group make a poster illustrating the process and present it to the class.

Figure 4-24 Processing iron ore into steel requires many steps.

Technology and society: What happens to all the scrap metal produced by society?

Community resources and services: Invite one or more people working in a primary processing plant to describe their work to the class.

Designing and making

List products that cannot be made from thermoplastics and explain why thermoplastics are not an option for these items.

Vocabulary

What is the difference between a *thermoplastic* and a *thermoset* plastic?

Alloy	Major Components	Uses
Brass	Copper and zinc	Electrical fittings, locks, hinges
Bronze	Copper and tin	Bells, castings, sculptures
Pewter	Tin and lead	Tableware, tankards
Solder	Tin and lead	General solder uses a 50-50 mixture. Electrical solder uses a 60-40 mixture.
Stainless steel	Steel and chromium	Cutlery, sinks, surgical instruments

Figure 4-25 Alloys are mixtures of two or more metals. Alloying makes a metal more useful. It combines good characteristics of the different metals used.

Plastics (Polymers)

Plastics are manufactured materials. The first plastics were made from natural substances such as animals, insects, and plants. Most plastics today are made from crude oil. Coal and gas are also used.

Plastics include two basic categories: thermoplastics and thermosets. To understand how a plastic is made, you must first remember that all matter is composed of minute particles called atoms. An atom is the smallest part of an element. For example, carbon, hydrogen, and oxygen are all elements. Atoms can combine with one another to form molecules. These molecules, joined together in chains, form the fundamental building blocks of a plastic material, **Figure 4-26.**

The scientific name for plastic is *polymer.* Polymeric materials are basically materials that contain many parts. "Poly" means many, and "mer" stands for monomer or unit. A polymer is a chain-like molecule made up of smaller molecular units.

Thermoplastics

The weak bonds that exist between the chains of thermoplastics can easily be broken by increasing their temperature, **Figure 4-27.** When this happens, the chains can move past one another. The polymer becomes softer and can easily be reshaped by applying pressure. *Thermoplastics* are materials that can be repeatedly softened by heating and hardened by cooling.

Thermoplastics are not used to make objects that must resist high temperatures. Examples of thermoplastics include acrylic, polyethylene, polyvinyl chloride (PVC), nylon, and polystyrene. Dentists use

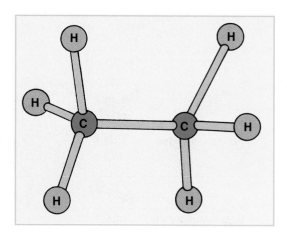

Figure 4-26 This is how a molecule of ethane looks. It has two atoms of carbon and six of hydrogen.

Technology and society: What impact will the increasing use of plastics have on the environment?

Community resources and services: Have students investigate the resources available in their community for disposing or recycling plastic products. How could these resources be improved?

thermoplastic to produce a detailed replica of your teeth and the shape of the tissues in your mouth. When the thermoplastic material is heated, it becomes soft and able to take up a new form. When it cools to mouth temperature, it hardens and can be removed, retaining the impression of the oral cavity.

Have you noticed that even one type of plastic can be sold in different forms? PVC has been used to make hard phonograph records and soft vinyl shower curtains. The pliability of the material is affected by combining it with various chemicals called *plasticizers*. These act as internal lubricants, allowing the long chains of PVC molecules to slide past one another.

Plastics have been criticized as major contributors to garbage. For example, polystyrene is molded into the food trays you see under meats or fast food containers. It is widely used because it does not support the growth of bacteria or fungi. Thus, it slows down spoilage and improves hygiene. The ever-increasing demand for polystyrene has added to the concern about its effect on landfill space. Some people claimed that polystyrene is not biodegradable and that it degrades into substances that contaminate water and the atmosphere.

Thermosets

You might think that plastics are modern materials, but one thermoset plastic was invented nearly a century ago. By the 1920s, Bakelite, made from phenol and formaldehyde molecules, was used in jewelry, telephones, pens, radios, ashtrays, and cameras. In fact, it is still used as an adhesive that glues plywood and particleboard. Thermoset plastics, **Figure 4-28**, can be heated only once. That is at the time they are made. Do you know how they got their name? "Thermo" means heat, and "set" means permanent. Heat causes the material to develop cross-links between the chains. Once connected, the individual chains form a rigid structure. The plastic is no longer affected by heat or pressure. Once a

Designing and making
How do the characteristics of thermoset plastics affect the design of products?

Activity
Collect and label examples of different plastics.

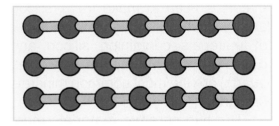

Figure 4-27 The molecular structure of a thermoplastic material shows that the chains have weak bonds to one another, which can easily be broken.

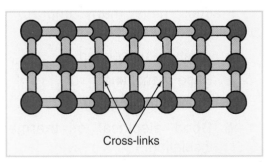

Cross-links

Figure 4-28 Molecular structure of a thermoset has cross-links between chains.

Useful Web site: Have students visit the Biodegradable Plastics Society Web site at www.bpsweb.net/02_english to gather information about the characteristics of Greenpla (biodegradable plastics).

Useful Web site: Have students investigate the properties and uses of Super-Absorbent Polymers at www.psrc.usm.edu/macrog.

thermoset has been heated and formed, its shape becomes permanent.

Articles made from thermosetting materials have good heat resistance. Examples of thermosets include polyester resins (often reinforced with glass fiber), phenol resins, urea resins, melamine resins, epoxy, and polyurethane.

Prior to the twentieth century, clothes were made of natural fibers, such as wool from sheep or silk from silkworms. Synthetic fibers are now used in most of the clothes we wear. They can be produced for specific purposes. Their color, strength, and resistance to heat, sunlight, and chemicals can be determined before they are made. Nylon is strong, hard-wearing, and used for lingerie and socks. Acrylic is soft and wool-like, making it ideal for sweaters and sleepwear. Polyester is strong, lightweight, and a favorite for shirts and leisurewear.

Today, there is no single material called plastic. Chemists can alter the "mix" each time to create plastics that look and behave very differently from each other, **Figure 4-29**.

Plastics have many important properties. These include:

◆ Ability to be colored.

◆ Ease of molding.

◆ Flexibility or rigidity.

◆ Good electrical or thermal insulation.

◆ Light mass (weight).

◆ Low cost.

◆ Resistance to rot and corrosion.

◆ Strength.

The characteristics and uses of typical thermoplastics and thermosets are summarized in **Figure 4-30**.

Ceramics

The word *ceramics* is from the Greek *Kermos* meaning "burnt stuff." The important characteristics of ceramics are hardness, strength, resistance to chemical attack, and brittleness. They are also impervious to heat. The space shuttle is covered with ceramic (silica) tiles to protect it from the heat of friction generated by a high-speed re-entry into the earth's atmosphere. Today, the term ceramics covers a wide range of materials, including abrasives, cement, window glass, and porcelain enamels on bathroom fixtures.

Most ceramics are thermosetting materials. Once they have been processed and hardened, they cannot be made soft and pliable again. The one exception is glass. It can be continuously reheated and reshaped. It is, therefore, thermoplastic in nature.

One of the great advantages of ceramics is the abundance of the raw materials used to make them: silicon and oxygen. Silicates (any number of materials containing silicon and oxygen, including sand, clay, and quartz) are the most common minerals on earth.

The two ceramic materials most familiar to you are glass and cement. Their manufacture is shown in **Figure 4-31** and **Figure 4-32**. The following are some basic characteristics of ceramics:

Links to other subjects: Have students investigate the use of ceramics in science and medical technology.

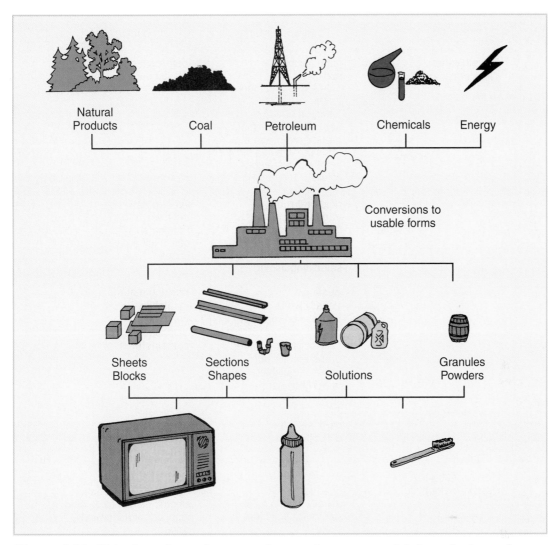

Discussion
Why has plastic become such a widely used material in the design and manufacture of mass-produced products?

Reflection
Do you prefer eating and drinking from ceramic or plastic dishes, bowls, and cups? Do you prefer to use metal or plastic utensils? Explain your answers.

Natural Products Coal Petroleum Chemicals Energy

Conversions to usable forms

Sheets Blocks Sections Shapes Solutions Granules Powders

Figure 4-29 This chart shows the steps of processing raw materials into plastic products. Note that "energy" is named as a raw material. Why?

- ◆ Strong and resistant to attack by nearly all chemicals.
- ◆ Withstand high temperatures.
- ◆ Lack malleability (are stiff, brittle, and rigid).
- ◆ Very stable (not likely to change shape because of heat or weather).
- ◆ High melting point.
- ◆ Hardest of all engineering (solid) materials.

- ◆ Raw materials are available worldwide and are consequently low in cost.
- ◆ Withstand outdoor weathering from the sun, moisture, environmental pollutants, and dramatic temperature changes.
- ◆ Poor electrical and thermal conductivity.
- ◆ Poor thermal shock resistance.

Technology and society: How could governments and manufacturers encourage people to use products made of Greenpla (biodegradable plastics)? Refer to www.bpsweb.net/02_english.

Designing and making
How might increasing the use of Greenpla (biodegradable plastics) change the design of products? Refer to www.bpsweb.net/02_english.

Activity
Have students collect examples of products made from the six types of plastics shown in Figure 4-30.

Resource
Reproducible Master, *Characteristics of Plastics*

	Plastic	Characteristics	Uses
Thermoplastics	Polyethylene (low density)	-fairly flexible -soft -cuts easily and smoothly -floats -"waxy" feel -not self-extinguishing -transparent when thin -translucent when thicker -can be dyed various colors	-detergent squeeze bottles -plastic bags -electrical wire covering
	Polyvinyl chloride	-rigid or flexible -transparent -fairly easy to cut -smooth edges -sinks -self-extinguishing	-water pipes -records -raincoats -soft-drink bottles -hoses
	Polystyrene	-opaque -usually white -tends to crumble on cutting -very buoyant in water -burns readily, not self-extinguishing -very lightweight	-packing material -insulation -ceiling tiles -disposable food containers
Thermosets	Polyester resin	-stiff -hard, solid feel -difficult to cut -brittle -burns readily	-repair kits -car bodies -boat hulls -garden furniture
	Urea formaldehyde	-opaque -usually light in color -stiff, hard, solid feel -sinks -flakes on cutting -burns with difficulty -good heat insulator	-light-colored domestic electric fittings (e.g., plug tops, adaptors, switch covers) -waterproof wood adhesives
	Melamine	-opaque -usually light in color -stiff -hard, solid feel -flakes on cutting -sinks -burns with difficulty -good heat insulator -resists staining	-tableware -surfaces for counters, tables, and cabinets -expensive electrical fittings

Figure 4-30   Note the characteristics and uses of plastics.

Useful Web site: Have students visit the Biodegradable Plastics Society Web site at www.bpsweb.net/02_english to gather information about products currently made using Greenpla (biodegradable plastics).

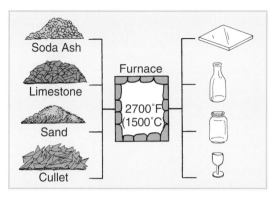

Figure 4-31 Raw materials and broken glass are used to make new glass products.

Figure 4-33 lists the characteristics and uses of common ceramics.

Composite Materials

When two or more materials are combined, a new material, known as a *composite*, is formed. For instance, concrete is a composite. It combines cement, sand, and gravel. Fiberglass is a composite of glass fibers and plastic resin.

Combining materials produces a new material. The individual components retain their physical identity, but, together, as a composite, they provide improved mechanical or other properties. Each material in the composite keeps its own properties, and combining them also adds new properties.

There are three types of composites: layered, fiber, and particle. Each will be explained.

Layered composites consist of laminations like a sandwich. Thin layers of material are tightly bonded.

How much do you know about the tires that support you when you ride in a car or bus, or land in an airplane? Tires are also layered composites, **Figure 4-34**.

Another common use of layered composites is in the construction of

Designing and making
How do the characteristics of composite materials affect the design of products?

Discussion
What is the difference between *layered*, *fiber*, and *particle composites*?

Vocabulary
What is a *kiln* and what is it used for?

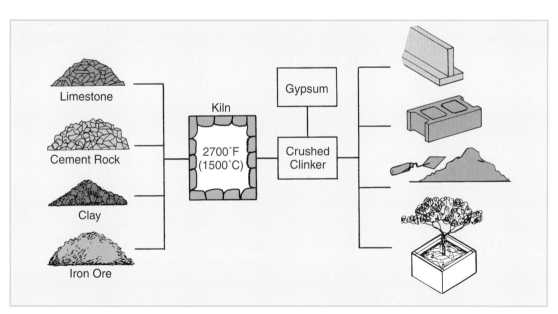

Figure 4-32 Cement is a common ceramic material. This is how it is made.

Useful Web site: Have students enter the term "composites" into a search engine and browse some of the Web sites to find examples of advanced composites used in bio-engineered products. A useful starting point is the "Bio-Engineered Products" link at www.composites.Sparta.com.

Designing and making
How is the visual effect of buildings made from stone and the visual effect of buildings made from concrete different?

Reflection
Do you prefer to sit on natural rocks or on concrete? Why do you prefer that material?

Activity
Have students create a collage showing products made from composites.

Resource
Reproducible Master, *Characteristics of Ceramics*

Type	Characteristics	Uses
Stone	-hardness varies from soft sandstone to hard granite -generally able to withstand effects of rain, spray, wind, frost, heat, and fire -color varies -surface appearance varies -resistant to corrosion -opaque	-building material -abrasives
Clay	-becomes plastic when mixed with water -has a bonding action on drying	-house bricks -tiles -earthenware -stoneware -china -porcelain
Refractories	-stability at high temperatures -can withstand pressure when hot -can withstand thermal shock -highly resistant to chemical attack	-lining of furnaces to produce glass, cement, metals, bricks -insulation on spark plugs -tiles on space shuttle -firebrick
Glass	-transparent -poor in tension -transmits visible light well -opaque to ultraviolet light	-fibers for insulation -fibers for filters -windows and doors -heat resistant cookware -light bulbs -drinking glasses -lenses
Cement	-correctly call Portland cement -made from limestone (80%) and clay (20%) heated in kiln to form a clinker that is ground to a fine powder -very durable -can withstand frequent freezing and thawing -can withstand wide range of temperatures	-the ingredient of concrete that binds the crushed rock, gravel, and sand together

Figure 4-33 Ceramics have characteristics that make them useful for many products and building materials.

hollow core doors, **Figure 4-35**. Sheets of plywood are glued to spacers of cardboard and solid wood.

The wings of some aircraft are made of layered composites. The center of the composite is an aluminum honeycomb. This honeycomb is covered with layers of fiberglass to form a sandwich, **Figure 4-36**.

Fiber composites consist of short or long fibers of a material such as glass or carbon embedded in a matrix

Useful Web site: Have students browse www.GreatBuildings.com to examine buildings in which the design is influenced by the characteristics of concrete. Have students print one image they find inspirational to share with the class.

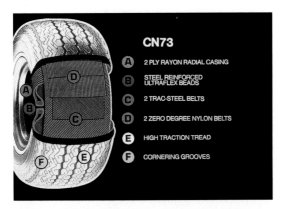

Figure 4-34 Tires are composites of steel, rubber, plastics, and fibers. (Pirelli)

Figure 4-36 This airplane wing is also a composite. The aluminum center looks like a honeycomb. It is covered with sheets of fiberglass. (Bombardier)

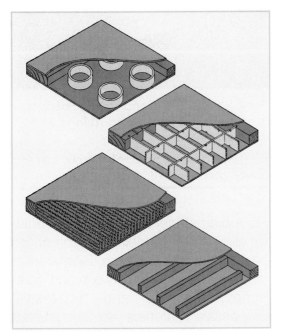

Figure 4-35 Here are sample sections of hollow core doors. Layers are made up of plywood, cardboard, and solid wood pieces.

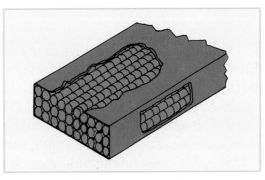

Figure 4-37 These are fibers in a matrix. The matrix could be a liquid such as resin. It is poured over the fibers and left to harden.

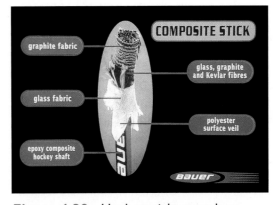

Figure 4-38 Hockey sticks must have strength and flexibility. (Canstar)

of another material such as resin, or other plastic, and metal, **Figure 4-37**. One use of a fiber composite is shown in **Figure 4-38**.

Particle composites, **Figure 4-39**, consist of particles held in a matrix.

Reflection
How would you feel if one or more parts of your body were replaced with bio-engineered products?

Vocabulary
What do the terms *prosthetics* and *implants* mean?

Activity
Have students use an Internet search engine to investigate the term "wetware" (technological products combined with the human body).

Technology and society: List ways in which biomaterials are currently used in medical technology. What ethical issues are involved in the increased use of biomaterials?

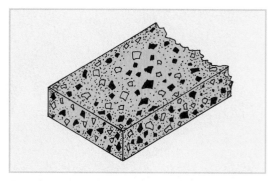

Figure 4-39 Here are particles in a matrix.

The most common particle composites are concrete and particleboard, **Figure 4-40**.

Biomaterials

The history of using artificial materials to substitute body parts dates back to early history. Gold was used to replace natural teeth, glass to replace eyes, and wood to replace limbs. In the last century, human-made materials and devices have been made that can replace parts of living systems in the body. These special materials, able to function in intimate contact with living tissue, are classified as a *biomaterial*. Metals, ceramics, polymers, and composites are the main classes of biomaterials. Many of these materials also have other common uses, but when they are used as biomaterials, they have a very high purity. They are also processed differently to ensure durability, wear-resistance, and biocompatibility. Being biocompatible means that the surrounding tissue in the body will not reject the new implanted material.

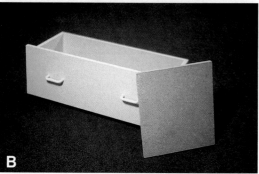

Figure 4-40 A—Concrete, a composite, is being poured into a form. (Morrison Knudsen Corp.)

B—This drawer is made of another composite, particleboard. (Ecritek)

A biomaterial is any material, natural or human-made, that comprises part of a living structure. Biomaterials can be classified into two broad categories: bioactive and biopassive. Biopassive materials will not interact with a person's biological system. For example, cobalt-chromium alloys are used as bearing materials for total joint replacements. Bioactive materials will interact with the biological system. For example, various polymers are used as sutures that degrade in the body.

Chapter 4 Review
Materials

Summary

Designers need to know about the properties of different materials so they can determine the material best suited for the object to be designed and built. Materials may be chosen for their physical, mechanical, thermal, chemical, optical, acoustical, electrical, or magnetic properties.

Most materials can be classified into five groups. These are woods, metals, plastics, ceramics, and composites. Within each group there are many varieties. Woods may be subdivided into softwoods and hardwoods. Metals include ferrous and nonferrous. Plastics are either thermoplastics or thermosets.

Most ceramics are thermoset materials, except glass, which is a thermoplastic. A composite is formed when two or more engineering materials are combined. Composites are increasingly important in the production of the objects we use today. Technologists must know the mechanical properties of each material they plan to use in a designed product. Exceeding the limits of the material may cause a material to fail unexpectedly. In the near future, materials used in situations where lives could be in danger may have computer chips embedded in them to warn of possible failure.

Modular Connections

The information in this chapter provides the required foundation for the following types of modular activities:

◆ Material Science

◆ Material Testing

◆ Plastics

Test Your Knowledge

Write your answers to these review questions on a separate sheet of paper.

1. What have some of the major periods throughout history been named after?

2. Name and define each of the seven properties of materials.

3. Set up a chart like the one below. Place each of the following materials in the appropriate column in the chart: polystyrene, cast iron, cedar, copper, glass, maple, aluminum, pine, cement, melamine, spruce, polyethylene, birch, porcelain, clay, steel.

Category of Material			
Woods	Metals	Plastics	Ceramics

4. List the characteristics of hardwood and softwood trees.

5. What happens to a tree once it has been transported to a sawmill?

6. Sketch the construction of the following manufactured boards: plywood, blockboard, and particleboard.

7. What is the difference between a ferrous and a nonferrous metal?

8. Is pure iron an alloy? Explain your answer.

9. Give three examples of both ferrous and nonferrous metals.

10. Copy the chart below and fill in the blank spaces.

Metal	Typical Use
Copper	
Aluminum	
High-carbon steel	
Cast iron	

11. The scientific name for plastic is _____.

12. Sketch the molecular chains of thermoplastics and thermosets.

13. Describe the fundamental difference between thermoplastics and thermosets in terms of the way their molecular chains are formed.

14. Plastics are generally manufactured from _____.
    a. crude oil
    b. wood chips
    c. mineral ores
    d. animal by-products

15. Copy the chart below. List four objects in your home that are made of plastic. In your opinion, what particular properties make plastic the most appropriate material for each of the objects you have chosen? Refer to Figure 4-30 in your text.

Object	Important Characteristics
detergent squeeze bottle	flexible, transparent, lightweight
1.	
2.	
3.	
4.	

16. The two ceramic materials most familiar to you are _____ and _____.

17. State three advantages and three disadvantages of ceramic materials.

18. What is a composite material?

19. Using sketches and notes, describe the differences between layered, fiber, and particle composites.

20. List six objects made using composite materials that can be found in your home. For each object, state the type of composite and the materials used.

Answers to Test Your Knowledge Questions *(continued)*
14. A. crude oil
15. Answers will vary; evaluate individually.
16. glass, cement.
17. Advantages (any order):
Strong and resistant to attack by nearly all chemicals
Withstand high temperatures
Very stable
High melting points
Raw materials available worldwide and, consequently, low cost
Withstand outdoor weathering
Disadvantages (any order):
Lack of malleability
Poor electrical and thermal conductivity
Poor thermal shock resistance
18. A material made by combining two or more materials.
19. For sketches, refer to Figures 4-34, 4-35, 4-37, and 4-39, pages 103–104.
Layered composites consist of layers like a sandwich. Fiber composites are formed of short or long fibers of one material embedded in a matrix of another material. Particle composites consist of particles held in a matrix.
20. Answers will vary; evaluate individually.

Answers to Test Your Knowledge Questions *(continued)*
10. Copper: water pipes in a home; Aluminum: cooking foil; High carbon steel: kitchen knives; Cast iron: engine block
11. polymer
12. Refer to Figures 4-27 and 4-28, page 97 of the text.
13. In a thermoset, the chains become cross-linked, and the plastic cannot be reshaped by heating. In a thermoplastic, the chains remain separated even after heating. The shape can be altered by reheating.
(continued)

Apply Your Knowledge

1. Choose three objects in your home that are made of different materials. State whether the materials used are appropriate. Explain your answer.

2. Describe the major processes used to change (a) a tree into finished lumber and (b) iron ore into standard stock.

3. State one advantage and one disadvantage of each of the five different types of materials.

4. Collect pictures of objects that are made of (a) layered, (b) fiber, and (c) particle composites. Label the diagrams to show the materials used in each composite.

5. Choose one property of materials. Collect samples of five different materials. Design and build an apparatus to test the materials for the property you have chosen.

6. Do you know that some metals can make you sick and have even caused madness? Research one of the following topics to find out more.

   A. The Romans loved sweet things but they didn't have any sugar. As a substitute they boiled grape juice in lead containers until it became sweet syrup. They then added it to various foods. In small quantities, lead can cause mental problems. In larger amounts, it can result in lead poisoning. Did this poisoning lead to the fall of the Roman Empire?

   B. George III was king of England at the time of the American Revolution. It is believed that he was at least partly insane. His madness may have been the result of lead poisoning from eating sauerkraut. It was one of his favorite foods, and it was cooked in lead pots. Did his madness lead to poor decisions resulting in loss of the American colonies?

   C. Lead carbonate was one of the main ingredients in paint until 1980. In the early to mid-1990s, babies and young children were found to have lead in their bodies. It was being ingested by biting their crib rails or by touching paint dust on windowsills and then sucking their fingers. What are the ingredients in modern paints and are they harmful to us?

7. Research one career related to the information you have studied in this chapter and state the following:

   A. The occupation you selected.

   B. The education requirements to enter this occupation.

   C. The possibilities of promotion to a higher level at a later date.

   D. What someone with this career does on a day-to-day basis. You might find this information on the Internet or in your library. If possible, interview a person who already works in this field to answer the four points. Finally, state why you might or might not be interested in pursuing this occupation when you finish school.

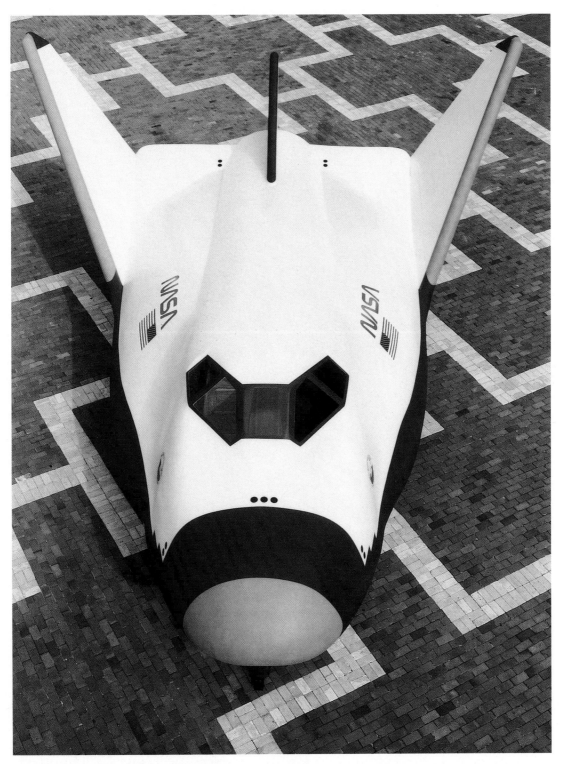

During re-entry, friction forces extremely hot temperatures on the outer surface of the space shuttle. Protection from the heat comes in the form of thermal blankets and tiles. The outer skin is made mostly of aluminum and graphite epoxy. (NASA)

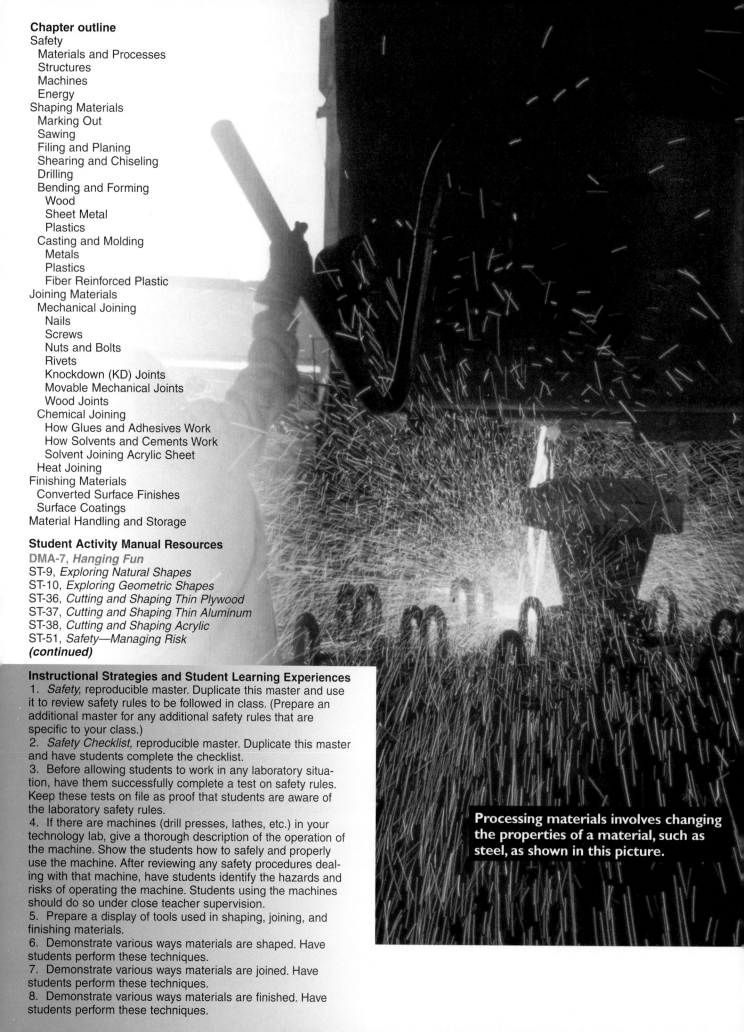

Chapter outline

Student Activity Manual Resources
DMA-7, *Hanging Fun*
ST-9, *Exploring Natural Shapes*
ST-10, *Exploring Geometric Shapes*
ST-36, *Cutting and Shaping Thin Plywood*
ST-37, *Cutting and Shaping Thin Aluminum*
ST-38, *Cutting and Shaping Acrylic*
ST-51, *Safety—Managing Risk*
(continued)

Instructional Strategies and Student Learning Experiences
1. *Safety*, reproducible master. Duplicate this master and use it to review safety rules to be followed in class. (Prepare an additional master for any additional safety rules that are specific to your class.)
2. *Safety Checklist*, reproducible master. Duplicate this master and have students complete the checklist.
3. Before allowing students to work in any laboratory situation, have them successfully complete a test on safety rules. Keep these tests on file as proof that students are aware of the laboratory safety rules.
4. If there are machines (drill presses, lathes, etc.) in your technology lab, give a thorough description of the operation of the machine. Show the students how to safely and properly use the machine. After reviewing any safety procedures dealing with that machine, have students identify the hazards and risks of operating the machine. Students using the machines should do so under close teacher supervision.
5. Prepare a display of tools used in shaping, joining, and finishing materials.
6. Demonstrate various ways materials are shaped. Have students perform these techniques.
7. Demonstrate various ways materials are joined. Have students perform these techniques.
8. Demonstrate various ways materials are finished. Have students perform these techniques.

Processing materials involves changing the properties of a material, such as steel, as shown in this picture.

5

Processing Materials

Key Terms

bending
casting
chemical joining
chiseling
coating
drilling
filing
finishing
forming
heat joining

jig
laminating
marking out
mechanical joining
molding
planing
safety
sawing
shearing

Objectives

After reading this chapter you will be able to:

- ◆ Demonstrate responsible and safe work attitudes and habits.
- ◆ Design and make a product using wood, metal, and plastic materials.
- ◆ Identify the correct tools and processes to be used to shape a material.
- ◆ Select the correct method for joining materials.
- ◆ Recall steps for applying a finish to a material.

Discussion
Why is it important that designers have a thorough understanding of how materials are worked using tools and equipment?

Career connection
Have students identify the careers in which using various materials to make a product is a central task. Students can browse the Web to find examples of work completed by people in these careers.

Student Activity Manual Resources (continued)
ST-5, *Writing a Design Specification*
EV-1, *Evaluating the Final Product*
EV-2, *Unit Review*

DMA-8, *Desk Lamp*
ST-3, *Using a Collage to Generate Ideas*
ST-8, *Exploring Existing Products*
ST-74, *What Is a Circuit?*
ST-76, *Building a Series Circuit*
ST-77, *Building a Parallel Circuit*
ST-5, *Writing a Design Specification*
ST-51, *Safety— Managing Risk*
EV-1, *Evaluating the Final Product*
EV-2, *Unit Review*

Designing and making
How can designers reduce the hazards and risks faced by consumers when using particular products and services?

Discussion
What types of hazards and risks should be considered when designing and making products? How can the risks be managed?

Reflection
What risk management strategies do you practice when using tools and equipment to cut and shape materials?

Vocabulary
What is the difference between a *hazard* and a *risk*?

Vocabulary
What does *risk management* mean?

Safety
What types of safety precautions must be observed by people working in environments where there are hazardous tools, equipment, and materials?

Resource
Activity 5-3, *Working Safely*

Resource
Reproducible Master, *Safety*

Safety

As you work through your technology course, you will use various materials. You will learn to select the correct hand and machine tools and practice safe techniques for each material. Tools can be dangerous unless you learn to use them correctly.

In addition, you will build models and prototypes. Each must be built with *safety* in mind. When testing models or prototypes or when you are involved in other experimental work, use great care. You must ensure that you, your friends, and the equipment are not harmed. The following notes cover some of the safety rules you should observe as you work through your Activity Book.

Materials and Processes

- Remove loose clothing (coats, jackets, and sweaters).
- Roll sleeves above the elbows. Remove all jewelry.
- Tie back long hair.
- Put on safety glasses before switching on a machine.
- Report all accidents, including minor cuts and scratches, to your teacher.
- Report unsafe conditions, such as damaged or worn equipment.
- Place rags containing oil, gasoline, paint, or solvents in approved metal containers.
- Use the right tool for the job, and use it correctly.

- Keep tools in proper condition and store in a safe place when not in use.
- Pass tools with the sharp edge toward you.
- Make all adjustments to a machine with the power off.
- Do not leave a machine until it has come to a complete stop.
- When using a portable electric drill, always make sure that the drill bit is secure before plugging in the drill.
- Avoid using dangerous materials such as lead or asbestos.

Structures

- Ensure that the structure will withstand the loads imposed on it.
- When testing a structure by loading it with weights, be prepared for the structure to collapse: protect yourself.
- When using a hot wire cutter to cut Styrofoam®, avoid noxious fumes by working in a well-ventilated area.

Machines

- Test all mechanisms. Move the parts by hand before connecting them to a power source.
- Check to ensure that all parts of a mechanism are correctly attached to the supporting structure.

Links to other subjects: What types of hazards and risks exist when participating in science labs, art class, and physical education? How are the risks managed?

Useful Web site: Students can use the dictionaries at www.techtionary.com and www.hyperdictionary.com to find the definition of any new technical and technological terms used in this chapter.

◆ Test vehicles only when you have sufficient space to permit the vehicle to move without danger to anyone.

Energy

◆ Ensure that all parts are fully insulated against leakage of electric current.

◆ If possible, use a 1.5 V cell or a 9 V battery as a power source for projects.

◆ When a 120 V power source is used, the project must be properly grounded.

Shaping Materials

When you have designed your product and have chosen the material, you are ready to make the product. After following several carefully thought-out steps, you will do this.

Marking Out

The first step in making a product is called *marking out.* Marking out involves measuring material and marking it to the dimensions on your drawing. You should do this carefully for two reasons. First, your marks must be accurate so that the pieces fit together. Second, materials are expensive. Making mistakes wastes time and money.

Most marking out starts from a straight edge. To make a straight edge on wood, a plane is used. Plastic and

metal are filed. When wood is used, this edge is called the face edge, and the best side is the face side, **Figure 5-1**. Lines should be easily seen. To be accurate, they must be thin. They should be marked as shown in **Figure 5-2**.

Sawing

Sawing removes material quickly. All saws have a row of teeth. They chip or cut away the material. Like all tools, the part that cuts must be harder than the material being cut. There are special saws for cutting wood and for cutting metal, **Figure 5-3**. Plastics and composites mostly use saws designed for cutting wood and metal.

Filing and Planing

Small amounts of material may be removed by *filing* and *planing*, **Figure 5-4**. Files are mainly used on metal. A special type of file, called a rasp, is used on wood. Various types of planes produce smooth, flat surfaces on wood.

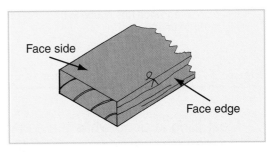

Figure 5-1 Mark lumber on both the face side and the face edge.

Discussion
What problems occur when making a product, if the materials are not marked out accurately?

Safety
Compare the hazards and risks involved when using a 9 V battery power with the hazards and risks in using a 120 V power source.

Resource
Activity 5-1, *General Safety*

Resource
Reproducible Master, *Safety Checklist*

Links to other subjects: Which tools are used for marking out materials in subjects other than technology education?

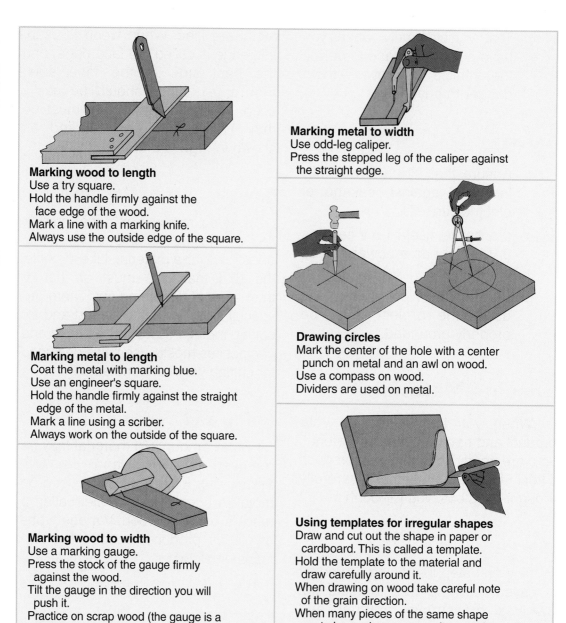

Marking wood to length
Use a try square.
Hold the handle firmly against the face edge of the wood.
Mark a line with a marking knife.
Always use the outside edge of the square.

Marking metal to length
Coat the metal with marking blue.
Use an engineer's square.
Hold the handle firmly against the straight edge of the metal.
Mark a line using a scriber.
Always work on the outside of the square.

Marking wood to width
Use a marking gauge.
Press the stock of the gauge firmly against the wood.
Tilt the gauge in the direction you will push it.
Practice on scrap wood (the gauge is a difficult tool to use).

Marking metal to width
Use odd-leg caliper.
Press the stepped leg of the caliper against the straight edge.

Drawing circles
Mark the center of the hole with a center punch on metal and an awl on wood.
Use a compass on wood.
Dividers are used on metal.

Using templates for irregular shapes
Draw and cut out the shape in paper or cardboard. This is called a template.
Hold the template to the material and draw carefully around it.
When drawing on wood take careful note of the grain direction.
When many pieces of the same shape are to be made, use masonite to make the template.

Figure 5-2 Read the instructions for using tools for marking out.

Most filing is done while the work is held in a vise. In straight or cross filing, you push the file across the work straight ahead or at a slight angle. Never run your fingers over a newly filed surface. Sharp burrs on the workpiece may cause a severe cut.

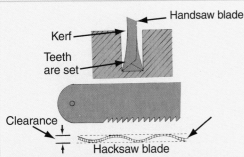

Handsaw blade

Kerf

Teeth are set

Clearance

Hacksaw blade

Saws need clearance

Teeth are bent in alternative directions; this is called set.

The kerf made by the teeth is wider than the blade thickness.

Teeth usually point away from the handle and material is cut on the push stroke.

The teeth on hacksaw blades are too small to be set. Clearance is achieved by stamping a wavy edge on the blade.

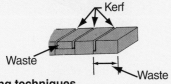

Kerf

Waste

Waste

Sawing techniques

Never cut on the line; the kerf is made touching the line but on the waste side.

Stand with your hand and arm in line with the saw cut.

Use your thumb at the side of the blade when you start the cut.

Always use the full lenght of the blade.

At the end of the cut support the work underneath.

Using a handsaw

Used for first cutting wood to approximate size.

There are two types: cross cut and rip.

A cross cut saw has finer teeth and is used for cutting across the grain.

A rip saw has larger teeth and is used for cutting parallel to the grain.

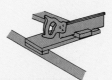

Sawing wood accurately

A tenon saw is used to make straight, accurate cuts in wood.

A bench hook, held firmly in the vise, should always be used.

Coping saw

Used to make curved cuts through wood and plastic.

Teeth point toward the handle; cuts on the pull stroke.

Hacksaw

Used to make straight cuts through metal and plastic.

Make sure at least three teeth are in contact with the material all the time.

Use small teeth for hard materials and large teeth for soft materials.

Junior hacksaw

Small and inexpensive.

Useful for cutting thin metals and light sections.

Abrafile (rod saw)

Used to make curved cuts through metal, plastic, and ceramics.

The blade is like a file and is held in a hacksaw frame.

The cutting edge is made of tungsten carbide particles bonded to a steel rod.

Hot wire cutter (an alternative to sawing)

Heated wire cuts straight and curved shapes in rigid foam plastic.

Must be used in a well-ventilated area as fumes are produced.

Figure 5-3 These tools are used for cutting operations.

Designing and making

If materials are not sawn accurately, what problems will occur when making a product?

Discussion

What problems did you encounter when using saws to cut materials?

Reflection

How do you ensure high quality results when cutting and shaping materials?

Safety

Review the hazards and risks involved in using saws. Discuss how working carefully and using the correct procedures can manage the risks.

Technology and society: What has been the environmental impact of replacing handsaws and axes with chain saws in the forestry industry?

Designing and making
What problems occur when making a product, if materials are not filed and planed accurately?

Discussion
What problems did you encounter when using files and planes to shape materials?

Safety
Review the hazards and risks involved in using tools for filing and planing. Discuss how working carefully and using the correct procedures can manage the risks.

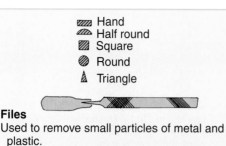

🮖 Hand
🮖 Half round
🮖 Square
⬤ Round
🔺 Triangle

Files
Used to remove small particles of metal and plastic.
Double cut files remove metal faster but make a rougher surface.
Single cut files produce a smooth surface.
Each shape is available in many sizes and degrees of coarseness.
Always use a file with a handle.

Using a file
For normal filing, hold the file at each end; it only cuts on the forward stroke (cross filing).
To produce a very smooth finish, push the file sideways (draw filing).

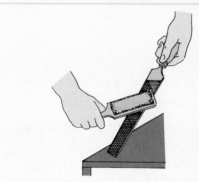

Cleaning a file
Small pieces of metal sometimes stick in the teeth of the file; this is known as pinning.
A special wire brush called a file card is used to clean the file.
Keeping the file clean is particularly important when filing plastic.

Rasps
Similar to coarse files.
Used for rough shaping (e.g.) free-form, sculptured shapes.

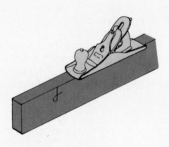

Planes
Used to remove a thin layer of wood (shaving).
A short plane, called a smoothing plane, is used on short pieces of wood.
A longer plane, called a jack plane, is used on longer pieces of wood.
Always plane in the direction of the grain.

Surforms
Are held like files and rasps but cut like planes.
Cut wood quickly but leave a rough surface.

Figure 5-4   Filing and planing tools are used to shape and smooth materials.

Planing also requires that the wood piece be held in a vise. When properly adjusted, a plane should take off fine shavings from the piece of wood. Planing removes the small ridges left by the power planer or saw. This saves having to remove them with sandpaper—a slow process.

Links to other subjects: How are files used in subjects other than technology education? How are the files different? For example, you can investigate the use of files in the care of horses' feet and teeth and how dentists use files on human teeth.

Shearing and Chiseling

Shearing and *chiseling* are other techniques used to shape materials. These tasks require shears and chisels. Shears, also called snips, cut thin metals. Chisels are designed to cut wood or metal, **Figure 5-5**.

Snips are designed for various kinds of cuts. A straight snips cuts straight lines and large curves. For cutting curves and intricate designs, an aviation snips or a hawk-billed snips is best.

Chisels must be designed for the type of material on which they are to be used. Wood chisels have very sharp cutting edges so they will cut rather than tear the wood. Metal-cutting chisels have thicker, tougher cutting edges. Refer to **Figure 5-5** once more. It shows how shears and chisels are to be used.

Drilling

Drilling is a process used to make holes in wood, plastic, metal, and other materials. A drill cuts while turning. Twist drills for cutting metals are made of carbon steel and

Designing and making
If materials are not accurately cut with shears and chisels, what problems will occur when making a product?

Discussion
What problems did you encounter when using shears and chisels to cut and shape materials?

Safety
Review the hazards and risks involved in using shearing and chiseling tools. Discuss how working carefully and using the correct procedures can manage the risks.

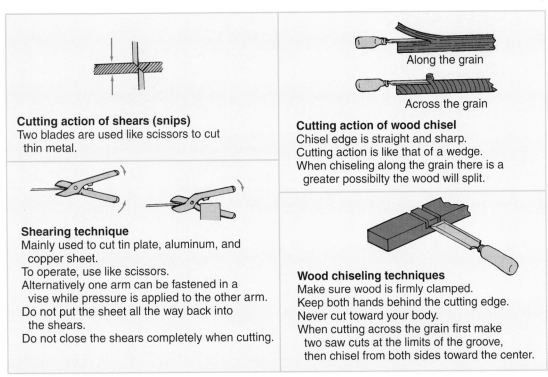

Cutting action of shears (snips)
Two blades are used like scissors to cut thin metal.

Shearing technique
Mainly used to cut tin plate, aluminum, and copper sheet.
To operate, use like scissors.
Alternatively one arm can be fastened in a vise while pressure is applied to the other arm.
Do not put the sheet all the way back into the shears.
Do not close the shears completely when cutting.

Along the grain

Across the grain

Cutting action of wood chisel
Chisel edge is straight and sharp.
Cutting action is like that of a wedge.
When chiseling along the grain there is a greater possibilty the wood will split.

Wood chiseling techniques
Make sure wood is firmly clamped.
Keep both hands behind the cutting edge.
Never cut toward your body.
When cutting across the grain first make two saw cuts at the limits of the groove, then chisel from both sides toward the center.

Figure 5-5 These are some common shearing and chiseling tools and techniques.

Community resources and services: Invite a sheet metal worker to class and discuss the design and installation of ductwork in a home.

high-speed alloy steel. Twist drills are also used to bore holes in woods and plastics. **Figure 5-6** shows some types of drills and how you are to use them.

Bending and Forming

Bending and *forming* are two different processes. Bending sheet material is like folding paper along a

Cutting action of drill
Drills cut by rotating a cutting edge into a material.
A twist drill has a cutting edge, a sprial groove (the flute) to release the chips, and a straight shank to hold the drill in a chuck.

Drilling technique
Mark the point you want to drill.
Use a center punch on metal.
Never hand hold work to drill; hold work in a vise or a clamp.
Place waste wood under the work to prevent damaging the bench after the drill has gone through; this also prevents the material cracking away.
When drilling deep holes remove the drill from the hole from time to time to avoid clogging and overheating.

Hand drill
Concentrate on keeping the drill vertical.
Turn the handle at a steady speed, trying not to wobble the drill.

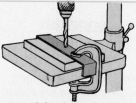

Portable electric drill
Clamp small pieces of material in a vise.
Center punch the location of the hole.
Place a small piece of wood on the underside to prevent splitting.
Do not bend the drill sideways or you will snap the drill bit.

Drill press
Remember that work is always held in a machine vise or clamp.
Always wear eye protection.
The speed can be adjusted: the larger the drill bit, the slower it should turn.
Place a piece of scrap wood under the workpiece to protect the table.

Countersink drill
Used to open out the end of a hole so that a flathead screw will fit flush with the surface.

Holesaw
Used to drill large holes in wood up to 3/4 in. (18 mm) thick.
Very useful for making discs or wheels.

Figure 5-6   This table shows the hole-drilling tools and how to use them.

Technology and society: Have students investigate the environmental impact of drilling on the ocean floor.

straight, sharp crease. It is also quite easy to bend it along a gentle curve.

However, forming sheet material is more difficult. Forming is the changing of the shape of a sheet material. It often employs the use of a mold. Take a flat sheet of paper and try to form it into a dome. That is much harder than bending, isn't it? The sheet must bend and curve in many directions. It creases and buckles, **Figure 5-7**.

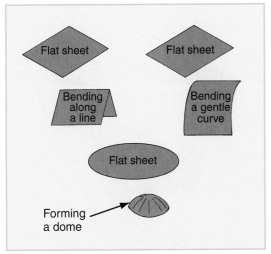

Figure 5-7 Sheet material will bend or curve easily in one direction. Bending in several directions at the same time is more difficult.

Wood

Wood does not bend easily. One way to create a curved shape from wood is to cut the curve out of a thick, solid block as shown in **Figure 5-8A**. The trouble is that the curved piece would break easily because of the short grain on the curved sections. The part made this way would be weak. But there is another way: laminating. It avoids the problem of short grain.

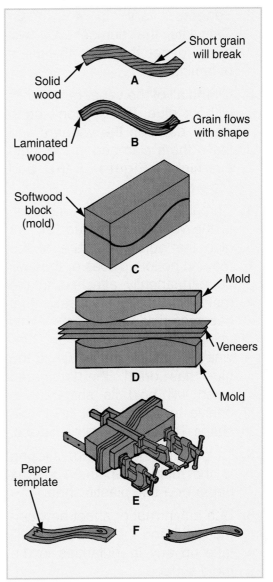

Figure 5-8 Laminating produces stronger parts than sawing the curved shapes out of solid wood.

Laminating is the gluing together of several veneers (thin sheets of wood). These can be easily bent, glued, and held in a mold until the glue dries. The steps for laminating are:

1. Make a mold, **Figure 5-8C**. The two parts must fit together exactly.

Designing and making
How does lamination change the way products can be designed and made?

Discussion
What problems did you encounter when laminating veneers?

Vocabulary
What does the term *laminating* mean?

Activity
Have students bring in products that include one or more laminated parts. Identify the materials used for the laminations.

Safety
Review the hazards and risks involved in laminating wood. Address how working carefully and using the correct procedures can manage the risks.

Useful Web site: Have students browse www.unalam.com to explore the use of laminated beams in large structures.

Designing and making
If materials are not bent accurately, what problems will occur when making a product?

Discussion
What problems did you encounter when bending sheet metal?

Reflection
How do you ensure high quality results when using tools and techniques for bending metal?

Safety
Review the hazards and risks involved in bending sheet metal. Address how working carefully and using the correct procedures can manage the risks.

2. Veneers vary in thickness. Calculate the number you will need to get the right thickness of laminate.

3. Clamp all of the veneers together without glue. If they don't bend easily, dampen the veneers and leave them clamped in the mold overnight, **Figure 5-8D** and **Figure 5-8E**.

4. Completely cover the surfaces of the veneers with glue. Do not glue the outside surface of the top and bottom pieces or they will be permanently attached to the mold! Use a resin adhesive that becomes rigid when it sets. Avoid using contact cement or PVA, which remains rubbery. Once the glue has dried, the laminated wood will hold its shape. It is unable to spring back to its old shape. The glue will hold it in place.

5. To prevent the veneers sticking to the mold, wrap them in wax paper or a thin plastic sheet.

6. Use a thin rubber sheet between the mold and the veneers. It will take up any irregularities in the mold surface.

7. Use bar clamps, **Figure 5-8E**, and squeeze the mold together until the glue is forced out along the edges of the veneer.

8. As shown in **Figure 5-8F**, make a template of the shape you want. Either glue it to the laminate or mark around it. Use it as a pattern for cutting out the shape.

Sheet Metal

Sheet metal can be bent and folded into three-dimensional objects. A pan is a good example. Before you begin to shape the metal, you must work out a development. A development is a pattern. It shows where the sheet metal must be cut and folded to make the object, **Figure 5-9A**. The development is marked on the sheet metal. You then use tin snips to cut the shape.

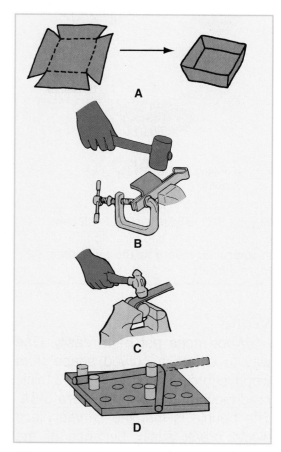

Figure 5-9   These are the steps for bending metal.

Useful Web site: Have students use the Internet to investigate how large sheets of metal are bent and folded. Helpful places to start are www.tpub.com/air/13-20.htm and www.sheetmetalworld.com.

In order to form straight bends, follow these steps:

1. Place the marked sheet metal into folding bars. The bend line should be touching the top edge of the bar.

2. Fasten the folding bars in a vise.

3. If the metal is wider than the vise jaws, add a C-clamp.

4. Use a mallet with a rawhide or nylon head. Bend the metal over the bar, **Figure 5-9B.**

5. For some shapes, you may find a block of wood more useful than folding bars.

Lengths of mild steel not more than $1/4''$ (6 mm) thick may be bent fairly easily. Follow these steps to bend strip steel.

1. Clamp the metal vertically in a vise.

2. Hammer the metal from one side to bend it to the needed angle.

A *jig* is a useful tool for bending metal.

◆ Small diameter rods can be bent on a peg jig, **Figure 5-9D.**

◆ Metal tubing can be bent using a similar jig.

Plastics

Thermoplastics can be bent or formed when heated between 300° and 400° F (150° and 200° C). Heat the plastic sheet or rod along the line of the bend. The narrower the heated line, the sharper the bend will be. Refer to **Figure 5-10A.**

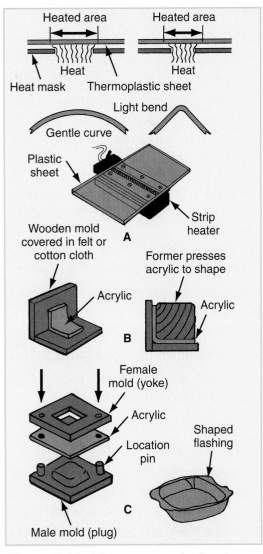

Figure 5-10 The processes for bending and forming thermoplastic sheet materials.

Using a strip heater provides heat along a straight, narrow line for a sharp bend.

To use the strip heater:

1. Place the sheet of plastic on the heater. The bend line must be exactly over the heat element.

2. For safety, wear gloves to protect hands from the hot plastic.

Designing and making
How are products designed using formed plastic different from those designed using wood or metal?

Discussion
What problems did you encounter when bending or forming thermoplastics?

Activity
Make a collection of products whose shape and form are dependent on the manufacturer's ability to bend plastic.

Safety
Review the hazards and risks involved in bending or forming thermoplastics.

Useful Web site: Use the Internet to investigate techniques for mass-producing large quantities of plastic items, such as the Anywayup Cup™ at www.mandyhaberman.com.

3. Heat both sides of the sheet.

4. Bend the plastic to the required shape. For a 90° bend, press the plastic into a mold, **Figure 5-10B**.

5. To produce a sharper bend, press a second mold into the corner, **Figure 5-10B**. Hold until cool.

NOTE: Wooden molds must be covered with cotton or felt material. This prevents the grain from marking the plastic.

Forming plastic in a mold calls for a two-part mold. To form the plastic:

1. Heat the acrylic sheet in an oven until pliable.

2. Place it over the plug. Press the yoke down on top, **Figure 5-10C**.

3. Allow the plastic to cool before separating the mold.

Casting and Molding

Pouring liquid or plastic material into a mold to shape it is called *casting* and *molding.* Do you realize that every time you make ice cubes you are actually casting? You are making a solid shape by pouring water into a tray. The liquid takes the shape of its container.

Casting is a method of making shapes that are almost impossible to produce by sawing, drilling, or filing. We use three basic materials for casting and molding. These are metals, plastics, and ceramics. When the material is poured into the mold, the process is called casting. When the material is forced into the mold, it is called molding.

Metals become liquid when heated above their melting point. As they cool, they solidify. Plastics are available in a liquid form. These liquids set hard through chemical action.

Ceramics are materials such as silica (sand), clay, and concrete. Silica must be melted to make glass and other products. Clays and concrete are not melted. They are mixed with liquid and poured into a mold.

Metals

Use the following steps to cast metal:

1. As shown in **Figure 5-11A**, produce a pattern. A pattern is just like the finished product. Usually it is made of wood, but it could be made of some other easily worked material.

2. Place the pattern on a flat surface. On the same surface, place a molding box, **Figure 5-11B**.

3. Pack the mold material around the pattern. The mold should be made from a material that will not burn as the hot metal is poured into it. Sand is often used because it is inexpensive.

4. Fill the molding box with sand. Tamp it tightly around and over the pattern. Fill the box completely.

5. Cover the box with another board and turn it over. Remove the board that was on the bottom.

6. Carefully remove the exposed pattern, **Figure 5-11C**.

Useful Web site: Have students visit www.architecturaliron.com to investigate decorative cast iron products. Students should print information they find interesting and share it with the class.

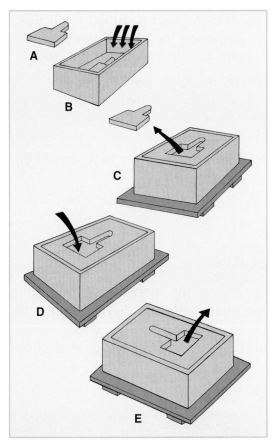

Figure 5-11 This is how to cast metals. A—Make a wooden pattern. B—Place pattern in a molding box. Pack sand around it until box is full. C—Place a flat wood cover on the box. Turn it over. Remove the pattern carefully. D—Pour molten metal into the cavity. E—When cooled and solid, remove the casting.

7. Pour molten metal into the cavity formed by the pattern, **Figure 5-11D**.

8. Allow the casting to cool and solidify. Then remove it, **Figure 5-11E**.

Plastics

Casting plastics has one big advantage over casting metal. The resins can be cast at room temperature.

Small articles, such as paperweights, can be cast. If you wish, you can also embed decorative objects in them. Follow these five steps:

1. Use a smooth mold. It will produce a smooth surface on the casting. Waxed drinking cups work well. Never use polystyrene (Styrofoam®) cups. The resin will dissolve them!

2. Measure the amount of resin you will need. Add the recommended amount of catalyst (hardener). Mix thoroughly.

3. Pour a layer of the mixed resin into the mold. Leave it to harden. This will form the top layer of the casting, **Figure 5-12A**.

4. Place the decorative object (coin, insect, or stamp) face down on the hardened layer of resin, **Figure 5-12B**.

5. Pour more resin around and over the object, **Figure 5-12C**. You can use clear resin throughout or you can add pigment to the last layer. This forms the base of the object.

When the resin has hardened, **Figure 5-12D**, remove the casting from the mold. Smooth rough edges and surfaces with wet or dry sandpaper. Then polish the casting with a polishing paste.

Fiber Reinforced Plastic

Fiberglass canoes, racing car bodies, and crash helmets may all be made from plastic resin reinforced with glass fiber. The reinforcing glass fiber makes the shells very tough,

Designing and making
If the material is not cast correctly, what problems will occur when making a plastic product?

Discussion
What problems did you encounter when casting plastic?

Vocabulary
What does *injection molding* mean?

Safety
Review the hazards and risks involved in casting plastics.

Resource
Activity 5-4, *Designing a Bathroom Rack*

Useful Web site: Have students investigate rapid injection molding at www.protomold.com/designguide. Students should print information they find interesting and share it with the class.

Discussion
What objects could be embedded when casting plastics?

Activity
Have students create a collage showing products made from fiber-reinforced plastics.

Safety
Review the potential hazards and risks involved in working with fiber reinforced plastic.

Resource
Activity 5-5, *Designing a Tape Storage Unit*

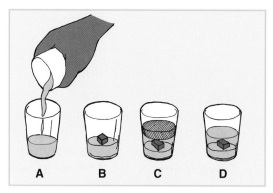

Figure 5-12 Steps for embedding an object in plastic.

while the thermosetting resin creates a hard surface. To make a fiberglass reinforced product:

1. Paint on a release agent over the surface of the mold, **Figure 5-13**.

2. Mix polyester resin and a catalyst in recommended proportions.

3. Brush on a gel coat of polyester resin.

4. Add a layer of fiberglass and coat it with more resin.

Figure 5-13 Molding fiberglass on a form, such as this boat, is called a "lay up."

5. Use a roller to make each layer take up the exact shape of the mold and remove air bubbles.

6. Add other layers of resin and fiberglass to produce the required thickness.

7. Allow the assembly to cure (set hard), and then remove it from the mold.

8. Color may be added to the resin.

Joining Materials

There are many ways of joining materials. **Figure 5-14** lists some common choices for wood, metal, and plastic. You will notice that some

	Mechanical	Chemical	Heat
Wood	Nails Joints for gluing Screws Nuts and bolts KD (knock-down) fasteners Wedges Hinges	Glues Adhesives	
Metal	Rivets Nuts/bolts/screws KD fasteners Hinges	Adhesives	Weld Braze Solder
Plastic	Rivets Nuts/bolts/screws KD fasteners Hinges	Solvents Cements	Weld

Figure 5-14 Joining wood, metal, and plastics. Nuts and bolts and screws are used for all three.

Useful Web site: Have students use the Internet to investigate the use of fiber-reinforced plastic in the manufacture of surfboards, sailboards, kayaks, and canoes.

Have students visit the Fiberglass Information Network at www.sustainableenterprises.com/fin to investigate the health issues related to the use of fiberglass. Students should print information they find interesting and share with the class.

methods, such as nuts and bolts, are good for all three. Others, such as soldering, may be used only for metal.

Mechanical Joining

Mechanical joining is the use of physical means to assemble parts. It can be done in one of two ways. One method is to use hardware such as nails, screws, or special fasteners. Another method is to shape the parts themselves, so they interlock.

Nails

Nails provide one of the easiest ways to join two pieces of wood. Nails hold the wood by friction between the wood fibers and the nail, **Figure 5-15**. Nails can be shaped many different ways. Each shape serves a special purpose, **Figure 5-16**. Remember these general points when using nails:

- ◆ Whenever it is possible, hold one of the pieces to be nailed in a vise, **Figure 5-17A**.
- ◆ Always nail through the thinner piece into the thicker piece.
- ◆ Avoid bending the nail. Strike it squarely with the face of the hammer.

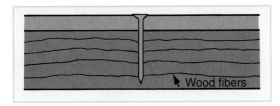

Figure 5-15 What holds the nail in the wood?

Type of Nail	Uses
Common	Structural or other heavy work where head will be exposed.
Finishing	Finishing work where nail head should not be exposed
Spiral	Building construction. Twisted shank causes nail to thread itself into wood, increasing its holding power.
Drywall	Fastening gypsum board to wood frames.
Concrete	Fastening to concrete. Nail is hardened to prevent bending.
Roofing	For wood, asphalt, and other roofing materials. Usually coated with zinc or galvanized to make rust resistant.

Figure 5-16 Nails are made in many different shapes to serve special fastening purposes.

- ◆ When using finishing nails, drive the nail below the surface using a nail set, as shown in **Figure 5-17B**.
- ◆ Stagger the nails. If you place them in a straight line, you may split the wood along the grain, **Figure 5-17C**.
- ◆ To remove nails, use a claw hammer, **Figure 5-17D**. Always use a block of waste wood to protect the surface of the wood.

Screws

Screws have a greater holding power than nails, **Figure 5-18**. They also rely on friction for their strength.

Designing and making
If nails are not inserted accurately, what problems will occur when making a product?

Discussion
What problems did you encounter when using nails to join materials?

Vocabulary
Explain the difference in how common nails and spiral nails enter a piece of wood.

Safety
Review the potential hazards and risks involved in using nails and screws.

Resource
Transparency 5-1, *Ways of Joining Materials*

Resource
Reproducible Master, *Uses of Different Nails*

Links to other subjects: Have students investigate the joining techniques, tools, and materials used in subjects other than technology education.

Designing and making
If screws are not inserted accurately, what problems will occur when making a product?

Discussion
What problems did you encounter when using screws to join materials?

Reflection
How do you ensure high quality results when using screws to join materials?

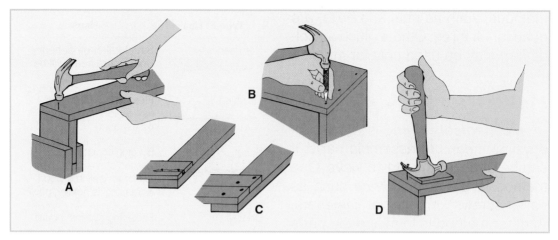

Figure 5-17 Methods for assembly and disassembly when using a hammer and nails.

When two pieces of wood are held together, the screw has only to grip the second piece. The head of the screw and the grip of the screw thread pull the two pieces together. Screws can be removed more easily than nails and without damaging the material.

Wood screws are used for fastening wood to wood and metal to wood (hinges to a door). They are also used for fastening all types of hardware to furniture (handles to doors).

In choosing the correct wood screw, you must decide the following items:

◆ Shape of head and type of slot.

◆ Length.

◆ Thickness or gauge.

◆ Material.

There are three shapes of screw heads, **Figure 5-19**.

◆ Flat head screws—used when the head of the screw must be flush with or below the surface of the wood.

◆ Oval head screws—used when the holding power of a flat head

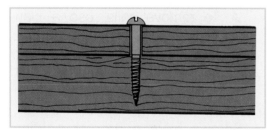

Figure 5-18 Wood fibers grip the threads of a screw. This gives the screw holding power.

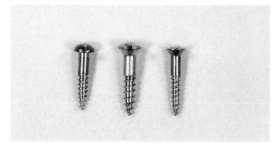

Figure 5-19 Types of wood screws are named for the shapes of their heads. (Christopharo)

Useful Web site: Have students investigate the use of screws in dental implants at www.dental-implants.com and medical implants at www.stanford.edu/group/esource/2c.html. Students should print information they find interesting to share with the class.

screw is needed. The screw head will show for decoration.

- ◆ Round head screws—used when the object that is being fastened by the screw is too thin to be countersunk.

There are several common head styles or types:

- ◆ Straight slot.
- ◆ Phillips.
- ◆ Allen.
- ◆ Robertson (square drive).

The types are shown in **Figure 5-20**. Screw lengths range from $1/4''$ to $6''$ (6 mm to 150 mm). The thickness of a wood screw is called its "gauge." Gauge is expressed as a number. Screw gauges range from 0 to 24.

Most screws are made of steel. They are very strong, but they can rust. Brass screws do not rust, but they are weaker than steel. Sheet metal screws, **Figure 5-21**, are used to join sheet metal, plastics, and particleboard. **Figure 5-22** shows the following three steps to fastening wood parts with a wood screw:

1. Hold the two pieces together. Drill a pilot hole the length of the screw.

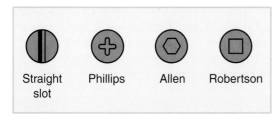

Figure 5-20 These are the common types of screw heads.

Figure 5-21 Sheet metal screws have threads that go all the way to the head. (TEC)

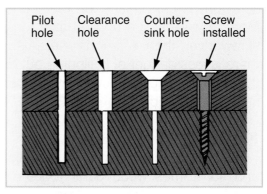

Figure 5-22 Special holes must be drilled when fastening hardwood with screws.

2. In the top piece, drill a clearance hole. This is a hole the same diameter as the screw shank (unthreaded part of the screw below the head).

3. If you are using a flathead screw or oval head screw, countersink the hole.

Now you are ready to install the screw. The diameter of pilot holes and clearance holes used most frequently are shown in **Figure 5-23**.

The two most common types of screwdrivers are the straight slot (standard) and Phillips head. Use the largest

Useful Web site: Have students research the inventors of the Robertson and Phillips screws at http://inventors.about.com.

Technology and society: History—Have students research techniques used for joining materials before the invention of screws.

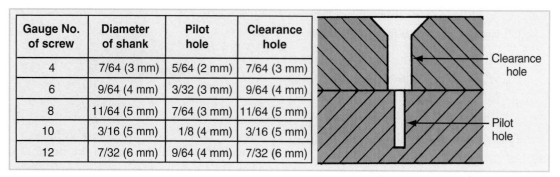

Gauge No. of screw	Diameter of shank	Pilot hole	Clearance hole
4	7/64 (3 mm)	5/64 (2 mm)	7/64 (3 mm)
6	9/64 (4 mm)	3/32 (3 mm)	9/64 (4 mm)
8	11/64 (5 mm)	7/64 (3 mm)	11/64 (5 mm)
10	3/16 (5 mm)	1/8 (4 mm)	3/16 (5 mm)
12	7/32 (6 mm)	9/64 (4 mm)	7/32 (6 mm)

Figure 5-23 These pilot and clearance holes are recommended for wood screws.

screwdriver convenient for the work. More power can be applied to a long screwdriver than a short one. Also, there is less danger of it slipping out of the slot. The tip of the screwdriver must fit the slot correctly, **Figure 5-24**.

Nuts and Bolts

Nuts and bolts fasten metal, plastic, and sometimes wood parts together. They are quite different from wood screws. Bolt threads do not depend on gripping the fibers of the material. Bolts go completely through a drilled clearance hole. (This is a hole large enough for the bolt to be pushed through.) A nut threads onto the bolt end. Tightening the nut squeezes the parts and holds them.

Washers are often used under the bolt head and the nut. This protects the surfaces. It distributes the load over a larger area, too. Lock washers prevent nuts from accidentally loosening due to vibration.

Sometimes, threads are cut into the hole in the second piece of material, **Figure 5-25**. This takes the place

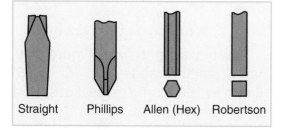

Straight   Phillips   Allen (Hex)   Robertson

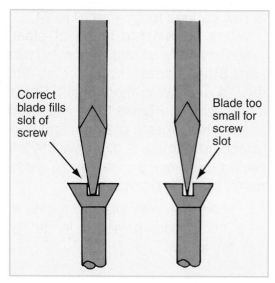

Correct blade fills slot of screw

Blade too small for screw slot

Figure 5-24 Top—There are several types of screwdrivers. Bottom—It is important to choose the correct size standard screwdriver.

Useful Web site: Have students browse www.precisionscrewandbolt.com to investigate the range of materials, sizes, and types of nuts and bolts available to manufacturers.

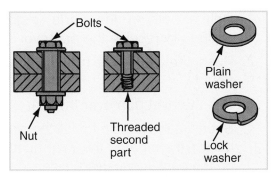

Figure 5-25   These are the two systems for fastening metal parts with bolts.

of the nut. Joints fastened with nuts and bolts can be taken apart and reassembled.

You need to make a number of decisions to choose the right nut and bolt. You have to decide on:

◆ Length.

◆ Diameter.

◆ Shape of head.

◆ Thread series. Some are coarse; others are fine. There is a standard and metric thread.

◆ Material.

Some of the choices are shown in **Figure 5-26**. Can you name any of the types shown?

Figure 5-26   Bolts and nuts are made with many different shapes for special purposes.

You tighten most nuts and bolts with wrenches. Machine screws are tightened with screwdrivers or Allen wrenches. A combination wrench is shown in **Figure 5-27**. It has one open end and one box end. The box wrench is usually preferred because it does not slip. Sometimes there is not enough room for a box end. Use an open end when clearance is a problem.

An adjustable wrench fits a range of nut sizes. As **Figure 5-28** shows, always pull on the wrench handle. Pushing can be dangerous. If the wrench should slip, you could injure your knuckles.

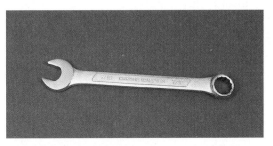

Figure 5-27   This is known as a combination wrench.

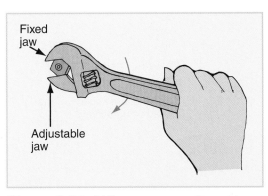

Figure 5-28   When using an adjustable wrench, always pull against the fixed jaw.

Useful Web site: Have students investigate the following Web site to look at the range of wrenches available to mechanics: www.snapon.com.

Rivets

Like nuts and bolts, rivets squeeze two or more pieces of metal or plastic together. They are either solid or pop type.

Solid rivets are usually made of mild steel. They may have round or flat heads. The four steps for installing a round head rivet are shown and described in **Figure 5-29**.

To use solid rivets, you must be able to reach both sides of the rivet. This is not always possible. Pop rivets are designed to be used from one side. They are made of a hollow aluminum head with a steel pin through it. Use the following procedure to install a pop rivet:

1. Drill a hole in the parts large enough to receive the pop rivet.
2. Push the pop rivet through the hole.
3. Slip the rivet gun over the pin.
4. Squeeze the handle to pull the pin back. This creates the rivet head on the back (concealed) side.
5. Continue squeezing until the pin breaks off, **Figure 5-30**.

Knockdown (KD) Joints

Some furniture is designed for "do-it-yourself" assembly. This requires special fasteners that are strong and

Align the parts to be joined and drill a hole.

Insert the solid rivet through the hole.

Peen (hammer) the end of the rivet to roughly shape the rivet head.

Complete the shape of the head using a rivet set.

Figure 5-29   Steps to fasten sheet metal when using solid rivets.

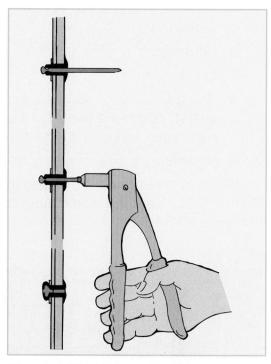

Figure 5-30   Using a pop riveter. Top—Insert rivet in predrilled hole. Middle—Slip tool over pin and squeeze. Bottom—Pin will break off when its sleeve has expanded on the blind side of the rivet.

easy to use. The fasteners require no special tools or skills. Known as KD, or "knockdown," joints, they can be taken apart and reassembled as needed. There are three common types, as shown in **Figure 5-31**.

Movable Mechanical Joints

Some joints are made so that the joined parts can move. Think of how a door is joined to its frame. A hinge is used. A knife switch in **Figure 5-32** is another example of a movable joint. Both are pin hinges. They can only move back and forth. We say that they "move through one plane only."

A second type of movable joint allows movement in more than one plane, **Figure 5-33**. The joystick of a video game uses a ball and socket joint. There are other common examples. A camera tripod uses a lockable ball and socket. It can move in three different directions. The drive shaft of a car uses a universal joint. It permits the joint to move up and down or left and right as the shaft spins.

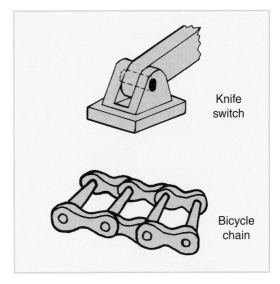

Knife switch

Bicycle chain

Door hinge

Figure 5-32 These are common examples of pin hinges. (TEC)

Designing and making
What problems will occur if movable mechanical joints are not installed accurately?

Discussion
What problems did you encounter when installing movable mechanical joints?

Safety
Review the hazards and risks involved when installing movable mechanical joints.

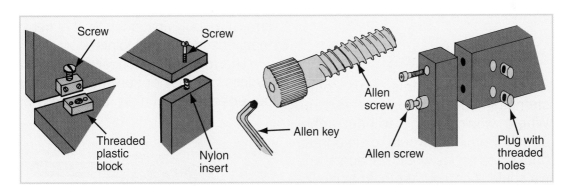

Screw

Screw

Allen screw

Allen key

Allen screw

Threaded plastic block

Nylon insert

Plug with threaded holes

Figure 5-31 Knockdown furniture uses three basic fasteners.

Links to other subjects: Have students investigate the materials used to make movable mechanical joints in soft materials, such as paper, cardboard, or fabric. Students can make a display board showing the techniques and materials in a variety of contexts.

Designing and making
How do the shape and form of furniture parts affect the wood joints used in its construction?

Discussion
Have students discuss advantages and disadvantages of the frame and box joints shown in Figure 5-37.

Vocabulary
What is the difference between a *tenon* and a *mortise*?

A third type of movable joint uses a flexible material, **Figure 5-34**. Polypropylene is used to make a variety of boxes and cases. The material itself acts as a hinge wherever it is folded. These are called "integral" or "living" hinges.

Wood Joints

Joints are designed to hold wood pieces together. A wood joint's strength depends on two things:

◆ The amount of mechanical interlocking (this means that the wood parts are cut so they cannot be pulled apart). See **Figure 5-35**.

◆ The surface area of the joint to be glued.

Wood joints can be grouped by type, **Figure 5-36**. One group is used on frames. The other group is used to make boxes. Frame joints are found on chairs, windows, doors, and similar products. Box joints are used on items like cabinets, drawers, and storage boxes. **Figure 5-37** shows eight different joints for constructing frames or boxes.

Chemical Joining

Mechanical joints are often strengthened by *chemical joining*. Where would we be without adhesives? Furniture would disintegrate, books and shoes would fall apart, and we couldn't cap our teeth. Imagine life without wallpaper, stamps, tape, or Post-it® notes. Glues, adhesives, solvents, and cements are all methods of chemical joining.

Figure 5-33   Top—Ball and socket joint is like the toggle on a video game. Bottom—A universal joint can turn up and down as well as side to side.

Glues and adhesives are used for joining woods and metals. Glues are made from natural materials. These include animal bones, hides, and milk. They are rarely used today. Still, we use the term "glue." It is more correct to say "adhesives."

Adhesives come from petroleum products. These adhesives are of two types: thermoplastic and thermosetting. One common thermoplastic adhesive is liquid white glue. Also known as polyvinyl acetate, it is commonly used in wood joints.

Thermoplastic adhesives harden by loss of water or solvent. They may be softened by heat and are not

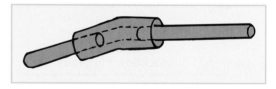

Figure 5-34   An example of a flexible joint that can move in any direction.

Links to other subjects: History—Have students investigate the types of wood joints used in furniture from previous eras, such as ancient Greece and early America. Students should create an annotated poster showing the evolution of wood joints.

Technology and society: Who has benefited from the invention of machines that mass-produce joints for wooden furniture? Who has not?

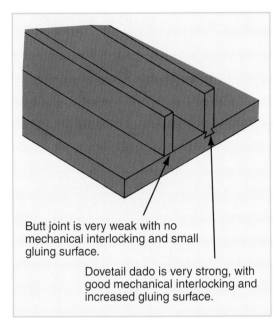

Butt joint is very weak with no mechanical interlocking and small gluing surface.

Dovetail dado is very strong, with good mechanical interlocking and increased gluing surface.

Figure 5-35 Joints are stronger when one piece is shaped to fit into the other. This is called an interlocking joint.

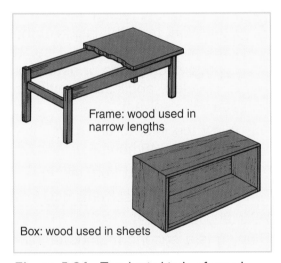

Frame: wood used in narrow lengths

Box: wood used in sheets

Figure 5-36 Two basic kinds of wood construction for furniture making are frame and box.

Frame joints	Box joints
Butt	Butt
Dowel	Rabbet
Mortise and tenon	Dado
Lap	Dovetail

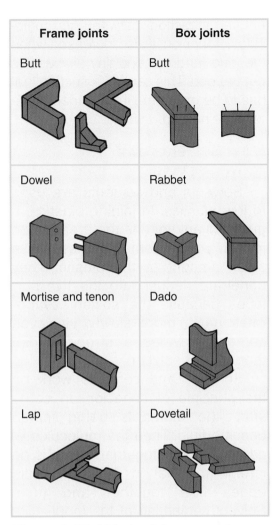

Figure 5-37 Which of these joints has the greatest mechanical strength? Which has the largest gluing surface?

How Glues and Adhesives Work

To act as an adhesive, the molecules that make up the glue must form strong links to one another, and the glue must stick to both surfaces being joined. This ensures they cannot be separated when the two surfaces are pulled apart. An adhesive must easily flow to coat both surfaces and have a

waterproof. Thermosetting adhesives include various types of resins. Heat will not soften them. They are waterproof.

Designing and making
What problems can occur if solvents and cements are not used correctly?

Discussion
What problems did you encounter when using solvents and cements to join materials?

Reflection
How do you ensure high quality results when using solvents and cements to join materials?

Vocabulary
What is the difference between a *solvent* and a *cement*?

Safety
Review the potential hazards and risks involved in using solvents and cements.

natural attraction between its molecules. Adhesion is also increased when the glue hardens, and tiny air bubbles get trapped. This causes a suction that has to be overcome if the surfaces are to be separated.

How Solvents and Cements Work

Solvents and cements are used to join plastics. In joining plastics, a pure solvent softens the areas to be joined, while cements dissolve a small amount of the plastic. They penetrate deeper into the two surfaces because the solvent evaporates much more slowly. However, cement provides a stronger joint than a pure solvent.

Solvents and cements work on the principle of cohesion. In cohesion, the materials being joined become fluid. Then the molecules of each piece mix together. There is no foreign material in the joint. Fluid edges flow together and fuse. Solvent cementing of thermoplastics is an example of cohesion fastening.

Figure 5-38 shows uses for different solvents, adhesives, and glues. For safety and good results, follow these general rules:

◆ Make sure the surfaces are clean and dry. Remove grease, paint, varnish, or other coatings.

◆ Carefully read the instructions on the packaging.

◆ Carefully read any safety instructions or cautions.

◆ Secure a good fit between the two surfaces.

◆ Work in a well-ventilated area especially when using solvents and cements.

◆ Clamp the joint until the adhesive or solvent dries.

Solvent Joining Acrylic Sheet

In solvent joining, the low-viscosity solvent travels through a joint area by capillary action. This is a force that causes a liquid to rise through a solid.

Properly done, solvent joining yields strong, perfectly transparent joints. It will not work at all if the parts do not fit together perfectly.

To solvent join plastic parts:

1. Hold the two pieces of acrylic in a jig, **Figure 5-39**.

2. Apply solvent along the entire joint. Work from the inside of the joint where possible. Use a hypodermic syringe or needle or a nozzled applicator bottle to apply the solvent.

3. Allow the joint to dry thoroughly (from 24 to 48 hours).

4. Remove the part from the jig.

SAFETY: Work with solvents only in a well-ventilated area!

Dipping is a second method of joining acrylic sheet. As shown in **Figure 5-40**:

1. Set up a tray of solvent. The tray must be larger than the plastic pieces.

2. Ensure that the tray is sitting level.

Technology and society: How does the increased use of solvents and cements to join plastics affect the natural environment? Do the advantages outweigh the disadvantages?

Class	Type	Uses	Comments
Glues	Animal	Interior woodwork	Difficult to use Must be used hot Not waterproof or heat proof
	Casein	Interior woodwork	White powder mixed with water Sets in six hours Heat and water resistant
Adhesives	Polyvinyl Acetate (PVA-white glue)	Wood, leather, paper	White liquid ready to use Hardens in under one hour Not waterproof
	Plastic resins (urea and phenol)	Wood	Urea: powder mixed with water Phenol: ready to use Hardens in approximately two–six hours Urea is water resistant: Phenol is waterproof Good strength
	Epoxy resin	Wood and metal	Two parts are mixed together Hardens in 12–24 hours Waterproof Very high strength
	Contact	Plastic laminates	Ready to use liquid Apply to both surfaces and let dry to touch Used in situation where clamps cannot be applied
	Instant or super glues (cyano acrylate)	Nonporous materials such as glass and ceramics	Ready to use liquid Hardens almost immediately Water resistant
Solvents	Pure solvent (methylene chloride and ethylene dichloride)	Acrylics	Colorless liquid, ready to use Bonds almost immediately Waterproof
	Solvent cement	Acrylics	Colorless, viscous liquid Sets in 12–24 hours Waterproof

SAFETY NOTE: Use solvents and cements in well-ventilated areas

Figure 5-38 Glues, adhesives, and solvents are designed for specific applications.

Vocabulary
Describe the difference between a *glue*, an *adhesive*, and a *solvent*.

Activity
Have students copy the first two columns of the chart in Figure 5-38 (Class and Type). Students should visit a local hardware store to investigate the availability and price of each type of glue, adhesive, or solvent.

Resource
Reproducible Master, *Glue Adhesive, and Solvent Applications*

Useful Web site: Have students investigate trends and innovations in adhesives by visiting www.specialchem4adhesives.com. Students should print information they find interesting and share it with the class.

Designing and making
What problems will occur if heat-joining materials are not used correctly?

Discussion
What problems did you encounter when using heat to join materials?

Reflection
How do you ensure high quality results when heat joining materials?

Vocabulary
What is the difference between *welding*, *brazing*, and *soldering*?

Activity
Have students type "welded sculpture" into an Internet search engine and investigate welded sculptures that use cut, bent, crushed, and painted metal. Sculptures may be stable or movable, figurative, humorous, or playful. Have students print images they find interesting and create an image board entitled "Welded Sculpture."

Safety
Review the hazards and risks involved in heat joining.

Figure 5-39 When you are solvent joining, apply solvent to the inside edges of the parts where possible.

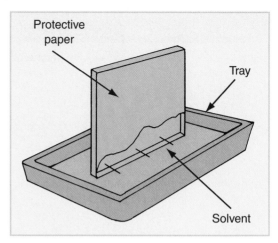

Figure 5-40 Solvent joining can be done by dipping, too.

3. Dip only the very edge of the plastic part into the solvent.
4. Use finishing nails in the bottom of the tray to keep the acrylic off the bottom.

Heat Joining

Heat joining is used mostly on metals. It is also used to some extent on plastics. Two types of heat joining are used on metals:

◆ Welding.
◆ Brazing and soldering.

Welding brings metals to their melting point. When they melt, the metals flow together. After they cool, they solidify, becoming one piece. The joint is as strong as the original metal.

Brazing and soldering work differently than welding. The heat melts the metal being used to join the parts. It does not melt metal in the parts themselves.

Brazing uses a brass alloy to join the parts. The alloy melts at 1650°F (900°C). Lead alloy is used in soldering. It melts at only 420°F (216°C). At the melting temperature, the metal runs through the joint. As the metal cools, it bonds the pieces.

The strength of the joining metal controls the joint's strength. The mild steel used for welding is the strongest. Brass, used for brazing, is much weaker, it is stronger than the lead alloy used for soldering.

Welding may also join plastics. This is possible with some thermoplastics such as PVC. A hot air torch heats the two parts of the joint. Heat fuses them.

Links to other subjects: Art—Have students investigate the welded sculpture of twentieth century artists, such as Julio Gonzalez, Nancy Graves, Mel Edwards, and Richard Hunt.

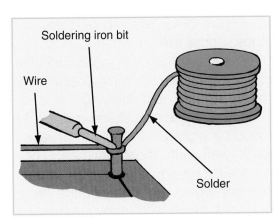

Figure 5-41 This is the proper way to use a soldering iron.

To solder tinplate, copper, brass, and mild steel:

1. Clean the tip of the soldering iron with an old file.
2. Heat an electric soldering iron.
3. Apply flux and solder to the tip. This is called tinning. It helps the solder to flow.
4. Hold the tip of the soldering iron against the joint. When it becomes hot enough, touch the solder to the metal. As the solder melts, allow it to run along the joint, **Figure 5-41**.
5. Allow the joint to cool slowly. Do not blow on it or move it in any way.

Finishing Materials

When a product is completed, its surface is usually finished. *Finishing* changes the surface by treating it or placing a coating on it. Finishing is done for several reasons:

- ◆ Protect the surfaces from damage caused by the environment.
- ◆ Prevent corrosion, including rust.
- ◆ Improve the appearance by covering the surface or treating it to bring out the natural beauty of the material.

Converted Surface Finishes

When the surface is treated to beautify or protect, it is called a converted surface. The material is chemically altered to change the way it reacts to elements in the environment. The reaction of the chemical and the atoms of metal on the product's surface provide the protective coating.

Some converted coatings are natural. Aluminum develops an oxide covering if exposed to the open air. The covering will resist the natural elements.

Surface Coatings

Materials applied to a surface are called *coatings.* The most common coatings are paints, enamels, shellac, varnish, lacquer, vinyl, silicone, and epoxy.

The first step in finishing is to prepare the surfaces. They should be clean and smooth. Surfaces can be made smooth using abrasive papers or abrasive cloths, which are made in

Discussion
What problems did you encounter when using finishing materials?

Reflection
How do you ensure high quality results when using finishing materials?

Vocabulary
What is the difference between a *surface coating* and a *converted surface finish*?

Resource
Reproducible Master, *Ways of Joining Materials*

Links to other subjects: Identify the finishing materials used in ceramics, painting, and graphic arts.

Community resources and services: Have students visit a hardware store and list the types of finishes and coatings available. Students should note the price of each.

Designing and making
What problems will occur when coated abrasives are not used correctly?

Discussion
What problems did you encounter when using coated abrasives to prepare the surface of a material?

Activity
Have students research the shapes, forms, and sizes of coated abrasives available to manufacturing industries. Students should print several examples they find interesting and summarize the characteristics and uses of each.

Safety
Review the potential hazards and risks involved when using coated abrasives.

Resource
Reproducible Master, *Abrasive Applications*

a wide range of grades and coarseness. Abrasive materials and their uses are described in **Figure 5-42**.

You should follow these three general rules when using an abrasive:

- ◆ Clean inside surfaces before assembling the project.
- ◆ Begin with a coarse abrasive. Then gradually work up to a fine grade.
- ◆ Support the abrasive whenever possible, **Figure 5-43**. A wood or cork block can be used for wood and plastic. Files can be used for holding abrasive papers while finishing metals, **Figure 5-44**.

Figure 5-45 lists ten different finishes. Some are for wood. Others are best used on plastics or metal. Finishes can be applied by wiping, brushing, rolling, dipping, and spraying.

You can apply stain and oil to wood by wiping with a cloth. Brushes work well with most finishes. They are best with liquid plastic and paint. A roller works well for painting large surfaces. Items with many curves and parts can be dipped.

Paint can be sprayed onto most shapes and materials. Aerosol spray cans are fast and easy to use but expensive. Spray guns use compressed air. They produce a high quality finish.

Material	Abrasive	Comments
Wood		Sandpaper was once the general name given to all abrasive papers used for smoothing wood. Today, the industry calls them coated abrasives.
	Flint paper	Crushed flint or quartz used as the abrasive Wears out quickly Cuts slowly Normally used in grades coarse to extra fine (50-320 grit)
	Garnet paper	Uses garnet as the abrasive More durable than flint paper Normally used in grades coarse to extra fine (50-320 grit)
Metal	Emery cloth	Uses the natural abrasive, emery Dull black in color Normally used in grades coarse, medium, and fine (3-3/0) Oil may be added to the fine grade to give a mirror finish
Wood and Metal	Aluminum oxide	An artificial abrasive Gray-brown in color Tough, durable, and resistant to wear Normally used in grades coarse, medium, and fine (40-180 grit) Used on steel and other hard materials
Wood, metal, and plastic	Silicon carbide paper (wet-and-dry paper)	An artificial abrasive Available in three common grades: coarse (50), medium (100), very fine (400) Paper is best used wet Creates a smooth, matte finish

Figure 5-42   Coated abrasives are designed for different materials. All prepare the surface of the material for finishing.

Correct   Incorrect

Figure 5-43 Always sand wood along the grain or you will see the scratches made by the abrasive.

Material Handling and Storage

Both raw materials and processed materials must be handled and stored properly. It is often necessary to transport materials several times during the manufacturing process. Materials may need to be stored between processing steps, as well. Improper handling or storage can result in damaged material and wasted processing effort.

Raw materials are usually moved to a storage facility prior to processing. For example, crude oil may be transferred by oil tanker and pipeline to a refinery. The crude oil must be handled and stored correctly so it is not exposed to water, dirt, or other contaminants.

Materials are often transported through a series of processing steps. For example, a conveyor in a food processing plant may transport plastic

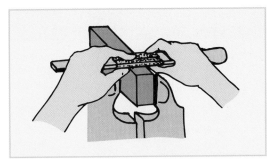

Figure 5-44 One way to use emery cloth on metals is to wrap it around a file.

containers to a filling station, a sealing station, a labeling station, and a packing station. Large and heavy materials may be handled by lift trucks or other similar vehicles.

Wood and steel products used in construction are processed to comply with standard specifications and are stored before being shipped to a building site. Lumber must be correctly stored to prevent rotting, insect infestation, and mold or fungus growth. Steel should be stored in a dry location to prevent excessive rusting. Building products also must be adequately supported to prevent excessive bending or deformation.

Finished products usually must be handled and stored even more carefully than raw materials. Relatively small products may be packaged in boxes with protective packing material. Boxes are often stacked and secured on reusable platforms called pallets. This makes the products easier to transport and helps protect them from damage while they are being handled. In addition to these packing safeguards, extra care must be taken when handling finished products.

Designing and making
How are finishes used to enhance the appearance of products and services?

Discussion
Under what circumstances would a finish NOT be applied to a surface?

Discussion
Does a finish always improve the appearance of a surface? Explain your answers.

Reflection
What type of finish do you prefer on the products you use?

Resource
Reproducible Master, *Finish Applications*

Material	Type	Comments
Wood	Liquid plastic (urethane)	Provides a clear coating Apply with a brush Gives a hard, water resistant, and long-lasting coating
	Stain	Changes the color of wood Cheaper wood can be stained to resemble the color of more expensive woods Applied with brush or cloth Another clear, protective finish must be applied later
	Paint	Two types: latex (water based) and oil based The surface must be primed with a primer coat Read and follow the manufacturers' suggestions
	Plastic laminate	Provides a decorative, durable surface The laminate is glued to a flat surface using contact cement
	Creosote and pressure treatment	Wood is immersed in a creosote or a preservative is forced into the wood under pressure Exterior use only
	Oil	Teak oil is preferred, as linseed oil requires preparation Used on handmade furniture
Metal	Paint	Surface must be completely free of oil and grease First apply a primer coat, then an undercoat, and finally a top coat
	Plastic coating	Metal is heated and dipped into fine particles of PVC that soften under the heat to form a smooth coating Useful for tool handles
	Enameling	A thin layer of glass is fused onto a metal surface For decorative work, copper is the metal used
Plastic	Polish	Surfaces are polished using very fine silicone-carbide paper followed by using a buffing attachment on a power drill Buff with light pressure to prevent melting the plastic
	Dye	Dip transparent plastic in a strong dye for a few minutes to give a tinted effect

Figure 5-45   Finishes are made for many different uses.

Links to other subjects: Have students create a chart similar to Figure 5-45, listing materials and finishes used in subjects other than technology education.

Chapter 5 Review
Processing Materials

Summary

The technologist must know how to select and use hand and machine tools safely. Marking out tools are used to measure and mark dimensions. Saws remove material quickly. Filing and planing remove small amounts of material. Materials may be shaped by shearing and chiseling. Shears are used to cut thin metal, and chisels are used to cut wood. Drilling is a process used to make holes in a material. Some sheet materials may be bent to shape. Others are formed or laminated. Casting and molding are methods of making shapes that are very difficult to produce by sawing, drilling, or filing.

Materials are joined together using mechanical devices, chemicals, or heat. Mechanical devices include nails, screws, nuts and bolts, rivets, and joints. Joints are frequently strengthened using chemicals such as glues, adhesives, solvents, or cements. Heat joining is used on metals and sometimes plastics. It includes welding, brazing, and soldering.

When the construction of an object is completed, a finish may be applied. Finishes may be sprayed, brushed, rolled, or dipped onto the surfaces.

Modular Connections

The information in this chapter provides the required foundation for the following types of modular activities:

- Manufacturing Technology
- Robotics and Automation
- Computer Numerical Control
- CNC Lathe
- CNC Mill
- CNC Wood Production
- Digital Manufacturing
- Manufacturing Design
- Manufacturing Prototype

◆ Manufacturing Concepts
◆ Materials Processing
◆ Welding Simulation
◆ Package Design
◆ Quality Control
◆ Quality Assurance (Service)
◆ Enterprise Production
◆ Safe Work Environment

Test Your Knowledge

Write your answers to these review questions on a separate sheet of paper.

1. All loose clothing should be fastened or removed _____.
   A. before operating any machine
   B. during the operation of the machine
   C. after operating a machine
   D. only when you are assisting the teacher

2. When passing hand tools to a friend, you should _____.
   A. slide them across the bench
   B. pass them with the sharp edge toward you
   C. pass them with the sharp edge away from you
   D. explain how they are to be used

3. Safety glasses or a face shield must be worn _____.
   A. on the drill press only
   B. on the band saw only
   C. every time you use a machine
   D. only by those people who wear prescription glasses

4. Before cleaning or adjusting any machine, you should _____.
   A. allow the machine to come to a complete stop
   B. run the machine at full speed
   C. allow the machine to coast slowly
   D. make sure the power is switched on

5. Before plugging in a portable electric drill, you should _____.
   A. place the chuck key into the chuck
   B. lock the drill switch in the ON position
   C. remove the drill bit
   D. make sure the drill bit is secure in the chuck

6. To mark a piece of wood to length, use a(n) _____ and _____.

7. A(n) _____ is used to make straight cuts through metal and plastic.

8. The edge of a piece of metal is made smooth and flat using a(n) _____.

9. The surface of a piece of wood is made flat using a(n) _____.

10. Tinplate, aluminum, and copper sheet are cut using _____.

11. The process used to cut a round hole in wood, metal, and plastics is called _____.

12. Laminating means _____.

13. A strip heater is used to _____.

14. Pouring liquid metal or plastic into a mold is called _____.

15. Describe the difference between mechanical, chemical, and heat joining. Give examples of each.

16. Why is a finish applied to the surface of a material?

17. Which of the following materials does NOT require a protective or decorative surface finish?
    A. Pine.
    B. Acrylic.
    C. Maple.
    D. Steel.

18. A finish may be applied to a surface of a material by _____, _____, _____, _____, or _____.

19. What are the three general rules to follow when using an abrasive to prepare surfaces for finishing?

20. If you were buying a basic kit of tools, what tools would be in your kit?

Apply Your Knowledge

1. Design and make a simple game for a young child.

2. List what you consider to be the five most dangerous activities in the technology room. What would you do to reduce the dangers?

3. Name five objects in your home. Describe how the parts of each object are joined.

4. Choose five objects each with a different type of finish. Make a chart to show (a) the material, (b) the finish, and (c) the reason why that finish has been used.

5. Make a list of the tools you have in your home. State the material and process for which each is designed.

6. Design and make an ergonomically correct chair using measurements of students in your class. It should do all of the following:

   ◆ Have a seat and back.

   ◆ Support a person weighing at least 200 pounds.

   ◆ Be comfortable to sit in.

   ◆ Be aesthetically pleasing.

   ◆ Be made from corrugated cardboard.

   ◆ Use only glue to connect the parts permanently.

   ◆ Use as little material as possible.

7. Research one career related to the information you have studied in this chapter and state the following:
   A. The occupation you selected.
   B. The education requirements to enter this occupation.
   C. The possibilities of promotion to a higher level at a later date.
   D. What someone with this career does on a day-to-day basis.
   You might find this information on the Internet or in your library. If possible, interview a person who already works in this field to answer the four points. Finally, state why you might or might not be interested in pursuing this occupation when you finish school.

Materials are processed by a number of different methods, ranging from low-technology for producing lumber to complex machines in high-technology facilities.

Chapter outline

What Structures Have in Common
Types of Structures
    Static Loads
    Dynamic Loads
Forces Acting on Structures
Designing Structures to Withstand Loads
    Reinforced Concrete

Student Activity Manual Resources

Instructional Strategies and Student Learning Experiences

1. Ask students to bring to class pictures of various types of structures. Make a display and label the various types of structures. Ask the students to determine which members were used to support the structure.
2. Take the class on a field trip or a walk. Point out both human-made and natural structures.
3. Help students to distinguish between static and dynamic loads by asking them to give examples of each.
4. *Forces Acting on Structures,* reproducible master. Use this master to help students visualize forces acting on structures. For a hands-on demonstration, obtain a piece of foam rubber with parallel lines on it. Then place stresses of compression and tension on it by bending it.
5. Ask students to describe or visit bridges. Have them identify the types of bridges. Ask students how the structures are designed to withstand loads.

Structures come in many different shapes and sizes, but they all have one common purpose: to support a given load. Skyscrapers support many floors of offices, restaurants, and businesses.

6

Structures

Discussion
Ask students to name and describe structures they have seen that left a lasting impression in their minds. Have students describe how they felt when they saw these structures.

Career connection
Have students identify careers associated with the design of structures, both large and small. Have students browse the Internet to identify world famous architects and view examples of their work.

Key Terms

abutment	static load
arch bridge	stay
cantilever bridge	structure
compression	strut
dynamic load	suspension bridge
load	tension
pier	tie
reinforced concrete	truss
shear	

Objectives

After reading this chapter you will be able to:

◆ Recognize many different types of structures, both natural ones and those made by humans.

◆ Recall that structures made by humans include bridges, buildings, dams, harbors, roads, towers, and tunnels.

◆ Identify the loads acting on structures.

◆ Analyze the forces acting on a structure.

◆ Demonstrate how structures can be designed to withstand loads.

◆ Design and make a product that incorporates structural principles.

Structures are all around us. We build them to live in or to cross a river. We build them to carry wires, to receive radio waves, and to transport people. Houses, bridges, and towers are not the only structures; airplanes, boats, and cars are structures, too.

The main purpose of a structure is to enclose and define a space. At times, however, a structure is built to connect two points. This is the case with bridges and elevators. Other structures are meant to hold back natural forces, as in the case of dams and retaining walls.

Everyone has built some kind of structure. Have you ever used a cardboard box to build a playhouse large enough to crawl inside? Have you ever constructed a ramp for a skateboard? Perhaps you built a treehouse from a variety of scrap materials. Maybe you have made a model crane, a dollhouse, a tunnel for a model railroad, or a sand castle on the beach. **Figure 6-1** shows two structures. What is the purpose of each?

Not all structures are made by humans. Living organisms, such as trees and our bodies, are natural structures. A giant redwood tree must be rigid enough to carry its own weight. Yet it is able to sway in high winds. Grass is flexible, because it springs back after it is stepped on. The bones of a skeleton have movable joints. They permit activities such as running and lifting. **Figure 6-2** shows both natural and human-made structures.

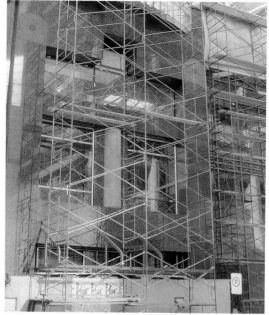

Figure 6-1 A scaffold supports workers while they build structures. Scaffolds are structures, too! They have connected parts and carry workers without collapsing. (Christopharo)

What Structures Have in Common

What do all structures have in common? They all have a number of parts, which are connected. The parts provide support so the structures can serve their purpose. One important job of all structures is to support a *load*. A load is the weight or force placed on a structure, **Figure 6-3**. For example, a load on a bridge would be a heavy vehicle crossing it, **Figure 6-4**. These vehicles must also carry loads, such as the weight from their own frame and the passengers they carry, **Figure 6-5**.

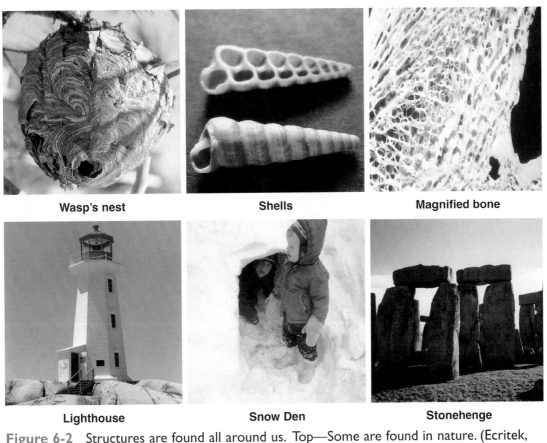

Wasp's nest **Shells** **Magnified bone**

Lighthouse **Snow Den** **Stonehenge**

Figure 6-2 Structures are found all around us. Top—Some are found in nature. (Ecritek, TEC) Bottom—Others are planned and built by humans. (Ecritek)

Discussion
What natural structures would you expect to see during a walk in a garden, through an urban park, or in the countryside?

Reflection
How do you feel when you walk in an environment where you are surrounded by natural structures?

Activity
Make a list of three animals. Make an illustrated poster showing the structures they create and the materials used to build each.

Tower **Platform**

Figure 6-3 Towers and platforms are important structures used every day. What other similar structures have you seen? What do they lift or support? (TEC, Seeds)

Links to other subjects: History—Have students choose one type of structure, such as dams, tunnels, or tables, and research its development during the past 100 years. Has the design changed? Have the materials used changed? Have the construction techniques changed? Students can create a collage to show the historical development.

Technology and society: How does the design of structures affect our enjoyment of both the natural environment and the made environment?

Designing and making
In what ways do the materials used in the design and building of structures affect its appearance?

Discussion
What structures do you see on your way to school? What materials have been used to build these structures? How many different purposes are served by these structures?

Reflection
What is your favorite human-made structure you have seen? Why is it your favorite?

Figure 6-4   Roads, sidewalks, and bridges are important structures. They help us travel from place to place. A—Pipes that supply water, electricity, and fuel are often built under sidewalks or roads. (Ecritek) B—Tunnels may provide a walkway under obstructions. (Ecritek) C—Bridges span other obstacles. D—Highways must be kept in good repair. (Jack Klasey)

Figure 6-5   These are a few structures for transportation vehicles. Both structures must carry people and support other parts of the vehicle. (Ford, Ecritek)

Useful Web site: Have students work in small groups to research the work of a famous architect. They can obtain a list of names by visiting the Great Buildings Web site at www.GreatBuildings.com. Have students create a poster giving a brief biography of the architect and showing three of his or her most famous works.

Community resources and services: Have the class make a photographic record of notable structures in their community. Identify the architect who designed each structure.

The load for a dam is the force of the water behind it. Both must also support the materials from which they are built. This is part of the load.

Types of Structures

Structures vary greatly in size and type. Look at the photographs in **Figure 6-6**. As you look, think about the loads that each of the structures must withstand. Think of the materials used in their construction. Think how the parts are connected together.

All structures must be able to support a load without collapsing. A roof must not only support its own mass but also a heavy blanket of snow. A dining chair must carry the load of a person sitting or fidgeting, **Figure 6-7**. There are two types of loads: static and dynamic.

High-rise building

Geodesic dome

Modern tent frame

House frame

Figure 6-6 How is the framework of each building like a skeleton? (Ecritek, TEC, Ecritek, Ecritek)

Vocabulary
Define the terms *static load* and *dynamic load*.

Designing and making
What general questions about the loads acting on a structure must designers ask when they develop design ideas for that structure?

Enrichment
Investigate the work of Marcel Breuer and Mies Van der Rohe, who designed tubular metal furniture similar to the chair shown in Figure 6-7.

Resource
Transparency 6-1, *Types of Loads*

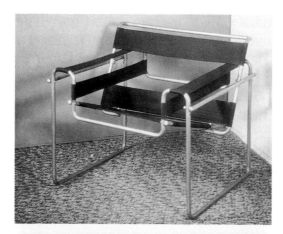

Figure 6-7 The structure of a chair must be such that it can carry the load of a person sitting on it. (Christopharo)

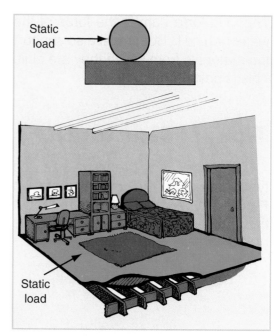

Figure 6-8 Objects at rest create static loads.

Static Loads

Static loads are loads that are unchanging or change slowly. They may be caused by the weight of the structure itself. Columns, beams, floors, and roofs are part of this load. They are also caused by objects placed in or on the structure. **Figure 6-8** is an example of such a load.

Dynamic Loads

A *dynamic load* is a load that is always moving and changing on a given structure. For example, the mass of a person walking across the floor creates a dynamic load. Other dynamic loads include the force of a gust of wind pushing against a tall building and a truck crossing a bridge, **Figure 6-9**.

Forces Acting on Structures

Both static and dynamic loads create forces, which act on structures. To understand these forces and what they do, imagine a plank placed across a stream, **Figure 6-10**. When you (the load) walk across the plank (the structure), what would you expect to happen? The plank bends

Useful Web site: Have students visit www.pbs.org/wgbh/buildingbig/lab/loads.html and complete the interactive laboratory.

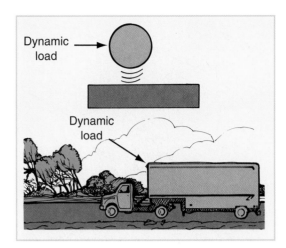

Figure 6-9 Moving objects create dynamic loads.

Figure 6-10 A person standing on a plank is a static load. Bending will cause compression on its top surface and tension on its bottom surface.

in the middle. The forces acting on the bridge may be shown by the foam rubber in **Figure 6-11**. Notice that parallel lines have been marked on it. Support the foam at each end. A vertical load applied to the center of the foam bends it, **Figure 6-12**.

Figure 6-11 Foam rubber with parallel lines drawn on it will show what happens when a load is placed on a beam.

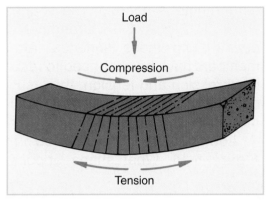

Figure 6-12 Bending causes compression and tension stress.

Notice what has happened to the parallel lines. At the top edge, the lines have moved closer together. The lines at the bottom edge have moved farther apart. The top edge of the plank is in *compression* (being squeezed) and the bottom edge is in *tension* (being stretched). Along the center is a line that is neither in compression nor in tension. It has no force acting along it. This line is called the neutral axis.

The design and construction of structures must minimize the effects of bending. Parts must be shaped so the forces of tension and compression are balanced. These energies are then said to be in a state of equilibrium, and there is little chance to bend.

Vocabulary
Define the terms *compression* and *tension*.

Designing and making
What activities in your home create compression and tension in structures? How are these structures designed to resist the forces?

Resource
Transparency 6-2, *Forces Acting on Structures*

Resource
Reproducible Master, *Forces Acting on Structures*

Useful Web site: Have students visit www.pbs.org/wgbh/buildingbig/lab/forces.html and complete the interactive laboratory.

Standards for Technological Literacy

7

Designing and making
How do the loads and forces acting on a structure affect the appearance of the structure?

Discussion
What geometric shapes can you identify in structures as you walk along a street? Which geometric shape appears more often than others?

Enrichment
Investigate the work of Buckminster Fuller, the inventor who designed the geodesic dome.

Designing Structures to Withstand Loads

As was shown by the foam rubber in **Figure 6-12**, the top and bottom surfaces of a beam are subject to the greatest compression and tension. These surfaces are where the greatest strength is needed. The shapes shown in **Figure 6-13** strengthen a beam along these surfaces. After members have been shaped to resist compression and tension, they must be connected in a way that minimizes bending.

Look at the structures in **Figure 6-14**. What shape appears

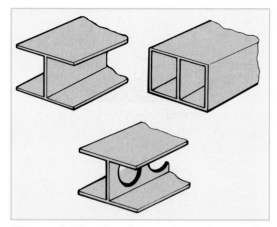

Figure 6-13 The shapes shown here will support heavy loads.

Pylon

Geodesic dome

Figure 6-14 Some shapes can support heavier loads better than other shapes can. What supporting shape appears most often in these two pictures? (TEC)

Useful Web site: Have students visit www.pbs.org/wgbh/buildingbig/lab/materials.html and complete the interactive laboratory.

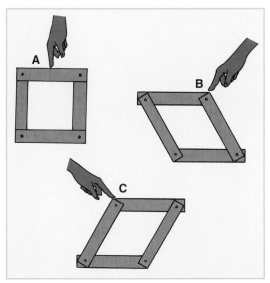

Figure 6-15 Why will frames B and C collapse when the load shown is applied?

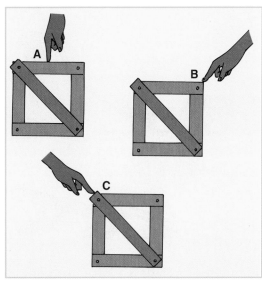

Figure 6-16 The frames retain their shape from loads at A, B, and C.

most often? You can see that the triangle appears most often. To understand why the triangle is important in structures, look at **Figure 6-15**. The frame is made of four connected members. If a load is applied at A, the frame retains it shape. However, if a load is applied at a corner (B or C), the frame will collapse. Now compare this frame to the one in **Figure 6-16**.

A rigid diagonal member (running from corner to corner) has been added, **Figure 6-17**. Once again, when a load is applied at A, the frame retains its shape. This time, however, it also retains its shape when a load is applied at corners B or C. At corner B, the load causes the diagonal to be in tension. A rigid member in tension is called a *tie*. When the load is

Figure 6-17 Why will a rope or chain work as a tie but not as a strut?

applied at corner C, the diagonal is in compression. A rigid member in compression is called a *strut*.

Standards for Technological Literacy

7

Vocabulary
Define the term *strut* as it applies to structures.

Discussion
What are the effects of building beam bridges with piers on the natural environment?

Enrichment
How do architects design buildings to resist wind shear?

Figure 6-18   Shear is a force that causes one part to slide over an adjacent part.

What would be the effect of replacing the rigid diagonal member with a nonrigid member such as a rope, chain, or cable? Would the frame retain its shape when loaded in each of the three positions? When is the rope in compression, and when is it in tension?

In addition to compression and tension, there is a third force acting on structures. This force is called *shear*. Shear is a multidirectional force that includes parallel and opposite sliding motions. To understand how shear takes place, imagine you are pulling the wagon in **Figure 6-18**. Suddenly, the wheels hit a rock. The effect is a sharp jolt on the pin. This force causes the material to shear.

Let us see how bending and the forces of compression, tension, and shear are resisted in the design of structures. Then we will see why bridges are built the way they are.

A major problem with bridges is that they bend, **Figure 6-19**. One common way to prevent a beam bridge from bending is to support the center with a *pier* as in **Figure 6-20**. However, it is not always possible to build piers under a bridge. Piers may not allow the passage of ships. Sometimes the river is too deep, runs too swiftly, or has a soft bed with no

Figure 6-19   A simple beam bridge bends easily.

Useful Web site: Have students visit www.pbs.org/wgbh/buildingbig/lab/forces.html, then click on the *sliding* button and complete the interactive laboratory.

firm foundation. Other ways have to be found to strengthen the beam bridge.

One solution is to make the beam much thicker. This, however, would make the beam very heavy. Its own mass would make it sag in the middle.

The beam could also be strengthened at the center where it is most likely to bend or break, **Figure 6-21**. Once again, notice that the strongest shape is the triangle. As we saw in Figure 6-16, a triangle does not have to be solid; it can be a frame and still be very rigid.

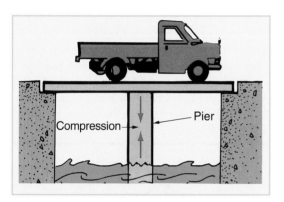

Figure 6-20 The pier of a beam bridge is compressed by the load on it.

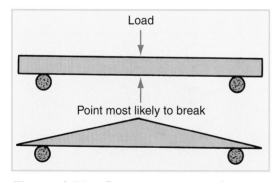

Figure 6-21 One way to strengthen a beam bridge is to make it thicker in the middle.

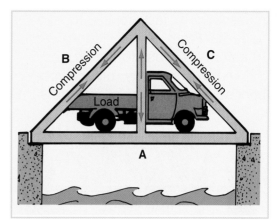

Figure 6-22 This diagram shows how a simple truss bridge works. Is the center (vertical) beam under tension or compression?

Truss bridges make use of the triangle in their design, **Figure 6-22**. As the truck crosses the bridge, its mass causes the bridge roadway to bend. Member "A" moves down. This pulls down on members "B" and "C," pulling them towards the end of the bridge and carrying the forces out to the bridge supports.

Most truss bridges are more complex than the simple truss. Many triangular frames are used to construct them, **Figure 6-23**. A bridge deck can also be supported from above. Cables, called *stays*, provide the support, **Figure 6-24**. Notice that the pylons are in compression and the stays are in tension.

The same principle is used for *suspension bridges*. Suspension bridges are the longest bridges, **Figure 6-25**. The bridge deck is suspended from hangers attached to a continuous cable. The cable is securely anchored into the ground at both ends. The cables transfer the

Useful Web site: Have students examine examples of truss bridges at www.richmangalleries.com/truss_bridges.htm. Have students select one of the bridges, print the image, and write a report on the bridge. The report should include details of the bridge's location, size, and materials used in the construction. Students can make sketches to show details of its construction.

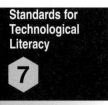
mass of the deck to the top of the towers. From there, compression transfers the mass to the ground.

There are many other types of bridges. Their design follows the same

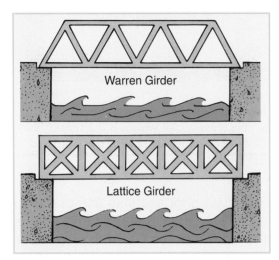

Figure 6-23 A truss is a long beam made up of shorter beams or girders that give strength to one another.

Figure 6-25 These examples of suspension bridges show how huge they can be. Top—Note the steel cables supported by the tower. Bottom—The Humber Bridge in England, the longest single-span suspension bridge in the world, stretches across the Humber Estuary.

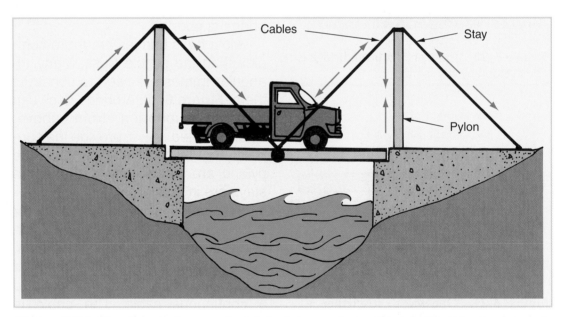

Figure 6-24 To where is the mass (weight) of the truck transferred, when the truck travels over the bridge?

Useful Web site: Have students examine examples of suspension bridges at www.richmangalleries.com/suspension_bridges.htm. Have students select one of the bridges, print the image, and write a report on the bridge. The report should include details of the bridge's location, size, and materials used in the construction. Students can make sketches to show details of its construction.

Technology and society: How did the Tacoma Narrows Bridge disaster change the way bridges were designed and built?

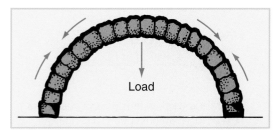

Figure 6-26 Top—Here is a simple arch bridge. Arch transfers load back to its ground supports. Bottom—World's longest arch bridge spans the New River Gorge in West Virginia. It is 3030′ (923.5 m)long. (Ecritek)

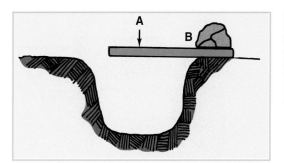

Figure 6-27 This shows the principle of the cantilever. Load at A is transferred to B.

Standards for Technological Literacy

7

Designing and making
What structures other than bridges make use of a cantilever beam?

Vocabulary
Define the term *abutment*.

Activity
Search the Internet for information about Fallingwater, a famous American home designed by Frank Lloyd Wright. How has Wright used cantilevers in the design of the home?

Resource
Activity 6-3, *Designing and Building a Bridge*

general principle: try to reduce bending. Two of the most common types are *arch bridges* and *cantilever bridges*.

In an arch bridge, the compressive stress created by the load is spread over the arch as a whole. The mass is transferred outward along two curving paths. The supports where the arch meets the ground are called *abutments*. They resist the outward thrust and keep the bridge up, **Figure 6-26**.

A beam can support a load at one end provided that the opposite end is anchored or fixed. This is known as a cantilever beam. The principle of a cantilever is seen in **Figure 6-27**. A cantilever bridge has two cantilevers with a short beam to complete the span, **Figure 6-28**.

Bridges are made from many materials. The most common are steel and concrete. Steel is fairly inexpensive, strong under compression and tension, but needs maintenance to prevent corrosion. Concrete is economical and resists fire and corrosion. It is strong under compression but weak under tension. However, it can be strengthened with steel rods.

Reinforced Concrete

Most modern bridges use steel and concrete. Steel cables made of wire rope are used to support the mass of the roadway and the traffic load on it. The towers of many bridges are made of steel. Steel trusses give rigidity to the bridge deck. They also resist bending.

Many bridges use concrete even though it is weak in tension. To overcome this weakness, the concrete is

Links to other subjects: History—Have students work in small groups to create a poster showing examples of Roman arch bridges. Students should include annotated sketches explaining the construction techniques used to build the bridges.

Useful Web site: Have students examine examples of both arch and cantilever bridges at www.richmangalleries.com/arch_bridges.htm and www.richmangalleries.com/cantilever_bridges.htm. Students should select one bridge of each type and write reports on each. Include sketches and details of the bridge's location, size, and construction materials.

Standards for Technological Literacy

7

Designing and making
How does the use of concrete in structures alter the visual environment?

Discussion
One of the most famous structures in Europe is the chapel of Notre Dame du Haut, designed by Le Corbusier. What natural structures does it remind you of?

Reflection
How do you feel when you look at pictures of structures that make use of the plastic qualities, ability to be shaped and molded, of concrete?

Vocabulary
What do the terms *hyperbolic* and *paraboloid* mean?

Enrichment
Have students work in small groups to investigate the use of reinforced concrete in the design and construction of hyperbolic paraboloid shells. A good resource is www.ketchum.org/-milo/photos-2.html.

Resource
Activity 6-1, *Designing a Paper Structure*

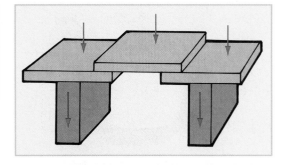

Figure 6-28 The theory used in the design of a cantilever bridge is shown on the left. An example of a cantilever bridge is seen in the photo on the right. (Ecritek)

reinforced with steel rods wherever it is in tension. The embedding of steel rods to increase the resistance to tension is the basic principle of *reinforced concrete*, Figure 6-29.

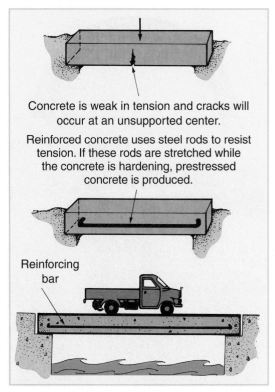

Concrete is weak in tension and cracks will occur at an unsupported center.

Reinforced concrete uses steel rods to resist tension. If these rods are stretched while the concrete is hardening, prestressed concrete is produced.

Reinforcing bar

Figure 6-29 Concrete is made stronger with steel-reinforcing rods.

Useful Web site: Have students investigate thin shell concrete structures. A useful starting point is www.ketchum.org/shellpix.html#candela, which shows structures designed by Felix Candela and Milo Ketchum.

Chapter 6 Review

Structures

Summary

All structures comprise a number of connected parts. These parts provide support and withstand a load without collapsing. There are two types of loads: static and dynamic. These loads create the forces of compression, tension, and shear. Individual members of a structure must be designed to minimize the effects of these forces. The members are then connected together in such a way as to minimize bending.

Bridges provide an example of how structures are designed to resist forces. A truss bridge uses the rigidity of the triangle to resist the forces of compression and tension. Cables and pylons in a suspension bridge resist these same forces. There are many other types of bridges. As with all structures, they are designed to withstand loads and minimize bending.

Modular Connections

The information in this chapter provides the required foundation for the following types of modular activities:

- Structural Engineering
- Applied Physics
- Bridge Design
- Tower Design
- Truss Design

Answers to Test Your Knowledge Questions
1. Answers will vary. Refer to Figure 6-1 through Figure 6-7, pages 148–152.
2. B. Bridge.
3. C. built to withstand a load.
4. Static: objects in a room. Dynamic: person or object moving. Examples may vary.
5. Top surface is in compression; bottom surface is in tension.
6. D. top and bottom surfaces.
7. D. Triangle.
8. A. strut.
9. B. tie.
10. truss
(continued)

Test Your Knowledge

Write your answers to these review questions on a separate sheet of paper.

1. Name three natural structures and three structures made by humans.

2. Which of the following is NOT a natural structure?
   A. Spider's web.
   B. Bridge.
   C. Tree.
   D. Beaver's dam.

3. All structures are _____.
   A. built to withstand heat
   B. made in factories
   C. built to withstand a load
   D. designed to house people

4. Name the two types of loads acting on structures. Give one example of each.

5. What forces are acting on the top and bottom surfaces of a beam loaded from above?

6. To strengthen a beam loaded on the top surface, it must be reinforced at the _____.
   A. top surface only
   B. bottom surface only
   C. center
   D. top and bottom surfaces

7. Which geometric shape gives the greatest rigidity to a structure?
   A. Square.
   B. Circle.
   C. Rectangle.
   D. Triangle.

8. A beam in compression is called a _____.
   A. strut
   B. tie
   C. post
   D. stay

9. A beam in tension is called a _____.
   A. strut.
   B. tie.
   C. post.
   C. stay.

10. A bridge that uses a series of triangular frames is called a(n) _____ bridge.

11. The world's longest bridges are _____ bridges.

12. Using notes and diagrams, explain how an arch bridge resists loads.

13. Using notes and diagrams, explain the principle of a cantilever bridge.

14. What are the most common materials from which bridges are built?

15. Concrete is weak in tension. How is this problem overcome?

Answers to Test Your Knowledge Questions *(continued)*
11. suspension
12. Refer to Figure 6-26, page 159.
13. Refer to Figure 6-27 and Figure 6-28, pages 159–160.
14. Steel and concrete.
15. By inserting steel rods to increase the resistance to tension.

Apply Your Knowledge

1. Look at the natural structures in the illustrations. Next look at the structures made by humans. For each of the structures made by humans, name the natural structure it most closely resembles.

2. Look at the structures in **Figures 6-4** and **6-5**. Write the location or address of a structure in your town that most closely resembles each one.

3. Name five different structures. For each structure, list the loads to which it is subjected. State whether each load is static or dynamic.

4. Draw a diagram of a plank bridge with a load on it. Label your diagram to show the forces of tension and compression.

5. Using only one sheet of newspaper and 4″ (10 cm) of clear tape, construct the tallest freestanding tower possible.

6. Using drinking straws and pins, construct a bridge to span a gap of 20″ (508 mm) and support the largest mass possible at midpoint.

7. Research one career related to the information you have studied in this chapter and state the following:
   A. The occupation you selected.
   B. The education requirements to enter this occupation.
   C. The possibilities of promotion to a higher level at a later date.
   D. What someone with this career does on a day-to-day basis.

   You might find this information on the Internet or in your library. If possible, interview a person who already works in this field to answer the four points. Finally, state why you might or might not be interested in pursuing this occupation when you finish school.

Chapter outline

Student Activity Manual Resources

DMA-11 *Pet House*
ST-11, *Investigating Pets*
ST-24, *Drawing to Scale*
ST-25, *Measuring Scale Drawings—Conventional*
ST-27, *Reading Scale Drawings—Conventional*
ST-52, *How Trusses Work*
ST-53, *Making Structures Rigid*
ST-29, *Modeling with Foam Board*
ST-37, *Cutting and Shaping Thin Aluminum*
ST-38, *Cutting and Shaping Acrylic*
ST-41, *Making a Wooden Candle Holder*
ST-51, *Safety—Managing Risk*
ST-5, *Writing a Design Specification*
EV-1, *Evaluating the Final Product*
EV-2, *Unit Review*

DMA-12, *Temporary Shelter*
ST-55, *Natural and Human-Made Structures*
ST-56, *Forces Acting on Structures*
ST-53, *Making Structures Rigid*
ST-54, *Struts and Ties*
ST-57, *Using Beams in Structures*
ST-58, *Strengthening Beams*
(continued)

High-rise construction projects frequently use structural steel or concrete reinforced with steel rods.

Instructional Strategies and Student Learning Experiences

1. Invite an architect to class to discuss various types of houses with students. Ask students to prepare a list of questions in advance.

2. *How Parts of a House Fit Together*, reproducible master. Use this master to identify the various components of a house and how they fit together.

3. As a class, visit the building sites of several types of houses. Ask students to write a report about their observations.

4. Have students write a paragraph describing their ideas of what their "dream house" would be like. Have them sketch the house and describe the various rooms. Also, have them list the types of materials they would use to construct various parts of the house.

5. Invite a plumber, HVAC technician, and electrician to class to discuss how these systems operate in a house. Ask students to prepare a list of questions in advance.

7

Construction

Key Terms

adobe
batter board
beam
closed-loop system
communication system
electrical system
floor plan
footing
foundation
heating system
insulation
joist

landscaping
modular construction
open-loop system
plumbing system
post and lintel
prefabrication
roof truss
site
subfloor
system
wall stud

Objectives

After reading this chapter you will be able to:

◆ Identify the principal types of residential buildings found in modern communities.

◆ Discuss the steps involved in selecting and buying a building site.

◆ Identify the component parts of a typical house and describe the function of each part.

◆ Identify the principal materials used in the construction of a house.

◆ Recognize that all systems have inputs, a process, and outputs.

◆ Use a systems model to explain an example of technology.

◆ Differentiate between an open-loop and a closed-loop system.

Discussion
In the late fifteen hundreds Sir Edward Coke wrote, "for a man's house is his castle." What do you think he meant? What does this famous quotation say to you?

Career connection
Have students identify the careers related to the construction industry. Have students use the Internet to find out the training and licensing requirements to enter one of these careers.

Student Activity Manual Resources (continued)
ST-32, *Making a 3-D Framework*
ST-51, *Safety— Managing Risk*
ST-29, *Modeling with Foam Board*
ST-4, *Identifying User Needs and Interests*
ST-5, *Writing a Design Specification*
EV-1, *Evaluating the Final Product*
EV-2, *Unit Review*

Modern communities rely on diverse people with different skills, educations, and jobs in order to be built well and maintained in good condition. Carpenters, electricians, cement masons, plumbers, sheet metal workers, bricklayers, and ironworkers build houses, stores, office buildings, and other commercial and industrial structures, **Figure 7-1**. Many of these tradesmen are licensed, certified, and bonded. They often undergo an apprenticeship and join a union, which ensures their skill and competency and also secures their wages and working conditions. Modern communities also rely on civil engineers, architects, and building, bridge, and home inspectors to design, plan, and ensure the keeping of certain government guidelines, regulations, building laws, and ordinances.

The Structure of a House

The structure most familiar to you is your home. Many materials have been used for different kinds of homes, **Figure 7-2**. Traditionally, people have built houses from the handiest material. They used what was most commonly available nearby. In North America, clay and straw, called *adobe*, were used in the Southwest. Thick walls kept the home warm in winter and cool in summer. Pioneers who homesteaded the Great Plains used sod cut by plow. The sod served as walls and sometimes roofs. In wooded regions of the north where lumber was plentiful, log cabins were

built. Wood is still a popular building material in North America. Treatment of the wood to retard fire and decay has made the frame house more durable than ever.

Building a home entirely on its site often takes a number of months. A new method is often used to speed up construction. It is called *prefabrication*. The term means building parts of the house in a factory. This is much faster because the parts can be made on an assembly line with power tools and heavy equipment. Workers are not affected by bad weather.

Prefabricated parts are moved to the building site for final assembly. Time is saved in the factory because of mass production methods, and time is saved on the site because much of the assembly has already been done.

A popular type of prefabrication is known as *modular construction*. Modules are basic units, such as rooms. Modules of different sizes and shapes can be combined on site.

The House Frame

The frame of a house provides a supporting structure. The simplest framed structure is a *post and lintel* or post and beam, **Figure 7-3**. The lintel is a beam simply supported on the posts. It carries the roof load. The posts are vertical struts compressed by the lintel. Post and lintel structures may be built one on top of another to frame multistory buildings.

Like the bridges you have read about, houses must also support

Useful Web site: Students can use the dictionaries at www.techtionary.com and www.hyperdictionary.com to find the definition of any new technical and technological terms used in this chapter.

Designing and making
How is a person's quality of life affected by the work done by construction workers?

Reflection
What job would you enjoy if you worked in the construction industry?

Activity
Describe using your own words what is happening in each of the photos in Figure 7-1.

Figure 7-1 A—Carpenters use wood to build and assemble the frame, roof, doors, trim, windows, stairs, and various other parts of a house. This carpenter is building a raised deck for a new house. (Tom Severson) B—Electricians install electrical and communication wiring, conduit, lighting fixtures and lighting circuits, outlets, switches, circuit breakers, motors, and various other circuits and electrical devices. This electrician is stripping wires for a wall outlet. (Tom Severson) C—Cement masons pour the cement for the sidewalks, driveways, foundations of houses, and decorative concrete work. D—Sheet metal workers install the ductwork for the heating, air conditioning, and ventilation systems. (Jack Klasey) E—Bricklayers set bricks, blocks, and stone for houses.

Links to other subjects: English—Have students select and research one construction technology career. Have them prepare a written report for the class. Discuss what information they should include in their reports. Also make sure they include a bibliography.

Designing and making
How has the development of steel framed buildings changed the made environment?

Discussion
What are the advantages and disadvantages of steel framed buildings?

Vocabulary
Define the terms *post*, *lintel*, *beam*, and *joist*.

Resource
Transparency 7-1, *Shape and Construction of Homes*

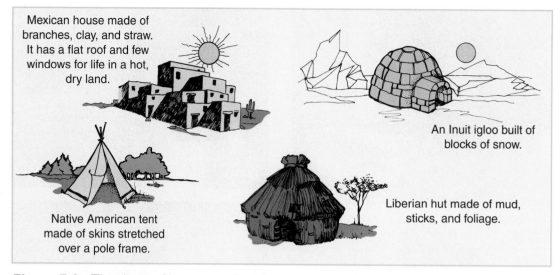

Mexican house made of branches, clay, and straw. It has a flat roof and few windows for life in a hot, dry land.

An Inuit igloo built of blocks of snow.

Native American tent made of skins stretched over a pole frame.

Liberian hut made of mud, sticks, and foliage.

Figure 7-2 The shape of homes is often the result of materials found locally.

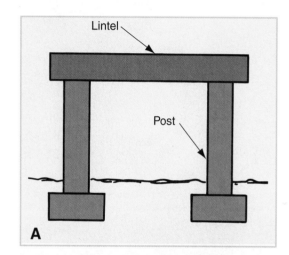

Lintel

Post

A

B

Figure 7-3 Here is the setup for post and lintel structures. A—Heavy posts support the horizontal lintel or beam. B—Steel framed buildings have characteristics similar to post and beam structures. (Ecritek)

loads. The structure of a typical house must support the static loads. These are the weight of the materials from which it is built and its contents. The house must also withstand dynamic loads created by weather conditions outside and the movement of people inside. To understand how these loads are supported in a house, look at **Figure 7-4**.

Loads are usually applied to horizontal members such as *joists* or *beams*. The total load then moves downward. It transfers to columns or bearing walls to the foundation. Then the load transfers to the soil. Therefore, most loads and forces tend to push downward. Horizontal forces are produced by wind.

Think about the loads placed on the floor of your home. If you and your friends are dancing, the mass of the people acts as a load. This load is transferred to a beam, then to a column. The load on the column is, in turn, transferred to the footing and to the underlying soil, **Figure 7-5**.

Useful Web site: Have students write a report about a famous steel frame structure. Students can search the Great Buildings web site at www.GreatBuildings.com. Click on *Advanced search*. Type *steel frame* into the *construction system* box and click *Search*. Have students choose one of the structures listed, such as the Bank of China Tower, Chrysler Building, or Empire State Building.

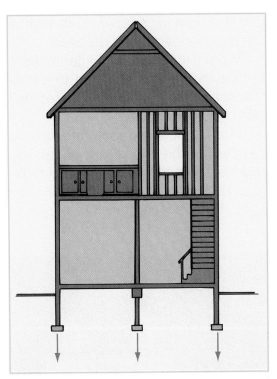

Figure 7-4 These are the principles of load transfer. Arrows show how load is carried through posts down to the foundation.

Planning for a Home

Before construction of a house can begin, two major decisions must be made. One is to decide on the

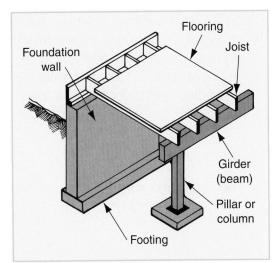

Figure 7-5 Loads on horizontal surfaces are transferred to the soil through vertical members of a building.

basic type of house. The other is to plan the house *site*. A site is the land where a house is built.

Four types of single-family houses are shown in **Figure 7-6** through **Figure 7-9**. The type one chooses must be based on a number of considerations. How much room is needed? How much land is available for a home? Which type has the most

Discussion
If you could choose to live in any type of single-family dwelling, which would you choose? Give reasons for your answer.

Reflection
What type of home do you dream about living in? What must you do to make your dream a reality?

Resource
Activity 7-1, *Designing Your Dream Home: Part I*

Figure 7-6 A bungalow has only one story. (Ecritek)

Technology and society: What is the impact on the environment of trying to provide a single-family dwelling for every family?

Community resources and services: What percentage of homes in your community are single-family, terraced, condominium, and apartments?

Designing and making
Why do houses look the way they do? What influences have shaped the neighborhood in which you live?

Activity
Take a walk around your neighborhood and look at the different styles of homes. What general observations can you make?

appeal? Will any family member have difficulty with steps? How much money do you wish to spend? Is the money available? Often banks or other lending institutions make loans for building.

Planning Inside Space

The second major task in the design and building of a home is to plan the interior spaces. This task is often performed by an architect.

What if there were no interior walls in your home and you had to divide the space into a number of rooms? How would you do it?

First of all, think about the spaces needed for the things in a home. You must be able to fit in all of your furniture and household equipment. You should also leave sufficient space to

Figure 7-7 A one-and-one-half-story home is shown here. The second-level rooms extend into the roof. (Ecritek)

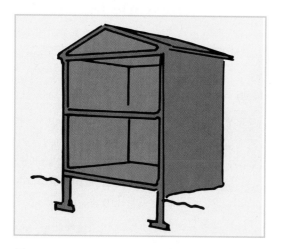

Figure 7-8 Houses may have two, three, or even more stories. The roof balcony on this house is called a "widow's walk." Do you know why? (Ecritek)

Useful Web site: Have students create a poster that illustrates and describes a house style they find interesting. Students can begin their research at http://architecture.about.com/library/bl-styles_index.htm. Discuss with students the information they can include on their poster.

Community resources and services: Are there bylaws in your community that restrict the type of homes that can be built?

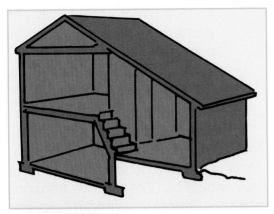

Figure 7-9 A split-level home is divided vertically. Floors of one part are located midway between the floors of the other part. (Ecritek)

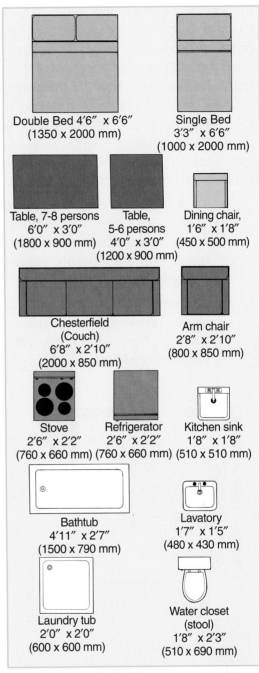

Figure 7-10 These drawings represent furniture and fixtures found in your home. They are used on floor plans so you can visualize how they will fit in a home.

move around. The sizes of most items of furniture and equipment are fairly standard. **Figure 7-10** shows a plan view and floor space required for common items.

Finding Space Needs

To determine the overall size and shape of any individual room, you can use small pieces of card stock. Cut the card stock to represent the size and shape of each piece of furniture. Position the shapes in different ways and study each arrangement. Continue until you find a suitable arrangement, remembering to leave space for people

Designing and making
What factors would you consider as you begin to plan the arrangement of spaces in a home?

Activity
Have students measure one room in their home and every piece of furniture it contains. Next have students draw a floorplan to scale and cut pieces of card to represent the furniture. Can the furniture be rearranged to improve the appearance and utility of the room?

Resource
Activity 7-2, *Designing Your Dream Home: Part II*

Technology and society: What factors determine the style of decoration and furniture in a home? How can people become better informed about well-designed interiors?

to circulate around the room. There must also be room to open doors and pass through doorways. The result would provide the size and shape of the room, **Figure 7-11**. Next, fit the rooms into the overall shape of the house. Generally, rooms are grouped according to their functions: living, sleeping, and working. There are a number of reasons for grouping rooms:

♦ To separate noisy and quiet areas so family members could work or play without disturbing others who are resting or studying.

♦ To place bedrooms and bathrooms close to each other for convenience in washing, bathing, and dressing.

♦ To give direct access between kitchen and dining area for convenience in carrying hot foods from the kitchen to the table and for clearing the table.

Rooms must be connected by hallways, stairs, and doorways. The ideal is to have good traffic patterns. You should be able to move from one area to another without passing through a third area, **Figure 7-12**.

Figure 7-11 Rooms are often planned around furniture. What furniture items are shown by symbols?

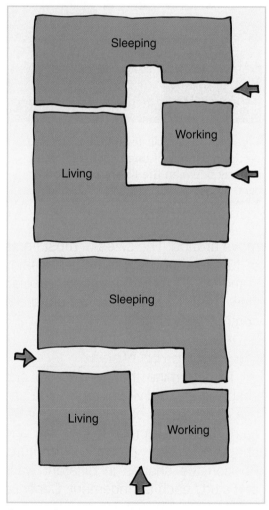

Figure 7-12 Hallways allow access to all rooms without passing through other rooms.

Technology and society: How does the arrangement of spaces in a home affect the quality of life of the people who live there?

For example, it should be possible to walk from the kitchen to the front door without going through several rooms.

When you feel you have a good arrangement of rooms you can make a drawing. An architect calls this a *floor plan*. The plan shown in **Figure 7-13** shows a two-bedroom apartment. **Figure 7-14** shows a three-bedroom bungalow.

Finding and Preparing a Site

The land on which you will build is called a site or lot. It can be any size. City lots are usually small. Those outside the city may be as large as several acres.

Planning the site is just as important as designing the home itself. It involves several important steps.

Selecting the Site

Where you locate a new home is important. You may want it to be in a certain community. Perhaps it should be close to your job, shopping centers, and schools. A quiet wooded area may be preferred.

Acquiring the Site

To purchase a building site, both the owner and you must agree on a price. If you are not willing to pay the asking price, you must negotiate. This means finding a price at which the owner will agree to sell.

Going into Contract

When a price is agreed upon, you sign a contract. This is a legal document, which sets down all the conditions of the purchase. Among other things, it lists the selling price agreed upon. When both parties sign the contract, it is binding. This means that

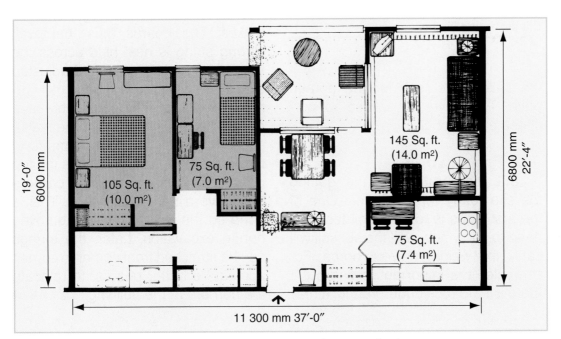

Figure 7-13 This floor plan shows an arrangement for a two-bedroom apartment.

Useful Web site: Have students use the Internet to research the way in which Frank Lloyd Wright used a dramatic natural site in the design of Fallingwater. One Web site to explore is www.wpconline.org/fallingwaterhome.htm.

Technology and society: To what extent should architects and town planners be required to design homes and communities that leave the natural environment undisturbed?

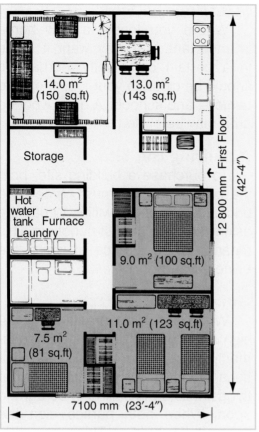

14.0 m² (150 sq.ft)

13.0 m² (143 sq.ft)

Storage

Hot water tank Furnace Laundry

9.0 m² (100 sq.ft)

11.0 m² (123 sq.ft)

7.5 m² (81 sq.ft)

12 800 mm First Floor (42'-4")

7100 mm (23'-4")

Figure 7-14 Here is a floor plan for a three-bedroom bungalow. See how the plan separates the sleeping and living areas from the dining and kitchen areas.

the seller must sell for the agreed price. The buyer is obliged to purchase at that price.

Site Preparation

Site preparation means getting the site ready for the home. One of the first steps is to do a soil test. You need to know how well the subsoil will carry the weight of your home. For example, it is important to find out if there is hard rock underground. It may be expensive to remove. There may be groundwater too close to the surface. This could cause flooding in the house. Other soils may be too light to carry loads well.

Once a soil engineer has determined that the site is suitable, a contractor will clear the site of boulders and excess soil. Grading may be needed to level a spot for the foundation. Lines and grades must be established to keep the work true and level. **Figure 7-15** shows how *batter boards* are used for this purpose. Small stakes are located at what will be the corners of the house. Nails driven into the tops of these stakes mark the four corners of the house. Straight lines between these nails indicate the outside edge of the foundation walls.

Once the four corners have been located, larger stakes are driven into the ground at least 4' (1.3 m) beyond the lines of the foundation. The batter boards are nailed horizontally to these stakes. The boards must be level. Strong string is next held across the tops of opposite boards and adjusted exactly over the tacks in the small corner stakes. A plumb bob may be used to set the lines exactly over the nails.

Next, a saw kerf (cut) is made to mark where the string crosses the top of each batter board. Some carpenters drive a nail at this point. This is done so the strings can be removed during excavation. Later, the strings can be stretched from corner to corner, across the batter boards, to relocate the corners of the building.

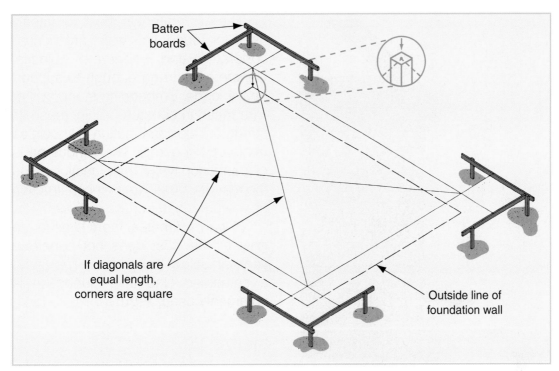

Batter boards

If diagonals are equal length, corners are square

Outside line of foundation wall

Figure 7-15 Batter boards support the lines set up to locate the building so excavating can begin for the foundation.

Standards for Technological Literacy

20

Discussion
Houses are built on different types of foundations, including slabs, piers, and foundation walls. What environmental factors influence the choice of foundation?

Vocabulary
Using your own words explain the terms *joist, jack post,* and *subfloor.*

Main Parts of a Home

Now let us look at the main structural components and materials used in house construction. There are five major components: foundation, floor, wall, ceiling, and roof, **Figure 7-16**. These separate structural components fit together to form a house.

Foundation

Most structures rest on a *foundation*, **Figure 7-17**. Normally a foundation lies below the surface of the ground. Notice that there are two major parts to this type of foundation: the *footing* and the foundation wall.

To understand the importance of the footing, think about the reason for wearing snowshoes. If you try to walk on deep, soft snow without them, you might sink down to your thighs. Snowshoes spread your body mass over a larger area of the snow's surface, **Figure 7-18**.

This prevents you from sinking. The same principle is used to build a foundation. The load of the building is first transmitted to the foundation wall. Then it is spread over a larger area by the footing. Thus, the building is prevented from sinking into the ground.

In most locations, it is necessary to drain away any subsurface water to avoid damp basements and wet floors. Tile laid around the wall footings serves this purpose. These are known as drain tile, perimeter tile, or weeping tile.

Community resources and services: Invite a local flooring expert into the class to discuss the advantages and disadvantages of various floor finishes.

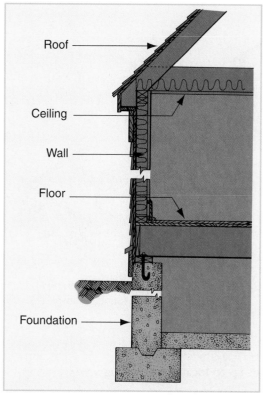

Figure 7-16 A typical section through a house shows its main structural parts.

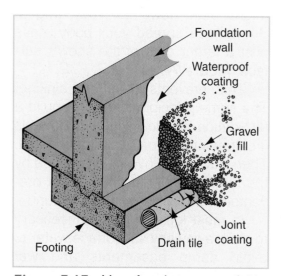

Figure 7-17 Many foundations are of this type. Explain why the footing is so wide.

The two materials most commonly used for foundation walls are poured concrete and concrete blocks. Concrete is strong enough to support heavy loads. Embedding steel rods or wire mesh in concrete will increase its strength. This combination is known as reinforced concrete. The various parts of foundation walls, their functions, and the materials used in their construction are summarized in **Figure 7-19**.

In warm climates, there is either no frost or the frost does not penetrate very far below the ground. Therefore, a combined slab and foundation is commonly used, **Figure 7-20**.

Floor

When the concrete for the foundation wall is poured, anchor bolts are set in the top. These bolts are used to fasten a sill to the foundation, **Figure 7-21**. Joists are nailed to the sill on edge, forming a framework.

Figure 7-18 Why doesn't the boy sink into the snow? (Ecritek)

Part	Function	Material
Footing	To transmit the superimposed load to the soil	concrete
Foundation wall	To form an enclosure for the basement and to support walls and other building loads	concrete concrete blocks
Weeping or perimeter tile	To provide drainage around footings	clay plastic
Gravel fill	To permit water to drain into the weeping tile	gravel

Figure 7-19 The component parts of a foundation.

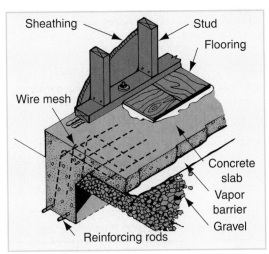

Figure 7-20 Where frost does not penetrate deeply, a combined slab and foundation can be used.

This framework is supported by a beam. Joists are usually made of wood nearly 2″ thick and 10″ wide or more.

When beams must span a long distance, they are supported in the

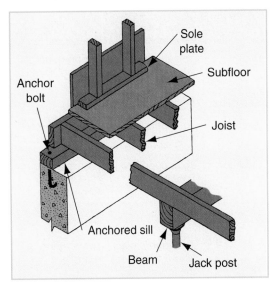

Figure 7-21 A floor is supported by joists.

middle by jack posts. The joists support a *subfloor*. A subfloor is a covering over joists. It supports other floor coverings. The various parts of floors, their functions, and the materials used in their construction are summarized in **Figure 7-22**.

Walls and Finish Flooring

The subfloor is fastened to the joists. Then the walls for the first floor are laid out and built. The many parts of this type of wall are shown in **Figure 7-23**.

Wall studs provide the framework for walls and partitions. The other various parts of walls, their functions, and the materials used in their construction are listed in **Figure 7-24**.

Ceiling and Roof

The construction of a ceiling often requires joists the same as for a floor.

Useful Web site: Have students investigate the structural components of vernacular architecture. They can begin with the Web sites at www.yvbsg.org.uk and www.hvbt.co.uk/morework.htm.

Discussion
Ask students to identify the materials used to build each structural part of a wall and floor, as shown in Figure 7-23.

Enrichment
Have students use the Internet to investigate the use of Aerogel™ in houses of the future at http://science.nasa. gov/newhome/ help/tutorials/ housefuture.htm.

Activity
Encourage students to look carefully at their home to identify the structural elements shown in Figures 7-20 to 7-23.

Part	Function	Material
Beam	To support joists when long distances are spanned	wood (pine or spruce) steel
Joist	To support a floor	wood (pine or spruce)
Subfloor	To support finish flooring	board or sheet material (e.g., tongue-and-groove pine, plywood)
Sill	To support joists where they meet the foundation	wood (pine or spruce)
Jack post	To support beams	wood (pine or spruce) steel

Figure 7-22   Parts of a floor frame.

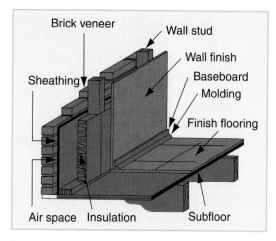

Brick veneer
Wall stud
Wall finish
Baseboard
Molding
Sheathing
Finish flooring
Air space   Insulation   Subfloor

Figure 7-23 This is a section of a wall and floor. Refer to Figure 7-24 for function of parts.

Part	Function	Material
Exterior surface	To provide protection and decoration to the outside of a building	brick aluminum siding wood
Air space	To provide a barrier against the passage of moisture	
Sheathing	Reinforce studs Provides insulation	wood fiberboard
Wall stud	To provide a framework for walls or partitions	wood (pine or spruce)
Insulation	To resist heat transmission	fiberglass polyurethane vermiculite
Vapor barrier	To retard the passage of water vapor or moisture	polyethylene sheet
Interior wall surface	To cover the interior wall framing	plasterboard wood paneling plaster
Finish flooring	To cover a subfloor and provide a decorative surface	parquet ceramic linoleum carpet

Figure 7-24. This table gives information on the parts of walls and finish flooring.

A roof is made up of sloping timber called rafters. Roofs are also built of a series of prefabricated trusses. Shaped like a triangle, the *roof truss* forms a framework to support the roof and any loads applied to it. Braces on the inside create triangles to support and strengthen the rafters.

A typical ceiling and roof construction is shown in **Figure 7-25**. The various parts of ceilings and roofs, their functions, and the materials used in their construction are given in **Figure 7-26**.

Links to other subjects: Geography—Have students investigate the ways in which geographic location affects the construction techniques used to build homes.

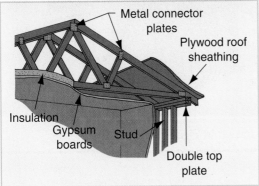

Figure 7-25. This picture and diagram show the construction of a truss roof. Gypsum board ceiling is attached to the underside. (TEC)

Part	Function	Material
Joist	To support a ceiling	wood (pine or spruce)
Insulation	To resist heat transmission	fiberglass polyurethane vermiculite
Vapor barrier	To retard the passage of water vapor or moisture	polyethylene sheet
Interior surface	To form the ceiling	plasterboard plaster
Roof truss	To form a framework for the roof and to support loads applied to it.	wood (pine or spruce)
Exterior finish	To provide protection from rain, snow, and other weather conditions	asphalt wood shingles tar and gravel

Figure 7-26 Study the parts of a ceiling and roof and remember their purpose and materials.

Designing and making
How does landscaping affect the way in which people can enjoy a neighborhood?

Discussion
How do the elements and principles of design (Chapter 2) apply to the design of urban landscapes presented at www.sustland.umn.edu?

Reflection
What features do you enjoy looking at in a garden?

Vocabulary
What does the term *landscaping* mean?

Figure 7-27 shows how all the many parts of a house fit together. The next time you see a building going up, see how many of the parts you can identify.

Finishing the House

The final stages in building a house include trimming, painting, decorating, and landscaping. Trimming involves covering rough edges and openings with moldings. For example, a baseboard is the trim used to cover the small space between a wall and a floor. Painting protects and improves the appearance of interior and some exterior surfaces. Wallpaper is the most common interior decorating material. Paneling or tongue-and-groove boards are also used to decorate walls.

Landscaping, **Figure 7-28**, is designing the exterior space that surrounds a home. It involves planning the location of lawns, hedges, trees, shrubs, and plants. The plan will also show the location of paths, such as driveways and walkways. It will also show special features such as patios, fences, walls, and plant boxes. After accesses and features have been built, topsoil is added to the site. Topsoil is a layer of rich earth. It is needed so that trees, shrubs, lawns, and plants can grow.

Links to other subjects: History—Have students write an illustrated report highlighting the work of famous American landscape designers. A useful starting place is www2.cr.nps.gov/hli/pioneer1.htm.

Technology and society: What is the environmental impact of reducing the amount of land devoted to gardens in a new subdivision?

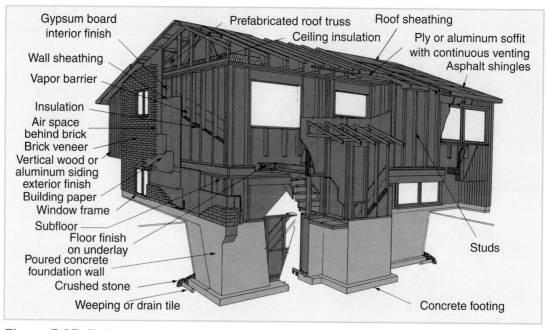

Figure 7-27 Find each of the five components shown in Figure 7-16.

Maintaining the House

Once your house is built, your work is still not done. As materials age and get used, they wear out and break down. That is why people must learn to maintain their homes. The following are a number of home maintenance jobs:

◆ Replace old carpet

◆ Repaint walls and ceilings

◆ Redo siding

◆ Repair roof leaks

◆ Replace blown fuses

◆ Rewire circuits for increased wattage

◆ Run conduit

◆ Repair concrete cracks

◆ Clean chimneys

◆ Change air filters in furnace

◆ Install more insulation in walls or ceilings

Many of these jobs one can learn without extensive training.

Learning basic home maintenance procedures has a number of advantages. You can improve the condition of your house. You can save money doing these jobs yourself (instead of hiring a

Figure 7-28 Part of a well-landscaped house is having flowers, shrubs, trees, grass, and walkways that are pleasing.

Technology and society: Should people be encouraged to do more of the maintenance in their home? Who would benefit and who would lose out from an increase in the number of Do-It-Yourselfers?

Community resources and services: Have students visit a local hardware store to explore the range of tools and materials available to anyone wanting to maintain their own home.

contractor). It can also bring you personal satisfaction for a job well done, whether working on your own house or helping a friend. There are other important reasons for keeping one's home well maintained. If you plan on selling your house, a home inspector will come and check everything from electrical, plumbing, dry wall, roof, carpentry, and furnace, hot water heater, and exhaust ducts. He will inform you what needs to be done in order to meet the code. You will then need to repair or replace whatever is not up-to-date with the code.

Systems in Structures

What happens when you telephone a friend? After lifting the receiver or pressing a button, you dial a number. Signals travel to a central location where automatic switching equipment sends your call to your friend's house. Your friend answers and your voices are carried over the lines. At the end of the conversation, the caller disconnects. The telephones, cables, and automatic switching equipment are part of a system.

Some systems are very large. Others are quite small. The sun and the planets that revolve around it form the solar system. The skeletal system of your body is made up of more than 200 bones. Together, they support the body's mass. They also give the body shape and protect important organs. The fuel system of a car pumps gasoline from a fuel tank, through fuel lines, to the injectors and into the cylinders of the engine.

What Is a System?

What do all systems have in common? How is a system defined? A *system* is a series of parts or objects connected together for a particular purpose. There are two types of systems: open-loop and closed-loop, **Figure 7-29**.

Open-Loop System

A portable space heater without a thermostat is an example of an *open-loop system*. When plugged in and switched on, the heating element warms the air passing over it. It continues to heat the room until switched off. There is no method of controlling whether there is too much or too little heat.

Closed-Loop System

In a *closed-loop system*, the same heater would be connected to a control mechanism such as a thermostat. When the room air reaches

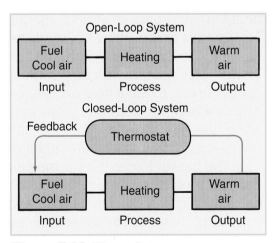

Figure 7-29 These diagrams compare open-loop and closed-loop systems.

Links to other subjects: Discuss with students how the systems concept is used in subjects other than technology education.

Community resources and services: Identify and describe systems that operate in your community that make life easier and safer.

Standards for Technological Literacy

2 3 20

Discussion
Discuss with the class the fuels used to operate the heating system in their homes. How does the type of fuel affect the design of the heating system?

Activity
Compare the cost of heating a home using fuel oil, natural gas, or electricity.

Safety
What safety features are included in your home heating system? Discuss your findings in class.

Resource
Transparency 7-2, *The Two Types of Systems*

Resource
Reproducible Master, *The Two Types of Systems*

the temperature you set on the thermostat, the heater shuts off. It will switch itself on again when the temperature falls below the set limit.

In both cases, the systems contain: input (cool air and fuel), process (burning the fuel), and output (warmed air). Input, process, and output are characteristics of all systems. However, in a closed-loop system there is also a feedback device, which provides control. Control of our environment is a major reason why technological systems have been created.

In our homes, there are four major systems: heating, electrical, plumbing, and communications. Each helps to control the environment in the home automatically according to instructions previously set.

Heating System

Figure 7-30 illustrates the major parts of a forced air *heating system*. Some heating systems use water to carry heat to rooms. However, forced air is a more popular way to move heat from a furnace to various rooms.

Cool air enters the bottom of the furnace. Here the filter traps dirt. A blower forces the filtered air up into a compartment, called a heat exchanger. The exchanger has passageways, which are heated by electricity or the combustion gases from burning oil or gas. The blower forces the warmed air through a network of ducts into each room. Cooler, heavier air sinks to the floor and flows through return air ducts leading back to the furnace. Control switches turn on the blower and the supply of heat. Thus, the furnace controls the temperature of the circulating air.

An important part of any heating system is the *insulation*. Insulation is material installed in walls and ceilings that helps control the heat loss in the winter and the heat gain in the summer. Some common forms of residential insulation are blanket, batt, rigid, and loose fill. Different types of insulation are commonly made from fiberglass, Styrofoam®, treated paper, and a variety of other materials.

This heating system is, in fact, composed of several subsystems. A subsystem is a smaller system that operates as a part of the larger system. The subsystems within the home heating system are:

◆ Heater to produce heat.
◆ Blower unit for pushing the heat through the ductwork.
◆ Network of ducts for carrying the heated air.
◆ Thermostat for providing continuous feedback.

Some heating systems may have other subsystems, including:

◆ Humidifier.
◆ Air conditioner.
◆ Electronic filter.
◆ Heat pump.

Electrical System

An *electrical system* supplies electricity for light, heat, and appliances. This electricity is carried throughout the home by a number of separate circuits. A circuit is a pathway for electrical current. Each circuit has two or three wires (two wires if run with conduit, three includes ground wire) running inside the walls and

---

Useful Web site: Have students investigate ways to reduce the heating costs in a home and the advantages of building R-2000 homes. Useful Web sites are www.homeenergysaver.lbl.gov and http://r2000.chba.ca/consumers/home.html.

Technology and society: Why should homeowners maximize the insulation in their homes?

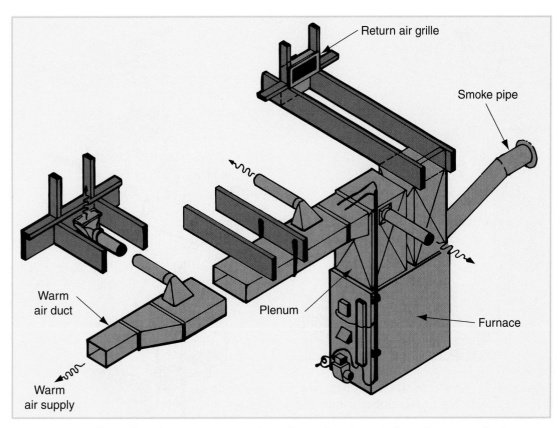

Figure 7-30 Forced air heating system. Air is heated and carried to all rooms of a house.

Labels in figure: Return air grille, Smoke pipe, Warm air duct, Plenum, Furnace, Warm air supply

ceilings. A circuit carries current from a power source, **Figure 7-31**. Electric current travels to lights, motors, or heaters and back to the source. To supply these circuits, electricity from a utility company's wires must pass through a meter and a service panel. A service panel distributes the power among the separate circuits. Lamps, television sets, and small appliances are connected to 120 volt, 15-ampere circuits. Appliances, such as a refrigerator, toaster, and power tools, are connected to 120 volt, 20-ampere circuits. Separate 240 volt, 30-ampere circuits are provided for a clothes dryer or an electric range.

Plumbing System

The *plumbing system* in a home is basically very simple. Potable (drinkable) water is brought into the house, **Figure 7-32**. It is piped directly to all the faucets and outlets, such as sinks, toilets, baths, and washing machine. It is also piped to the hot water heater. From this heater, heated water goes to all hot-water faucets, the washing machine, and the dishwasher. The used water is drained from the house and disposed of by a separate system.

Community resources and services: Have students visit local merchants and compare the cost of new gas, electric, and oil-fired furnaces.

Community resources and services: Have students investigate the electrical system in their community that enables electricity to be delivered to their homes.

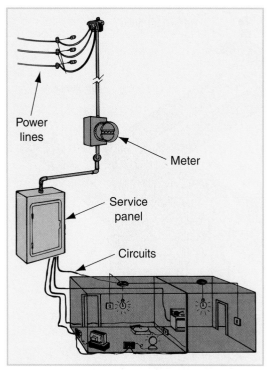

Figure 7-31 An electrical system has only a few key parts. Try to remember names of all the parts.

Communication System

The *communication systems* in your home include the telephone system, the radio and television broadcasting system, Internet connection, home PC network, and the cable or satellite television system.

Telephone service is provided to most homes by copper wires or fiber optic cables. A nationwide switching system enables your telephone to be connected to any other telephone. The same copper or fiber cables can be used for data transmission so that any home can have access to a national or international computer network.

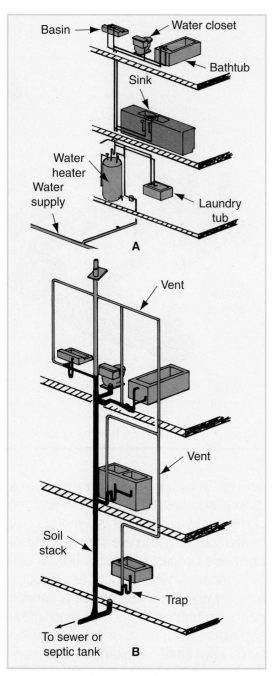

Figure 7-32 A—Water distribution system brings fresh water to different rooms of the home. B—Drain system carries away wastewater.

Links to other subjects: Social studies—Have students investigate the effects on populations who do not have an adequate supply of clean water.

Radio and television signals are received in each home. The programs may be broadcast from local stations. Satellites and antennas allow you to receive live and instantaneous radio and television coverage of events from around the world, **Figure 7-33**.

New Automated Systems

In the early 1970s, the microprocessor, often called a microchip, was developed. Complete electronic circuits were etched on a tiny slice of silicon. These "chips" became smaller and smaller and more and more powerful. The microchip is now so small that it can be embedded into any device. This enables automatic control, programming, and connection to just about anything. Think about the center of your home, the kitchen. A refrigerator could suggest what meal might be cooked based on what is stored inside and display the ingredients on its video screen. Cupboards could be designed so that when food is consumed, it is automatically reordered from an online grocer. Microwave ovens could scan the bar code on the food packaging and set themselves to the correct power level and cooking time. Such devices are not fantasies. They could be made right now with existing technologies. Imagine a Kitchen Command Center that would include a microwave with a flat computer screen enabling on-line shopping, banking, and e-mail!

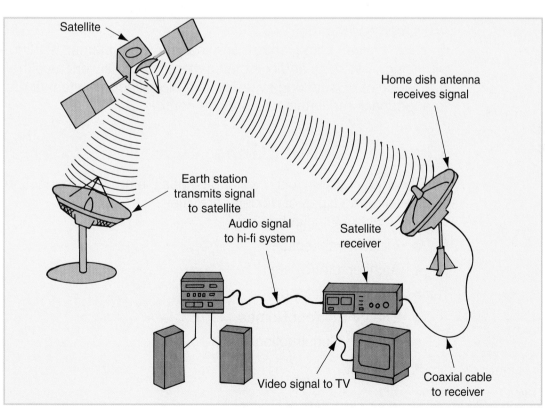

Satellite

Home dish antenna receives signal

Earth station transmits signal to satellite

Audio signal to hi-fi system

Satellite receiver

Video signal to TV

Coaxial cable to receiver

Figure 7-33 A telecommunication system can be linked to a satellite in space. Signals can be received in an instant in spite of great distances.

Useful web site: Have students visit the Intel Web site to investigate how computer chips are made at www.intel.com/education/makingchips/index.htm and how microprocessors work at www.intel.com/education/mpworks/index.htm.

Technology and society: How does the increase in automated communication systems affect people negatively?

Chapter 7 Review

Construction

Summary

The structure of a house must support the loads of the materials from which it is built plus its contents. It must also support the dynamic loads created by the weather conditions outside and the movement of people inside.

There are four types of single-family houses. These are bungalow, one-and-a-half-story, two-story, and split-level. Space within a house must accommodate all furniture and equipment and still leave space to move around. Rooms are grouped according to their function: living, sleeping, and working.

The first step in building a house is to check the subsoil conditions. Next, a contractor clears the site, establishes lines, and grades to ensure that the construction will be level. There are five major components of a house. These are the foundation, floor, walls, ceiling, and roof.

In a typical home, there are four major systems: heating, plumbing, electrical, and communication. These all work as either open-loop or closed-loop systems. In a closed-loop system, there is a feedback mechanism to provide control.

Modular Connections

The information in this chapter provides the required foundation for the following types of modular activities:

- ◆ Construction Technology
- ◆ Virtual Architecture
- ◆ Architectural Design
- ◆ Architectural Model
- ◆ Energy Efficient Homes
- ◆ Building Construction
- ◆ Masonry

- ◆ Plumbing
- ◆ Residential Wiring
- ◆ Wall Covering
- ◆ Construction Systems
- ◆ Electrical Systems
- ◆ Electronic Systems

Test Your Knowledge

Write your answers to these review questions on a separate sheet of paper.

1. List the materials used in the construction of the following different types of homes:
   A. Mexican house.
   B. Native American tent.
   C. Inuit igloo.
   D. Liberian hut.

2. What factors determined the use of a particular material when early settlers built their homes?

3. What characteristics do reinforced concrete buildings have in common with post and beam structures?

4. Sketch and name the parts of a post and lintel structure.

5. List two static and two dynamic loads that a typical house must withstand.

6. List four types of single-family houses.

7. How could you plan alternative arrangements of furniture in a room?

8. In the floor plan of a house, rooms are grouped according to their function.
   A. What are the three functions?
   B. State three reasons why rooms are grouped by function.

9. The five major structural components of a house are walls, floors, ceilings, foundation, and roof. List the order in which these are built on a construction site.

10. Describe the similarity between the footing of a foundation wall and a pair of snowshoes.

Answers to Test Your Knowledge Questions
(continued)

8. A. Living, sleeping, and working. B. To separate noisy areas from quiet areas. To position bedrooms and bathroom close together. To provide direct access between kitchen and dining area.

9. Foundation. Floors. Walls. Ceilings. Roof.

10. They both spread mass over an increased surface area to prevent sinking.

11. Refer to Figure 7-17 and Figure 7-20, pages 176 and 177.

12. sill, anchor bolts. joists. walls or partitions. insulation. plasterboard or wood paneling. framework for the roof.

13. Each has a series of parts connected together for a particular purpose. They are either of an open-loop or closed-loop type.

14. It continues to heat the room until it is switched off. There is no method of controlling whether there is too much or too little heat.

15. Refer to Figure 7-29, page 181. Closed-loop systems send a feedback to the input, which helps control the output. Open-loop systems do not have a feedback mechanism.

16. Heating. Electrical. Plumbing. Communication.

17. Student responses will vary.

11. Draw and label the parts of a house foundation.

12. Complete the following sentences:
    A. A foundation wall is connected to a(n) _____ by means of _____ _____.
    B. A subfloor is supported by a number of _____.
    C. Wall studs provide a framework for _____.
    D. Heat transmission is resisted by the use of _____.
    E. Interior wall surfaces are usually made of _____.
    F. The function of a roof truss is to form the _____.

13. What do all systems have in common?

14 Why is a portable space heater without a built-in thermostat an example of an open-loop system?

15. Using notes and diagrams, describe the difference between an open-loop system and a closed-loop system.

16. List the four major systems in a home.

17. Make a list of household devices that are not currently automated. Describe which of these might be automated and how their function could be changed.

Apply Your Knowledge

1. Collect five pictures to illustrate the different types of residential buildings. Name each type of building.

2. Make a simplified copy of Figure 7-15. Label the five major components and state the function of each.

3. Take a close look at the inside and outside of your home. List all the materials used in its construction.

4. Use the systems model in Figure 7-28 to describe how a spaghetti dinner is prepared.

5. Give one example of a device in your home that uses an open-loop system and one example that uses a closed-loop system. Draw a system diagram of each.

6. Construct a scale model of the floor and walls of one room in your home. Think of the furniture you would like to choose if the room were empty. Cut blocks of Styrofoam® or cardboard to represent the furniture. Position them in the room.

7. More and more new houses and apartments are "smart homes," loaded with appliances connected to the Internet. Which devices can be connected to a local network, and for what purpose?

8. Describe ways to bring existing homes up to R2000 standards, thereby making the homes more energy efficient?

9. Research one career related to the information you have studied in this chapter and state the following:
   A. The occupation you selected.
   B. The education requirements to enter this occupation.
   C. The possibilities of promotion to a higher level at a later date.
   D. What someone with this career does on a day-to-day basis.
   You might find this information on the Internet or in your library. If possible, interview a person who already works in this field to answer the four points. Finally, state why you might or might not be interested in pursuing this occupation when you finish school.

Chapter outline
Simple Machines
  Levers
    Mechanical Advantage
  Moments and Levers
  The Pulley
  The Wheel and Axle
  The Inclined Plane
    Calculating Inclined Planes
  The Wedge
  The Screw
Gears
Pressure
  Hydraulics and Pneumatics
Mechanism
Friction

Student Activity Manual Resource

Instructional Strategies and Student Learning Experiences
1. *Six Simple Machines*, reproducible master. Use this master as a basis of discussion about the six simple machines.
2. Have students design a machine to perform a simple task. Ask students to present their machines to the class.
3. Invite a mechanical engineer to class to discuss the principles behind the operation of various types of machines. Have students prepare a list of questions in advance.
4. Have students distinguish between the concepts of hydraulics and pneumatics. Ask them to give examples of how these are used in machines.
5. Have students explain why friction between moving parts must be minimized in most machines and how this can be accomplished.

Machines are designed to change or transmit energy. These gears change the speed and direction of rotating shafts.

Machines

Key Terms

friction	pneumatics
gear	power
hydraulics	pressure
inclined plane	pulley
lever	screw
linkage	torque
machine	wedge
mechanical advantage	wheel and axle
mechanism	work
moment	

Objectives

After reading this chapter you will be able to:

◆ Identify different types of simple machines.

◆ Recognize where these simple machines exist in everyday objects.

◆ List the advantages of simple machines.

◆ Identify systems having mechanical operations and explain their function.

◆ Describe the principle of hydraulic and pneumatic systems.

◆ Design and make a product that incorporates one or more mechanisms.

Discussion
Have students name machines they use every day that help them do work, make their lives safer, and help them enjoy their leisure time.

Career connection
Have students investigate the careers associated with the design and development and use of new machines. Students could focus their research on the careers associated with the production of a particular product. Alternatively, students could find out what products result from the work of people in one career sector.

Designing and making
Why is it important that designers make machines not just functional, but also reliable, pleasing to look at, and easy to service?

Reflection
What would your life be like if there were no machines?

Vocabulary
What is the definition of the term *machine*?

Vocabulary
What is a *shadoof*?

Safety
How do designers reduce the hazards posed by the use of machines?

Resource
Activity 8-1, *Building an Ancient Machine*

A *machine* is a device that does some kind of work by changing or transmitting energy. Our society relies heavily on machines. Machines help us to produce televisions in factories and food on the land. They help us to dig trenches and bore tunnels. Around the home, the automatic washing machine, the lawn-mower, and the vacuum cleaner lighten our workload. The dentist's drill removes decay from our teeth. The jet plane speeds us to our vacation site. Soon our whole houses will be automated and controlled by a single computer.

Machines do not need to be large or complicated. A knife, a bottle opener, and a claw hammer are also machines.

Machines have been used for thousands of years to make work easier. In ancient Egypt, a shadoof used a lever to raise a leather bucket filled with water, **Figure 8-1**.

Early civilizations used a *pulley*, **Figure 8-2**. A pulley is a simple machine made of a wheel with a groove for a rope, chain, or belt. It is used to change the direction of applied force, which aids us in lifting heavy objects. The first pulley may have been developed by throwing a rope over a tree branch in order to pull in a downward direction. A further development of the pulley was the windlass. This was often used to raise water from wells, **Figure 8-3**. People in ancient times used sections of tree trunks to move heavy loads. They found that it took less effort than sliding them, **Figure 8-4**. Later, this idea

Figure 8-1 The shadoof from ancient times used a lever and leather bucket to lift water to a trough. The trough carried the water to fields.

Figure 8-2 A pulley is designed to change the direction of applied force. As the woman pulls down on the rope, the pail moves upward.

Links to other subjects: History—Have students investigate how the invention of the pulley contributed to the development of ancient civilizations.

Useful Web site: Students can use the dictionaries at www.techtionary.com and www.hyperdictionary.com to find the definition of any new technical and technological terms used in this chapter.

Figure 8-3 The windlass is a wheel and axle machine. It is also somewhat like a pulley.

Figure 8-5 Over time, wheels became better made. Roman chariots used wheels with wooden spokes. How has the wheel changed since then?

led to the development of the wheel and axle, **Figure 8-5**.

Blocks of stone used to build the pyramids were raised to great heights by means of a ramp called an *inclined plane*, **Figure 8-6**. The inclined plane allowed lifting of heavy loads with minimal effort.

Two further developments of the inclined plane were the wedge and screw. See **Figures 8-7** and **8-8**. One of the most important uses of the wedge was the plow.

Simple Machines

Today, we refer to the lever, pulley, wheel and axle, inclined plane, wedge, and screw as simple machines. These

Standards for Technological Literacy

7

Discussion
Discuss with students the daily jobs around the home that are made easier by the use of machines.

Activity
Have students locate examples of machine diagrams from home (such as instructions provided by manufacturers with bicycles, appliances, and tools). Which are easiest to understand? What drawing techniques have illustrators used to make ideas clearer?

Safety
How has the concept of safety in the use of machines changed from ancient times to modern times?

Figure 8-4 What machines are used today to move a heavy load?

Links to other subjects: History—Have students investigate and write a report to explain how massive historical structures (such as the Egyptian pyramids or Stonehenge) were built.

Useful Web site: Have students investigate the mechanical drawings of Leonardo da Vinci at www.kfki.hu/~arthp/html/l/leonardo/12engine/index.html.

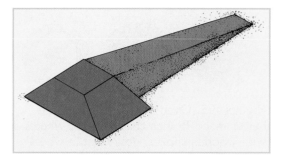

Figure 8-6 The Egyptian pyramids could not have been built without the inclined plane.

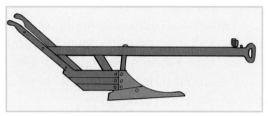

Figure 8-7 A plow is a type of inclined plane called a wedge.

Figure 8-8 Archimedes was a Greek inventor and mathematician. His invention could lift more water faster than any other method used at that time.

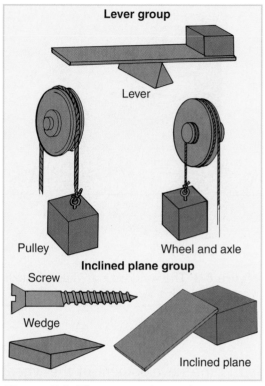

Figure 8-9 The six simple machines were important inventions.

six simple machines can be divided into two groups. One is based on the lever and the other on the inclined plane, **Figure 8-9**.

Levers

You have probably played or seen a game of baseball, **Figure 8-10**. A baseball bat is a lever. A *lever* has a fulcrum, effort, and resistance. The fulcrum is the point where the bat is held. The batter's muscles supply the effort, and the resistance is the ball, **Figure 8-11**.

To understand the principle of the lever, look at the boy in **Figure 8-12**. He is using a branch to move a heavy rock. The branch is the lever. The mass of the rock is the resistance (R). The boy's muscle power pushing down on the lever provides the effort (E). The rock on which the lever is pivoting is the fulcrum (F). These three elements—

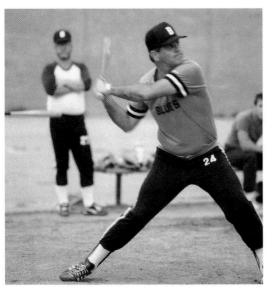

Figure 8-10 Baseball players use a bat as a lever to strike the baseball with greater speed. (TEC)

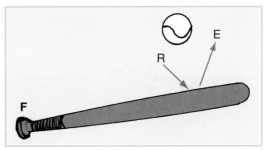

Figure 8-11 The baseball bat is an example of a lever. F-Fulcrum R-Resistance E-Effort.

Figure 8-12 This is an example of a lever being used to move a heavy load.

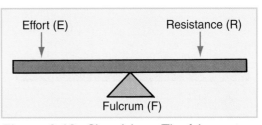

Figure 8-13 Class 1 lever. The fulcrum is between effort and resistance.

Discussion
What sports equipment other than a baseball bat makes use of a lever?

Vocabulary
Name the three elements common to all levers.

Resource
Transparency 8-1, *The Six Simple Machines*

Resource
Reproducible Master, *The Six Simple Machines*

resistance, effort, and fulcrum—are always present in a lever. However, they can be arranged in different ways to create three different classes of levers.

In a Class 1 lever, the fulcrum is placed between the effort and the resistance, **Figure 8-13**. Some applications of Class 1 levers are illustrated in **Figure 8-14**.

In Class 2 levers, the resistance is placed between the effort and the fulcrum, **Figure 8-15**. Some applications of class 2 levers are illustrated in **Figure 8-16**.

In Class 3 levers, the effort is applied between the resistance and the fulcrum, **Figure 8-17**. Some applications of Class 3 levers are illustrated in **Figure 8-18**.

From the many examples shown, you can see that some levers are designed to increase the force available. Examples are a wheelbarrow, a bar used to move a crate, and a garden spade. Other levers are designed to increase the distance a force moves or the speed at which it moves. Examples are a fishing rod and a human arm.

Useful Web site: Have students visit www.btinternet.com/~hognosesam/bitsofstuff/bitsofstuff/levers.htm and watch animations of levers in action.

Designing and making
What tasks do you complete in the home that make use of the mechanical advantage provided by a lever?

Discussion
Have students identify examples where machines are used to multiply force.

Enrichment
Have students investigate how cantilever brakes on a bicycle use the concept of mechanical advantage at www.sheldonbrown. com/cantilever-geometry.html.

Rowing with oars

Claw hammer

Scales

Lifting a crate

Scissors

Seesaw

Figure 8-14 These are familiar examples of the use of Class 1 levers.

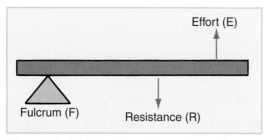

Effort (E)

Fulcrum (F)   Resistance (R)

Figure 8-15 Here is an illustration of a Class 2 lever. Do you see how it is different from a Class 1 lever?

Mechanical Advantage

When a small effort is applied to a lever to move a large resistance, there is obviously an advantage. This is called the *mechanical advantage* of the lever.

Mechanical advantage is equal to the resistance divided by the effort. The greater the resistance that can be moved for a given effort, the greater the mechanical advantage is. The formula is:

$$\text{Mechanical Advantage (M.A.)} = \frac{\text{Resistance}}{\text{Effort}}$$

For example, if a lever can make it possible to overcome a resistance of 90 newtons (N) when an effort of 30 N is applied, the mechanical advantage

Links to other subjects: Social studies—Have students investigate nonmechanized methods of moving large quantities of topsoil in Third World countries.

Technology and society: How has the invention of ever-larger earth moving machines affected the natural environment?

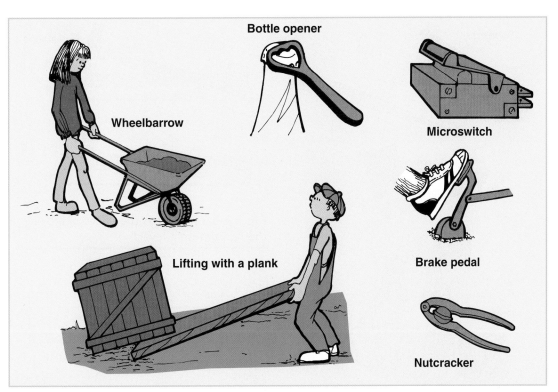

Bottle opener

Wheelbarrow

Microswitch

Lifting with a plank

Brake pedal

Nutcracker

Figure 8-16 Do you recognize these examples of Class 2 levers?

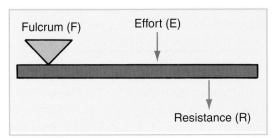

Fulcrum (F)   Effort (E)

Resistance (R)

Figure 8-17 How is a Class 3 lever different from a Class 1 lever?

will be 3. The newton (N) is the metric unit of force or effort. The formula is:

$$\text{M.A.} = \frac{90N}{30N} = 3$$

In other words, the human effort applied is being multiplied by the machine. In this case, it is the lever. The effort required becomes less as the fulcrum and resistance are brought closer together. However, the

disadvantage is that as the fulcrum and resistance are moved closer, the load moves a shorter distance, **Figure 8-19**.

Moments and Levers

Imagine a lever with a fulcrum in the middle. On one side is an effort and on the other a resistance. When at rest, this lever is said to be balanced. If the effort is increased, the lever will turn in a counterclockwise direction. If the resistance is increased, the lever will turn in a clockwise direction, **Figure 8-20**. The turning force is called a *moment*.

The moment depends on the following two things: the effort and the distance of the effort from the fulcrum.

Moment = Effort × Distance

Links to other subjects: History—Have students search the Internet to gather information about how people used levers to erect the statues on Easter Island.

Vocabulary
What is the definition of the word *linkage*? Explain the similarities in a "link" between two people and a mechanical "link."

Discussion
How are the linkages in a mechanical system similar to the human body?

Activity
Make a list of products in your home that include one or more linkages.

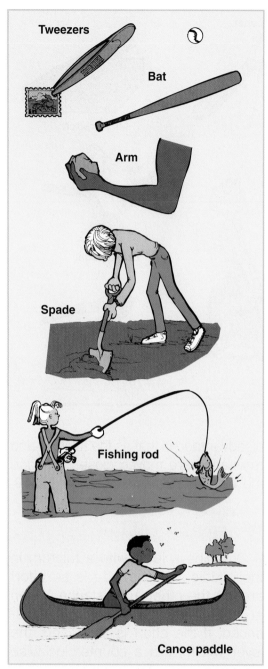

Figure 8-18 No doubt you have used some, if not all, of these types of Class 3 levers.

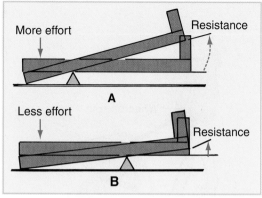

Figure 8-19 What is the effect of moving the fulcrum on a lever?

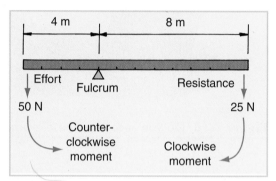

Figure 8-20 Forces acting on a lever are called moments.

Levers discussed so far have been used to increase force, distance moved, or speed. Levers can also be used to reverse the direction of motion.

Think of a lever with a fulcrum in the center. If it pivots about its fulcrum, the ends move in opposite directions. One end moves down, and the other end moves up, **Figure 8-21**. A single lever with a pivot in the center reverses an input motion.

This idea is used in linkages. A *linkage* is a system of levers used to transmit motion. **Figure 8-22** illustrates a reverse motion linkage.

If a beam is in balance, the clockwise moments are equal to the counterclockwise moments.

$$4 \times 50 = 8 \times 25$$

Links to other subjects: How is the term *linkage* used in science? How is it used in history? How do these relate to the term's meaning in technology education?

Useful Web site: Have students work through the interactive demonstrations on pulleys at www.flying-pig.com/mechanisms/pages/linkage.html and www.technologystudent.com/cams/link1.htm.

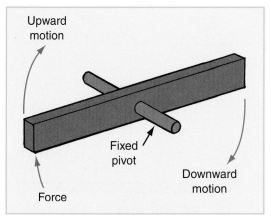

Figure 8-21 Some levers are designed to change motion.

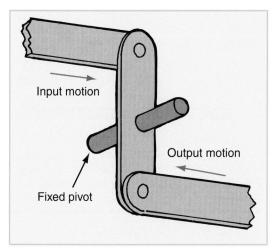

Figure 8-22 Look at this example of a reverse motion linkage. The pivot is fixed at the center of one lever. Input force equals output force.

The input force and output force are equal. If the pivot is not at the center, the input force is increased or decreased at the output. This is shown in **Figure 8-23**.

The Pulley

The pulley is a special kind of Class 1 lever, **Figure 8-24** and **Figure 8-25**. Its action is continuous. The resistance

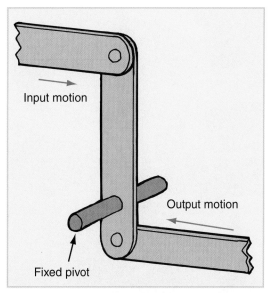

Figure 8-23 In this example, is the input force increased or decreased at the output?

arm is the same length as the effort arm. The length of each arm is the radius of the pulley, **Figure 8-26**.

Pulleys are used for lifting heavy objects. A bale of hay can be lifted into a hayloft using a single pulley suspended from a beam. Car engine hoists enable one person to lift a car engine having a mass of over 450 lb. (200 kg). Cranes use pulleys to lift enormous loads.

Figure 8-24 The simple pulley changes direction once.

Discussion
What is the advantage of using a single fixed pulley to lift a mass?

Vocabulary
Explain the difference between a *fixed pulley* and a *movable pulley.* Use diagrams and notes if necessary.

Safety
What hazards may be present when pulleys are used?

Links to other subjects: History—Investigate the use of block and tackle gear on old sailing ships.

Useful Web site: Have students work through the interactive demonstrations on pulleys at www.flying-pig.com/mechanisms/pages/pulley.html.

Vocabulary
What is the difference between a *simple pulley* and a *block and tackle*?

Safety
What risk management strategies should be observed when using a pulley or block and tackle to lift a heavy object?

Figure 8-25 A chain hoist is a pulley designed to lift heavy loads. (Ecritek)

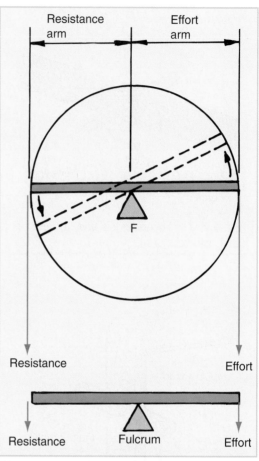

Figure 8-26 Do you see why a pulley is a special kind of Class I lever?

Two types of pulleys are used to lift heavy objects. One is called fixed, and the other is called movable.

In a single, fixed pulley system, **Figure 8-27**, the effort is equal to the resistance. There is no mechanical advantage. It is easier, however, for the operator to pull down instead of up. There has been a change in direction of force. The distance moved by the effort (effort distance) is equal to the distance moved by the resistance (resistance distance).

A single, movable pulley system has a mechanical advantage of two. Both ropes support the resistance equally. The amount of effort required is half that of the resistance. The disadvantage is that the operator must pull upward. Also the effort distance is two times the resistance distance. In all pulley systems, as the effort decreases, the effort distance increases, **Figure 8-28**.

To have the advantages of change of direction and decreased effort, movable and fixed pulleys can be combined as shown in **Figure 8-29**. In both examples, the mechanical advantage is two. However, the effort must be exerted over twice the distance.

Links to other subjects: Science—What is the difference between what you learn about pulleys in science classes and what you learn about pulleys in technology classes? Why are they both important?

Useful Web site: Have students work through the interactive demonstrations on pulleys at www.phy.ntnu.edu.tw/java/wheelAxle/pulley.html.

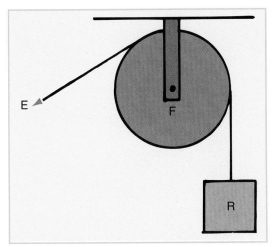

Figure 8-27 In a single fixed pulley, the effort is equal to the resistance, and there is no mechanical advantage.

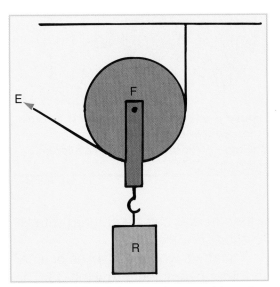

Figure 8-28 With a single, movable pulley, effort equals half the resistance and the effort moves twice the distance of the resistance.

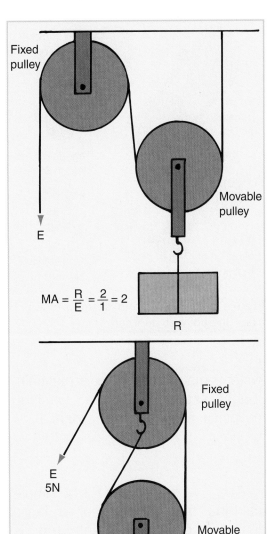

$$MA = \frac{R}{E} = \frac{2}{1} = 2$$

$$MA = \frac{R}{E} = \frac{10}{5} = 2$$

Figure 8-29 A fixed pulley changes direction of the effort. A movable pulley decreases effort.

Pulleys may also be used to transmit motion, increase or decrease speed, reverse the direction of motion, or change motion through 90°, **Figure 8-30**. These types of pulley

Discussion
Which products used in your home incorporate one or more wheels and axles?

Enrichment
Have students investigate the design of the original Ferris Wheel at http://users.vnet.net/ schulman/Columbian/ ferris.html and compare it to the design of the London Eye at www.londoneye.com.

Safety
What hazards and risks are present in the use of wheels and axles?

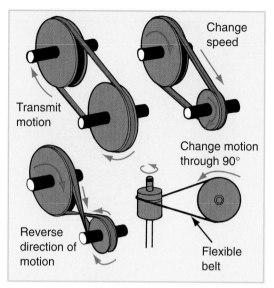

Figure 8-30 Pulleys can transmit motion from one point to another.

Figure 8-31 Top—A doorknob is a common example of the wheel and axle. Bottom—Where are the effort, fulcrum, and resistance?

systems may be used in cars (fan belt), upright vacuum cleaners, washing machines, and electrical appliances.

The Wheel and Axle

The *wheel and axle* is a simple machine that consists of a large diameter disk or wheel that is attached rigidly to a smaller diameter bar (axle). Effort applied to the outer edge of the wheel is transmitted through the axle. Think of it as a special kind of Class 1 lever, **Figure 8-31**.

Like levers, the wheel and axle contains three elements: effort, fulcrum, and resistance. In the case of the doorknob, the effort is applied to the rim of the wheel (knob). The knob multiplies the effort and transmits it through the axle (bar). The resistance is the door latch.

A similar application is a car steering wheel. The driver's effort, applied to the wheel, is increased. The result

is that the car can be steered with little effort, **Figure 8-32**.

The drive wheel and axles of a car are other examples, **Figure 8-33**. The effort is supplied by the engine through the axle to the circumference (rim) of the wheel. The resistance is the force of the road on the wheel.

The Inclined Plane

There are two ways of loading the motorcycle onto the truck, **Figure 8-34**. One way is simply to lift it. However, it is much easier and

Useful Web site: Have students work through the interactive demonstration on the wheel and axle at www.flying-pig.com/mechanisms/pages/wheel.html and www.usoe.k12.ut.us/curr/science/sciber00/8th/machines/sciber/machine7.htm.

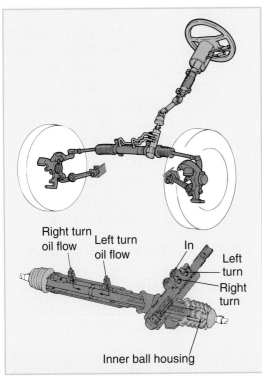

Figure 8-32 A car's steering wheel multiplies the effort applied by the driver to steer the vehicle.

safer to push the motorcycle up a sloping plank. The sloping surface formed by the plank is called an *inclined plane* or ramp.

There are many uses for an inclined plane. Many mountain roads zigzag up the mountainside rather than running straight up to the top.

Common uses are shown in **Figure 8-35** through **Figure 8-37**:

Figure 8-33 Can you trace the path of the effort from the engine to the rim of the wheel?

Discussion
Give examples of inclined planes in objects used every day.

Vocabulary
What is the difference between the *mass* of an object and its *weight*?

Safety
What hazards and risks are involved in using an inclined plane? How can these risks be managed?

Resource
Activity 8-4, *Applications of the Inclined Plane*

Figure 8-34 Left—It would be impossible for one person to lift the motorcycle onto the truck. Right—One person could push it up a ramp onto the truck. Would a shorter ramp make the job harder or easier? (Ecritek)

Links to other subjects: History—Have students search the Internet to gather information about how ancient Egyptians used inclined planes when building the pyramids.

◆ Loading and unloading a car transporter.

◆ Reaching different levels of a multistory parking lot.

◆ Replacing steps so persons with strollers or wheelchairs have access.

Calculating Inclined Planes

How do people who build ramps decide about their length? Guessing about it is too costly. They might have to tear down the ramp and start over. Therefore, they use mathematics.

In general, the length of a ramp depends on the mass of the object to be moved. Heavier objects need longer ramps or more effort to move them upward. The more gradual the rise, the less effort required. We can say the length of the inclined plane is directly related to the mass of the object. This means that as mass increases, ramp length increases. **Figure 8-36** shows this relationship. So does the following formula:

$$\text{Effort (E)} \times \text{Effort Distance (ED)}$$
$$= \text{Resistance (R)} \times \text{Resistance Distance (RD)}$$

Figure 8-35 We use inclined planes for many purposes. Top left—Unloading cars from a transporter. Bottom left—Parking cars on different levels of buildings. Right—Helping persons in strollers or wheelchairs move from one level to another. (Ecritek)

Now, let us see how we might use the formula to work out a problem. Suppose, for example, that a mass of 450 lb. (200 kg) needs to be raised 6′ (2 m). Then suppose that the most force that can be exerted is 100 lb. (50 kg). How long must the ramp be so the force is able to move the object?

$$ED = R \times \frac{RD}{E} =$$

$$450 \times \frac{6}{100} = \frac{2700}{100} = 27'$$

$$ED = R \times \frac{RD}{E} =$$

$$200 \times \frac{2}{50} = \frac{400}{50} = 8m$$

The Wedge

The *wedge* is a special version of the inclined plane, **Figure 8-37**. It is two inclined planes back-to-back, **Figure 8-38**.

The shape is effective because the force exerted pushes out in two directions as it enters the object, **Figure 8-39**. Do you see the difference between the wedge, **Figure 8-38**, and the inclined plane in **Figure 8-36**?

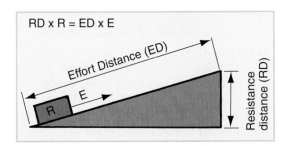

Figure 8-36 The mass of an object and the distance it is raised are related to the length of the inclined plane.

Figure 8-37 An axe is a wedge, which is a special kind of inclined plane. However, its uses are much different from the inclined plane.

Figure 8-38 A wedge is like two inclined planes joined together.

Figure 8-39 Because of its shape, an axe enters material easily and can split it apart.

Discussion
Which products used in your home incorporate one or more wedges?

Safety
Are there any safety hazards involved in using a wedge? How can the risks be managed?

Resource
Transparency 8-2, *Formulas for Levers and Inclined Planes*

Resource
Reproducible Master, *Formulas for Levers and Inclined Planes*

Links to other subjects: History—Have students investigate how the invention of the plow contributed to the development of ancient civilizations in Egypt, Greece, and Mesopotamia.

Discussion
Which products in your home make use of one or more screws?

Activity
Have students make a card model to show how a screw thread is a modified inclined plane.

Safety
Are there hazards and risks involved in using screws? What safety precautions do you take when using screws?

Other applications of the wedge include the plow, a doorstop, the blade of a knife, and the prow of a boat, **Figure 8-40**.

The Screw

It is not hard to find examples of how we use the screw. When you see someone using a scissor jack to lift a car, you are seeing a screw at work, **Figure 8-41** and **Figure 8-42**.

A *screw* is an inclined plane wrapped in the form of a cylinder. To illustrate how this works, take a rectangular piece of paper and cut it along a diagonal. This triangle will remind you of an inclined plane. Now wrap it around a pencil. Roll from the short edge of the triangle to shape it like a screw, **Figure 8-43**.

When we examine the inclined plane, we find that the time taken for the load to be pushed up a longer inclined plane increases while the effort decreases. The same is true in the case of the screw threads on a nut and bolt. The greater the number of threads, that is, the shallower the slope, the longer it takes to move the nut to the head of the bolt. However, it will be easier to move the nut against a resistance.

The wedge-shaped section of a tapering wood screw reveals another application of the wedge, **Figure 8-44**. It allows the screw to force its way into the wood.

Screw threads may be used in two quite different ways. They may be used to fasten as with wood screws, machine screws, and lightbulbs. They may also transmit motion and force.

Figure 8-40 We use the wedge in many products. Top—Wedge-shaped prow (front) of a boat cuts through the water easily. (Carnival Corporation) Middle—The wedge used in this hydraulic splitter is like two inclined planes joined together. Bottom—Bucket of a backhoe is wedge-shaped to enter soil more easily. (Hewitt)

Links to other subjects: Social studies—Have students write an illustrated report to show how an Archimedes' water screw is used for irrigation. The following Web sites provide useful starting points: www.wikipedia.org/wiki/Archimedes'_screw and www.mcs.Drexel.edu/~crorres/Archimedes/Screw/ScrewEngraving.html.

Figure 8-41 A scissor jack lifts heavy loads with less effort applied. (Christopharo)

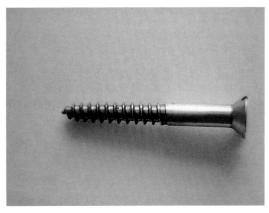

Figure 8-44 A wood screw also has a wedge shape to push aside the wood fibers as it enters the wood.

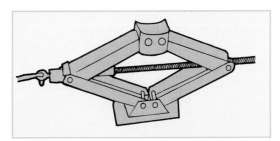

Figure 8-42 As the screw of the jack is turned, the two ends of the jack are forced together and the car is raised.

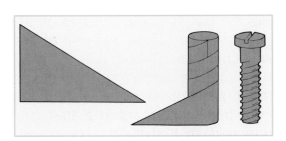

Figure 8-43 A screw is an inclined plane wrapped into a cylinder.

Examples are C-clamps, vises, and car jacks. Note also that a screw converts rotary motion into straight-line motion, **Figure 8-45**.

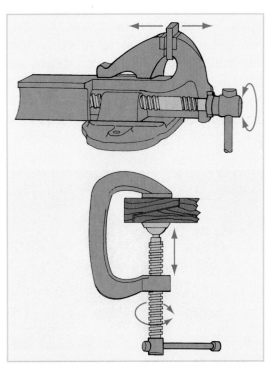

Figure 8-45 Screw threads are used on vises and C-clamps. They convert rotary motion to straight line motion.

Standards for Technological Literacy

7

Discussion
What common household products that use gears can you name?

Reflection
How many gears do you have on your bicycle? Do you use them all?

Safety
What hazards and risks are present when gears are being used? How can these risks be managed?

Resource
Activity 8-3, *Identifying Simple Machines*

Useful Web site: Have students work through the interactive demonstrations on gears at www.flying-pig.com/mechanisms/pages/gears.html.

Gears

Gears, Figure 8-46 and Figure 8-47, are *not* classified as one of the six simple machines. They are similar to pulleys in that their motion is usually circular and continuous. Gears have an advantage over pulleys. They cannot slip.

Many bicycles have gears to help make pedaling as effortless as possible. When climbing hills, the cyclist selects a low gear to make pedaling easier. When descending a hill, a higher gear provides a high speed in return for slower pedaling.

A mechanical clock contains many different sized gear wheels. They are arranged so that they rotate the clock hands at different speeds.

Gears, like pulleys, are modified levers. They transmit rotary motion. They increase or decrease speed, change the direction of motion, and transmit a force. This force, known as *torque*, acts at a distance from the center of rotation. To understand this concept more easily, think of torque as

Figure 8-46 In what ways are gears similar to pulleys?

Figure 8-47 Gears cannot slip because of the teeth. (Ecritek)

a measure of turning effort. It is similar to the use of a wrench to tighten a bolt, Figure 8-48 and Figure 8-49.

An effort (E) is applied at a distance (R) from the center of the nut. The torque on the nut is calculated by multiplying the effort (E) by the distance (R). The applied force is measured in pounds (newtons) and the distance from the center of rotation is measured in feet (meters). Therefore, torque is measured in ft.-lb. (N-m).

As you look at a gear, you can consider the center of the gear to be like the nut. Consider the end of the gear tooth to be like the end of the wrench, Figure 8-49.

Gears are used in groups of two or more. A group of gears is called a gear train. The gears in a train are arranged so that their teeth closely interlock or mesh. The teeth on gears that mesh are the same size and of equal strength. The spacing of the teeth on each gear is the same.

When two gears are of the same size mesh, they act as a simple

---

Useful Web site: Have students investigate torque specifications for automobiles sold in the US at www.torquespecs.com.

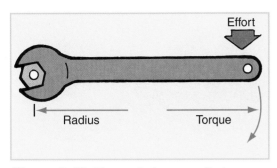

Figure 8-48 Torque is force applied to a radius.

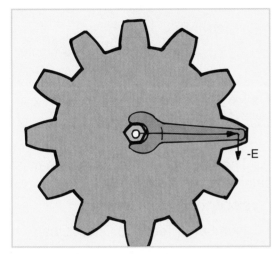

Figure 8-49 Compare a gear to a wrench turning a nut.

The amount of torque delivered by a gear is described as a ratio. For example, suppose that a gear of 10 teeth meshes with a gear of 30 teeth. The small gear will make three revolutions for each revolution of the larger gear. As the small gear makes one revolution, its 10 teeth will have meshed with 10 teeth on the larger gear. The large gear will have turned through ten thirtieths or one third of a revolution. The small gear will have to make three revolutions to turn the large gear through one revolution. The gear ratio is, therefore, 3:1, **Figure 8-52**.

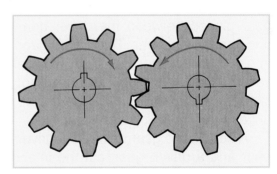

Figure 8-50 On a gear train, as one gear moves, its torque is transmitted to another gear.

torque transmitter. They both turn at the same speed but in opposite directions, **Figure 8-50**. The input motion and force are applied to the driven gear. The driven gear transmits the output motion and forces.

When two gears are of different size mesh, they act as torque converters, **Figure 8-51**. The larger gear is called a wheel. The smaller gear is called a pinion. The pinion gear revolves faster, but the wheel delivers more force.

Figure 8-51 Two gears act like levers to convert torque.

Standards for Technological Literacy

7

Discussion
Which products used in your home incorporate a simple gear train?

Vocabulary
What is the difference between a *wheel gear* and a *pinion gear*?

Activity
Have students use thin card to make a model of a simple gear train.

Links to other subjects: Mathematics—Have students calculate the gear ratio of various combinations of gear wheels.

Useful Web site: Have students work through the interactive demonstrations on gear trains at www.technologystudent.com/gears1/gears2.htm.

Figure 8-52 Here is a gear train with a 3:1 ratio. If the smaller gear is driving the larger one, the input torque will be multiplied by three. (TEC)

Gear trains are either simple or compound. In a simple gear train, there is only one meshed gear on each shaft. **Figure 8-53** shows an idler gear placed between the driver gear and the driven gear. The driver gear and the driven gear now rotate in the same direction. The idler gear does not change the gear ratio between the driver gear and the driven gear.

A compound gear train has a driver gear and a driven gear, but the intermediate gears are fixed together on one common shaft, **Figure 8-53**. The gear wheels on the intermediate shaft are not idlers, for one is a driven gear and the other is a driver gear. They do affect the ratio of the gear train.

Just like the six simple machines, gears provide mechanical advantage. This advantage is calculated in the following way:

Mechanical Advantage (MA) =

Number of Teeth on Driven Gear
Number of Teeth on Driver Gear

Figure 8-53 Here are two compound gear trains. (TEC)

The velocity of the driven gear is calculated as follows:

Velocity of Driven Gear

= Number of Teeth on Driver Gear

× Velocity of Driver Gear / Number of Teeth on Driven Gear

An example of these calculations for the gears is shown in **Figure 8-54**.

Gears are designed in a variety of types for a variety of purposes. The five most common gear types are spur, helical, worm, bevel, and rack-and-pinion.

The spur gear is the simplest and most fundamental gear design, **Figure 8-55**. Its teeth are cut parallel to the center axis of the gear. The strength of spur gears is no greater

Links to other subjects: Mathematics—Have students calculate the mechanical advantage of various compound gear trains.

Useful Web site: Have students work through the interactive demonstrations on compound gear trains at www.technologystudent.com/gears1/gears3.htm.

$$MA = \frac{30}{15} = \frac{2}{1} = 2$$

$$Velocity = \frac{15 \times 60}{30} = 30 \text{ rpm}$$

Figure 8-54 You can calculate the mechanical advantage and velocity of compound gears by using the formula above. (TEC)

than the strength of an individual tooth. Only one tooth is in mesh at any given time.

To overcome this weakness, helical gears are sometimes used, **Figure 8-56**. Since the teeth on helical gears are cut at an angle, more than one tooth is in contact at a time. The increased contact allows more force to be transmitted.

Figure 8-55 On spur gears, the teeth are cut straight across the width of the gears.

A worm is a gear with only one tooth. The tooth is shaped like a screw thread. A wormwheel meshes with the worm, **Figure 8-57**. The wormwheel is a helical gear with teeth inclined so that they can engage with the threadlike worm. This system changes the direction of motion through 90°. It also has the ability to make major changes in mechanical advantage (M.A.) and speed. Input into the worm gear system is usually through the worm gear. A high M.A. is possible because the helical gear advances only one tooth for each complete revolution of the worm gear. The worm gear in **Figure 8-57** will rotate 40 times to turn the helical gear only once. This is an M.A. of 40:1. Worm gear mechanisms are very quiet running.

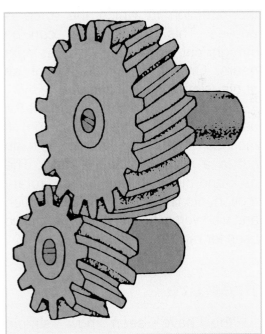

Figure 8-56 Helical gears, cut at an angle, allow several teeth to engage at one time.

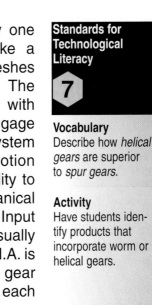

Useful Web site: Have students work through the interactive demonstrations on worm gears at www.flying-pig.com/mechanisms/pages/worm.html.

Standards for Technological Literacy

Vocabulary
What is the advantage of *bevel gears* over *spur gears*?

Activity
Have students identify products that incorporate bevel gears.

Activity
Have students identify products that incorporate a rack and pinion.

Figure 8-57 Here is a worm and wormwheel. The worm, right, has only one tooth. It spirals like a screw thread. (Franke)

Bevel gears, **Figure 8-58**, change the direction in which the force is applied. This type of gear can be straight cut in the same way as spur gears. Or, they may be cut at an angle, similar to helical gears.

Rack-and-pinion gears, **Figure 8-59**, use a round spur gear (the pinion). It meshes with a spur gear that has teeth cut in a straight line (the rack). The rack-and-pinion transforms rotary motion into linear (straight-line) motion and vice versa. **Figure 8-60** shows two uses for rack-and-pinion gears.

Pressure

You have seen how simple machines and gears are able to move a greater resistance with a smaller

Figure 8-58 Top—Bevel gears are designed to change the direction of the applied force. Bottom—A hand drill uses bevel gears.

effort. Now you will see that pressure can increase the effort applied. **Figure 8-61A** shows a diagram of a simple system for multiplying force through use of pressure. **Figure 8-61B** shows a simple pressure device used to lift an automobile.

Useful Web site: Have students work through the interactive demonstrations on gears at www.flying-pig.com/mechanisms/pages/rackandpinion.html and www.flying-pig.com/mechanisms/pages/bevel.html.

Useful Web site: Have students work through the interactive demonstrations on rack-and-pinion-gears at www.technologystudent.com/gears1/gears4.htm.

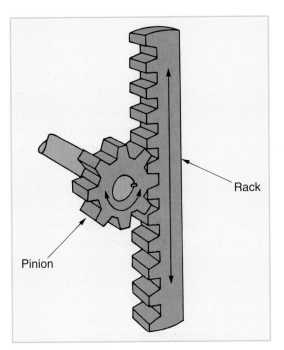

Figure 8-59 Rack-and-pinion gears convert rotary motion to straight-line motion.

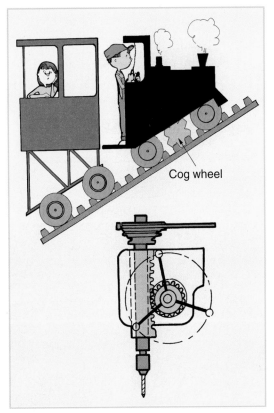

Figure 8-60 An inclined railroad and a feed mechanism for a drill press are both good examples of rack-and-pinion gears.

Discussion
What is the effect of air pressure on the human body? A useful starting point for students is http://kids.earth.nasa.gov/archive/air_pressure/.

Enrichment
Have students research and prepare an oral report on Blaise Pascal.

Vocabulary
What does the term *barometric pressure* mean?

To understand how pressure can be increased, think about the area on which the effort presses. Would you rather a woman step on your foot with a small, pointed heel or with a larger, flat heel, **Figure 8-62**?

When a surface area is small, a little effort produces a large pressure. *Pressure* is the effort applied to a given area. To calculate, divide effort by area. The formula is:

$$\text{Pressure} = \frac{\text{Effort}}{\text{Area}}$$

For example, if a 120 lb. woman rests her mass on a 4in.² heel, the pressure is 30 psi. On the other hand, if she rests her mass on a 1/4 in.² heel, the pressure increases to 480 psi.

In the metric system, effort is measured in newtons. Area is measured in meters squared (m²).

Pressure is calculated in newtons per meter squared (N/m²). The metric unit of measure for pressure is the pascal (Pa). Since a pascal is small, kilopascals are generally used (1 kPa = 1000 Pa).

Consider another example of how area affects pressure. A knife has a sharp edge. Pressed against a surface, it takes up a very small area. That is why it cuts: the material offers little resistance to such a tiny surface. A dinner fork works in a similar way. The narrow prongs place enough pressure on the food to pierce it easily.

Links to other subjects: Language—Have students discuss how the meaning of the term *pressure* when used in science and technology is different from the use of the term in everyday use. Social studies—Have students discuss the effects of positive and negative peer pressure.

Useful Web site: Have students investigate how atmospheric pressure is used in vacuum forming plastics. A useful starting point www.technologystudent.com/equip1/vacform1.htm.

Discussion
Which products used in your home make use of hydraulics or pneumatics?

Vocabulary
What is the difference between *hydraulic* and *pneumatic*?

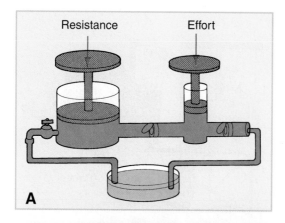

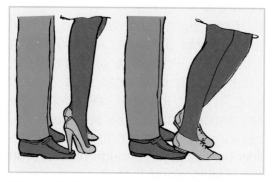

Figure 8-61 Hydraulics use pressure to increase force. A—In a hydraulic system, a fluid transmits effort from where it is applied to where it is used. B—A hydraulic lift is used to support the mass (weight) of an automobile.

Figure 8-62 Imagine someone stepping on your foot. Which will hurt more, the pointed heel or the flat heel?

Hydraulics and Pneumatics

The study and technology of the actions and reactions of liquids at rest and in motion is called *hydraulics*. *Pneumatics* is the study and technology of the characteristics of gases. Unlike solids, liquids and gases flow freely in all directions. Pressure, therefore, can be transmitted in all directions. For example, water will flow in a garden hose even when the hose is bent in many directions.

Figure 8-63 shows a model of a hydraulic lift. The effort is being applied to a piston. Pressure produced by the effort is being transmitted by the liquid to a second piston. This piston moves the resistance. The second piston has a larger area and so the pressure presses on a larger area. This produces a larger effort. If the resistance piston has four times the area of the effort piston, the effort on it will be four times greater.

From this example, you can see that the effort acting on a piston from a liquid under pressure depends on the area of the piston. The larger the area is, the larger the effort will be. However, the distance moved by the larger piston will be less than the distance moved by the smaller piston. In **Figure 8-63**, the smaller piston moves four times the distance of the larger piston.

The hydraulic brake system on cars operates using the same principles as the hydraulic lift, **Figure 8-64**. Using the brake pedal, a car driver applies a small effort to the piston. Hydraulic fluid is transmitted through the brake lines to a larger piston. The

Useful Web site: Have students work through the interactive demonstrations on pistons at www.flying-pig.co.uk/mechanisms/pages/piston.html.

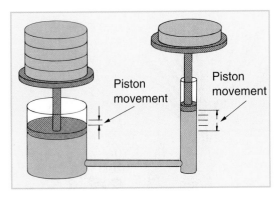

Figure 8-63 This is a model of a hydraulic lift. Large movement in one piston will create a smaller movement in the other piston. Why?

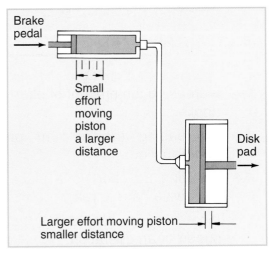

Figure 8-64 This simplified diagram of an automobile brake system is a good example of a hydraulics system.

larger piston forces brake pads and shoes against the disks and drums. The brake systems of large trucks, buses, and trains are often pneumatically operated.

Among the many common applications of hydraulic power are dentist and barber chairs, door closers, and power steering. Common applications of pneumatic power include a variety of tools such as air drills, screwdrivers, and jackhammers. Sometimes hydraulic and pneumatic systems are combined. For example, air pressure forces hydraulic fluid to raise the lift in a garage, **Figure 8-61B**.

Because of their many advantages, most industries use hydraulic and pneumatic systems. These advantages include the ability to:

◆ Multiply a force using minimal space.

◆ Transmit power to wherever pipe, hose, or tubing can be located.

◆ Transmit motion rapidly and smoothly.

◆ Operate with less breakage than occurs with mechanical parts.

◆ Transmit effort over considerable distance with relatively small loss.

Mechanism

A *mechanism* is a way of changing one kind of effort into another kind of effort, **Figure 8-65**. For example, a C-clamp holds two pieces of wood together while glue sets. Rotary motion is changed to linear motion to apply pressure, **Figure 8-66**.

Mechanisms can be combined to form machines. Their advantages include:

◆ Changing the direction of an effort.

Figure 8-65 How many simple machines make up this Rube Goldberg mechanism?

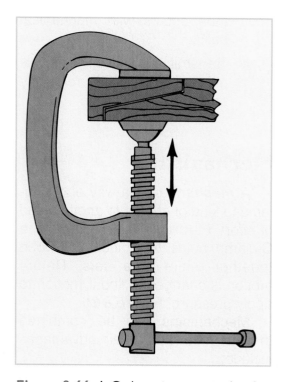

Figure 8-66 A C-clamp is an example of a mechanism.

♦ Increasing the amount of effort applied.

♦ Decreasing the amount of effort applied.

♦ Applying an effort to a place otherwise hard to reach.

♦ Increasing or decreasing the speed of an operation.

Machines change one kind of energy into another and do work. The amount of *work* done is approximately equal to the amount of energy changed.

Work = (Energy Change)
= Effort × Distance Moved in Direction of Effort

For example, how much work is done to move a 50 lb. resistance a distance of 5′?

Work = 50 × 5
= 250 ft.-lb.

Useful Web site: Have students research the work of cartoonist Rube Goldberg and examine his "inventions" at www.rube-goldberg.com.

In metric, how much work would be done to move a 50 N resistance through 4 meters?

Work = 50 × 4 = 200 joules

Machines make it easier to do work. However, no machine does as much work as the energy put into it. If a machine did the same amount of work as the energy supplied, it would be 100% efficient. Most machines lose energy as heat or light. Approximate efficiencies of some common machines are listed in **Figure 8-67**.

Another important term associated with work is power. *Power* is the rate at which work is done or the rate at which energy is converted from one form to another or transferred from one place to another. It is basically work along with the time it takes to do that work. Power can be expressed in this formula:

$$\text{Power} = \frac{\text{Work}}{\text{Time}} \qquad P = \frac{W}{T}$$

Mechanisms use or create motion, **Figure 8-68**. The four basic kinds of motion are:

Watt's Steam Engine	3%
Modern Steam Engine	10%
Gasoline Engine	30%
Nuclear Power Plant	30%
Aircraft Gas Turbine	36%
Diesel Engine	37%
Rocket Engine	48%
Electric Motor	80%

Figure 8-67 Scientists, engineers, and manufacturers are always trying to design more efficient engines, motors, and machines.

- Linear – straight-line motion.
- Rotary – motion in a circle.
- Reciprocating – backward and forward motion in a straight line.
- Oscillating – backward and forward motion like a pendulum.

Mechanisms are often used to change one kind of motion into another kind. Some examples are shown in **Figure 8-69**.

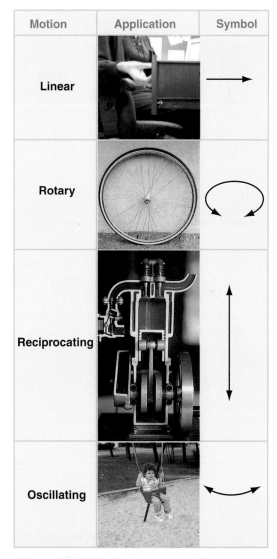

Motion	Application	Symbol
Linear		
Rotary		
Reciprocating		
Oscillating		

Figure 8-68 Mechanisms are used to create different kinds of motion. (Jack Klasey, Ecritek)

Standards for Technological Literacy

16

Discussion
List products in your home that use or create linear, rotary, reciprocating, and oscillating motion.

Vocabulary
What is the definition of the term *power* when applied to mechanisms?

Resource
Transparency 8-4, *Machine and Energy Formulas*

Resource
Reproducible Master, *Machine and Energy Formulas*

Useful Web site: Have students work through the interactive demonstrations on mechanisms at www.flying-pig.com/mechanisms/index.html.

Designing and making
How do designers reduce the amount of friction acting on vehicles that travel at very high speeds?

Discussion
Have students investigate and discuss the types of friction acting on bodies at http://scienceworld. wolfram.com/physics/ Friction.html.

Activity
List products in your home that involve the following changes of motion (a) from rotary to linear, (b) from rotary to reciprocating, and (c) from reciprocating to rotary.

Resource
Activity 8-6, *Mechanical Coin Sorter*

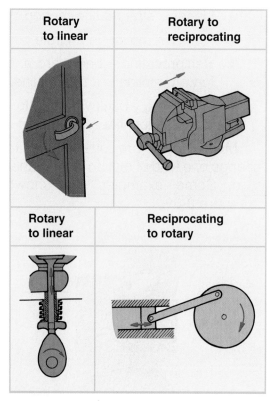

Figure 8-69 Mechanisms change motion from one form to another.

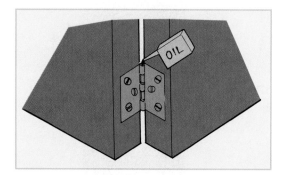

Figure 8-70 Oil lubricates parts to reduce friction.

Friction

As you pedal your bicycle, you are working against *friction*. Friction is a force that acts like a brake on the movement of moving objects. Your finger will slide without much effort on a pane of glass. But if you do the same thing on sandpaper, you can feel a resistance slowing up your movement.

The moving parts of mechanisms do not have perfectly smooth surfaces. The tiny projections on the surfaces rub on one another. This creates friction and results in heat. The friction between the moving parts must be minimized so that:

◆ Less energy will be needed to work the machine.

◆ Wear and tear will be reduced.

◆ Moving parts will stay cooler.

Friction may be reduced in several ways as shown in **Figure 8-70** through **Figure 8-73**.

◆ Oiling – oil separates two surfaces that would otherwise touch, rub, and wear each other away.

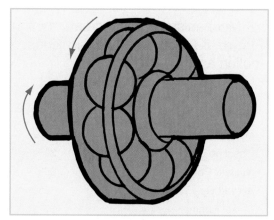

Figure 8-71 Ball bearings allow parts to "roll" over one another, reducing friction.

Useful Web site: Have students conduct the experiment on using bearings to reduce friction described at www.smm.org/sln/tf/f/friction/friction.html.

Figure 8-73 One of the effects of streamlining is to reduce air drag, which is a form of friction. (NFB)

Figure 8-72 Hovercraft use a cushion of air to reduce friction. (Canada Coastguard)

◆ Ball bearings – steel balls enable surfaces to roll over one another instead of sliding.

◆ Air or water cushions – compressed air or water separates moving parts.

◆ Streamlining – the shape of a fast-moving object can be changed to reduce its resistance to air or water.

Designing and making
How might smart materials change the design of machines in the future?

Enrichment
Have students use the Internet to investigate very big machines (such as earth moving equipment) and very small machines (such as an analogue wrist watch).

Useful Web site: Have students research the use of simple machines in the design of artificial limbs. Three useful web sites are: www.waramps.ca/nac/limbs.html, www.animatedprosthetics.com, and www.alatheia.com.

Chapter 8 Review
Machines

Summary

Machines can be small like a bottle opener or large and complex like a bulldozer. The six simple machines are the lever, pulley, wheel and axle, inclined plane, wedge, and screw.

The lever has four advantages. The lever may be used to increase force, increase the distance a force moves, increase the speed at which it moves, or reverse the direction of motion.

A pulley is a special kind of lever that acts continuously. It is used for lifting heavy objects vertically. It may also be used to transmit motion, increase or decrease speed, reverse the direction of motion, or change motion through 90°.

The wheel and axle is also a special kind of lever. Effort applied to the outer edge of the wheel is transmitted through the axle.

The inclined plane is most commonly seen in the form of a ramp. The more gradual the slope, the less force needed to push an object upward.

The wedge is a special version of an inclined plane. It is two inclined planes back-to-back. A screw is an inclined plane wrapped in the form of a cylinder.

Gears are modified levers used in the transmission of rotary motion. They increase or decrease speed, change the direction of motion, and transmit a force.

Simple machines and gears are able to move a greater resistance with a smaller effort. Pressure can also be used to increase an effort applied. Pressure in liquids is called hydraulics, and in gases, pneumatics.

Mechanisms change one kind of effort into another kind of effort. They can be combined to form a machine. Machines change one kind of energy into another kind and do work.

Modular Connections

The information in this chapter provides the required foundation for the following types of modular activities:

- ◆ Simple Machines
- ◆ Weights and Measures
- ◆ Precision Measure
- ◆ Mechanisms
- ◆ Mechanical Systems
- ◆ Applied Physics
- ◆ Engineering Design
- ◆ Dragster Design
- ◆ Auto Demolition Derby
- ◆ Pneumatics
- ◆ Fluid Power Technology
- ◆ Fluid Power Systems

Test Your Knowledge

Write your answers to these review questions on a separate sheet of paper.

1. List four machines used by early civilizations to make work easier.

2. Identify objects in or around your home that contain one or more simple machines.

3. When you hit a ball with a baseball bat, the bat is an example of a lever. Explain why.

4. Draw three diagrams to illustrate the difference between a Class 1, Class 2, and Class 3 lever.

5. Which of the following objects is a Class 1 lever?
   A. Nutcracker.
   B. Wheelbarrow.
   C. Hockey stick.
   D. Scissors.

Answers to Test Your Knowledge Questions
1. Lever, wheel, inclined plane, and pulley (any order).
2. Answers will vary. (Evaluate individually). Examples: Bottle opener (lever). Pizza cutter (wheel and axle). Corkscrew (screw). Sloping driveway (inclined plane). Knife (wedge). Garage door mechanism (pulley).
3. There is a fulcrum, an effort, and a resistance.
4. Refer to Figure 8-13, 8-15, and 8-17 on pages 195–197.
5. D. Scissors.
(continued)

6. Which of the following objects is a Class 2 lever?
   A. Nutcracker.
   B. Seesaw.
   C. Fishing rod.
   D. Baseball bat.

7. Which of the following objects is a Class 3 lever?
   A. Scale.
   B. Tweezers.
   C. Nutcracker.
   D. Claw hammer.

8. A laborer using a Class 1 places the load the same distance from the fulcrum as the effort. If the fulcrum is moved closer to the load, the mechanical advantage of the machine _____.
   A. increases
   B. decreases
   C. remains the same
   D. approaches one

9. A lever is used to move a load of 1500 newtons with a force of 300 newtons. What is the mechanical advantage of the lever?
   A. 30.
   B. 15.
   C. 5.
   D. 3.

10. State the four advantages of levers.

11. The advantage of a single, fixed pulley system is that it _____.
    A. increases mechanical advantage.
    B. decreases the effort required.
    C. changes the direction of force.
    D. makes the work harder.

12. The disadvantage of a single, movable pulley system is that it _____.
    A. decreases the mechanical advantage.
    B. makes the operator pull down on the rope.
    C. increases the effort needed to move a load.
    D. makes the operator pull up on the rope.

13. What force would be required to raise a load of 15 newtons using a single, fixed pulley?
    A. 30.
    B. 15.
    C. 7.5.
    D. 5.

14. What force would be required to raise a load of 15 newtons using a single, movable pulley?
    A. 30.
    B. 15.
    C. 7.5.
    D. 5.

15. List three examples of a wheel and axle.

16. List three practical applications of an inclined plane.

17. A 100 N load has to be moved to the top of an inclined plane 10 m long and 2 m high. What effort is required?
    A. 10 N.
    B. 20 N.
    C. 50 N.
    D. 100 N.

18. A wedge is composed of _____.

19. What is the connection between an inclined plane and a screw thread?

20. State four advantages of simple machines.

21. A gear is a modified form of _____.

22. State three reasons why gears are used.

23. The largest gear in a two-gear train is called the _____.
    A. driver
    B. driven
    C. pinion
    D. wheel

24. The wheel in a two-gear train has 60 teeth and the pinion 15 teeth. What is the ratio of the gear train?
    A. 4:1.
    B. 4:15.
    C. 6:15.
    D. None of the above.

25. Sketch a simple gear train in which the first and last gears are rotating in the same direction. Use arrows to show the direction of rotation.

26. Pressure is defined as _____. It is calculated by _____.

27. The study of pressure in liquids is called _____. The study of pressure in gases is called _____.

28. Give two examples of an object operated by hydraulics and two examples of an object operated by pneumatics.

29. List five advantages of using machines.

30. Name the four basic kinds of motion and give one example of each.

Apply Your Knowledge

1. Identify three tasks in the home that would be impossible to complete without the assistance of simple machines. Name the simple machines used.

2. Design and build a robotic arm that will move an AA dry cell from one location to another. Use hypodermic syringes and tubing.

3. Design and build a mechanism that will make a loud noise. Your solution must contain at least two simple machines.

4. Design and build a method of measuring the mass of a series of weights from 1 oz (25 g) to 16 oz (500 g).

5. Consult your library or the Internet to learn about the mechanical inventions of Leonardo da Vinci. Which of his mechanical inventions are somewhat similar to the machines and vehicles we see today?

6. Check www.rubegoldberg.com and sketch one of the designs that fascinates you.

7. Research one career related to the information you have studied in this chapter and state the following:
A. The occupation you selected.
B. The education requirements to enter this occupation.
C. The possibilities of promotion to a higher level at a later date.
D. What someone with this career does on a day-to-day basis. You might find this information on the Internet or in your library. If possible, interview a person who already works in this field to answer the four points. Finally, state why you might or might not be interested in pursuing this occupation when you finish school.

The pulley is one of the most important simple machines. It can be used to change the direction of applied force and give the operator a mechanical advantage based on the pulley configuration. Notice the groove in the center. This helps reduce rope slippage.

Chapter outline

Student Activity Manual Resources

Instructional Strategies and Student Learning Experiences

1. Have students list as many modes of transportation as possible.
2. Have students clip pictures of various modes of transportation and make a collage.
3. *Modes of Transportation Comparison Chart,* reproducible master. Use this master as a basis of discussion about the advantages and disadvantages of various modes of transportation.
4. Invite an automotive technician to class to discuss the differences between the four-stroke gasoline engine and the four-stroke diesel engine.
5. Have students visit the library, search the Internet, and/or write to NASA to research current developments in space travel. Ask students to write a report and present it in class.
6. Ask students to describe local impacts of transportation and the advantages and disadvantages of each. Have students try to come up with solutions to counter the disadvantages.

Transportation is the moving of people and things from one place to another. Transportation into space requires tremendous energy due to the weight of the vehicle and the force of gravity.

Transportation

Discussion
During the twentieth century, our society became increasingly dependent on automobiles for transportation. What are some advantages and disadvantages of the automobile? What are some alternatives to automobile transportation?

Career connection
Have students identify careers in transportation industries and the qualifications of people who work in them.

Key Terms

AGV
diesel engine
electric motor
engine
external combustion engine
fuel cell
gasoline engine
HPV
internal combustion engine

jet engine
on-site transportation
thrust
transportation
transportation system
turbine
turbofan
turboshaft

Objectives

After reading this chapter you will be able to:

◆ State the advantages and disadvantages of various modes of transportation.

◆ Explain the principles of various types of engines and motors.

◆ Recall what industries are affected by transportation systems.

◆ Identify what processes are involved for the entire transportation system to work.

◆ Explain how transportation technology influences everyday life.

◆ Discuss the environmental impact of transportation systems.

Standards for Technological Literacy

Vocabulary
Define the term *hydrocarbon*.

Safety
Make a list of the various safety features built into different forms of transportation.

Resource
Activity 9-1, *Transportation Words*

Modes of Transportation

Transportation forms a vital part of our lives, giving us access to education, recreation, jobs, goods, services, and other people. Different transportation systems enable people to travel from house to house, town to town, country to country, and even the earth to the moon. See **Figure 9-1**.

Suppose you want to send a package to a friend who lives in a distant city. You might transport the parcel to the post office using your bicycle. From the post office, a truck might take the parcel to the airport. An airplane delivers it to a city across the continent. There, another truck takes it to a central depot for sorting. Finally, a mail carrier might use a bus or mail truck to deliver the parcel to your friend's house. The bicycle, truck, airplane, and bus are individual modes of transportation. Together, they form a transportation system.

You are probably familiar with the millions of miles of public roads, but you might not know about other transportation systems, including the commercially navigable waterways and hundreds of ports on our inland waterways, the Great Lakes, and our coastal regions. You have probably never thought about the millions of miles of pipelines in North America, but they too are part of our transportation system, as they transport oil and natural gas.

Land Transportation

Transportation burns most of the world's supply of petroleum. Hydrocarbon fuels produce carbon dioxide, a greenhouse gas thought to be the main cause of global climate change. Although regulations for our vehicles have ensured that the vehicles pollute less than they used to, the number of vehicles has increased, and we use the vehicles more often. For example, over 90 percent of workers in North America travel to work by car.

By contrast, public transportation is the main way most people in the world travel. While we are used to traveling by car, most people in the world cannot afford a car. Walking, cycling, and taking minibuses or trains are their means of getting around.

Even in North America, public transportation can often be the fastest way to travel. When city streets are congested, buses traveling in dedicated bus lanes, metro cars moving underground, and high-speed trains going between major urban centers can all travel at much higher speeds than private vehicles can. What would happen if all means of transportation were public? We would have more people in each vehicle, fewer cars would be on the roads, and traffic jams might be eliminated!

When large numbers of passengers are transported, we call it mass transit. Mass transit is a good description because underground systems, such as those in London and Hong

Links to other subjects: History—Sketch or make a model of an early machine used to harness the energy of humans or animals (such as, the treadmills for pumping and grinding used in Mesopotamia).

Useful Web site: Students can use the dictionaries at www.techtionary.com and www.hyperdictionary.com to find the definition of any new technical and technological terms used in this chapter.

Mode of Transportation	Advantages	Disadvantages
Bicycle	• cheap to operate • flexible • nonpolluting	• transports only one person or small articles. • limited to short travel distances
Car	• flexible and convenient • provides privacy	• transports no more than six passengers • causes high levels of pollution • road networks use arable land
Truck	• flexible means of moving freight short and medium distances • goes directly from point of origin to destination	• not suitable for very long distances • causes high levels of pollution
Bus	• relatively flexible and convenient • carries at least 50 people • cheap form of urban mass transit	• causes pollution • not suitable for very long distances
Train	• economical and efficient for large loads of freight over long distances • can carry hundreds of passengers • creates little traffic congestion • very safe • causes very little air pollution	• not as convenient as cars • requires a track system that is expensive to built and maintain
Subway train	• can carry hundreds of passengers • does not cause congestion or pollution	• expensive to build
Ship	• can carry huge loads of freight or large numbers of passengers over long distances	• special facilities needed for loading and unloading • can only be operated on water
Airplane	• can carry passengers and freight over long distances quickly • can pass over all types of terrain	• expensive to operate • airport located away from urban centers; needs support of other modes of transportation
Space shuttle	• reusable form of space travel	• extremely expensive • only used for scientific and technological experiments in space

Figure 9-1   Here are nine different modes of transportation and how they compare.

Discussion
Discuss different types of bicycles and their uses. Include the materials used for construction and the cost in the discussion.

Enrichment
Explain how the four basic forces (lift, weight, drag, and thrust) acting on an aircraft operate.

Activity
Research Leonardo da Vinci's fifteenth century drawings. What examples of transportation vehicles did he predict? Sketch one of them.

Resource
Reproducible Master, *Modes of Transportation Comparison Chart*

Resource
Activity 9-2, *Elastic-Powered Vehicles*

Kong, carry millions of passengers each day. Underground systems might solve the worst traffic problems, especially at rush hour, as weather and street-level congestion do not affect them. They are, however, extremely costly to build. Many mass transit railways include not only underground tracks, but surface and elevated tracks. San Francisco's Bay

Useful Web site: Have students investigate one aspect of early attempts to fly at www.inventingflight.com and report their findings.

Standards for Technological Literacy

2

Designing and making
Design and make a small-scale fantasy vehicle powered by an electric motor and battery.

Vocabulary
What does the term *funicular* mean?

Activity
Divide the class into small groups. Have each group create a multimedia presentation about a famous train and present it to the class. Two excellent Web sites are www.infoplease.com/spot/trains1/html and www.factmonster .com/ipka/A0855836 .html.

Area Rapid Transit (BART) includes 30 km of tunnels, 40 km of surface track, and 50 km of elevated track.

Road Transportation

Roads carry an extraordinary variety of motor vehicles. In addition to cars, buses, and many sizes of delivery trucks, there are fire engines, ambulances, police vehicles, mixer trucks, crane carriers, and garbage trucks, to name only a few types of vehicles. There are also vehicles designed for off-road use, including earthmovers designed for shifting earth and rock during construction and mining operations.

Tractor-trailers are the most common vehicles for long-distance hauling. The tractor part has a cab, an engine, and a transmission with a turntable (or fifth wheel) on top of the rear axle, where it hooks to the trailer. Rules mandate the number of hours tractor-trailer drivers can drive and how much rest or sleep time is necessary. Long-distance trucks are, therefore, often equipped with sleeper cabs, where the drivers can rest while not driving. These are equipped like small apartments, with good sound systems, stoves, refrigerators, storage places, and wash basins. They are air-conditioned and insulated against noise and vibration, so the drivers stay fresh and alert.

Rail Transportation

Railroads are part of North America's history and folklore. In the nineteenth century, Native Americans fought in vain to defend their land against the "steel horses" bringing settlers westward to the Midwest and California. The Trans-Siberian railroad in Russia is the longest in the world, stretching 9300 km, much of that over frozen wasteland. The Orient Express in Europe has been the setting of several films and novels. There are also fictional trains, such as the Hogwarts Express, which transports Harry Potter to Hogwarts Academy.

The diesel-electric system is a common type of fueling system for locomotives, but there is a movement toward electric trains, as these trains are nonpolluting. Electric locomotives are generally faster than diesels. For example, the French TGV (*train à grande vitesse*, or high-speed train) can travel up to 380 km per hour. Electric trains pick up their power from overhead power lines or a third rail alongside the ordinary track. Trains make efficient use of fuel because there is not a lot of friction between the steel wheels of the train and the steel rails on which they run. There are also special kinds of trains that run on unique guideways, such as monorails, maglevs, rubber-tired subways, funiculars, and cog trains.

Water Transportation

For over 5000 years, people have used ships to travel, trade, and fight in far-off lands. Nowadays, planes have largely replaced ships as people carriers, but every day, there are thousands of merchant ships plying the seas. These ships vary in size, from simple cargo carriers to giant oil

Links to other subjects: History—Have students investigate and describe the development of the railroads in North America. A useful starting place is www.infoplease.com/ipa/.

Useful Web site: Have students access the Web sites http://members.tripod.com/shumsw/ and www.geocities.com/uksteve.geo/boxship to investigate very large containerships.

tankers. Although they differ in size, all of them have double hulls, which are double steel plates with a space in between, for safety reasons. While a ship's advantage lies in the amount of goods it can transport, its disadvantage is its slow speed, due to the resistance, or drag, of water on the hull.

Hydrofoils and hovercraft successfully overcome this disadvantage by reducing the drag of water on the hull. When at rest or moving slowly, a hydrofoil looks like any other boat, but when it increases its speed, the entire hull lifts out of the water. The only parts that remain submerged are the rudder and part of its wings, or foils. This lift is made possible by the shape of the foils. A hovercraft flies a few centimeters above the surface on a cushion of air. This means it is free of friction from the ground or water underneath. In the early days of hovercraft, transportation regulators could not decide whether hovercraft are boats or planes. Hovercraft are used as passenger and cargo ferries and by the military as patrol craft and troop carriers.

The most popular method of transporting goods by sea is in containerships. Modern containerships are relatively fast, as they can cross the Atlantic Ocean in seven days. Other ships transport loose cargo, such as coal, grain, and mineral ores in holds below deck. Others specialize in transporting mail, cargo, and passengers to smaller communities, which sometimes are not even connected by roads. Cruise ships create

a "fun ship" image by transporting vacationers to islands in the Caribbean, the Mediterranean, and the South Seas.

In the nineteenth century, steam powered ships. Robert Fulton's *North River Steamboat* (also known as the *Clemont*) was an early paddle steamer. Nowadays, most ships use diesel power, with a few using nuclear reactors to make steam to drive turbines. The screw propeller has replaced the paddle wheel.

Air Transportation

There is a huge variety of transportation vehicles that fly. We are most familiar with planes, including regularly scheduled passenger planes, which operate according to timetables, and charter planes, which companies hire to fly passengers. Helicopters are often called on to help in rescue operations, such as lifting a pet to safety from the side of a cliff, helping people who are lost in remote areas, or rescuing earthquake survivors.

Sports enthusiasts use many other means of air travel, including gliders, hang gliders, parachutes, and ultralight aircraft. Gliding is a sport that enables people to experience the exhilaration of flying in a vehicle that looks like a small plane but has no engine. The most common way to launch a glider is to pull it into the air by means of a steel cable connected to a powerful winch. Hang gliders are quite different, as they do not even need a tow. The riders launch themselves from high places and return to

Standards for Technological Literacy

Enrichment
Make a labeled sketch showing the parts of a sailboat. Include the mast, mainsail, boom, jib, bow, stern, hull, keel, and rudder in the sketch. Add notes explaining the function of each part.

Activity
Ask students to bring pictures of the vehicles their parents or grandparents used to drive. Have students report on what their parents or grandparents remember about the vehicles.

Activity
Build a model plane.

Links to other subjects: History—Write an illustrated report explaining how waterwheels have been used in the past 2000 years.

Useful Web site: Investigate the history of flight at www.centennialofflight.gov. Students should print a page they find interesting and share it with the class.

Standards for Technological Literacy

2

Designing and making
What design decisions do designers of containers and tractor-trailers have to make?

Vocabulary
What does the phrase *intermodal transportation* mean?

Activity
Investigate the life of Malcolm McLean, the inventor of container shipping.

the ground using movements of their bodies to turn and land. Skydivers jump out of planes, and some do acrobatic movements, before landing on predestined targets. Ultralight aircraft are like hang gliders with very small engines. They operate at a maximum speed of 55 knots per hour and only between sunrise and sunset.

Other flying vehicles include hot air balloons and blimps. In a hot air balloon, passengers ride in a basket suspended under the balloon. Blimps are often seen at major sporting events, and they are usually used for photography or advertising. The strangest aircraft is the ornithopter, an engine-powered vehicle in which the mechanical flapping of its wings creates the thrust and nearly all the lift!

Intermodal Transportation

Airplanes, ships, trains, and trucks all have particular advantages, but no single system is better than the others. Ships can move the largest loads from country to country. Trains carry large loads efficiently over long distances on land, but they cannot collect and deliver goods to your door or to the store around the corner. The amount of cargo that trucks carry is limited, and while planes are the fastest means of travel, the cost of planes is the highest.

In recent years, there has been a willingness to integrate separate transportation systems through an intermodal chain. Freight might be transported by ship to the West Coast from China, packed in containers.

Each container can be loaded onto a flatbed railcar when it arrives at a port. When the containers arrive at a large city, each one might be loaded onto a tractor-trailer and taken to a warehouse. Finally, smaller trucks deliver the items to local stores.

Technology has facilitated intermodal transportation by inventing new techniques to transfer freight from one mode to another. In some countries, trucks can drive straight onto a train and drive off again when they reach their destination. Some intermodal vehicles have two sets of wheels, one for use when coupled to a train and the other for road use. The most important development was a simple one: the container. A container is a box measuring 20′ long, 8′ high, and 8′ wide. As its size has been standardized worldwide, ships and lifting equipment are available in all ports and rail terminals. Goods are shipped in bulk, so costs are reduced and cranes can handle the containers quickly. The contents are protected from damage and theft, and containers can be stacked to minimize storage space.

Now that shippers can think in terms of a variety of transportation methods, they can move away from the dominance of trucks and address concerns about congestion, safety, and environmental pollution. Regulations can even mandate the use of intermodal transportation. The law requires freight crossing Switzerland to be placed on railcars, so as to reduce air pollution.

Useful Web site: Have students access the Web site www.alteich.com/tidbits/t060401.htm and print material about container shipping they find interesting, to share with the class.

Technology and society: What are the impacts on the environment and on employment of using containers?

On-Site Transportation

On-site transportation transports people and materials from one spot to another in a defined location. A few examples of sites are a mine, gravel pit, building project, factory, or building. Elevators, escalators, and conveyors move people or materials in a building. Robots in a factory move materials cheaply and quickly from one spot to another. Some are stationary and move parts short distances. Others are called automated guided vehicles, or *AGVs*. They follow a set path and transport parts and materials over longer distances. Conveyors, trucks, and pipelines move materials at many sites, including mines, refineries, wells, and construction sites.

Human Powered Vehicles (HPVs)

An *HPV* is any vehicle powered solely by one or more humans. Rowboats, canoes, and bicycles are common examples. The shape of the bikes you see everyday has changed very little in the past 100 years. However, there is more choice in the materials, including steel, aluminum, carbon fiber composite, and titanium. Top of the line bikes now include features such as hydraulic disk brakes and dual suspension for riding off-road terrain that has drops, big roots, logs, and other obstacles.

Conventional bikes designed for road use have several disadvantages. There is a limited amount of downward pressure that can be placed on the pedals. Also, as much as 80% of a rider's energy is used just to push away the air in front of the bicycle. Wind resistance can be reduced in several ways. By making the tubing oval shaped and by hiding parts, such as cables, within the frame, all surfaces become smoother. A second way is to lower the rider into a recumbent position, lying on his or her back or stomach. For a third method, the machine and rider can be partially or fully covered by a streamlined shape that resembles an aircraft wing or teardrop, **Figure 9-2.**

Engines and Motors

Most forms of transportation need engines to make them move. *Engines* are machines made up of many mechanisms. They convert a form of energy into useful work.

An engine needs a constant supply of this energy to keep it working.

Figure 9-2 What three things have been done to these bicycles in order to lower wind resistance? (Easy Racer)

Standards for Technological Literacy

2

Discussion
Name and describe the function of simple machines found on a bicycle.

Discussion
Describe two differences between internal combustion engines and external combustion engines.

Vocabulary
Define the terms *aerodynamics* and *human powered vehicle*.

Vocabulary
Define the terms *turbine engine* and *turbofan engine*.

Activity
Make annotated sketches to show how the shape of bicycles has changed since they were invented.

Activity
Investigate and report on the inventions of Nikolaus Otto, particularly his contribution to the internal combustion engine.

Community resources and services: Contact your local university or technical college to find out what experiments are in progress related to engines, vehicle design, or transportation.

Standards for Technological Literacy

Discussion
Why can't the firing order of an internal combustion engine be changed from intake, compression, power, and exhaust?

Resource
Transparency 9-1, *How the Four-Stroke Gasoline Engine Works*

Resource
Transparency 9-3, *Four-Stroke Gasoline Engine*

The energy may come from burning fuel such as gasoline, diesel, or kerosene. Engines that burn this fuel inside them are called *internal combustion engines*. Internal combustion engines depend on hot, expanding gases for power. Another type of engine burns its fuel outside the engine. This kind of engine is called an *external combustion engine*. The fuels used as energy sources for such engines include coal and oil. External combustion engines may heat water or gases to produce power.

Engines also fall into another kind of grouping: reciprocating piston engines and turbines. Reciprocating engines include gasoline engines and diesel engines. These are most commonly used in motorcycles, cars, trucks, and buses. Turbine engines include turbofan engines and turboshaft engines. They are used in aircraft, hovercraft, and helicopters.

Four-Stroke Gasoline Engine

Imagine trying to pedal the family car at 55 mph (90 km/h). The driver and one passenger would have to pump their legs up and down about 60 times a second.

Although it is impossible for legs to move this fast, pistons in a gasoline engine slide up and down at this speed. The pistons are short metal drums moving inside cylindrical holes in a metal block. The power to move the pistons comes from a mixture of air and gasoline. This mixture is ignited by an electric spark.

Expanding gases push the pistons down to rotate the crankshaft. The crankshaft then transmits turning power to the drive train. The four-stroke *gasoline engine* is an internal combustion engine, **Figure 9-3**.

A piston in a four-stroke cycle travels down and up the cylinder four times (down twice and up twice) to produce a complete sequence of events (cycle), **Figure 9-4**. The following is what happens during one complete cycle of the engine:

1. First Stroke—Intake

   The intake valve opens. The rotating crankshaft pulls the piston downward. The piston sucks fuel and air into the cylinder.

2. Second Stroke—Compression

   The intake valve closes. The crankshaft pushes the piston upward. The mixture is squeezed into a small space to make it burn with an explosion. Just before the piston reaches the top of the stroke, the spark plug ignites the mixture.

Figure 9-3 The four-stroke gasoline engine is a popular power source for transportation. (Nissan)

Technology and society: Have students debate the advantages and disadvantages of the internal combustion engine.

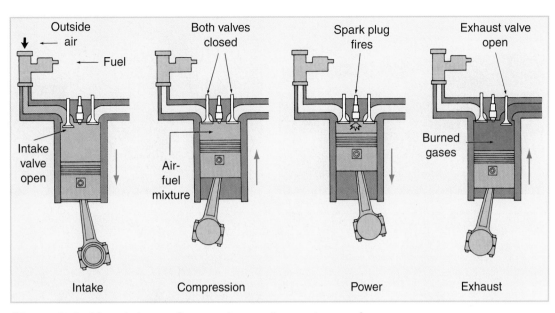

Figure 9-4 Here is how a four-stroke gasoline engine works.

Standards for Technological Literacy

Enrichment
Although no spark plugs are used in diesel engines, some models use glow plugs to assist in starting.

Activity
Draw a diagram to illustrate the interior of the internal combustion engine. Label the piston, cylinder, intake and exhaust valves, spark plug, and air/fuel mixture.

3. Third Stroke—Power

Both valves are closed, so expanding gases cannot escape. The hot, expanding gases force the piston downward with great force. The piston pushes hard on the crankshaft, making it rotate faster.

4. Fourth Stroke—Exhaust

At the bottom of the power stroke, the exhaust valve opens. The piston comes up again and pushes the gases from the burned fuel out of the cylinder. The cycle starts over again.

The four-stroke gasoline engine, Figure 9-5, has the following advantages:

◆ It adapts easily to speed changes.

◆ It provides good acceleration.

◆ It has sufficient power for medium size machines.

The gasoline engine also has the following disadvantages:

◆ It is not strong enough for heavy work.

◆ It pollutes the air.

◆ It burns a relatively expensive fuel.

The Four-Stroke Diesel Engine

Unlike the gasoline engine, the *diesel engine* does not need a spark to ignite its fuel. It uses the heat from compressed air to ignite the fuel. Pure air is drawn into the cylinders. This air is squeezed to a much higher pressure than the air-fuel mixture in a gasoline engine. At the top of the compression stroke, a fine spray of diesel fuel is injected into the cylinder. As the air is compressed, its temperature increases. When the fuel spray

Links to other subjects: History—Using the Internet, investigate the life and death of Rudolf Diesel, the inventor of the diesel engine.

Figure 9-5 Gasoline engines power all sorts of vehicles. (Ford, Ford, Polaris, Laser)

meets the hot compressed air, the fuel ignites. The four-stroke diesel engine is an internal combustion engine. Its operation is similar to the four-stroke gasoline engine.

1. As in a gasoline engine, there is a pair of valves on each cylinder. However, in a diesel engine, one valve controls the air intake, while the other valve allows exhaust gases to escape from the cylinder.

2. The upward stroke of the piston compresses the air.

3. An injector squirts fuel into the cylinder.

4. The fuel ignites spontaneously and the expansion of gases forces the piston downward.

The four-stroke diesel engine has the following advantages:

◆ It uses less fuel.

◆ It lasts longer than a gasoline engine.

◆ It stands up well to long, hard work.

The following are the disadvantages of the diesel engine:

◆ It is noisier.

◆ It has slower acceleration.

◆ It pollutes the air more.

Useful Web site: Write a report describing how Harley Davidson is using computer simulation to reduce costs.

◆ It must be more heavily built to withstand the high pressures in the cylinders.

Despite the diesel engine's disadvantages, the future is still bright. Modern electronically controlled diesels provide better fuel efficiency than most other cars on the road, and research and development experiments are working to reduce the levels of noise and exhaust emissions.

Diesel engines are sometimes used in cars. However, they are more commonly used to power large machines and vehicles, such as tractors and buses, **Figure 9-6**.

The Steam Turbine

The term turbine originally described machines driven by falling water, **Figure 9-7**. They were also known as waterwheels. There are two types. The overshot wheel is driven with water from above. The wheel rotates clockwise. An undershot wheel is driven with water from below so that the wheel turns counterclockwise. Later, the term *turbine* was also given to heat engines powered by steam. Engineers discovered that a strong jet of steam could turn a large wheel with blades on it. A steam turbine's hundreds of blades are set on a long shaft completely enclosed in a strong metal case. The blades on the rotor spin to drive the shaft. Blades on the stator are fixed to the outer casing and cannot spin, **Figure 9-8**. Each stator fan guides the steam flow so that as it moves along the turbine it has plenty of thrust to move the next rotor fan in its path. The spinning fans

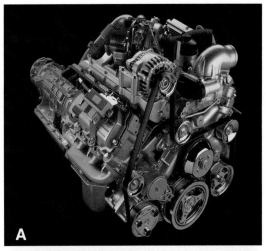

Figure 9-6   A—Notice the cutaway of a diesel engine. These engines are found in many kinds of vehicles and machines. B—Snowcat. C—Ship. D—Bus. (Nissan, TEC, Ecritek, Ecritek)

Discussion
What are some advantages of an engine, such as a steam turbine, that operates with rotary motion compared to a gasoline engine that uses reciprocating motion?

Activity
Write a report describing what each of the following inventors discovered in relation to the use of steam: Hero (about 130 BC), Leonardo da Vinci (late fifteenth century), Thomas Savery (late seventeenth century), and James Watt (early eighteenth century).

Resource
Transparency 9-5, *Steam Turbine*

Links to other subjects: History—Investigate the working and living conditions of people who labored in early factories powered by waterwheels.

Standards for Technological Literacy

2

Discussion
Which of the two types of water-wheels provides the most power and will work harder?

Discussion
Compare the differences between the three jet engine types (turbojet, turbofan, and turbo prop).

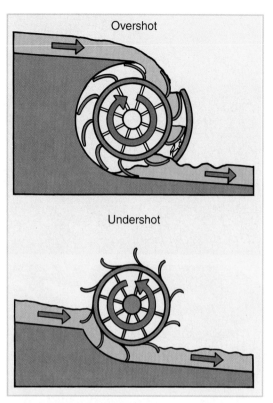

Overshot

Undershot

Figure 9-7 Overshot—An overshot wheel turns clockwise from the mass of the water passing over its blades. Undershot—An undershot wheel is turned counter-clockwise from the force or motion of the water striking the blades underneath.

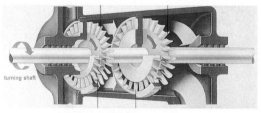

turning shaft

Figure 9-8 This is a cutaway of a steam turbine. How is it like a water wheel? The stator blades (colored blue) do not move. They guide the steam so it passes over the rotor blades, turning them. Steam turbines are a type of external combustion engine.

turn the drive shaft. The shaft, in turn, drives a propeller.

After its journey through the turbine, the steam cools and turns into water. This water then flows back to the boiler. Heated to steam again, it returns to work the turbine. The water is heated either by burning fuel or by a nuclear reactor outside the turbine.

The steam turbine has several advantages. It runs smoothly, has a long life, and is powerful and suited to slow, large machines that require an engine bigger than a diesel. This is shown in **Figure 9-9**.

One disadvantage is that it needs a lot of room since space is also needed for a boiler. Another is that it does not adapt as easily to changes in speed as piston engines.

Jet Engines

To get an airplane into the air, you need pushing power or *thrust*. What is thrust? Think of stepping forward off a skateboard, **Figure 9-10**. As you go forward, the skateboard moves backward.

Figure 9-9 Steam turbines power large seagoing vessels. (Ecritek)

Figure 9-10 One example of thrust is leaping off a skateboard. This creates thrust to push the skateboard in the opposite direction.

Have you ever held a garden hose when someone turned it on full? Suddenly, a jet of water bursts out, and the hose jumps out of your hand. As the water shoots forward, the hose goes backward. Firemen are sometimes pushed over by this backward force. Blow up a balloon and let it go without tying the neck. The balloon is driven in much the same way as the hose and skateboard.

These are just three examples of one of the basic laws of nature. Isaac Newton discovered it 300 years ago. He said that for every force in one direction there is always an equal force in the opposite direction.

All jets work on this principle—for every action there is an equal and opposite reaction. The reaction to the rush of gases out of a *jet engine* is a thrust that drives the airplane forward. The engine sucks in air at the front, squeezes it, and mixes it with fuel. The mixture ignites and burns quickly. This creates a strong blast of gases. These hot gases expand and rush out

of the back of the engine at great speed. Many people believe that the gases "push" against the air to propel the plane forward. This is not true. As the gases shoot out backward, the jet goes forward.

The *turbofan* is one kind of jet engine that powers aircraft, **Figure 9-11**. A jet engine produces a very noisy gas stream. In a turbofan engine, however, the gas stream drives a large fan located at the front of the engine. This creates a slower blast of air. Thrust is as great as a simple jet, but the engine is quieter.

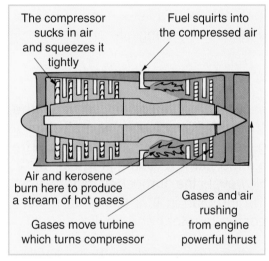

The compressor sucks in air and squeezes it tightly

Fuel squirts into the compressed air

Air and kerosene burn here to produce a stream of hot gases

Gases move turbine which turns compressor

Gases and air rushing from engine powerful thrust

Figure 9-11 Top—This is how a turbofan engine operates. Bottom—The Concorde aircraft, using a turbofan engine, is capable of speeds greater than 1300 mph (2100 km/h). (British Airways)

Standards for Technological Literacy

2

Reflection
What were you doing when you experienced Newton's third law of motion?

Enrichment
The turbofan engine is used in most commercial aircraft. It uses less fuel than a turbojet to produce the same power.

Resource
Transparency 9-6, *Turbofan Engine*

Useful Web site: Visit www.centennialofflight.gov to learn more about jet aircraft and jet engines.

Standards for Technological Literacy

2 18

Discussion
What are Isaac Newton's first and second laws of motion, and how do they relate to rockets?

Reflection
Why would your weight be only one-sixth as much on the moon, but your mass remains the same?

Enrichment
Sir Isaac Newton lived from 1641 to 1727.

Vocabulary
Define the terms *vacuum* and *zero gravity.*

Resource
Transparency 9-7, *Turboshaft Engine*

Air is drawn into the engine by the compressor fans. As pressure increases, the compressed air mixes with fuel. Ignition takes place, and temperature and pressure increase. The burned mixture leaves the engine through the turbine, which drives the compressor and the fan at the front of the engine. Pressure thrusts the engine forward, while the exhaust gases rush out of the back in a jet stream. The turbofan is an internal combustion engine.

The turbofan engine has several advantages. It is relatively lightweight, very powerful, and uses less fuel than other jet engines. Its disadvantage is that although quieter than other jet engines, all jet engines are noisy.

The Turboshaft Engine

Like the turbofan engine, the *turboshaft* engine uses a stream of gases to drive turbine blades. The blades turn a shaft, **Figure 9-12**. However, the turboshaft differs from the turbofan in that this shaft is connected to rotors or propellers.

The Rocket Engine

A firework rocket, an inflated balloon that zooms around the room when it is released, a jet engine, and a rocket engine all work on the same principle, **Figure 9-13**. The compressed air that is forced into a balloon rushes out of the nozzle. This creates thrust, resulting in forward motion. Inside a rocket, a fuel is burnt to produce hot, high-pressure gas. This escapes from the rocket to provide the thrust.

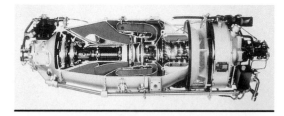

Figure 9-12 The turboshaft engine is lightweight and powerful. It is used on all but the smallest helicopters. (Top—Pratt & Whitney, Canada. Bottom—Bell Helicopter)

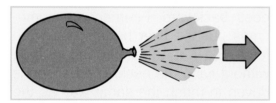

Figure 9-13 Rocket engines work on the same principle as compressed air rushing out of a balloon. Unequal pressure in the balloon and in the rocket causes them to move.

What pushes rockets forward? Imagine that the fuel is burning and the rocket's exhaust is closed. As the high-pressure gas burns, it pushes out in all directions against the inside of the rocket, **Figure 9-14A**. The rocket does not move because the force is equal in all directions. Now imagine that the exhaust is opened. Hot gas will rush through the opening. There is little or no downward force on the bottom of the combustion chamber, but there is

Useful Web site: Have students access the National Aeronautics and Space Administration (NASA) multimedia Web site at www.nasa.gov/multimedia/podcasting/vodcast_page_regular.html and view one or more of the short videos showing recent work at NASA.

upward force on the top. The rocket is pushed up, **Figure 9-14B**.

The difference between a jet engine and a rocket engine is in where they get oxygen to burn their fuel. The jet uses the oxygen in the air. A rocket must carry its own oxygen. It operates outside the earth's atmosphere where there is no air.

Rocket engines burn a variety of fuels called propellants. Some propellants are solid, while others are liquid. Nearly all space rockets, **Figure 9-15**, use liquid propellants. One of the propellants is the fuel, such as kerosene or liquid hydrogen. The other is liquid oxygen or some substance that can provide the oxygen for combustion.

Alternative Motors and Engines

Electric motors change electrical energy into mechanical energy. They provide smooth turning power to drive a shaft. Sometimes the electricity comes from the generators in power

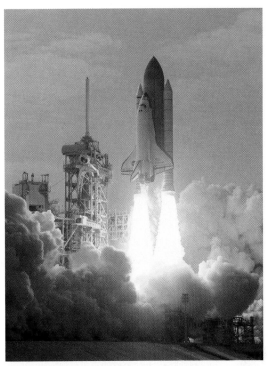

Figure 9-15 Because space has no oxygen, rockets must carry both fuel and oxygen for combustion. (NASA)

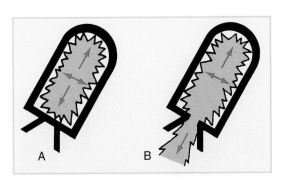

Figure 9-14 This is how rockets move. A—This rocket will not move because the pressure inside it is pushing equally in all directions. B—This rocket will move because the pushing force is greatest forward.

stations. This electricity is sent either by overhead cables to trains and trolleys or to the electrified rails used for subway trains, **Figure 9-16**.

Electric cars and small delivery vehicles have been in use for 100 years. They have many advantages, including little maintenance and no oil changes. They have one big disadvantage. Most use batteries that provide a driving range of less than 100 miles (160 km). Recharging the batteries takes several hours. This limitation means they are most suited to forklift trucks, local delivery vehicles, and golf carts.

Universities and technical colleges have been making experimental solar cars for many years. Usually,

Technology and society: Why do some countries subsidize the development of hybrid electric cars?

Figure 9-16 Electric motors power several types of transportation vehicles. A—Subway trains use electric motors because they do not pollute enclosed areas. B—Lift truck motors get their electricity from a battery. C—Many trolley cars run on electricity. (TEC)

solar cells that convert sunlight to electricity totally cover the upper surfaces of these cars. Each solar cell produces only about 1/2 volt of electricity, so hundreds of them are needed to power the car. The sun's energy not only powers the car's motor, but it also charges a battery that can supply the energy needed when the sun is hidden behind clouds. This method of powering a car is very interesting because the energy source is inexhaustible and no pollution results from its use. The 2005 North American Solar Challenge race made history. This special international edition began in Austin, Texas and finished in Calgary, Alberta, for a total of 2500 miles (approximately 4000 km). See **Figure 9-17**.

Fuel cell cars may solve the problem of distance, while also having the same low emissions level of electric vehicles. The *fuel cell* vehicle creates power by combining oxygen from the air with hydrogen from an onboard tank, **Figure 9-18**. The power is used

Figure 9-17 Eclipse 5 was a competitor in the Formula Sun Grand Prix American Solar Challenge. The solar cells, which cover most of the upper body surface, convert the energy in light into electrical energy.

Technology and society: Should governments subsidize the development and production of vehicles that use alternatives to the gasoline engine?

Ford's Focus FCV (Fuel Cell Vehicle)

A

B

C

Figure 9-18 A—This diagram shows the components of a fuel cell vehicle. B—Which parts do you think are the same as those in gasoline engine automobiles? Which parts look like they are specifically made for an FCV? C—The hydrogen tanks, wiring, air compressor inverter, and the water compressor inverter are located in the trunk of this FCV. (Ford)

to turn electric motors that drive the wheels. The only material leaving the exhaust pipe is water vapor. When methanol is used instead of hydrogen a small amount of carbon dioxide is produced. There are several different designs for fuel cell vehicles. Look at **Figure 9-19** and compare it with Figure 9-18. What are the advantages and disadvantages of each design?

Other types of engines and hybrids are being developed. The gasoline-electric hybrid combines a gasoline engine with electric motors. The gasoline engine is used mainly for power applications, such as accelerating and driving uphill. The electric motors usually take over when the car is coasting or going a set speed, **Figure 9-20**.

Other alternatives are the CNG, which uses compressed natural gas as fuel, and liquefied petroleum gas, which burns a mixture of mostly propane with some butane, **Figure 9-21**.

Transportation Systems

Modern vehicles, such as cars and buses, are complex machines composed of a number of subsystems. Automobiles have systems that control the emissions, lights, speed, and many other functions.

Systems in Vehicles

The modern car is composed of the following subsystems: electrical, emission control, computer, fuel, axles and drive train, steering, suspension, brake, heating and air conditioning, and engine. If one of these systems is

Standards for Technological Literacy

2 18

Discussion
Cars are usually owned or leased for a period of years, but are rarely used for more than a few hours a day. For the rest of the time, they sit in a driveway or parking lot. What kind of system could be designed so that cars can be borrowed for a short time to go shopping or to visit a friend?

Enrichment
All new vehicles in Delhi, India that transport paying passengers now operate using compressed natural gas (CNG). The pollution levels in the city have been drastically reduced.

Vocabulary
Figure 9-18 shows a car that uses "drive by wire" technology. What does that mean?

Safety
List the parts of a car that make up its safety system.

Resource
Transparency 9-9, *Electric Motor*

Resource
Activity 9-4, *Balloon-Powered Hovercraft*

Technology and society: How are the priorities of car companies different from the priorities of consumer protection agencies?

Discussion
List the products carried by road (highway), rail, and continuous flow transportation systems.

Vocabulary
Define the term *system.* Explain how transportation can be thought of as a system.

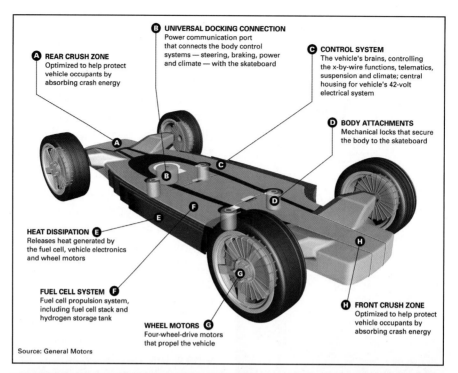

B UNIVERSAL DOCKING CONNECTION
Power communication port that connects the body control systems — steering, braking, power and climate — with the skateboard

A REAR CRUSH ZONE
Optimized to help protect vehicle occupants by absorbing crash energy

C CONTROL SYSTEM
The vehicle's brains, controlling the x-by-wire functions, telematics, suspension and climate; central housing for vehicle's 42-volt electrical system

D BODY ATTACHMENTS
Mechanical locks that secure the body to the skateboard

E HEAT DISSIPATION
Releases heat generated by the fuel cell, vehicle electronics and wheel motors

F FUEL CELL SYSTEM
Fuel cell propulsion system, including fuel cell stack and hydrogen storage tank

G WHEEL MOTORS
Four-wheel-drive motors that propel the vehicle

H FRONT CRUSH ZONE
Optimized to help protect vehicle occupants by absorbing crash energy

Source: General Motors

Figure 9-19 The AUTOnomy runs on hydrogen and oxygen. Each wheel has its own electric motor. There are no engine, drive train, axles, and pedals, so any shape of body can be fitted. (General Motors)

not working, the car's performance and efficiency will suffer. For example, if a car has a leak in its vacuum lines, it could affect the transmission, electrical, and power brake systems. Power brakes use the vacuum to help you push the pedal. The timing of the ignition system is also affected negatively. This example shows how one small problem can eventually cause a number of other troubles.

Useful Web site: Access your state or provincial Department of Transportation's Web site and identify the transportation systems that operate in your area.

Community resources and services: What transportation systems operate in your area? Describe the vehicles involved in the transportation systems and how they are coordinated. If possible, have a dispatcher from a transportation company meet with the class.

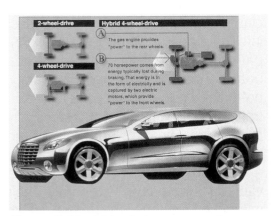

Figure 9-20 Hybrid electric cars often employ a regenerative braking process, which can provide power to the battery or directly to the electric motors. (DaimlerChrysler)

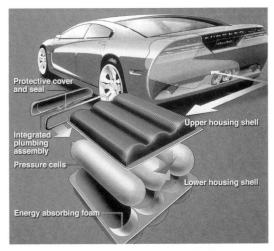

Figure 9-21 Automobile manufacturers are developing cars that use compressed natural gas as fuel. (DaimlerChrysler)

Systems in Society

Transportation systems are not those systems confined to the physical components of vehicles. *Transportation systems* are organized sets of coordinated modes of travel within a set area that are usually run by the local government. For example, in Chicago, the Chicago Transit Authority (CTA) has schedules for all three of its different modes of transportation—bus, subway, and "L" train. All three of these parts of the CTA must run on time. People are depending on these forms of transportation for timely arrivals to their destinations.

Imagine that your commute to work each morning starts with a bus ride from a local bus stop to the exchange station. Then you take the 8:30 bus to the subway station in order to ride the 8:50 subway train. After getting off at the next subway stop, you walk the rest of the way to your job. If the 8:30 bus were running late, you would miss the 8:50 subway. In the end, you would arrive late to work. If one part of the CTA is late, it is the public that suffers.

Transportation systems are not limited to ones at the local level, such as the CTA. They also include national and international systems that utilize planes, helicopters, boats, barges, semitrailers, trucks, and trains to deliver passengers, foods, and other goods. Schedules are made for each of these systems and businesses. Several processes help the entire transportation system run smoothly. They are the following: receiving, holding, storing, loading, moving, unloading, evaluating, marketing, managing, communicating, and using conventions. These are all necessary for the successful day-to-day operation of our modern transportation systems.

Standards for Technological Literacy

2 **18**

Discussion
What is the purpose of the cameras installed beside the highway?

Enrichment
Vertical takeoff and landing vehicles (VTOL), such as helicopters, can ascend and descend vertically into and out of small areas. Which of the six sectors of the economy listed on this page could make use of their services?

Activity
Have students select one of the six sectors of the economy listed on this page; ensure that each is covered. Ask them to observe any vehicles operating in that sector over a period of a few days and make note of their origin and destination.

Resource
Activity 9-5, *A Monorail Vehicle*

Community resources and services: Invite a long-distance trucker to meet the class. Ask about the routes traveled and the calculations made for distance covered and fuel used.

Like all other technological systems, transportation systems include inputs, processes, outputs, and feedback. Inputs include the people and materials that are transported; the people who develop, operate, and maintain the system; and the machines and structures within the system. Processes include managing and organizing the system, along with the actual transport of materials or people from one location to another. Outputs include the successful transportation of materials or people, the impacts on society, and the impacts on the environment. Feedback involves periodic checks comparing the location of cargo within the system to the transportation schedule. Another example of feedback is the evaluation of the completed transportation process—did the cargo arrive at the intended location at the intended time without unintended incident?

Another part of our transportation systems is our state Departments of Transportation and our National Department of Transportation. State Departments of Transportation are in charge of regulating state transportation laws and working with several other bureaus and divisions. The National Department of Transportation develops plans and programs to help improve the nation's transportation systems.

The Division of Highways is in charge of the construction, maintenance, and design of state highways. Highways and other paths of travel (sidewalks, bicycle paths, etc.) can be classified into two different categories: intelligent and nonintelligent. Intelligent pathways have electronic message boards that display lane capacity, potential congestion problems, and traffic flow. Unintelligent pathways rely on natural beauty and interesting designs to attract users.

Transportation systems play a vital role in today's economy. Their influence reaches to nearly all of today's industries:

◆ Agriculture—Trucks, trains, and planes deliver fruit, vegetables, and other agricultural goods to vendors and grocery stores.

◆ Manufacturing—Coal, iron ore, steel, and lumber are taken to mills and plants by boat, train, and semitrailer.

◆ Construction—Trucks take lumber, bricks, mortar, and concrete to job sites.

◆ Communication—News vans and trucks go on-location with camera operators and news anchors to broadcast stories. Helicopters fly over the city and report the traffic to TV and radio stations.

◆ Health—Vans, ambulances, and helicopters take organs from donors to recipients and take injured people to the hospital.

◆ Safety—Police and government agencies use squad cars, vans, buses, planes, and helicopters to enforce laws, respond to accidents.

Useful Web site: Access the NASA and Canadian Space Agency Web sites to find information about space industries that have been formed, or are being formed, as a result of space exploration, such as mining minerals from asteroids and extracting gases from the atmosphere for energy and life support.

Transportation systems are designed by civil engineers and city planners. Roadways are heavily influenced by the purpose of the transportation system, government regulations, and local ordinances.

Quality control in transportation involves inspection and maintenance of three separate components: transportation vehicles, transportation structures, and transportation processes.

Transportation vehicles such as automobiles, locomotive engines, and boats are manufactured using a quality control process that produces an acceptable product. These vehicles also require periodic maintenance to ensure proper operation.

The construction of transportation structures includes quality control methods typical of the construction industry. These include material testing and on-site inspections. Following construction, structures such as roads, bridges, tunnels, and railways are continually inspected and repaired as needed.

The quality of a transportation process is based on the effectiveness of the system in transporting goods or people in a timely and safe manner to the intended destination. The process is controlled by monitoring the system and making adjustments as needed. For example, overcrowded roads can produce unacceptable delays and unsafe conditions. To remedy this situation, the city planner may recommend adding more lanes to the road, improving traffic control systems, developing additional routes, or providing mass-transit options.

Exploring Space

What is it like to go into space? Can you imagine being one of seven people strapped to rockets the height of a 20-story building? When the solid rocket boosters (SRBs) are ignited there is no going back. The fuel cannot be extinguished. The booster rockets propel the orbiter to a speed of 3,500 miles per hour (5,700 km per hour). Once the vehicle has left earth's atmosphere, large thrusts of energy are not needed. The spacecraft is no longer trying to overcome the earth's gravity. Therefore, after two minutes of flight, the SRBs separate from the orbiter at an altitude of 30 miles (49 km). They fall to the earth with the aid of parachutes and are picked up by retrieval ships.

The orbiter's main propulsion unit then takes over. It consists of three engines burning liquid hydrogen and oxygen. After about eight minutes of flight, the orbiter's main engines shut down and the small Orbiting Maneuvering System (OMS) engines take over. To return to earth, the orbiter turns around and fires its OMS engines. After descending through the atmosphere, it lands like a glider. On-board computers control all the functions. There are three Space Shuttle Orbiters in operation. They are Discovery, Atlantis, and Endeavour.

There was a time when the destination of flights into space was the Moon. Today, many flights with humans aboard are destined for the International Space Station (ISS), in order to deliver component parts of the

Standards for Technological Literacy

Reflection
What experiences have you had with vehicles that pollute the air? What kinds of vehicles pollute the most?

Activity
Use the Internet to investigate the living conditions for astronauts living on the ISS. Comment on the accommodations for sleeping, eating, showering, and working.
Use the Internet to investigate the types of experiments that are being conducted on the ISS.

Links to other subjects: Literacy—Describe how space stations you have seen or read about in science fiction compare with the ISS.

Community resources and services: Contact NASA or Canadian Space Agency to determine if your class may be able to communicate with an astronaut by phone, e-mail, or in person.

space station, **Figure 9-22**. The ISS is a sprawling assembly of laboratories, living space, service areas, and solar panels the size of two football fields. Sixteen countries are cooperating in its construction.

Astronauts and cosmonauts travel to the International Space Station to conduct experiments in space. For example, one experiment studies the effects of blood pooling and poor blood pressure regulation caused by reentry and landing. Both of these conditions can inhibit cerebral activity and influence judgment and reason.

No large cranes are available in space to lift the pieces of the ISS into place. In their place, there is a "space arm" known as the Space Station Remote Manipulation System (SSRMS), **Figure 9-23**. The SSRMS not only assembles the station but also maintains the station and maneuvers equipment and payloads. Its two hollow booms, which are the diameter of telephone poles, are made of 19 layers of carbon fiber. They are joined at an aluminum elbow and end in a "hand" that is like sockets from a mechanic's ratchet set.

Space station technology involves initiatives to develop and transfer technology to other sectors of our economy. Space arm mechanisms

Figure 9-22 The International Space Station (ISS) is in use as an engineering and science laboratory. (NASA)

Technology and society: Think of a destination to which a large number of people in your community travel. For example, if you live in a suburb, many people will travel to a nearby city each day to work. Describe an ideal but practical transportation system that would get them to their destination, as rapidly as possible, with the least amount of pollution.

Community resources and services: Contact your local transit authority/corporation. Find out what their future plans are for your area.

Figure 9-23 The Space Station Remote Manipulation System (SSRMS) was a contribution the Canadian Space Agency made to the International Space Station (ISS). (NASA)

are being modified to control prosthetic hands for children. Space arm hardware and software are used in robotic devices.

Living on the ISS will be like going to the Arctic or to a drilling rig in the Atlantic Ocean. People will stay there for one to three months, then return to Earth for a long rest. Since the ISS is expected to have a 15-year life span, there could be 100 visits from space shuttles over this period in order to deliver personnel, supplies, and experimental cargo.

Both European and US space agencies have plans to go further into space. In particular, there are Mars programs, including plans for robots to bring back samples to Earth. Mars is an exciting prospect as a NASA research team has found evidence that strongly suggests primitive life may have existed on Mars more than 3.6 billion years ago.

The Impact of Transportation

In general, transportation moves people and materials as rapidly as possible, **Figure 9-24.** But vehicles cause pollution, noise, and, sometimes, traffic tie-ups. Even if you only have a week to spend with friends in another country, you can get there in less than a day. Products can be delivered from all points of the globe to your local supermarket. Fresh fruit and vegetables are available at any time of the year.

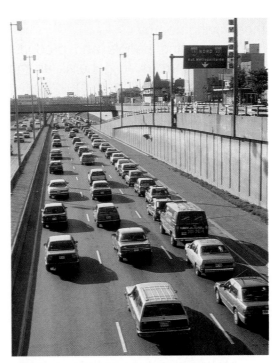

Figure 9-24 Transportation has both good and bad effects. It is needed to move people and materials, but vehicles cause pollution, noise, and, sometimes, excessive traffic. (Ecritek)

Technology and society: Visit your local supermarket and investigate the origin of fruits and vegetables. What means of transportation do you think were used to bring them to the store?

Useful Web site: Have students access the following Web sites and share information they find interesting with the class:
http://www.bp.com/extendedsectiongenericarticle.do?categoryId=9006010&contentId=7012265 and http://www.carbonfootprint.com/.

Enrichment
Carbon dioxide levels are steadily increasing. The current increase is twice as fast as it was in the 1970s. In 1960, it was at a level of 315 parts per million; by 2006, it reached 381 parts per million.

Activity
Investigate what steps are being taken by your government either to reduce carbon dioxide emissions of motor vehicles or to alter gasoline so fewer harmful pollutants are emitted.

Transportation systems make our lives much more convenient, but they do have negative impacts. Several modes of transportation pollute the environment. The pollutants in exhaust emissions from cars, trucks, buses, and airplanes cause many problems. Air pollution causes a bad smell in the air. The pollution level may be so high that people have difficulty breathing. Their eyes water, and they experience a burning sensation in their lungs. Air pollution can also corrode the surface of buildings. The features of historical buildings are eaten away. Noise pollution is another problem. The noise from highways and airports increases the stress level of those living nearby.

Transporting people and products can be dangerous. In general, as the number of vehicles and their speed increases, so does the number and severity of accidents. One way to reduce these problems is to make use of mass transit, **Figure 9-25**. Most cars carry a driver and, occasionally, a few passengers. In contrast, one bus will carry more than 50 passengers, and a commuter train can seat hundreds.

When there are no local buses or trains available, people should be encouraged to drive to a train station or to carpool to a station and then to take mass transit from that point. In some cities, drastic measures have been taken to reduce traffic problems. Special lanes have been marked on the roads to be used only by full cars or by buses. In some countries, cars are banned from city centers.

Figure 9-25   One way to avoid some of the disadvantages of car transportation is to use mass transit, such as trains. Top—These commuters walk or drive to the train stations and ride to their jobs or shopping. Bottom—The Las Vegas monorail system is the first fully automated, large-scale monorail in the United States. (Bombardier)

Many large communities are examining ways to improve mass transit systems and encourage their use. At the same time, restrictions on the unlimited use of private cars are being considered. Each day tens of thousands of cars stream into and out of many major cities. Each provides convenience and privacy to its occupants but at considerable cost to the environment. Can we continue to pay this cost?

Useful Web site: Access the following Web site and compare the mass transportation vehicles used in your neighborhood with those shown: http://en.wikipedia.org/wiki/Public_transport.

Chapter 9 Review
Transportation

Summary

Transportation systems move people or freight from a point of origin to a point of destination. These systems rely on both the vehicles and the people that operate, maintain, and repair the vehicles. A system may use one or more modes of transportation: bicycle, car, truck, bus, train, subway, ship, airplane, or space shuttle.

Most modes of transportation need an engine to make them move. An engine is a machine composed of many mechanisms that converts a form of energy into useful work. Engines are classified into two groups: internal combustion and external combustion. Internal combustion types include reciprocating piston engines and turbines. Reciprocating piston engines are used in motorcycles, cars, trucks, and buses. Turbine engines are used in aircraft, hovercraft, and helicopters.

The most common form of reciprocating piston engine is the gasoline engine. In a gasoline engine, a mixture of air and gas is ignited. Pistons are pushed down to drive a crankshaft in a rotary motion. The crankshaft then transmits turning power to the drive train. In a diesel engine, air is squeezed to very high pressure. A fine spray of diesel fuel is injected and spontaneous ignition occurs.

In a turbine engine, a large wheel with blades is turned by water, a jet of steam, or hot gases. A water wheel uses the force of falling water. A steam turbine uses a jet of steam. A turbofan engine uses a stream of hot gases.

A rocket engine works in a similar way to a turbofan engine. Forward motion is the result of the thrust of high pressure gas escaping from the rocket.

Electric and fuel cell vehicles are being developed as alternatives to gasoline engines. Electric motors change electrical energy into mechanical energy. Fuel cells combine oxygen from the air with hydrogen from tanks to create power.

Transportation systems are used to transport people and goods and affect the following industries: agriculture, manufacturing, construction, communication, health, and safety. The types of processes involved with transportation systems are receiving, holding, storing, loading, moving, unloading, evaluating, marketing, managing, communicating, and using conventions. Government organizations, such as the state and National Departments of Transportation, regulate the various systems.

Transportation systems make our lives more convenient. However, they also have negative impacts on the environment due to air and noise pollution. Also, as the number of vehicles and their speed increase, so do the number of accidents and the traffic congestion. One way to reduce these problems is to increase the use of public transportation.

Modular Connections

The information in this chapter provides the required foundation for the following types of modular activities:

- Transportation Systems
- Transportation Technology
- Transportation Modeling
- Automotive Design
- Watercraft Design
- Rocket Design
- Aerodynamics
- Automotive Safety
- Land Transportation
- Automobile Technology
- Engines
- Small Gas Engines
- Four-Stroke Small Engine
- Flight Technology
- Flight Simulator
- Flight Trainer
- Rocketry and Space
- Aerospace
- Alternative Energy

- ◆ Solar Energy
- ◆ Dragster Design
- ◆ Auto Demolition Derby
- ◆ Radio-Controlled Transportation
- ◆ Catapults
- ◆ Digital Transportation

Test Your Knowledge

Write your answers to these review questions on a separate sheet of paper.

1. Describe the ways in which the modes of transportation affect the way your family lives.

2. State the best mode of transportation for each of the following situations:
   A. A student traveling five blocks to the sports field.
   B. Carrying a huge load of grain from one continent to another.
   C. Transporting passengers across town without congestion or pollution.
   D. Providing a cheap and flexible form of urban mass transit.

3. List the modes of transportation that you have used in the past. List the advantages and disadvantages for each.

4. Name, in the correct sequence, the four strokes of an internal combustion engine.

5. What is the function of an intake valve?

6. What is the function of a spark plug?

7. What is the function of an exhaust valve?

8. What is the major difference between a gasoline engine and a diesel engine?

9. Make a sketch to show how a jet of steam can be used to turn a shaft.

10. What is the basic principle of a jet engine?

11. Turbofan engines are used to power _____ and turboshaft engines are used to power _____.

12. Use sketches to show how rockets are pushed forward.

Answers to Test Your Knowledge Questions
1. Student responses will vary. Refer to the introduction to Chapter 9.
2. A. Bicycle. B. Ship. C. Subway train. D. Bus.
3. See Figure 9-1, page 229.
4. Intake. Compression. Power. Exhaust.
5. It allows a mixture of gasoline and air to enter a cylinder.
6. It ignites a mixture of gasoline and air in a cylinder.
7. It allows the burnt gases to escape from the cylinder.
8. In a gasoline engine, a mixture of gasoline and air is compressed and ignited by a spark. In a diesel engine, pure air is compressed; diesel fuel is then injected and ignition occurs spontaneously.
9. See Figure 9-8, page 238.
10. For every force in one direction, there is an equal force in the opposite direction.
11. jet airplanes, helicopters.
12. See Figure 9-14, page 241.
(continued)

13. Explain why a rocket engine must carry its own oxygen.

14. Electric motors change _____ energy into _____ energy. They do not pollute the environment because _____.

15. Several modes of transportation pollute the environment. State one way in which this pollution can be minimized.

Apply Your Knowledge

1. List the modes of transportation that you use in a period of one week. State the advantages and disadvantages of each.

2. Name the most appropriate mode of transportation to move:
   A. a prize bull from a farm in Texas to a farm in Alberta.
   B. 10,000 commuters from a suburb to a city center.
   C. a transplant organ from one city to another, 500 miles (800 km) away.
   D. a large supply of fresh fruit from South America to North America.

3. Identify one problem in your town caused by a mode of transportation. Describe the problem and suggest ways to resolve it.

4. Design and build a vehicle that will travel 15′ (5 m) across a flat surface. The only power source permitted is a large elastic band and a propeller.

5. Design and build a balloon rocket using a balloon, drinking straw, clear tape, and string. The rocket must travel in a straight line for a distance of at least 15′ (5 m).

6. Future possibilities for powering spacecraft include:
   A. reusable single-stage-to-orbit vehicles.
   B. magnetically levitated catapults.
   C. light sail aircraft powered by lasers.
   D. microwave beamed energy.
   E. space-tethers carrying thousands of volts.
   Draw a picture of one of these methods, describe its operation, and comment on its advantages and disadvantages.

7. Create an attractive poster to show the relative distances of Earth, Mars, Venus, and Jupiter. Add interesting information about each of the planets to your poster.

8. If we are ever able to make contact with a civilization that is 250 light-years away and ask a question, it would take 500 years to get a reply. What question would you ask?

9. Create an extraterrestrial character, either as a drawing or in 3-D form, for use in a new movie. Keep in mind that there have been many such characters in movies including *The Man Who Fell to Earth*, *Close Encounters of the Third Kind*, *Star Trek: The Motion Picture*, *Alien*, *E.T. the Extra-Terrestrial*, and *Galaxy Quest*.

10. What negative effects does an astronaut experience by being in space for prolonged periods of time? What happens to his muscles, ears, sense of touch, heart, spine, bones, lungs, stomach, and eyes?

11. Research one career related to the information you have studied in this chapter and state the following:
    A. The occupation you selected.
    B. The education requirements to enter this occupation.
    C. The possibilities of promotion to a higher level at a later date.
    D. What someone with this career does on a day-to-day basis.
    You might find this information on the Internet or in your library. If possible, interview a person who already works in this field to answer the four points. Finally, state why you might or might not be interested in pursuing this occupation when you finish school.

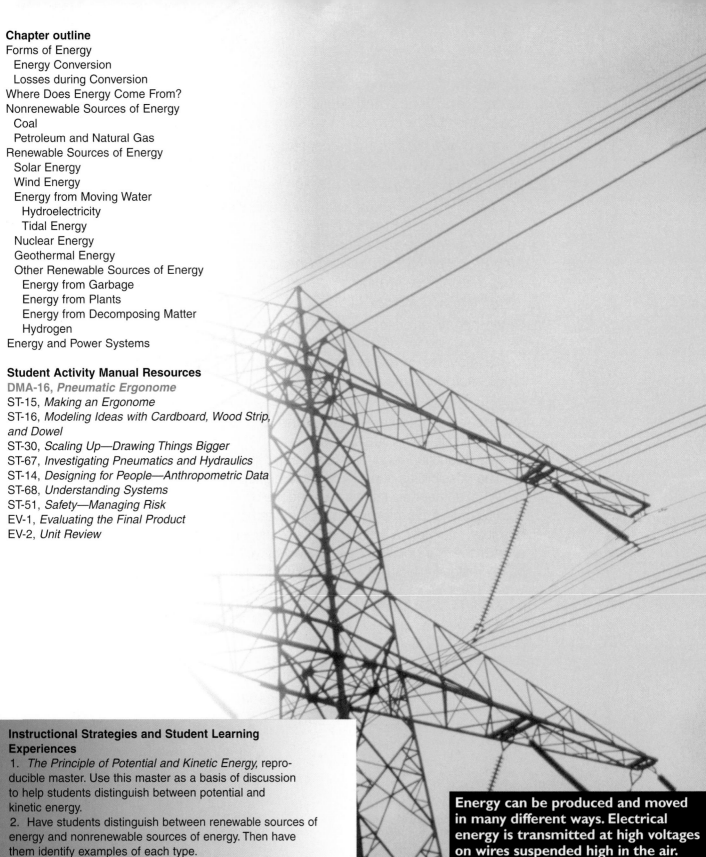

Chapter outline

Student Activity Manual Resources

DMA-16, *Pneumatic Ergonome*
ST-15, *Making an Ergonome*
ST-16, *Modeling Ideas with Cardboard, Wood Strip, and Dowel*
ST-30, *Scaling Up—Drawing Things Bigger*
ST-67, *Investigating Pneumatics and Hydraulics*
ST-14, *Designing for People—Anthropometric Data*
ST-68, *Understanding Systems*
ST-51, *Safety—Managing Risk*
EV-1, *Evaluating the Final Product*
EV-2, *Unit Review*

Instructional Strategies and Student Learning Experiences

1. *The Principle of Potential and Kinetic Energy,* reproducible master. Use this master as a basis of discussion to help students distinguish between potential and kinetic energy.

2. Have students distinguish between renewable sources of energy and nonrenewable sources of energy. Then have them identify examples of each type.

3. If possible, visit a home or business that uses solar panels. Ask students to prepare a list of questions in advance to ask the owner about the effectiveness of these panels.

4. Ask students to write paragraphs predicting what types of energy sources will and will not be used 50 years from now.

5. Have students divide into groups to debate the advantages and disadvantages of the following energy sources: solar energy, wind energy, energy from moving water, hydroelectricity, tidal energy, wave energy, nuclear energy, geothermal energy, and biomass energy.

Energy can be produced and moved in many different ways. Electrical energy is transmitted at high voltages on wires suspended high in the air.

Energy

Discussion
Discuss the importance of energy in the daily life of each student and his/her personal uses of energy.

Career connection
Have students identify careers associated with the production, monitoring, and maintenance of energy supplies.

Key Terms

biomass energy
chemical energy
convection
electrical energy
energy
geothermal energy
gravitational energy
heat energy
hydroelectricity
hydrogen
kinetic energy
light energy
mechanical energy
nonrenewable energy
nuclear energy

nuclear fission
nuclear fusion
photovoltaic
potential energy
radiant energy
radiation
renewable energy
solar energy
sound energy
strain energy
thermal energy
tidal energy
wave energy
wind energy

Objectives

After reading this chapter you will be able to:

◆ Explain society's dependence on energy.

◆ Describe the difference between potential and kinetic energy.

◆ Identify the various forms of energy and their applications.

◆ Describe how energy can be changed from one form to another.

◆ Distinguish between nonrenewable and renewable sources of energy.

◆ List the advantages and disadvantages of each source of energy.

◆ Design and make a pneumatically controlled robot.

Discussion
Ask students to give examples of how springs and elastic bands are used in toys, games, furniture, and machinery. Identify each as an example of potential energy or kinetic energy.

Enrichment
The word *kinetic* comes from the Greek *kinema* meaning motion. What English word does this remind you of where motion is involved?

Vocabulary
Define the term *energy*. State the difference between *potential energy* and *kinetic energy*.

Resource
Reproducible Master, *The Principles of Potential and Kinetic Energy*

Every day we use energy in one form or another. When you ride your bicycle to school, you are using your own physical energy to turn the pedals. When you fly a kite, wind gives energy to keep the kite in the air. The family car uses the energy in gasoline.

At home, the sun shines in your windows and warms up your house. The sun provides the energy. In winter, the energy stored in wood, coal, or other fuels heats your home. When you turn on a light, you are using electrical energy.

What is energy? *Energy* is the capacity to do work. To understand it more fully, we need to look at two categories: potential energy and kinetic energy.

Think about lifting a sledgehammer to drive a post into the ground. As you hold the hammer in the air, it has potential energy. Energy is stored in the hammer until it is dropped, and the energy is released to hit the post. When you wind up a spring in a toy, you are giving the spring potential energy. Once again, energy is being stored. A stretched elastic band also has potential energy. *Potential energy* is often called "stored energy."

When the head of the hammer is dropped to hit the post or the spring is released to drive the toy, both the hammer and the toy gain kinetic energy. *Kinetic energy* is the energy an object has because it is moving, **Figure 10-1**.

It is important to realize that energy is neither created nor destroyed. We

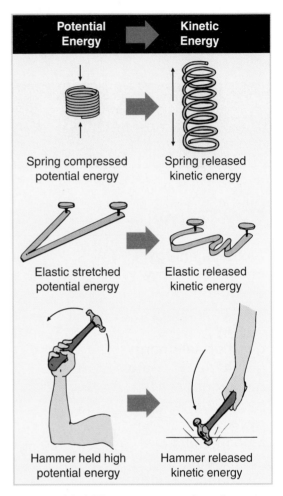

Figure 10-1. These are examples of potential and kinetic energy.

are simply changing its form. The first law of thermodynamics states: "The total amount of energy in the universe remains constant. More energy cannot be created; existing energy cannot be destroyed. It can only be converted from one form to another." You will see examples of conversion in the next section.

Links to other subjects: Science—Describe some of the experiments you have conducted in science classes to investigate potential and kinetic energy.

Useful Web site: Students can use the dictionaries at www.techtionary.com and www.hyperdictionary.com to find the definition of any new technical and technological terms used in this chapter.

Forms of Energy

Energy is available in many different forms. Each picture describes a different form of energy and the way energy gets things moving so that work is done.

A great deal of *chemical energy*, **Figure 10-2**, is locked away in different kinds of substances. Such energy is found in the molecules making up food, wood, gasoline, and oil. The energy often is released by burning the chemical. Burning rearranges the substance's molecules and releases heat.

Objects always tend to move toward the lowest possible level. This is due to the *gravitational energy* (attraction or pull) of the earth. It causes objects to fall. It is why water runs or objects roll downhill, **Figure 10-3**.

Energy of motion, often associated with or caused by a machine is *mechanical energy*. However, mechanical energy is not always caused by a machine. Two good examples of mechanical energy are a waterfall (natural mechanical energy), **Figure 10-4**, and a hydroelectric power plant (machine-related mechanical energy). The power plant harnesses the falling water's mechanical energy by using it to turn turbines. These turbines are connected to generators, which produce electricity.

Certain materials that can be stretched or compressed have a tendency to return to their original shape. This is known as *strain energy* or the energy of deformation. It is the kind of energy found in a rubber band, a fishing pole, or a bow used to shoot arrows, **Figure 10-5**.

- A skateboard at the top of a hill has gravitational energy. This energy is available because of the pull of gravity.
- Gravitational energy is potential energy before the object moves and kinetic energy when the object is moving.

Figure 10-3 Gravitational energy has energy because of its position.

- Chemicals in the body provide the muscles with energy to do work.
- Gasoline provides the chemical energy to keep the motorcycle moving.
- Chemical energy can be stored in batteries.

Figure 10-2 Chemical energy is used in natural (human body) and artificial (batteries) systems.

Community resources and services: Invite an engineer to the class to describe how chemical, gravitational, mechanical, or strain energy is used in his/her work.

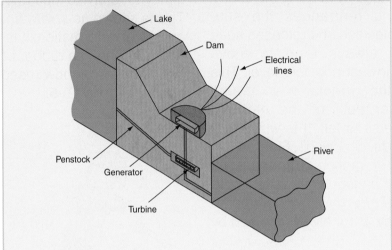

Figure 10-4 Mechanical energy is demonstrated in two ways: naturally by the falling water and machine-related by the movement of the turbines.

Electrical energy is the movement of particles from one atom to another. Atoms are made up of several particles: neutrons, protons, and electrons. Neutrons are neutrally charged; they are neither positive nor negative. Protons are positively charged, while electrons are negatively charged. Electrons will jump from one atom to another if attracted by protons. This energy is enough to provide light and operate electrical devices, such as motors and heaters, **Figure 10-6**.

Heat energy occurs as the atoms of a material become more active. Another name for heat energy is *thermal energy*. If you could look at atoms under a strong enough microscope, you would see that the atoms move about. The faster they move, the warmer the material, **Figure 10-7**.

Light energy is related to heat energy. Another name for it is

Links to other subjects: Science—What is the second law of thermodynamics and how is it used?

- Strain energy can be stored in a material that stretches, such as elastic, nylon, or a spring.
- When an arrow is fired from a bow, the potential (strain) energy of the bow and string changes into the kinetic energy of the arrow.

Figure 10-5 An example of strain energy or energy of deformation is the bow and arrow. Some materials are elastic and try to return to their original shape or size.

- Electrical energy is a flow of electrons from one atom to another within a conductor.
- Electrical energy can move from place to place and readily changes into other forms such as heat, light, and sound.
- Electrical energy can be stored in batteries or produced in a generating station.

Figure 10-6 How many of the devices you use every day need electrical energy?

radiant energy. It travels as a wave motion. This is a type of energy that is coming to us from the sun. Since we use the sun to power certain devices, we often refer to light energy as solar energy. Light is part of the electromagnetic spectrum. It travels at 186,000 miles (300,000 km) per second, **Figure 10-8**.

Sound energy is a form of kinetic energy, **Figure 10-9**. It moves at about 1100′ (331 m) per second. This is much slower than light energy.

Nuclear energy occurs as atoms of certain material are split. This action, called nuclear fission, creates huge amounts of energy, **Figure 10-10**. Most of it is in the form of heat. Another way of producing nuclear energy is through fusion, which is the joining together of atoms.

Energy Conversion

One of the most important laws of science states that energy can neither be created nor destroyed. It can only be changed from one form to another. The energy you use to pedal your bicycle comes from the food you eat. Your body has made a change in the form the energy takes.

There are many examples of energy changing its form. In the example of a cyclist, the chemical energy in the muscles is changed to kinetic energy of the bicycle. When the

Discussion
Some bikes have a small generator that rubs against a wheel to provide power for its lights. What kind of energy conversion is taking place?

Activity
Refer to Figure 10-7. What other examples of conduction, convection, and radiation can you think of?

Examples
Figure 10-7 shows examples of energy conversion by changing electrical energy to heat energy. List as many other examples of energy conversion as you can.

Resource
Activity 10-2, *Changing the Form of Energy II*

Heat energy travels through matter in three ways:
• Convection occurs when expanded warm liquid or gas rises above a cooler liquid or gas. When liquid is heated, it expands and its volume increases. The amount of material (its mass) does not change. Since its mass is more spread out, hot liquid is less dense than cold liquid around it. In a mixture of hot and cold liquid, the cold liquid will sink to the bottom and the hot liquid will rise to the top.

• Conduction occurs when heat energy passes from molecule to molecule in a solid. There is movement of heat energy without any obvious movement of the material.

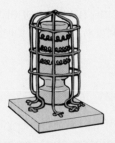

• Radiation occurs when heat energy is moving in the form of electromagnetic waves. For example, when you stand in the sunshine or in front of an electric heater, heat is transmitted without involving a material between you and the source of the heat.

Figure 10-7 Heat energy can move through matter by convection, conduction, and radiation.

brakes are applied, this kinetic energy is changed into heat energy as a result of friction between the brake shoes and the wheel. When a flashlight is

• Light travels in straight lines.
• TV pictures, lamps, and the sun are seen because of the light they send out.
• Most other objects are seen because they reflect light.

Figure 10-8. Most objects have no light of their own. We see them by the light they reflect.

switched on, the chemical energy in the battery is changed to electrical energy. Electrical energy is changed to light energy when the bulb is lit. A bungee jumper has gravitational energy because of the height above the ground. This energy is changed to kinetic energy during the dive. If the horn of a car blows, electrical energy becomes sound energy.

As you can see from the example of the cyclist and the flashlight, energy is used in a variety of natural and artificial power systems to provide the energy needed to drive actions, processes, and other power systems.

Links to other subjects: Science—Fireflies emit a bright spark of light at night. How is this light created?

- Sound is produced when matter, such as a tuning fork or human vocal cords, vibrates.
- The vibrating object has kinetic energy due to movement.
- The string on a guitar has potential energy when pulled back. When released, it has kinetic energy.

Figure 10-9 Sound waves carry vibrations from a source to our ears and make our eardrums vibrate.

Losses during Conversion

When switching on a lightbulb, you may expect to change all of the electrical energy to light energy. Not so! Only a portion of the electrical energy is converted into light energy. The rest is converted into heat energy. You can feel the heat produced by holding your hand near to the lightbulb, **Figure 10-11**. In all energy changes, some of the energy is used as intended, but some is wasted. However, remember that although there has been a change in the form of energy, the total

Figure 10-10 Uranium is used in nuclear power stations to produce heat to turn water into steam. (Michael Rennhack)

Figure 10-11 A lightbulb is one example of energy conversion. Electrical energy converts into light energy and heat.

amount of energy remains the same: energy is neither created nor lost. Engines follow this principle. No one can build an engine that performs its work and does not give off thermal energy to the surroundings.

Useful Web site: Use the Internet to investigate the sound level (in decibels) of various activities. At what decibel level does temporary or permanent damage to human hearing occur? Helpful Web sites to begin a search are www.infoplease.com and www.ask.com.

Discussion
Why will nonrenewable sources of energy NOT be replaced?

Enrichment
Describe differences between the various forms of coal (peat, lignite, bituminous, and anthracite) and identify which is the most common.

Vocabulary
What is the difference between a *renewable* and a *nonrenewable* energy source?

Activity
Sketch a diagram illustrating the interior of an underground mine. Include the main shaft, airshaft, coal seams, preparation plant on the surface, and layers of other rock (such as sandstone, shale, and limestone).

Resource
Transparency 10-1, *Sources of Energy*

Resource
Reproducible Master, *Sources of Energy*

Where Does Energy Come From?

Much of the energy we use comes from the sun. The sun's heat keeps us warm. Heat from the sun causes wind and rain. Most plants need light energy from the sun for growth. These plants provide humans and animals with the energy they need to do work. Over millions of years, some of these plants have been changed into petroleum and coal. These fossil fuels may be used to provide energy for machines.

All sources of energy make up two groups: nonrenewable and renewable. *Nonrenewable energy* sources will eventually be used up and cannot be replaced. They include coal, oil, and natural gas. *Renewable energy* sources will always be available. They include the sun, wind, and water. These two major groups are summarized in **Figure 10-12**.

Nonrenewable Sources of Energy

The most important nonrenewable sources of energy are coal and petroleum. All were most probably formed from the remains of living matter.

Coal

Coal developed from the remains of plants that died millions of years ago. For this reason, it is often referred to as a fossil fuel. The coal-forming plants probably grew in swamps. As the plants died, they gradually formed a thick layer of vegetable material.

Renewable	Nonrenewable
Water	Coal
Sun	Petroleum
Wind	Gas

Figure 10-12 Energy comes from a source, which is either renewable or nonrenewable.

Sometimes, ancient seas covered this layer. Sediments buried the plant layers. As this process was repeated, the layers of vegetable material became squeezed under great pressure and heat for a long time. The result was coal, **Figure 10-13**.

Removing coal from the ground is called mining. Coal mines are of two types: surface mines and underground mines. Surface mining involves stripping away the soil and rock that lie over a coal deposit. The coal can then easily be dug up and hauled away, **Figure 10-14**. Underground mining involves digging

Technology and society: What hazards do miners face and what precautions can be taken to reduce the risks?

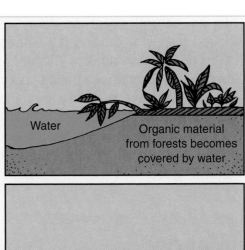

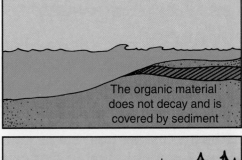

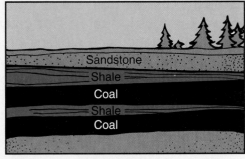

Water

Organic material from forests becomes covered by water.

The organic material does not decay and is covered by sediment

Sandstone

Shale

Coal

Shale

Coal

Figure 10-13 After millions of years of heat and pressure, organic material becomes coal.

tunnels into the coal deposit. Miners go down a shaft in a large elevator and then ride through the tunnels in cars. The cars take them to the coal face where large machines rip coal from its million-year-old home, **Figure 10-15.** Coal is primarily used as a fuel for electrical power generating stations. It is also used to power industrial processes, particularly those manufacturing steel.

Figure 10-14 In open pit or surface mining, a dragline removes soil and rock to expose a coal deposit. (Seeds)

Figure 10-15 In an underground coal mine, the tracks are for coal cars that carry coal and miners to the coal face. (Seeds)

Petroleum and Natural Gas

Petroleum was formed from the bodies of countless billions of microscopic plants and animals that lived in the seas millions of years ago. As these animals and plants died, they

Discussion
Discuss how coal mining can alter the landscape. What measures can be taken to reclaim the land after mining?

Enrichment
What must be done to coal so that it can be pumped through pipelines?

Activity
Investigate and describe the methods used in transporting coal to industries and generating stations.

Links to other subjects: Geography—Where are coal reserves located in North America?

Standards for Technological Literacy

Discussion
Discuss how oil is transported from off-shore drilling platforms to a refinery. What are the dangers involved and the precautions that must be taken?

Vocabulary
Describe the process of *distillation* or *cracking*. What petroleum products are derived from this process?

Activity
What types of products use each of the following natural gases: butane, ethane, methane, and propane?

sank to the bottom to mix with mud and sand on the sea floor. As more and more layers were added, pressure increased and the remains of the sea life in the deeply buried layers slowly transformed into oil and gas, **Figure 10-16**.

Oil and gas are removed from the ground by drilling deep holes. Holes are made either by drilling rigs located on land or by drilling platforms situated on the ocean, **Figure 10-17**.

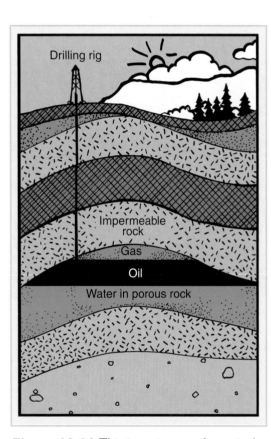

Figure 10-16. This is a picture of a typical rock formation containing oil and natural gas deposits. When a well is sunk, the oil may come out in a gusher. This is due to underground pressure. Sometimes, however, it must be pumped to the surface.

In its natural or crude form, oil removed from the ground is useless. The crude oil is processed in an oil refinery. The process consists basically of heating the crude oil. This separates the components. The lighter components rise to the top and the heavier components remain at the bottom, **Figure 10-18**.

Renewable Sources of Energy

Renewable energy is energy that does not get used up. The energy provided by the source can be renewed as it is used. In the past, most energy has been obtained from burning nonrenewable fossil fuels. Currently, hydroelectric generating stations produce a great deal of electrical energy. But as the nonrenewable sources of energy become scarce, alternative sources (solar, wind, tidal, geothermal, and biomass) are being developed.

Figure 10-17 Offshore drilling platforms are needed to drill oil wells under the oceans. (Seeds)

Technology and society: The 1997 Kyoto Protocol requires nations, who signed the agreement, to reduce greenhouse gas emissions an average of 5.2% from the 1990 levels by the year 2010. How can countries achieve this reduction?

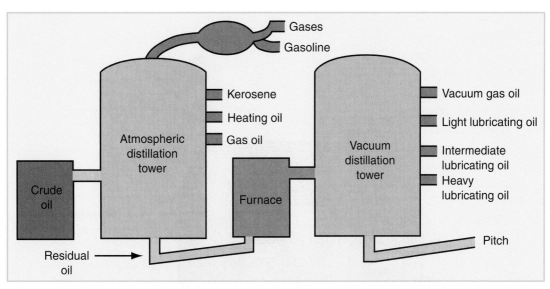

Figure 10-18. By distillation or "cracking," crude oil can be converted into more than 800 products.

Solar Energy

Solar energy, energy from the sun, is the most important of the alternative sources of energy. The idea of collecting energy from the sun may seem to be a very good one, but the main drawback is that this type of energy is not always available. In the winter and on cloudy days, there may be too little. At night, there is none. Yet these are the times when it is most needed; however, energy from the sun can be collected and stored for later use.

A solar panel collects heat from the sun's rays. The heat is carried away to provide hot water or to heat buildings. In one kind of solar panel, water flows through pipes or channels under a plate of glass. These pipes or channels are painted black to absorb heat better. This heat transfers to the water. Pipes carry it to the hot water system where the heat is released.

Solar panels are usually placed on the roof of a building, **Figure 10-19.**

Using solar cells, we can convert the sun's energy directly into electrical energy. These cells are silicon wafers. They generate electricity by the *photovoltaic* method. In this method, solar energy dislodges electrons from the wafers. The loose electrons move through conductors. This creates an electric current.

Solar photovoltaic power is best known for its use in space. The International Space Station (ISS) has eight solar panels. Each panel has 16,200 solar cells, **Figure 10-20.** During its 90-minute orbit, approximately 40 minutes is without sunlight, hence an electrical power storage system is necessary with rechargeable batteries.

Uses of solar power on earth vary from portable devices, such as pocket calculators, to recreational use in

Links to other subjects: Geography—Find the geographical latitude of where you live. A solar collector must be placed at an angle approximately equal to the latitude where you live. Which direction should the panels face in order to collect maximum sunlight?

Standards for Technological Literacy

Discussion
The advantage of generating energy from wind turbines is that the wind energy is free. What are the disadvantages in developing wind turbine installations?

Vocabulary
What is the meaning of the term *convection*?

Activity
Have students make a series of pinwheels with different size and shape blades to investigate the effects of these changes on the speed of rotation.

Resource
Activity 10-4, *Solar Hot Water Heater*

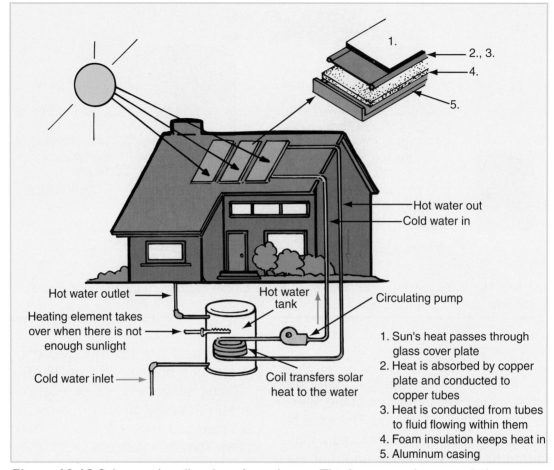

Hot water out
Cold water in

Hot water outlet
Heating element takes over when there is not enough sunlight
Cold water inlet
Hot water tank
Circulating pump
Coil transfers solar heat to the water

1. Sun's heat passes through glass cover plate
2. Heat is absorbed by copper plate and conducted to copper tubes
3. Heat is conducted from tubes to fluid flowing within them
4. Foam insulation keeps heat in
5. Aluminum casing

Figure 10-19 Solar panels collect heat from the sun. This heat is used to provide hot water.

distant locations. A remote home can be virtually self-sufficient with solar power. An inverter that converts direct current (DC) to alternating current (AC) can be used to run most domestic appliances.

Wind Energy

As long as the sun continues to shine, the wind will continue to blow. Wind currents occur because of differences in temperature between different parts of the earth. More heat from the sun reaches the equator than the poles, **Figure 10-21**. This is because the sun's rays hit the earth directly at the equator (A). Near the poles (B), they hit at an angle. The heat is spread over a wider area.

The air above the equator expands most and rises by convection. When it reaches about 30° latitude, the warm air cools and falls. At about 60° latitude, it meets cold air from the poles and has to rise over it. This air movement creates more convection currents, **Figure 10-22**.

Convection currents are also created as a result of the difference in temperature between the land and the sea. During the day, the land heats up more quickly than the sea.

Community resources and services: Have students browse the Internet for universities that have built experimental cars operating on solar power (such as Iowa State University). If possible, invite a representative from a university to talk to the class.

Figure 10-20 The International Space Station has solar panels that produce approximately 110 kW of power. (NASA)

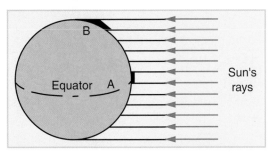

Figure 10-21. More solar heat is delivered at the equator than at the poles. Can you explain why?

Warm air rises from the land. Cooler air from the sea moves in to take its place, **Figure 10-23**. At night, the land cools more quickly than the sea. The process reverses, and warm air rises from the sea. Cooler air from the land moves in to replace it, **Figure 10-24**.

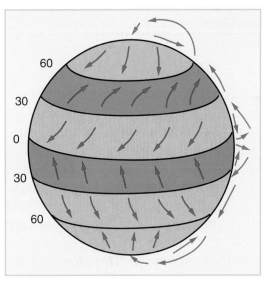

Figure 10-22 Air currents follow a certain pattern over the entire earth.

Links to other subjects: Geography—Have students browse www.hornsrev.dk to investigate the wind farm at Horns Rev off the coast of Denmark. Draw a map to show its location. Describe the turbines. How many turbines have been built and how much power is generated?

Useful Web site: Have students visit www.windelectric.com to learn more about the use of wind energy.

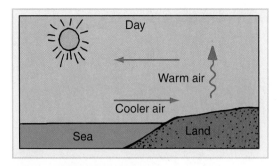

Figure 10-23 Sea breezes occur as warmer air over land rises and cooler sea air moves in to replace it.

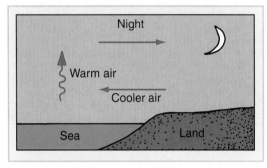

Figure 10-24 At night, landmasses cool faster than oceans, because the breezes blow from land to sea.

Wind energy is one of the oldest sources of energy. For many centuries, wind has turned wheels to grind grain and pump water. Today, wind is being looked at to do other work. It can spin wind turbines that will generate electricity, **Figure 10-25**.

The wind is free; however, it is also unreliable. The wind does not blow constantly. No one can predict when it will blow. Therefore, some method is needed to store some of the electricity generated during times of high winds. Usually a number of batteries serve this purpose. On windless days, the batteries provide electrical energy.

Energy from Moving Water

The energy of moving water is found in rivers (*hydroelectricity*), estuaries (*tidal energy*), and oceans (*wave energy*). Why do rivers flow? Look at **Figure 10-26** for the answer.

The sun evaporates water mainly from the sea and also from rivers, lakes, and plants. This water vapor rises to form clouds. The clouds move with the land breezes. When they

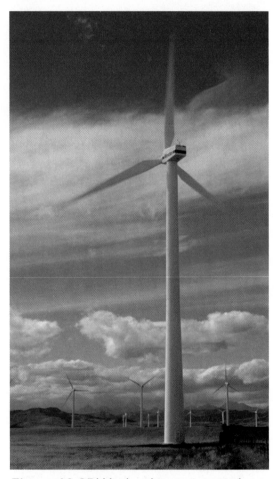

Figure 10-25 Wind turbines on a wind farm capture the kinetic energy in surface winds and convert it into electrical energy in the form of electricity.

Links to other subjects: Geography—Draw a map of the world showing the locations of the largest hydroelectric projects.

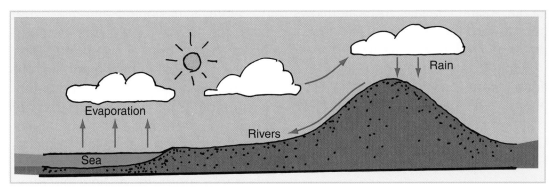

Figure 10-26 Rivers flow because of gravitational energy. Rains deliver the water to high ground. The water then flows to lower ground.

reach high ground, they are forced to rise. This causes them to cool, and they cannot hold as much water. Water falls as rain on high ground and forms rivers.

Hydroelectricity

Water from rivers can be stored behind a dam, **Figure 10-27**. To generate electricity, water flows through very large pipes called penstocks. The penstocks direct water onto turbine blades, spinning them. The turbines are connected to generators as **Figure 10-28** shows.

Hydroelectricity is more economical compared to other electric power systems. After the initial expense of building the dam and generating station, the cost of producing electricity is small. No fuel is needed apart from the energy provided by the sun.

Tidal Energy

Tides from the sea are yet another alternate source of energy. The tides are regular and inexhaustible. The force of tidal currents can be used to produce electricity. The method is much the same as the way waterfalls or streams and rivers are used. The force of tidal water, however, can be captured when it is rising as well as when it is falling.

To understand this method, think of a dam-like structure being placed across the mouth of a bay. As the tide rises, the water flows through a tunnel in the dam. It turns a turbine inside the

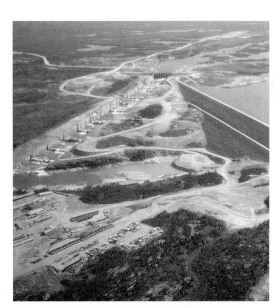

Figure 10-27 Dams store water, which will be used to produce electricity.

Standards for Technological Literacy

2

Discussion
Which hours of the day and which months of the year are peak demand times for electricity?

Enrichment
Hydroelectric dams normally have between 100′ to 1000′ (30 m to 300 m) of water behind the dam.

Vocabulary
What is a *penstock*?

Resource
Transparency 10-2, *How Hydroelectricity Is Produced*

Links to other subjects: Geography—Tidal power is used to generate electricity in following places: Bay of Fundy, Nova Scotia, Canada; Kislaya Guba, Barents and Okhotsk Seas, Russia; La Rance River, and Northwestern France. Draw a map marking each of these locations.

Technology and society: Should people be forced to move from their homes because their land is needed for the lake behind a dam?

Standards for Technological Literacy

2

Discussion
What is the difference between nuclear fusion and nuclear fission? Which can be used as a power source on Earth and which is responsible for the power of the Sun?

Enrichment
Discuss the effects of radiation on the human body.

Vocabulary
Define the term *radiation*.

tunnel. As the tide falls, the water flows back toward the ocean. Once again, it turns the turbine, **Figure 10-29**.

Only a few places on earth have tides great enough to make tidal energy practical. Tidal electricity generating plants are working in France. One is working experimentally in the Bay of Fundy, Nova Scotia.

Nuclear Energy

There are two types of nuclear energy: nuclear fission and nuclear fusion. *Nuclear fusion* is the same energy source that powers our sun and the stars. It requires high temperatures. So far, technologists have not been able to produce it on earth. At present, nuclear power stations use only the fission process.

What is fission? Remember that all solids, liquids, and gases are composed of chemical elements. The smallest unit of each element that still retains the properties of that element is an atom. Although atoms are very small, they are made of even smaller subatomic particles called protons, neutrons, and electrons. At the center of each atom is a tiny nucleus, **Figure 10-30**.

Most atoms have a stable nucleus, which means they do not change. In a few atoms, the nucleus is unstable. These unstable nuclei try to become stable. They throw off particles or rays. These rays are called *radiation*. The atoms are radioactive.

Uranium is a metal. Its atoms have very large nuclei. Very large nuclei are often particularly unstable.

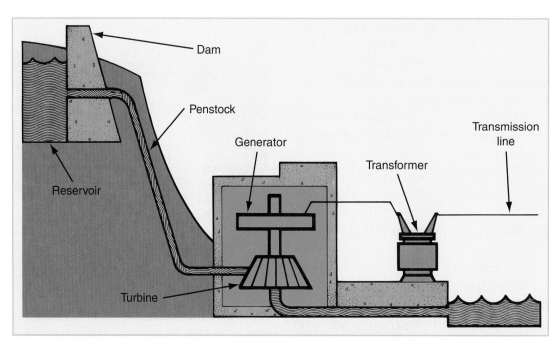

Figure 10-28 Stored water runs through the turbine with great force causing it to spin rapidly. The turbine drives an electric generator, which produces electricity.

Links to other subjects: Science—What type of uranium is used to generate atomic power and how is it refined?

Useful Web site: Have students visit the International Atomic Energy Agency's Web site at www.iaea.int to identify the countries that favor using nuclear power. Why do they prefer nuclear power over the many alternatives?

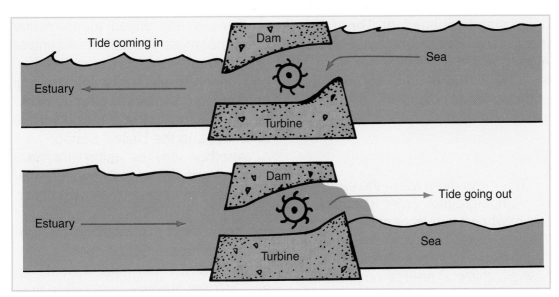

Figure 10-29 Tides can be used to produce electricity. Whether flowing into or out of the estuary, the water spins a turbine.

Discussion
Discuss the advantages and disadvantages of nuclear power. How is the chain reaction in a nuclear reactor slowed or shut down?

Enrichment
When nuclear generating plants are shut down permanently, robots are often used to cut up the reactor and other highly radioactive parts of the plant into pieces that can be hauled away.

Vocabulary
What is the meaning of the term *chain reaction*?

Activity
Draw a diagram illustrating the chain reaction that occurs when neutrons, from a split atom, hit and split other atoms.

When a neutron hits an atom of uranium, the nucleus of the atom splits, **Figure 10-31**.

The atom splits into two parts, called fission fragments. Together, the fragments weigh slightly less than the original atom. The loss in mass turns into energy. On average, a fissioned atom produces one, two, or three neutrons. These neutrons may hit other uranium nuclei. When they do, these nuclei may also split. This gives out more neutrons, and a chain reaction occurs. *Nuclear fission* is the energy produced by splitting atomic nuclei.

The heat produced by nuclear fission is used to heat water, which turns into steam. This steam is used to drive turbines. Generators attached to the turbines produce electricity.

The fission process is noted for the large amount of heat energy released. The fissioning of 2.2 lb. (1 kg) of uranium produces about the same amount of heat as burning 2.9 million lb. (1.3 million kg) of coal!

It was once thought that nuclear energy would become the major energy source. However, a number of nuclear generating stations have been shut down because of technical and other problems. The general public is concerned about the possibility of major disasters, such as the

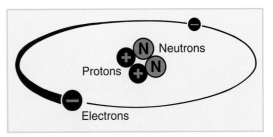

Figure 10-30 The nucleus of an atom contains protons (positively charged) and neutrons (no charge). Electrons (negatively charged) orbit around outside the nucleus in a "cloud."

Links to other subjects: Geography—Draw a map to show the location of Three Mile Island and Chernobyl, the sites of two major disasters with nuclear power stations.

Technology and society: Should governments build more nuclear reactors to accommodate an ever-increasing demand for electricity?

Standards for Technological Literacy

2

Enrichment
Temperatures in mines usually rise an average of 100°F per mile (35°C per km) of depth. At the molten center of the Earth, it is believed that temperatures are about 6000°F to 8000°F (3500°C to 4500°C).

Activity
Investigate the use of geothermal energy in Iceland. When did they start using this energy? How is it transported to homes? How deep are the wells? What is the temperature of the steam and water? How many people receive geothermal water?

ones that happened at Three Mile Island (US) and Chernobyl (Russia). Also, costs per kilowatt-hour are at least twice that of other conventional sources. Safe disposal of nuclear waste is vital.

Geothermal Energy

The rocks deep within the earth are hot. Underground water changes to steam when it comes near these rocks. This is called *geothermal energy*. Technologists are not absolutely sure how geothermal energy originated. It is generally assumed that about 80% of the heat is due to the decay of naturally occurring radioactive materials, such as uranium. The remainder is heat left over from the original formation of the earth.

Hot springs and geysers are evidence of the tremendous heat still in the earth's core. Most of the houses in Reykjavik, Iceland are heated by water that has been warmed by the natural heat of the earth.

If wells are drilled in an active area, steam can rush to the surface with enough force to drive turbines and generate electricity. Geothermal generating stations are already working in Iceland, Japan, Mexico, New Zealand, and the United States.

Another way of using the heat in the earth is to drill two parallel holes into the hot rock. An explosive charge is placed at the bottom of the holes, and the resulting fracture joins the holes. Cold water is then pumped down one hole. It heats up and returns through the other hole as steam. The steam generates electricity, **Figure 10-32**.

One problem with geothermal power concerns pollution. Geothermal steam brings strong chemicals with it. If this condensed steam is dumped into streams and rivers, their temperature can rise to dangerous levels. The chemicals will also pollute the water.

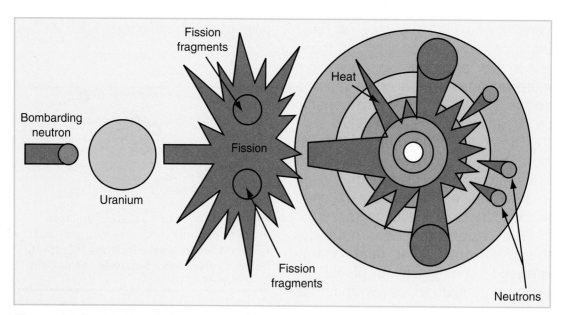

Figure 10-31 Nuclear fission occurs when an atom of uranium splits.

Links to other subjects: Geography—Where is the geyser named "Old Faithful" located? Where is the hot springs area in the Salton Sea-Imperial Valley region located?

Useful Web site: Have students visit the Pacific Gas and Electric Web site at www.pge.com to learn about the use of geyser power.

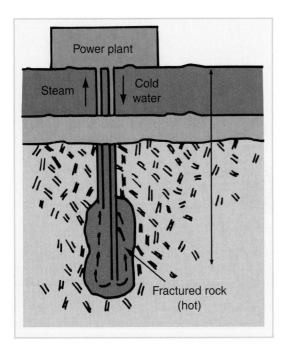

Figure 10-32 This shows a cross section of a geothermal well with a power station. Cold water is sent down through one well to be heated by the super hot rocks. Steam rises up the other well.

Other Renewable Sources of Energy

Other renewable sources of energy are being developed in different parts of the world. The most important are the following:

◆ Energy from garbage.

◆ Energy from plants.

◆ Energy from decomposing matter.

◆ Hydrogen.

Together, these four sources produce only a small part of our energy needs.

Energy from Garbage

People produce a lot of garbage. The average North American throws away about 4 lb. (2 kg) of garbage a

day. Think of the population of your town or city. How much garbage do you suppose is produced in one week?

About 50% of garbage is combustible. These combustible materials may be used as fuel. They will produce steam to turn the turbines in an electric generating plant. This is a partial solution to the disposal of large amounts of urban garbage.

Energy from Plants

Energy from plants is called *biomass energy*. Some fast-growing plants can be burned as fuels. Wood has been used as a fuel for thousands of years. It is still the most commonly used fuel in the developing world, where four out of five families depend on it as their main energy source.

When trees are harvested in North America, about 50% of the tree is converted into lumber or pulpwood. The remaining 50%, mainly branches, twigs, and bark, is discarded. This is wasteful. All parts of the tree contain stored solar energy.

Recently, technologists have tried to use this previously wasted energy. Branches, twigs, and bark are ground up into wood chips and wood pellets. This waste wood is then used as fuel for a steam boiler. Many pulp and paper mills use wood-fired generators to make their own electricity, **Figure 10-33**.

Biomass fuels may also be in liquid form. For example, sugar cane grows quickly in many regions. Sugar produced from the cane can be fermented to make alcohol, which can be mixed with gasoline to produce gasohol, **Figure 10-34**.

Technology and society: What problems are created when all, or most, of the trees in an area are cut down to provide fuel for heating, cooking, and wood for construction?

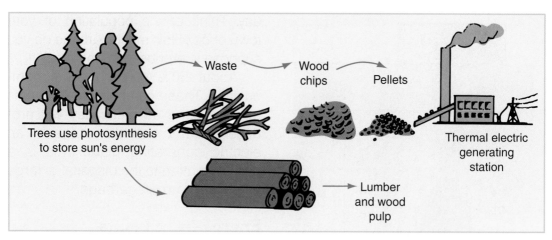

Figure 10-33 Waste wood products provide energy for producing electricity. Wood is a biomass source of energy.

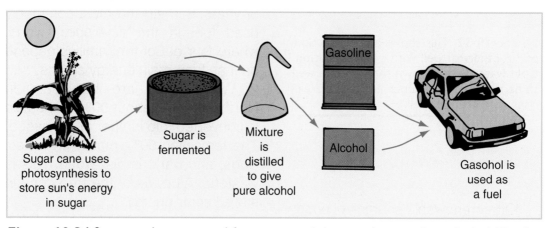

Figure 10-34 Sugar can be processed from cane and then used to produce alcohol. Blends of alcohol and gasoline are used to fuel automobiles.

Energy from Decomposing Matter

On farms, manure can be collected. Farms also have plant wastes. There are pasture plants that have not been eaten, leftover feedstock, fruits, vegetables, and grains that are damaged or unsold.

When manure and organic wastes are put into closed tanks, bacteria will digest them. This produces methane gas, which scientists call "biogas." This can be used for cooking, lighting, and running engines. It is a common method of producing energy in many parts of the world. In China, there are over 7 million biogas digesters supplying energy for 35 million people.

Digesters are designed to be batch load or continuous load. In a

Useful Web site: Have students browse the Montreal Protocol Web site at www.unep.org/ozone/montreal to gather information about reducing substances that deplete the ozone layer. Students should summarize the recommendations and discuss in class.

Technology and society: During some years, the US and Canada grow too much grain to be used domestically. What issues should governments consider when deciding whether to use this excess grain in (a) making gasohol, (b) feeding hungry people in third world countries, and (c) making electricity?

batch load type, the digester is loaded with a soupy mix of wastes. The mix is called slurry. The digester is then sealed and not emptied until the materials stop producing gas.

A continuous load digester accepts a small amount of fresh slurry continuously. This type produces gas as long as slurry is being fed into it. **Figure 10-35** diagrams two different kinds of continuous load digesters.

Hydrogen

The name *hydrogen* comes from the Greek word meaning "water generator," as water is produced when hydrogen burns in the presence of oxygen. One of the first successful uses of this fuel was in the Saturn V rocket that took men to the moon. The Space Shuttle has a huge external tank filled with liquid hydrogen and liquid oxygen. This fuel not only lifts the shuttle into orbit but also produces the electricity needed during a mission. The power to make the two elements combine comes from a fuel cell. A fuel cell is a device that allows hydrogen and oxygen to combine without a combustion process. Reaction occurs when electrons are released from the hydrogen and travel to the oxygen through an external circuit. As electrons travel through the circuit, they generate a current that can power any electrical device. The reaction benefits the astronauts in another way. The only by-product is water, which is the water that astronauts drink.

If fuel cells produce energy without any toxic by-products, why don't we use them in other vehicles? The main reason is cost. Fuel cells use

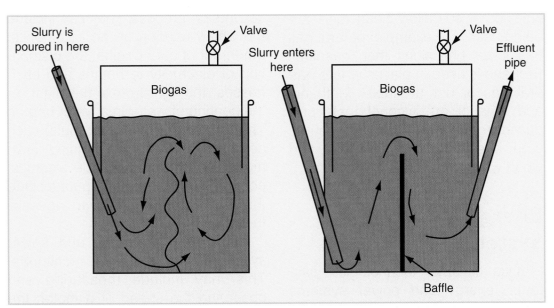

Figure 10-35 These are the two types of biogas digesters. Waste is made into a soupy mixture called "slurry."

platinum, a very expensive metal. Currently hydrogen, as a fuel, cannot compete economically with petroleum. Other issues remain. The hydrogen fuel tank takes up most of the trunk space. Also, there are very few fueling stations across North America. In addition, temperatures below freezing point present a major problem when the by-product of the motor is water. Some environmentalists also point out that, in order to be pollution free, hydrogen must be made by using renewable energy. If the hydrogen is manufactured by using fossil fuels, we are only shifting pollution from vehicle tailpipes to hydrogen production plants. We will have done nothing to reduce air pollution and greenhouse gases, if this happens. Nonetheless, there are experimental vehicles on the road. Both Chicago and Vancouver have three buses powered by hydrogen fuel cells. In the meantime, other so-called clean fuels are being used. For example, natural gas has lower emissions than gasoline. Propane is used in places where natural gas is not available. Methanol and ethanol are alcohol fuels produced from biomass matter, such as corn, beets, sugar cane, wheat, and wood wastes.

Energy and Power Systems

Like all other technological systems, energy and power systems include inputs, processes, outputs, and feedback. Understanding these individual components is critical to understanding the system as a whole.

There are many types of inputs in energy and power systems. Wind energy is an input for windmills. Sun energy is an input for solar power systems. Gravity and water are inputs for waterwheels and turbines within dams. Other inputs include the people who develop, operate, and maintain the systems and the materials and machines that compose the systems.

Processes within energy and power systems are typically conversion processes. The conversion types include mechanical-to-electrical (for example, an electrical generating station), electrical-to-mechanical (electric motors), chemical-to-mechanical internal combustion engines), and chemical-to-electrical (batteries and fuel cells). Other processes include the management and operation of the system.

Outputs of energy and power systems are typically the outputs of the conversion process. They can include the wheel rotation on an automobile, the thrust of a jet engine, the distribution of electricity to homes and businesses, the light from a flashlight, or the heating of water in a water heater. There are also societal outputs, such as the convenience of riding lawn mowers or readily available electrical power. Environmental outputs include the smoke produced by burning fossil fuels.

Feedback of energy and power systems includes system monitoring. This may include measuring and observing the speed of a motor, the amount of current flowing through an electrical wire, or the temperature of water in an aquarium.

Links to other subjects: Science—What are the chemical formulas for methanol, ethanol, and natural gas?

Chapter 10 Review

Energy

Summary

Energy is available in many different forms: chemical, gravitational, strain, electrical, heat, light, sound, and nuclear. Energy can neither be created nor destroyed. It can only be changed from one form to another.

Almost all energy originates from the sun. Sources of energy can be divided into two groups: renewable and nonrenewable. Nonrenewable energy—coal, oil, and natural gas—cannot be replaced. Renewable energy—sun, wind, and water—will always be available.

Coal developed from the compressed remains of plants that died millions of years ago. It is removed from the ground by surface mining or underground mining. Petroleum is formed from the compressed layers of plants and animals that once lived in the seas. Oil and gas are removed from the ground by drilling wells.

Solar energy uses energy from the sun. Solar panels are usually placed on the roof of a building. They can extract heat from the sun's rays and use it to provide hot water or to heat buildings. The sun's energy can also be converted directly to electrical energy using solar cells.

Wind is one of the oldest forms of energy used to do work. In the past, wind was used to turn windmills. Today, it powers wind turbines.

Currently, most cars and trucks are using petroleum products to power their engines. Some alternative "clean" fuels such as natural gas, propane, methanol and ethanol are available. However, most experts agree that hydrogen and electricity are the fuels of the future. What we need are clean, sustainable energy sources that will prevent climate change, air pollution and further damage to the environment.

The water from a river can be stored behind a dam and used to drive turbines that generate electricity. The force of tidal currents can be used to generate electricity in much the same way. Experiments are also being conducted to harness the energy of waves.

Nuclear power stations use nuclear fission. In this process, the nucleus of an atom of uranium is split into two parts. Together, these parts weigh slightly less than the original. The loss in mass turns into energy.

Answers to Test Your Knowledge Questions
1. potential, kinetic
2. Chemical—battery
Gravitational—skiing downhill
Strain—stretched elastic band
Mechanical—falling water turning a waterwheel
Electrical—flow of electrons in a wire
Heat—fire (or element on an electric stove)
Light—flashlight
Sound—vibrating string on a guitar
Nuclear—fission
3. Riding a bicycle—chemical energy in muscles is changed to mechanical energy by the rider's legs pumping the pedals.
Braking—kinetic energy of the bicycle wheel is changed to heat energy due to friction
Lighting—electrical energy is changed to light energy and heat energy in a bulb
(Evaluate individually; there are many other examples.)
4. coal, oil, natural gas
sun, wind, water
5. D. Solar energy
6. Heat from the sun is stored in solar panels filled with a liquid. The sun's energy is converted to electricity by means of photovoltaic cells.
7. They are unreliable since they depend on wind, which does not always blow.
8. See Figure 10-28, page 272.
9. Tidal energy, since it also uses water to drive turbines.
(continued)

Geothermal energy is produced when underground water comes into contact with hot rocks deep within the earth and is changed into steam. Wells are drilled and the steam rushes to the surface with enough force to drive turbines.

Modular Connections

The information in this chapter provides the required foundation for the following types of modular activities:

◆ Applied Physics

◆ Energy and Power

◆ Forces

◆ Alternative Energy

◆ Solar Energy

◆ Heat and Energy

Test Your Knowledge

Write your answers to these review questions on a separate sheet of paper.

1. A wound spring has _____ energy, and a falling boulder has _____ energy.

2. List the nine forms of energy and give one example of each.

3. Give three examples to show that energy can neither be created nor destroyed. It is only changed from one form to another.

4. Three examples of nonrenewable energy sources are _____, _____, and _____ _____. Three examples of renewable energy sources are _____, _____, and _____.

5. Which of the following is a renewable source of energy?
   A. Coal.
   B. Gasoline.
   C. Natural gas.
   D. Solar energy.

6. Describe two methods of collecting energy from the sun.

7. What is the major disadvantage of wind turbines as a source of electrical energy?

8. Make a sketch to show how hydroelectricity is produced. Indicate, using arrows, the direction of flow of water.

9. Which alternative energy source is similar to hydropower and why?

10. What are the major problems in using nuclear energy to produce electricity?

11. Describe how geothermal energy is used to produce eletricity.

12. How can energy be produced from garbage?

13. Energy from plants is called _____ energy.

14. Why is biogas a useful form of energy in a rural area?

15. Give one example of each of the three types of heat transmission.

Apply Your Knowledge

1. Think back to a time when a power failure occurred in your area. List the devices you were unable to use. Imagine that the power stayed off for 48 hours. What alternative sources of energy could you use?

2. Describe three situations that involve potential energy and three that involve kinetic energy.

3. Make a list of nine forms of energy shown in **Figure 10-2** through **Figure 10-10**. For each, describe a further example of how the energy form is used to do work.

4. List five devices in your home that use energy. Describe the energy change(s) that take place when each is used.

5. List the ways in which you use energy each day (a) around the house, (b) in traveling, (c) at school, and (d) for leisure. State whether the energy comes from a renewable or nonrenewable source.

6. Research the Internet, your library, or your public service utility to find ways to save energy. Create a poster that illustrates conservation ideas relating to your yard or garden, the vehicles your friends or parents drive, the types of food you buy, or the amount of recycling you do.

7. Research one career related to the information you have studied in this chapter and state the following:
   A. The occupation you selected.
   B. The education requirements to enter this occupation.
   C. The possibilities of promotion to a higher level at a later date.
   D. What someone with this career does on a day-to-day basis.
   You might find this information on the Internet or in your library. If possible, interview a person who already works in this field to answer the four points. Finally, state why you might or might not be interested in pursuing this occupation when you finish school.

Answers to Test Your Knowledge Questions *(continued)*
10. Locating sites where large amounts of cooling water are available. Danger of radioactive leaks. Finding geologically stable locations. Disposal of highly radioactive wastes.
11. If wells are drilled in an area where hot springs and geysers are evident, steam will rush to the surface. This steam will have enough force to drive turbines to generate electricity.
12. Garbage can be burned to produce steam that turns turbines in an electrical generating plant.
13. biomass
14. It can be produced from manure and organic wastes.
15. Answers may vary. Refer to Figure 10-7 on page 262.

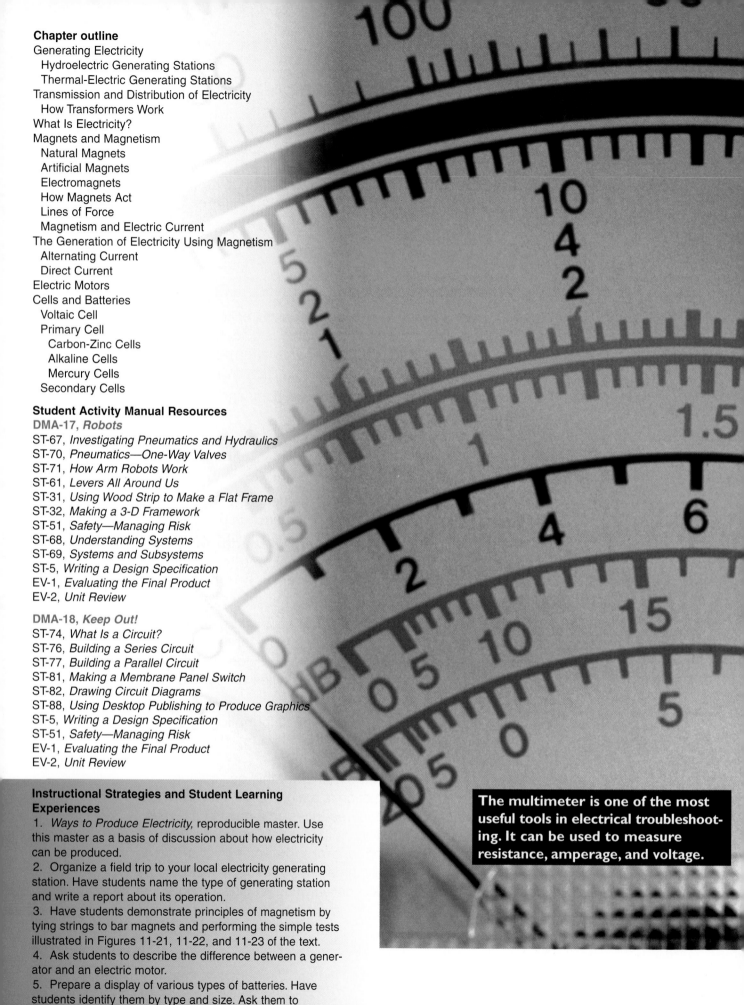

Chapter outline

Student Activity Manual Resources

Instructional Strategies and Student Learning Experiences

1. *Ways to Produce Electricity*, reproducible master. Use this master as a basis of discussion about how electricity can be produced.

2. Organize a field trip to your local electricity generating station. Have students name the type of generating station and write a report about its operation.

3. Have students demonstrate principles of magnetism by tying strings to bar magnets and performing the simple tests illustrated in Figures 11-21, 11-22, and 11-23 of the text.

4. Ask students to describe the difference between a generator and an electric motor.

5. Prepare a display of various types of batteries. Have students identify them by type and size. Ask them to describe various uses for each battery.

The multimeter is one of the most useful tools in electrical troubleshooting. It can be used to measure resistance, amperage, and voltage.

Electricity and Magnetism

Key Terms

alternating current
amperage
battery
cell
circuit
conductor
direct current
distribution line
electric current
electromagnet
electromotive force

electron
generating station
generator
magnetism
primary cell
secondary cell
transformer
transmission line
voltage
voltaic cell

Objectives

After reading this chapter you will be able to:

◆ Describe the different ways electricity can be produced.

◆ List the advantages and disadvantages of each method of generating electricity.

◆ Describe the principal components of a network for the transmission and distribution of electricity.

◆ Explain the nature of electricity by referring to the movement of electrons.

◆ State the laws of magnetism.

◆ Explain how the laws of magnetism are used in the generation of electricity.

◆ Describe how an electric motor operates.

◆ Explain what type of electrical current a battery produces and from what other form of energy is electricity produced.

◆ Design and make a product that uses electrical control.

Imagine your town or city without electricity. It can happen. At seventeen minutes after five o'clock in the afternoon of November 10, 1965, the lives of 30 million people were suddenly interrupted.

- ◆ 800,000 riders were trapped in the New York subway.

- ◆ All nine television channels in the metropolitan area of New York were forced to go off the air.

- ◆ Kennedy International and LaGuardia airports were shut down and airplanes found themselves circling, unable to land.

- ◆ 5,000 off-duty police officers were summoned to duty.

- ◆ 10,000 National Guardsmen were called up to help protect the city.

- ◆ Militiamen were alerted in Rhode Island and Massachusetts.

- ◆ Broadway theaters and movie houses were closed.

- ◆ Thousands of people hiked across the Brooklyn and Queensboro bridges.

- ◆ Highways were jammed with traffic for more than five hours.

- ◆ Thousands of New Yorkers were trapped in elevators in the city's skyscrapers.

In less than 15 minutes the power failure spread across more than 49,000 square miles (about 128,000 km²). New York state, New England, and parts of New Jersey, Pennsylvania, Ontario, and Quebec had no electricity.

It was the largest power failure in history. The first signs of trouble appeared at the power company's control center. An operator noticed problems on the center's interconnecting system with upstate power companies. By then, the blackout was only seconds away. It was too late for action. The demands for reserve power went so high that automatic switches shut the system down to protect it.

In our daily lives, we take electricity for granted. To most people, it is merely something that always arrives at the home. Only when it is gone do we realize how we depend on it, **Figure 11-1**.

Generating Electricity

The electrical energy supplied to your home comes from a generating station. A *generating station* uses energy from a source of power to turn turbines, which produce electricity. The making of electricity is called generating.

Figure 11-1 How would a blackout this evening affect you?

Useful Web site: Students can use the dictionaries at www.techtionary.com and www.hyperdictionary.com to find the definition of any new technical and technological terms used in this chapter.

Technology and society: How are the interactions between people affected during a major blackout?

There are several types of generating stations. They are named after the power source used. Remember the names: hydroelectric and thermal-electric.

Hydroelectric Generating Stations

"Hydro" is another word for water. Hydroelectric generating stations use the energy of flowing or falling water. The station is located at a waterfall or at a dam, **Figure 11-2**. As the water drops to a lower level, its mass spins a turbine, **Figure 11-3**. A turbine is a finned wheel. When the falling water strikes the fins, the turbine turns rapidly. The turbines are connected to generators. A *generator* is a device that produces an electric current as it turns.

Thermal-Electric Generating Stations

Thermal-electric generating stations use steam to drive turbines. A heat source produces the steam. The steam is directed onto the blades of a turbine. The turbine spins rapidly. As in hydroelectric systems, the turbines drive generators. The spinning generators produce electricity.

Heat for powering thermal-electric turbines comes from one of two sources. The first is by burning fossil fuels, **Figure 11-4**. Fossil fuels come from once-living animals and plants. They include coal, oil, and natural gas. The second source is nuclear fission. Fission is the splitting of uranium

Standards for Technological Literacy

2 16

Discussion
Once a hydroelectric dam is built, the cost of producing electricity is low and damage to the environment is minimal. So, why aren't more hydroelectric plants built?

Reflection
What measures did you take during a blackout to continue your activities safely?

Figure 11-2 Dams are built to store water. (TEC)

Figure 11-3 This is the turbine room of a hydroelectric generating station. (TEC)

Useful Web site: The Three Gorges Dam Project in China is one of the largest dams for hydroelectric power generation ever constructed. Visit www.power-technology.com to find out more about the area of land that was flooded, the people who had to move from their homes, and the amount of electricity that will be produced. Also, search the Internet to learn about the potential impact on the environment and the local climate.

Discussion
Describe the differences between and similarities in the two methods of generating electricity shown in Figures 11-4 and 11-5. Which of the three methods of driving generators shown in Figures 11-4, 11-5, and 11-6 could be restarted in the shortest time? Which would take the most time to restart? Explain your answers.

Activity
Figure 11-6 illustrates the potential energy behind a dam that is converted into kinetic energy when it spins turbine blades. What other types of potential and kinetic energy did you learn about in Chapter 10?

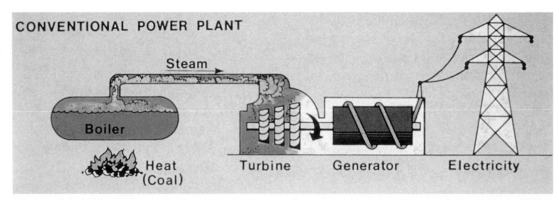

CONVENTIONAL POWER PLANT

Steam

Boiler

Heat (Coal)   Turbine   Generator   Electricity

Figure 11-4 This is a simple diagram of a system for generating electricity with fossil fuels. (AEC)

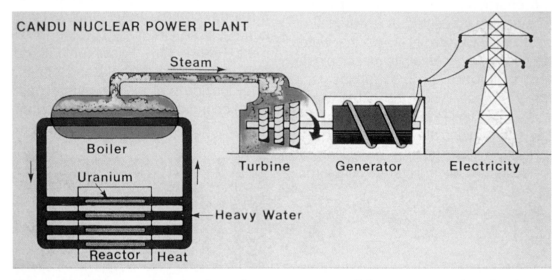

CANDU NUCLEAR POWER PLANT

Steam

Boiler

Uranium

Turbine   Generator   Electricity

Heavy Water

Reactor Heat

Figure 11-5 Here is a diagram of a nuclear power station. Splitting atoms, rather than burning fuels, creates the heat. (AEC)

atoms. The process releases enormous amounts of heat. In a nuclear station, **Figure 11-5**, the nuclear reactor does the same job as the furnace in fossil-fuel stations.

Any device that changes one form of energy to another is called a converter, **Figure 11-6**. Hydroelectric generating stations change the potential energy of water behind a dam. As it falls into the turbine, it becomes kinetic energy. Thermal-electric generating

stations convert the heat energy stored in fossil fuels and uranium into kinetic energy. In both cases, the kinetic energy is converted to electrical energy by generators.

Each method of generating electricity has advantages and disadvantages. Look at **Figure 11-7**. If a generating station had to be built near your home, which would you choose? Why?

Most of the electricity used in homes and factories is produced either

Technology and society: Should governments pass legislation that forbids building generating stations close to urban areas? What would be the environmental and social impacts of such legislation? If a generating station must be built close to an urban area, which type would you recommend?

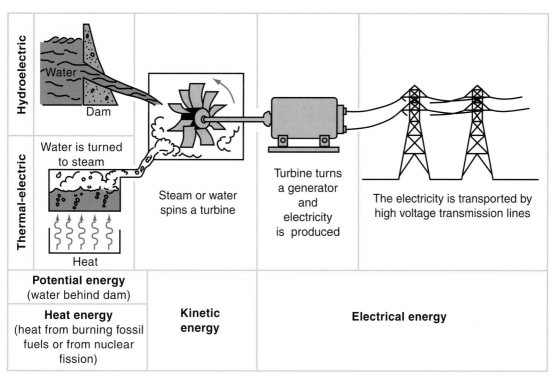

Figure 11-6 All generating stations are energy converters.

Standards for Technological Literacy

2 **16**

Discussion
Why are generating stations rarely located close to where the electrical energy is used?

Activity
Divide the class into two groups. After investigating the claim that electromagnetic radiation from high voltage transmission lines can negatively affect one's health, have the class debate the issues involved.

Safety
What types of leisure, work, and home renovation activities should NOT be conducted near a transmission line?

Resource
Transparency 11-1, *Ways of Generating Electricity*

in hydroelectric or thermal-electric generating stations. There are, however, other methods. Friction, chemical action, light, heat, and pressure also generate electricity, **Figure 11-8**.

Transmission and Distribution of Electricity

Generating stations are rarely found close to where the electrical energy is used. The electricity that comes to your home may have traveled a great distance.

After leaving the generating station, the electricity is fed into a network of *transmission lines* and *distribution lines*. These lines

transport the electricity to wherever it is needed, **Figure 11-9**.

The transmission lines, **Figure 11-10**, resist the flow of electrical energy. Thus, some of the energy is lost along the way. Increasing voltage and reducing amperage greatly reduces this loss. *Voltage* is a measure of electrical pressure. *Amperage* measures the amount of current.

A *transformer* changes voltage and amperage. It consists of a pair of coils and a core.

Neither your home nor a factory can use electricity at the high voltage carried by transmission lines. It would destroy the wiring, appliances, and machines. The voltage must be reduced before current enters distribution lines. Once again, a

Useful Web site: Have students investigate the largest underground hydroelectric power house, located at Churchill Falls, Labrador, by visiting www.ewh.ieee.org/reg/7. Students should print information they find interesting and share it with the class.

Standards for Technological Literacy

2 16

Discussion
Where are used radioactive materials buried? How are they stored? How long do they remain radioactive?

Vocabulary
What is the meaning of the term *radioactive*?

Safety
What measures have miners taken in the past to ensure their safety? What safety precautions are routine today?

	Advantages	Disadvantages
Hydroelectric Station	Cheapest method overall. Most environmentally safe method. Cheap to operate — raw material is free. Low maintenance and operational costs. No harmful combustion products. No harmful wastes of any kind. Water used is not polluted. Flow of water and, therefore, the amount of electricity generated is adjustable.	Sites are normally a long way from cities so long wires are needed to bring the electricity to the consumers. Transmission towers are unsightly. Large amounts of land are taken up by transmission corridors. Transmission lines emit electromagnetic radiation that may be a health hazard. Dams disrupt rivers and, therefore, the marine life. The reservoir behind the dam covers a large expanse of land, thereby displacing people and animals.
Fossil-Fuel Station	Uses fuel that is often available locally or can be easily transported. Small generators can be built to supply local needs.	Uses a nonrenewable resource. Oil-fired thermal plants are becoming too costly and must be converted to natural gas or coal. Coal is bulky, heavy, costly to move, and dirty. Burning coal produces a large amount of ash. Rain that penetrates ash heaps or buried ashes will pollute streams or groundwater. Harmful particles and gases are released into the air and combine with water vapor in the air to form acid rain that damages trees, lakes, and buildings. Mining coal is a dangerous occupation. Oil extracted from under the sea may sometimes leak into the seawater.
Nuclear Station	Large amounts of electricity can be generated using a small amount of material. Can be built wherever there is a supply of water for cooling. No acid rain is created.	Uses a nonrenewable resource. Radioactive nuclear wastes must be disposed of. Deep burial sites must be found for waste that will remain radioactive for thousands of years. Mining the uranium fuel is expensive; it is also hazardous to miners who are exposed to cancer-causing gas. Reactors are expensive to build and maintain. A reactor could overheat and release radioactive substances into the environment.

Figure 11-7 There are three types of electricity generating stations. Which type would you select?

transformer is used. The first drop in voltage occurs when transferring electrical energy to distribution lines, **Figure 11-11**. Another reduction occurs when transferring electrical energy from distribution lines to service lines. This transformer may be located on a pole or on the ground, **Figure 11-12**. Electricity enters your home through a service line. It also

Technology and society: What are the risks to the environment of using nuclear power, coal, natural gas, and hydroelectricity? Which method causes the least long-term damage to the environment?

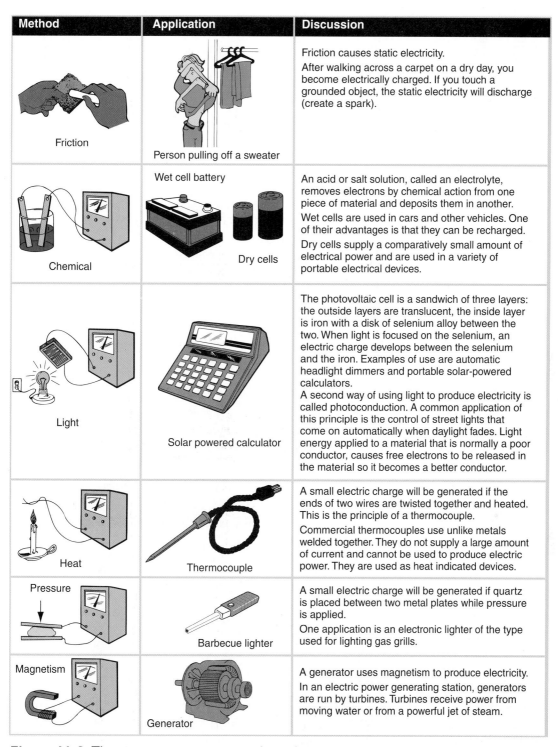

Method	Application	Discussion
Friction	Person pulling off a sweater	Friction causes static electricity. After walking across a carpet on a dry day, you become electrically charged. If you touch a grounded object, the static electricity will discharge (create a spark).
Chemical	Wet cell battery Dry cells	An acid or salt solution, called an electrolyte, removes electrons by chemical action from one piece of material and deposits them in another. Wet cells are used in cars and other vehicles. One of their advantages is that they can be recharged. Dry cells supply a comparatively small amount of electrical power and are used in a variety of portable electrical devices.
Light	Solar powered calculator	The photovoltaic cell is a sandwich of three layers: the outside layers are translucent, the inside layer is iron with a disk of selenium alloy between the two. When light is focused on the selenium, an electric charge develops between the selenium and the iron. Examples of use are automatic headlight dimmers and portable solar-powered calculators. A second way of using light to produce electricity is called photoconduction. A common application of this principle is the control of street lights that come on automatically when daylight fades. Light energy applied to a material that is normally a poor conductor, causes free electrons to be released in the material so it becomes a better conductor.
Heat	Thermocouple	A small electric charge will be generated if the ends of two wires are twisted together and heated. This is the principle of a thermocouple. Commercial thermocouples use unlike metals welded together. They do not supply a large amount of current and cannot be used to produce electric power. They are used as heat indicated devices.
Pressure	Barbecue lighter	A small electric charge will be generated if quartz is placed between two metal plates while pressure is applied. One application is an electronic lighter of the type used for lighting gas grills.
Magnetism	Generator	A generator uses magnetism to produce electricity. In an electric power generating station, generators are run by turbines. Turbines receive power from moving water or from a powerful jet of steam.

Figure 11-8 There are many ways to produce electricity.

Standards for Technological Literacy

2 **16**

Reflection
When have you noticed sparks from friction or felt the effects of static electricity?

Activity
Set up experiments to reproduce those shown in Figure 11-8. Show students examples of the devices shown in Figure 11-8.

Resource
Activity 11-2, *Static Electricity*

Resource
Reproducible Master, *Ways to Produce Electricity*

Links to other subjects: Science—Demonstrate the large charges and very high voltages produced by a Van de Graaff generator. The generator can be used to perform electrostatic tricks, such as placing a wig on the dome will cause the hair to stand upright.

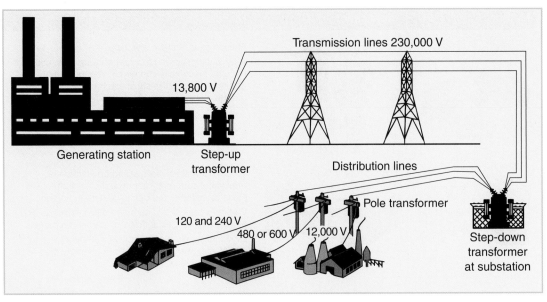

Figure 11-9 Shown here is an electrical power transmission and distribution system. Voltage is greatly increased before the electricity is transmitted over long distances.

Figure 11-10 Transmission lines are carried by triangular towers. Insulators support the lines where they attach to the tower. (TEC)

Figure 11-11 Distribution substations step down the voltage. (TEC)

passes through a meter and a main switch, **Figure 11-13**.

How Transformers Work

There are two types of transformers. Step-up transformers increase voltage, and step-down transformers reduce it.

Basically, a transformer consists of two coils of wire around a core. One coil has more turns than the

Figure 11-12 A pole transformer further reduces the voltage in the distribution lines. (TEC)

Figure 11-13 Electricity enters your home through a service line, a meter, and a main switch. (TEC)

other. **Figure 11-14** shows the construction of step-up and step-down transformers.

What Is Electricity?

When you flip a light switch, electricity flows through wires and lights a bulb. What exactly is it that flows through the wire when the switch is turned on? There is no perfect answer. A scientist would say that electric current consists of a flow of electrons. What does this mean?

One explanation is called the electron theory. Everything around us is made from very small particles called atoms. Atoms are made of even smaller particles called protons, electrons, and neutrons. The protons have a positive charge, and the *electrons* have a negative charge. Neutrons have no charge and play no role in electricity.

An atom normally has the same number of electrons as protons. The negative and positive charges cancel each other. Such atoms are electrically neutral, **Figure 11-15**.

In metals, electrons can detach themselves from the nucleus. Therefore, metals can conduct electricity because the electrons that are detached are free to move. When a metal wire is connected in a circuit, the free outer electrons can all be pushed in the same direction. This flow of electrons is called an *electric current*, **Figure 11-16**.

Suppose that electrons are made to flow from one end of a wire to the other. The end that loses electrons becomes positively charged—the positive terminal. The end that gains electrons becomes negatively charged—the

Designing and making
How do designers protect buildings from the damage that may be caused by a lightning strike?

Discussion
What climatic conditions cause lightning? What disasters are caused by lightning?

Vocabulary
What is an *electron*?

Examples
Use a simple circuit that includes a dry cell, wires, and a bulb to explain the flow direction of electrons.

Safety
How can you minimize the danger to yourself from a lightning strike?

Resource
Activity 11-5, *Detecting an Electric Current*

Useful Web site: To learn more about electron flow and magnetic fields visit www.unis.org/UNIscienceNet/Electricity_and_magnetism.html.

Useful Web site: To learn more about atoms and electrons visit www.unis.org/UNIscienceNet/ELECT_Knowledge.html.

negative terminal. See **Figure 11-17**. Electrons can move through the wire, which acts as a conductor. A *conductor* is a material that will allow an electric current to flow easily. For example, when free electrons pass into the filament (thin wire) in a lightbulb, collisions between electrons and atoms are more frequent. This increases the temperature of the wire, causing the wire to become white-hot and emit light.

Why are electrons able to move through a wire? There are two reasons.

◆ A force pushes them along a path.

◆ There is a closed path, called a *circuit*, in which they can move.

Think of the force as like the force created by a pump pushing water through pipes. In a circuit, an *electromotive force* (EMF) pushes electrons through a conductor. We also call this force voltage.

Just as we cannot say what an electron really is, we cannot describe an electromotive force. However, it is possible to describe some of the devices that are capable of producing electromotive force.

The two most common sources of EMF are generators and chemical reactions. Generators employ magnetism and mechanical energy to produce electricity, and dry cells and batteries use chemical reactions to produce direct current electricity.

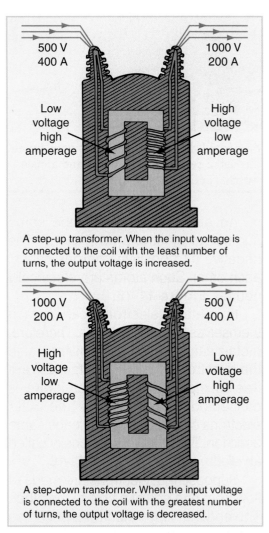

A step-up transformer. When the input voltage is connected to the coil with the least number of turns, the output voltage is increased.

A step-down transformer. When the input voltage is connected to the coil with the greatest number of turns, the output voltage is decreased.

Figure 11-14 The "inside" of step-up and step-down transformers helps explain how they work.

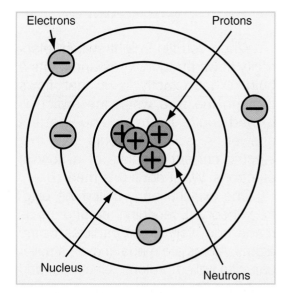

Figure 11-15 Protons and neutrons make up the nucleus of the atom. Electrons orbit around the nucleus.

Links to other subjects: Science—Use a model to illustrate the composition of an atom.

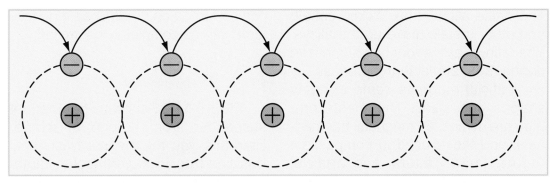

Figure 11-16 The movement of electrons occurs when a force is applied to one end of a wire. The free electrons move from one atom to the next. This process is repeated along the whole length of the wire. This causes what we call an electric current.

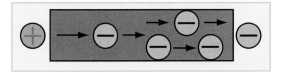

Figure 11-17 The terminal with a surplus of electrons is called negative. The terminal with a scarcity of electrons is called positive.

These two sources provide most of the electricity that we use. Other sources of EMF are friction, light, heat, and pressure. (Look back at Figure 11-8.)

Magnets and Magnetism

The production of electricity depends upon magnets and magnetism. *Magnetism* is the ability of a material to attract pieces of iron or steel.

Materials that are attracted by magnets are called magnetic materials. Among them are iron, steel, and nickel. Magnets do not attract nonmagnetic materials, such as aluminum, copper, glass, paper, and wood, **Figure 11-18**. Magnets fall into three different groups: natural magnets, artificial magnets, and electromagnets.

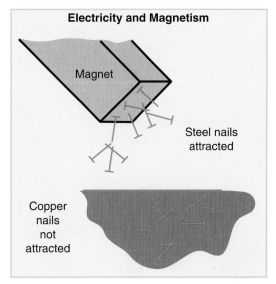

Electricity and Magnetism

Magnet

Steel nails attracted

Copper nails not attracted

Figure 11-18 Which materials are magnetic and which are nonmagnetic?

Natural Magnets

Natural magnets, such as lodestone, occur in nature. Lodestone is a blackish iron ore (magnetite). Its weak magnetic force varies greatly from stone to stone.

Artificial Magnets

Artificial magnets, also called permanent magnets, are made of hard

Links to other subjects: Geography—Demonstrate how to use a compass and explain the difference between the North Pole and the Magnetic North.

Designing and making
What were some of the problems that designers of Maglev trains had to overcome?

Discussion
If a bar magnet were to be cut into two halves, would the two bars each have a single pole or would each part have two poles?

and brittle alloys. Iron, nickel, cobalt, and other metals make up the alloys. The alloys are strongly magnetized during the manufacturing process.

Artificial magnets come in many shapes and sizes. The most common are horseshoe magnets, bar magnets, and those used in compasses, **Figure 11-19**. Natural and artificial magnets can retain their magnetism indefinitely.

Electromagnets

Electromagnets are so named because they are magnetized by an electric current. They consist of two main parts. One is a core of special steel, while the other is a copper wire coil wound on this core, **Figure 11-20**. Unlike permanent magnets, electromagnets

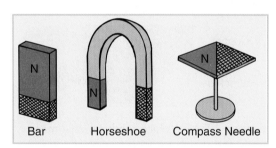

Bar   Horseshoe   Compass Needle

Figure 11-19 There are three common artificial magnets.

Figure 11-20 An electromagnet can be turned on and off.

can be turned on or off. Their magnetic force can be completely controlled.

How Magnets Act

Figure 11-21 shows a bar magnet suspended from a loop of thread. Held this way, the magnet twists until it is lined up in a north-south direction. The end that points toward the north is called the north-seeking pole. The end that points toward the south is called the south-seeking pole.

Suppose that two magnets are suspended so that the north pole of one is brought close to the south pole of the other. The magnets attract one another, **Figure 11-22**. This is the first law of magnetism. It states that unlike magnetic poles attract each other.

Suppose that the north poles come close to one another. The magnets push or move away from each other.

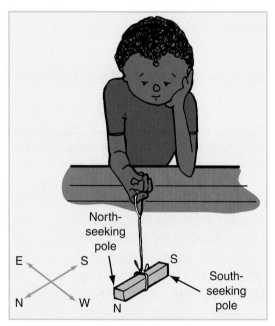

North-seeking pole

E   S

N   W

S

South-seeking pole

Figure 11-21 The north pole of a bar magnet that is free to swing will always point north.

This is the second law of magnetism. It states that like magnetic poles repel one another, **Figure 11-23**.

Lines of Force

Unlike poles attract, and like poles repel. This suggests that around a magnet there are invisible lines of force. Although you cannot see them they can be shown to exist. Place a sheet of paper over a magnet and sprinkle iron filings on the paper. When the paper is tapped gently, the small iron particles form a distinct pattern, **Figure 11-24**. The lines of force shown by the iron filings take the shape shown in **Figure 11-25**.

Now suppose that two magnets are laid end to end, and the experiment with iron filings is repeated. The lines of force demonstrate the two laws of magnetism, **Figure 11-26**.

Magnetism and Electric Current

An electric current passed through a wire also creates a magnetic field

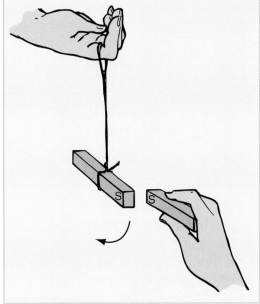

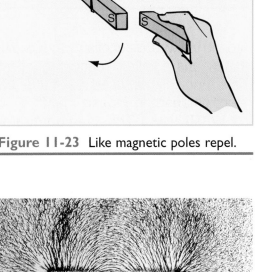

Figure 11-23 Like magnetic poles repel.

Figure 11-24 Iron filings show the lines of magnetic force around a magnet.

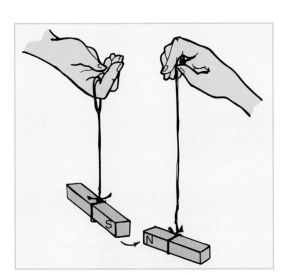

Figure 11-22 Unlike magnetic poles attract.

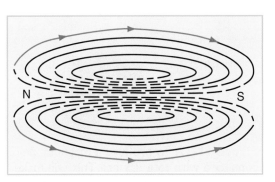

Figure 11-25 Note the pattern and direction of magnetic lines of force.

Activity
Have students repeat the experiments shown in Figures 11-22 and 11-23.

Activity
Have students use iron filings to observe the lines of magnetic force around a magnet.

Links to other subjects: History—Who was Michael Faraday (1791-1867) and what work did he do related to electromagnetism?

Discussion
Why is it correct to say that electrical energy is *converted* from another form of energy rather than saying it is *created*?

Activity
Demonstrate how to produce a magnetic field and how to make a simple elec-tromagnet, as shown in the Figures 11-27 and 11-29. Ask students to explain why the iron filings are attracted to the wire or to the nail.

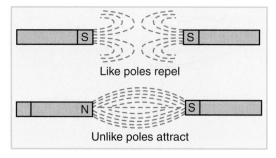

Figure 11-26 Try this experiment with iron filings and two magnets.

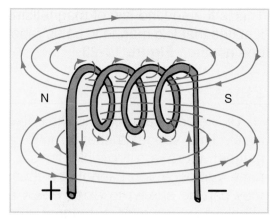

Figure 11-28 Coiling the wire shown in Figure 11-27 creates a magnet with poles.

around the wire. Magnetism produced by this means is called electromagnetism. This principle is used to make the electromagnet in **Figure 11-20**.

WARNING: The demonstration shown in **Figure 11-27** should only be done by your teacher. A carbon-zinc cell should be used. NEVER use an alkaline cell. It may explode.

If the wire in **Figure 11-27** is wound to form a coil, it becomes a magnet with poles, **Figure 11-28**. The magnetic strength of this coil can be controlled. It depends on the strength

of the current and the number of loops in the coil.

If a wire is coiled around a core of magnetic material, such as a soft iron nail, the nail becomes an electromag-net. It remains strongly magnetic only as long as there is current in the wire, **Figure 11-29**.

The Generation of Electricity Using Magnetism

So far, you have learned that electric current is the flow of electrons

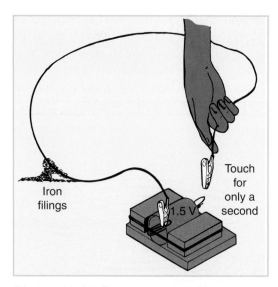

Figure 11-27 To produce a magnetic field, connect a wire to a dry cell. This will make the wire a magnet without poles.

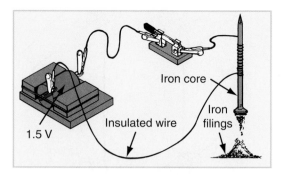

Figure 11-29 Here is a simple electro-magnet. The nail remains magnetized as long as there is current in the coil.

Useful Web site: Have students visit the NASA Web site to learn about the history of magnetism.

in a circuit. Causing electrons to flow is called "generating electricity." Electrical energy is not created. It is converted from other forms of energy.

A generator is the most practical and economical method today of producing electricity on a large scale. It uses magnetism to cause electrons to flow.

To see this in action, connect a length of copper wire to a milliammeter. As shown in **Figure 11-30**, move part of the wire loop through a magnetic field. A small current flows while the wire is cutting across the magnetic field.

The strength of the current depends on two things. One is the strength of the magnetic field. The other is the rate at which the lines of force are cut. The stronger the magnetic field or the faster the rate at which the lines of force are cut, the greater the current.

The direction of electron flow depends on the direction in which the lines of force are cut. Look at **Figure 11-30** again. When the wire moves down through the lines of force of the magnet, electrons flow in one direction. When the wire moves up, electrons flow in the opposite direction.

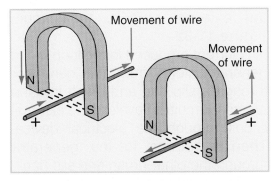

Figure 11-30 Moving a wire loop through a magnet creates a small current in the wire.

The end that loses electrons becomes positively charged. The end that gains electrons becomes negatively charged.

Alternating Current

Alternating current (AC) is electron flow that reverses direction on a regular basis. It is the type of current you use in your home. It is the type of current produced by power stations.

How is it produced? This will become clear as the basic operation of a generator is explained.

Figure 11-31 shows a simple generator. It is no more than a loop of wire turning clockwise between the poles of a magnet. Remember what was explained earlier. Current is produced only when a wire cuts through lines of magnetic force.

Now refer once more to **Figure 11-31**. With the loop (wire) in position A, no lines of force are cut. The generator produces no current. As the loop continues turning, it reaches position B. At this point, one side of the loop moves downward through the lines of force. At the same time, the other side of the loop is moving up through the lines of force. Because the wire is a closed loop, current travels through it in one direction.

As the loop reaches position C, half a revolution is completed. As in A, there is no current. Why? No lines of force are being cut.

The loop continues to turn. It reaches position D. The two sides once more cut lines of force. There is a difference, however. The side that moved downward before is now moving upward. Likewise, the side that moved

Links to other subjects: Science—Have students investigate how Joseph Priestly, Henry Cavendish, C. A. Coulomb, and G. S. Ohm contributed to our understanding of electricity and magnetism.

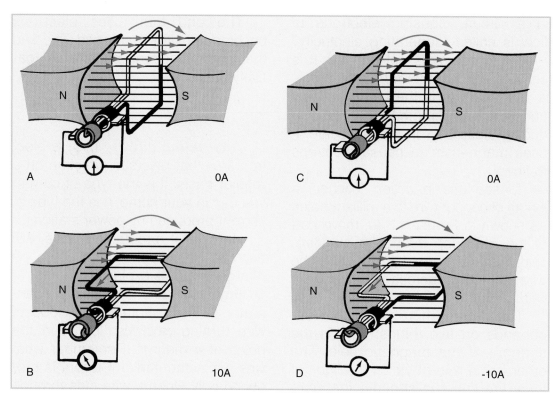

Figure 11-31 This is a basic AC generator. Note that an ammeter is connected across the terminals of the wire loop. At positions B and C, the loop or wire is cutting across lines of force to create current in the wire. The current moves through the wire into the ammeter. At B, it is going one direction. At D, current reverses. Notice that when the current is reversed the current is read as negative current. In reality, AC current is changing so quickly that a negative voltage or current cannot be read.

upward before is now moving downward. What happens? The electron flow reverses. Because the direction of flow alternates as the loop turns, the current produced is called alternating current.

Electricity produced by the generator must have a path along which it can flow. You know the path as a circuit. Therefore, the terminals (ends) of the loop must always be in contact (in touch) with an outside wire. This outside wire is stationary. The contact is made with slip rings and brushes.

A separate slip ring is permanently fastened to each terminal of the wire loop. Each slip ring turns with the loop. A brush is placed against each slip ring. As the slip rings turn, the brushes maintain rubbing contact with them. The wire forming the stationary part of the circuit is attached to the brushes. Electrical devices, such as a lightbulb, are connected to the external part of the circuit, **Figure 11-32.**

Current produced in the loop of the generator flows from the generator through a slip ring and brush into the external circuit. It travels in the external circuit through the electrical device. Then it returns to the generator through the other brush and slip ring.

As already noted, the direction of electron flow keeps changing or alternating. About 90% of the electricity

N
S
Slip ring
Brush
Stationary
or outside
circuit

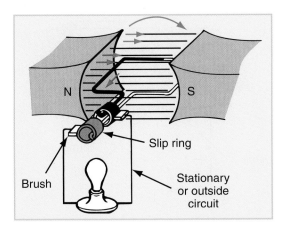

Figure 11-32 Look at this simple generator and its external circuit. What would happen if one end of the wire becomes detached from its brush?

produced in the world today is alternating current. It is easier to generate in large quantities than direct current. Even more importantly, it is easier to transmit from one place to another.

In North America, with few exceptions, alternating current makes 60 complete cycles each second. A cycle is a flow or pulse in one direction and a pulse in the opposite direction. In many European countries, the alternating frequency is 50 cycles per second. Cycles are given in Hertz rather than cycles per second. A Hertz is equal to one cycle per second.

Generators at any of the large generating stations are more complex than the simple loop generator shown in this chapter. However, their basic principle is the same. Generated current can be increased in the following two ways:

- Increasing the rate at which the lines of force are cut.
- Strengthening the magnetic field.

Therefore, many loops of wire are used instead of one. Powerful electromagnets supply the magnetic field.

For practical reasons, the loops are mounted around the inner surface of the generator housing. They remain stationary and are called the stator. The electromagnets are mounted around a rotating shaft. This assembly is called a rotor. It is placed inside the stator. In this way, current is created by having lines of force cutting across conductors instead of by having a conductor cutting across lines of force, Figure 11-33.

Direct Current

Direct current (DC) is current that does not change direction in an external circuit. The direct current generator uses a single, split ring. It replaces the two slip rings of an alternating current generator. Current in the loop still alternates. However, the split ring, called a

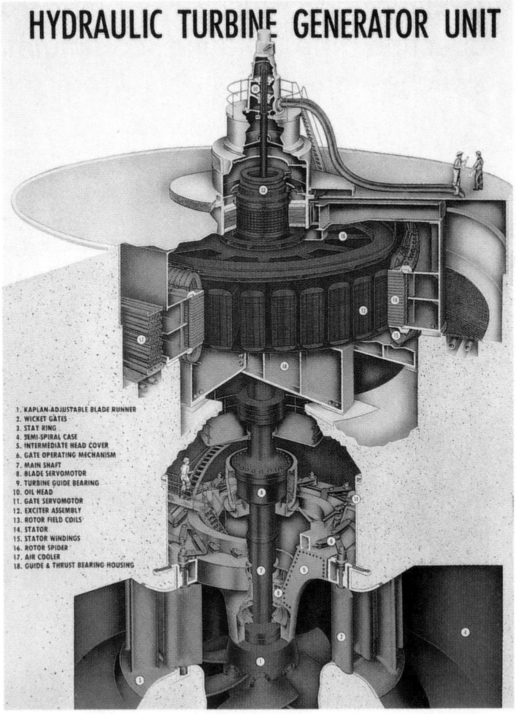

HYDRAULIC TURBINE GENERATOR UNIT
1. KAPLAN-ADJUSTABLE BLADE RUNNER
2. WICKET GATES
3. STAY RING
4. SEMI-SPIRAL CASE
5. INTERMEDIATE HEAD COVER
6. GATE OPERATING MECHANISM
7. MAIN SHAFT
8. BLADE SERVOMOTOR
9. TURBINE GUIDE BEARING
10. OIL HEAD
11. GATE SERVOMOTOR
12. EXCITER ASSEMBLY
13. ROTOR FIELD COILS
14. STATOR
15. STATOR WINDINGS
16. ROTOR SPIDER
17. AIR COOLER
18. GUIDE & THRUST BEARING HOUSING

Figure 11-33 Look at this large commercial generator. Which part is the rotor and which the stator? (Transalta)

Discussion
Why do power companies generate large quantities of AC electricity, rather than DC electricity? Why are some appliances made for North America not usable in Europe?

Vocabulary
Describe the function of the *rotor* and the *stator* in an electric generator.

Useful Web site: Have students type the phrase "DC current" into a search engine to investigate the uses of DC in portable and mobile equipment.

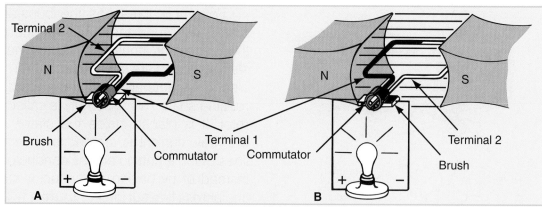

Figure 11-34 This figure explains how a commutator produces direct current. A—At this point of rotation, terminal 1 is contacting the brush connected to the negative side. B—Half a turn later, as current changes direction, terminal 1 is in contact with the brush connected to the positive side. Current through the external circuit continues in the same direction.

commutator, sends current only one way through the circuit. The brushes and commutator of a DC generator are shown in **Figure 11-34**.

Each half of the commutator is attached to one of the wire loop's terminals. As the current changes direction, the rotating commutator switches the terminals from one brush to the other every half revolution. **Figure 11-35** shows a small DC generator.

Direct current is used in portable and mobile equipment, such as flashlights and car accessories. It is also used in electronic and sound reproduction equipment. A disadvantage of DC current is that it is difficult to transmit over long distances.

Figure 11-35 A bicycle's light uses direct current supplied by the small generator. What advantage does this generator have over a light operated by a battery? (Ecritek)

Electric Motors

In many ways, an electric motor is like a generator. However, while the two have similar parts, their purposes are different. A generator converts kinetic energy to electrical energy. An electric motor changes electrical energy to kinetic energy.

Both a generator and an electric motor apply the laws of magnetism, and both contain magnets and a rotating coil of wire. The coil of wire of an electric motor is placed in a magnetic field, **Figure 11-36**. The motor spins when a current is applied to the coil of wire.

What makes an electric motor run? Electrons flowing through the coil of

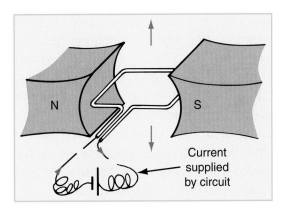

Figure 11-36 A simple electric motor is much like a generator.

wire of an electric motor cause a magnetic field around the coil. Remember the laws of magnetism? Unlike poles attract, and like poles repel. When current is introduced in the coil, the coil's magnetic field reacts with the magnets in the motor. The coil spins as it is either attracted or repelled by the motor's permanent magnets.

Unfortunately, the coil spins for only part of a turn from the effects of magnetism. The rotation would stop except for the effects of a split copper ring that rotates with the coil. For an instant, the current stops and the coil coasts. When current starts up again, magnetic force keeps the coil turning.

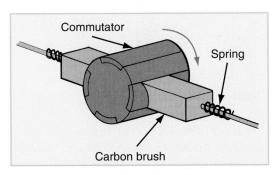

Figure 11-37 Brushes of an electric motor are always rubbing against the spinning commutator. This allows the electricity to flow into the commutator and into the coil.

The split copper ring is called a commutator, **Figure 11-37**. Current passes into and out of the coil through brushes that press against the commutator. In this way, current always passes down on the right side and back on the left side of the coil. The effect of this is to switch the poles in the coil's magnetic field. The rotation then continues in one direction.

The brushes also serve a second purpose. Since they do not rotate, they prevent the wires from twisting.

Brushes are usually made from carbon. It is a good conductor and produces less friction than metal. The brushes are spring-loaded. Pressure from the spring ensures continuous contact with the commutator.

Cells and Batteries

What most of us call a battery is not a battery at all. It is really a cell. The energy source for a car, however, is rightly called a battery. It is made up of several cells. How they differ and how they work are explained in the following sections.

Voltaic Cell

A *cell* has a single positive electrode, a single negative electrode, and an electrolyte. A *battery* is a package containing several cells together. Cells are classified as either primary or secondary.

The simplest of cells is the *voltaic cell*. Two rods, one copper and one zinc, are immersed in a container filled with a solution of water and sulfuric acid. (The mixture is known as an electrolyte.) The acid attacks and corrodes both of the metals, **Figure 11-38**.

Designing and making
How has battery technology changed consumer products?

Vocabulary
What is the difference between a *cell* and a *battery*?

Activity
Show the class a set of carbon brushes and commutator.

Resource
Activity 11-4, *A Mini Power Pack*

Links to other subjects: Science—Why is sulfuric acid used as the electrolyte in batteries?

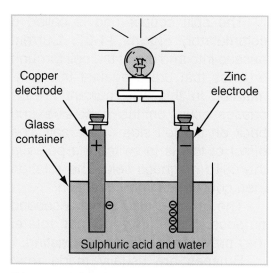

Figure 11-38 Notice on this voltaic cell that the zinc electrode is negative and the copper electrode is positive.

Some of the atoms from the metals pass into the solution. Each atom leaves behind a pair of electrons. However, the zinc rod tends to lose atoms to the solution faster than the copper rod. Since the zinc rod builds up more electrons than the copper rod, it becomes negative. If the two electrodes are connected by a conductor, excess electrons tend to flow along the conductor from the zinc to the copper. This flow of electrons produces an electric current, which will illuminate a lightbulb. Since this flow of electrons is in one direction only, cells and batteries produce DC voltage.

Primary Cell

A *primary cell* is one whose electrode is gradually consumed during normal use. It cannot be recharged. Primary cells are used in flashlights, digital watches, and cameras.

The primary cells used today employ the same principles as the voltaic cell in **Figure 11-38**. There are many different types of primary cells. They all have three main parts: the electrolyte and two electrodes. The electrolyte is usually a very active chemical such as an acid or an alkali. (Acids are compounds that react with a base, such as metal. An alkali is a substance capable of neutralizing acids.) Inside the battery, **Figure 11-39**, two chemical reactions take place. One is between the electrolyte and the negative electrode (cathode). The other is between the electrolyte and the positive electrode (anode). These reactions change the chemical energy stored in the cell into electrical energy. When the chemicals have all reacted, the cell has

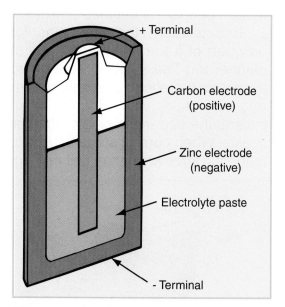

Figure 11-39 This cross section of a dry cell shows how it provides chemical storage of electricity.

Links to other subjects: Science—What chemical reaction takes place at the negative and positive poles in a battery, resulting in the production of an electric current?

Useful Web site: Have students browse the Web to find out which fruits and vegetables make the best batteries. A useful Web site for this information is www.pa.msu.edu/sci_theatre/ask_st/020393.html.

no chemical energy left, and it can give no more electricity.

Carbon-Zinc Cells

The carbon-zinc cell is the most common primary cell. It is also the least expensive, but it is short-lived. Carbon-zinc cells are produced in a range of standard sizes, **Figure 11-40**. These include 1.5 V AA, C, and D cells, as well as 9 V rectangular batteries. Six 1.5 V cells are put together to form a 9 V battery, **Figure 11-41**.

Alkaline Cells

Alkaline cells are produced in the same sizes as carbon-zinc cells. However, they can supply current longer.

Mercury Cells

Both the carbon-zinc and alkaline cells are much too large for some uses, such as digital watches, hearing aids, calculators, and miniature electronic equipment. For these applications, mercury cells are used, **Figure 11-42**. Mercury cells develop 1.34 V.

Secondary Cells

A *secondary cell*, **Figure 11-43**, is one that can store electrical energy fed into it. Then, as needed, the electricity can be drawn from the cell in the form of an electric current. In other words, it can be recharged time and time again when the electrical energy is used. Lead plate electrodes are placed in a solution of sulfuric acid. A current passing through the lead plates produces chemical changes. The sulfuric acid solution gets stronger and the cell becomes capable of producing an electric current. This is called charging a cell. When charged, the cell can produce a current in a circuit.

As electricity is drawn from the cell, the chemical change that took place during charging reverses. However, the materials in the cell are not used up; they are only changed. Therefore, the entire process can be repeated.

Each pair of electrodes in a secondary cell can produce about 2 V.

Figure 11-40 These are typical primary cells. Shown from left to right are AA, C, D, and 9 V sizes. (TEC)

Figure 11-41 This cutaway of a 9 V battery shows that six 1.5 V cells are connected. What other sizes (voltages) of batteries have you used? (TEC)

Designing and making
What products use rechargeable batteries? How do the design specifications of the battery change according to the function of the product?

Discussion
What are the differences between primary cells and secondary cells?

Activity
Have students visit a hardware store to identify the various types and sizes of cells and batteries available. Compare them with the ones described in this chapter. Students should compare the prices of the different battery types.

Links to other subjects: History—Investigate the work of Volta who, around the year 1800, invented the first battery. What were the component parts of this battery?

Technology and society: Should government legislate that all batteries sold be rechargeable? How should communities help people recycle old batteries in an effort to minimize environmental damage caused by the chemicals in them?

Discussion
Electric vehicles were in operation in the early 1900s. They are now starting to be used again. Why have electric vehicles never become very popular?

Activity
An electric vehicle has a consumption of 26 kw/h for 100 km and one kw/h costs 6 cents. A regular car takes 10 liters of gasoline for 100 km and a liter of gasoline costs 85 cents. Using the above criteria, which option is the most cost effective and by how much?

Activity
Have students browse the Internet to learn about the design of batteries for electric vehicles. Students should investigate the distance traveled on one charge, time required to recharge the battery, life span of the battery, and cost per kilowatt to recharge. Research the latest developments in secondary cell batteries, including nickel-metal, hydride, lithium-ion, and ZEBRA.

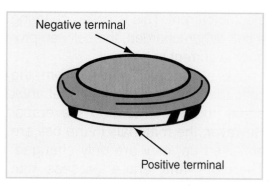

Figure 11-42 Mercury cells are smaller and shaped differently from carbon-zinc and alkaline cells. (TEC)

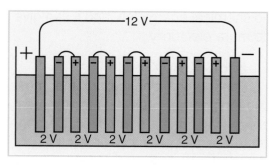

Figure 11-44 A modern automobile battery is made by connecting six cells to produce 12 V.

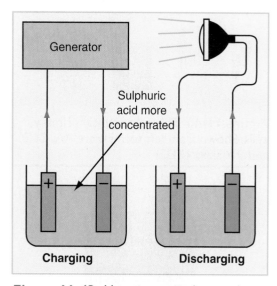

Figure 11-43 Here is a typical secondary cell. Charging—The cell can store energy fed into it during charging. Discharging—As needed, the stored energy discharges and produces an electric current.

Most motor vehicles require 12 V to operate the starter motor. Therefore, six pairs of electrodes, or cells, must be connected together. **Figure 11-44** shows a number of cells connected together to form a battery commonly called a lead-acid battery. These

batteries are used to power lights, radios, and all accessories in most cars. Rechargeable power packs are popular as emergency power and light sources. They can jump-start cars, inflate tires, run power tools, be used on campsites, and power back-seat televisions or video games. A power pack can also provide a stable power source in emergency medical situations and in locations such as the International Space Station (ISS).

Recently, there has been a lot of interest in electrically powered vehicles, which help to reduce air pollution in urban centers. Lead-acid batteries can be used, but they only provide enough energy for a short travel distance of 50 miles (80 km) before they need to be recharged. A nickel-cadmium battery gives slightly better performance and an increased range, up to 60 miles (100 km). Nickel-metal hydride is currently the highest performing battery with more than twice the power and range of a lead-acid battery. Other batteries in experimental phase include lithium-ion, lithium metal-polymer and ZEBRA batteries.

Technology and society: Builders of electric vehicles (EVs) in France receive financial aid to reduce the price of the vehicles. In Switzerland, financial incentives are given according to the fuel efficiency of a car. In the United States, several states have specific financial arrangements to encourage EVs. Why are these governments helping car manufacturers?

Chapter 11 Review
Electricity and Magnetism

Summary

Most of the electricity we use is produced at hydroelectric or thermal-electric generating stations. Other ways to produce electricity include friction, chemical action, light, heat, and pressure.

Electric current consists of a flow of electrons. Electrons move because an electromotive force pushes them. This force is provided by a generator or by chemical change in a dry cell or battery.

Most electricity is produced by generators using the principles of magnetism. An electric current flows in a wire if it is moved through a magnetic field. The current generated may be alternating or direct.

An electric motor is similar to a generator. However, a generator converts kinetic energy to electrical energy; an electric motor changes electrical energy to kinetic energy.

Small amounts of electricity may be stored in cells or batteries. Primary cells are consumed gradually and cannot be recharged. Secondary cells can be recharged.

Modular Connections

The information in this chapter provides the required foundation for the following types of modular activities:

◆ Technological Systems

◆ Electrical Systems

◆ Electronic Systems

Test Your Knowledge

Write your answers to these review questions on a separate sheet of paper.

1. If the electricity supply were cut off at 6:00 p.m. tonight, how would your community be affected?

2. What is the meaning of the word "hydro" in the phrase "hydroelectric generating station?"

3. Compare hydroelectric generating stations and thermal-electric generating stations by stating:

   A. In what way they are similar.

   B. In what way they are different.

4. After each entry below, describe where in a transmission and distribution system you would find the voltages listed.

   A. 230,000 V.

   B. 13,800 V.

   C. 240 V.

   D. 120 V.

5. A transformer can increase or decrease _____ and _____.

6. Electric current may be described as _____.

7. Name the three different groups of magnets.

8. What is the main difference between a permanent magnet and an electromagnet?

9. State the two laws of magnetism.

10. The most practical and economical method of producing electricity is _____.

11. If a wire is moved through a magnetic field created by a horseshoe magnet, _____ are caused to flow.

12. The alternating frequency in North America is _____ cycles per second (Hertz).

13. Describe the difference between alternating current and direct current.

14. How is the purpose of an electric motor different from that of a generator?

15. What are the disadvantages of most dry cells?

16. How does a dry cell produce electrical energy?

17. Over a period of a year a portable stereo system uses a large number of dry cells (often referred to as batteries). In order to reduce the amount of money you spend on power, what kind of cells could be used and why?

18. Describe the difference between a dry cell and a battery.

19. To produce 24 V, a lead acid battery needs _____ cells.

Apply Your Knowledge

1 Where and how is the electricity used in your home produced?

2. Make a model to illustrate one method of generating electricity.

3. Describe the components of a network for the transmission and distribution of electricity. How many of these components can you see in your neighborhood?

4. Make sketches with notes to illustrate (a) an atom and (b) electron flow.

5. Describe how the laws of magnetism are used to generate electricity.

6. Repeat the experiment illustrated in **Figure 11-24**. Iron filings can be made by cutting steel wool into tiny pieces using an old pair of scissors. You can fix the pattern of iron filings in place using hair spray.

7. Make a sketch to show how you would make an electromagnet.

8. List five objects in your home that use an electric motor.

9. Research one career related to the information you have studied in this chapter and state the following:
   A. The occupation you selected.
   B. The education requirements to enter this occupation.
   C. The possibilities of promotion to a higher level at a later date.
   D. What someone with this career does on a day-to-day basis. You might find this information on the Internet or in your library. If possible, interview a person who already works in this field to answer the four points. Finally, state why you might or might not be interested in pursuing this occupation when you finish school.

Answers to Test Your Knowledge Questions *(continued)*
13. Direct current continues to flow in the same direction through a wire (conductor), but in alternating current the flow continually reverses.
14. A generator converts mechanical (kinetic) energy into electrical energy. An electric motor converts electrical energy into mechanical (kinetic) energy.
15. They cannot be recharged.
16. Through the conversion of chemical energy.
17. Nickel cadmium cells because they can be recharged.
18. A battery is formed when two or more dry cells are connected.
19. 12

Chapter outline

Electric Circuits
  Protecting Circuits
  How Fuses and Circuit Breakers Work
  Direction of Current
  Types of Circuits
    Series Circuits
    Parallel Circuits
    Series-Parallel Circuits
AND and OR Gates
Conductors, Insulators, and Semiconductors
  Superconductivity
Measuring Electrical Energy
  Ohm's Law
  Watt's Law
Electronics
  Resistors
  Diodes
  Capacitors
  Transistors
  Integrated Circuits

Student Activity Manual Resources

DMA-19, *Electronic Jewelry*
ST-8, *Exploring Existing Products*
ST-74, *What Is a Circuit?*
ST-76, *Building a Series Circuit*
ST-77, *Building a Parallel Circuit*
ST-83, *Using Light Emitting Diodes*
ST-36, *Cutting and Shaping Thin Plywood*
ST-38, *Cutting and Shaping Acrylic*
ST-51, *Safety—Managing Risk*
ST-81, *Making a Membrane Parallel Switch*
ST-4, *Identifying User Needs and Interests*
ST-5, *Writing a Design Specification*
EV-1, *Evaluating the Final Product*
EV-2, *Unit Review*

DMA-20, *Game of Chance*
ST-8, *Exploring Existing Products*
ST-83, *Using Light Emitting Diodes*
ST-84, *Using Diodes*
ST-85, *Using Push-to-Make and Push-to-Break Switches*
ST-31, *Using Wood Strip to Make a Flat Frame*
ST-51, *Safety—Managing Risk*
ST-86, *Using Desktop Publishing to Produce Instruction
    Sheets*
ST-5, *Writing a Design Specification*
EV-1, *Evaluating the Final Product*
EV-2, *Unit Review*

Instructional Strategies and Student Learning Experiences

1. Invite an electrician to speak to the class about various tools and materials used when wiring a house or repairing electrical appliances.
2. Invite a firefighter to class to discuss how to prevent electrical fires and what to do in case of an electrical fire.
3. *Reading Resistor Color Bands,* reproducible master. Use this master when demonstrating how to read the values of resistor color bands.
4. Demonstrate how to read an electric meter. Have students read their home meters or the school meter for a week and record their readings. Then, have them calculate the electrical energy used.
5. Bring various types of electronic components to class. Have students identify, label, and display them. Then, have students describe the purpose of each component.

Manufacturers are continually reducing the sizes of electronic components and circuits. This creates an opportunity to add more components and circuits (resulting in more functionality and features) to a device.

Using Electricity and Electronics

Key Terms

ampere	parallel circuit
capacitor	resistance
conductor	resistor
diode	schematic
electric circuit	semiconductor
electronics	series circuit
insulator	series-parallel circuit
integrated circuit	short circuit
load	transistor
ohm	volt
Ohm's Law	watt
overload	Watt's Law

Objectives

After reading this chapter you will be able to:

◆ Design, draw, and build different types of circuits.

◆ List examples of insulators and conductors.

◆ Use Ohm's Law to calculate current, voltage, or resistance.

◆ Recall Watt's Law to calculate power, current, voltage, or resistance.

◆ Name and state the function of common electronic components.

◆ Design and make a product that uses electronic control.

Discussion
Show students a dry cell and identify the positive and negative ends. When several cells are connected, should they be connected positive (+) to positive (+) or positive (+) to negative (-)?

Vocabulary
What is the difference between a *pictorial drawing* and a *schematic*? Which would you find easier to draw? Explain your answer.

Activity
Have students select one room in their home and list all the electric and electronic appliances it contains. Using the cell holder from the previous chapter (or a commercially available one), make a simple circuit by connecting two leads to the cell and a lamp with its base.

Safety
Review the hazards and risks involved in working on electric circuits. Examine how working carefully and using the correct procedures can manage the risks. Working with electricity requires considerable care. Due to the danger of injury or death, all of the examples in this chapter use only low voltage cells or batteries. It is never advisable to allow students, who are being introduced to electric and electronic circuitry, to use 120 V.

Think about the things that electricity does. It operates motors found in many large and small appliances. The motors run electric mixers, blowers, pumps, dishwashers, washing machines, and many other appliances. They power subway trains, trolleys, and golf carts. Perhaps you have had an electric toy with a small electric motor.

Remember that every time you work with electricity, you must keep safety in mind. Water and electricity are a particularly dangerous combination. Keep all electrical devices away from swimming pools and other bodies of water. Do not fly kites near power lines. If a ball or other object gets stuck in a tree near a power line, do not try to get it down. Use only 1.5 V cells or 9 V batteries to power your experiments in this course.

Electricity is used to help us communicate. We often use telephones, radios, tape recorders, televisions, and computers.

Factories use electricity and electrical circuits for starting and stopping machines automatically. Electricity controls assembly lines and robots. Whole factories can be run electrically. A few people working at computers can control machines, lights, assembly lines, packaging, and loading of products. Indeed, it is hard to imagine what we would do without electricity.

Electric Circuits

An *electric circuit* is a closed path for electric current. The path starts from a source, such as a cell or battery. It continues through a resistance (load), such as a lamp, before it returns to its source. All power systems, including electrical circuits, must have a source, a process, and at least one load. To make a simple circuit, connect a lamp, two pieces of wire, and a 1.5 V cell. In the simple circuit shown in **Figure 12-1**, a wire is connected to the negative terminal of the cell. Electrons flow from the negative terminal and continue through the wire and the lamp to the positive terminal.

Figure 12-1 is a pictorial of this circuit. However, it takes too long to make a picture of each component. This is especially true when the circuit is complicated. Therefore, it is easier to use something simpler. The circuit diagram in **Figure 12-2** uses symbols rather than pictures. Symbols may be drawn quickly and are understood everywhere. Diagrams of circuits using symbols are called *schematics*.

The circuit in **Figure 12-2** has voltage, resistance, and current. Voltage is supplied by the cell. The resistance

Figure 12-1 A pictorial drawing of a simple circuit showing a lamp lit by a 1.5 V cell.

Useful Web site: Students can learn more about simple circuits at www.aspire.cs.uah.edu/textbook/ohms.html.

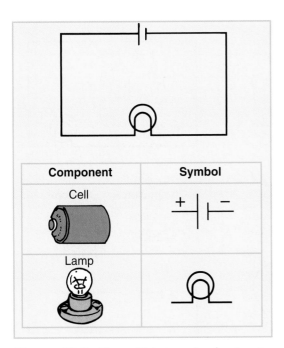

Figure 12-2 Top—This is a simple circuit schematic of the circuit pictured in Figure 12-1. Bottom—These are the components and their symbols.

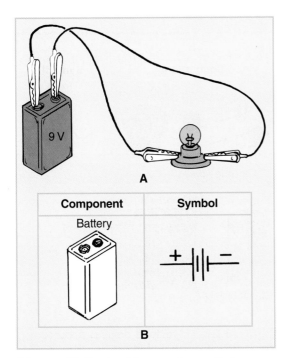

Figure 12-3 A—You can turn off the lamp by releasing any of the four clips. B—A 9 V battery and its schematic symbol.

is the glowing element in the lamp. The current is the flow of electrons in the circuit. If there is only one resistance, such as a lamp in the circuit, it is a simple circuit.

If you wished to turn the lamp on and off, you could do it by connecting and disconnecting a wire at any one of the four places shown in **Figure 12-3**. However, an easier method would be to add a switch to the circuit.

A switch is a device that enables the circuit to be turned on and off. With the switch closed, current flows, and the lamp illuminates. When the switch is open, the flow stops, and the lamp turns off. Mechanical switches are also used to direct current to various points. The simplest switch is a single-pole, single-throw (SPST) switch, **Figure 12-4**. There are many types of switches. **Figure 12-5** shows six.

Protecting Circuits

Too much current can overheat and damage circuits. To prevent this, a fuse or a circuit breaker is added, **Figure 12-6**. These are devices that open the circuit when current is too high. The fuse "blows" or burns out. The circuit breaker trips a contact. In either case, when the circuit is open, it is no longer complete, and current stops.

When applied to using electricity, the term *load* means something in the circuit that uses up the electric current. The bigger the load, the more current it needs.

High current can be caused by an overload or a short circuit. An *overload* occurs when lights or appliances in the circuit demand

Useful Web site: Visit www.pge.com/microsite/PGE_dgz/power/breaker.html to learn more about fuses, circuits, and breakers.

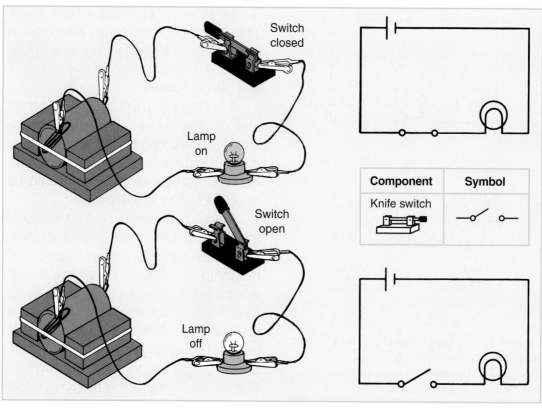

Figure 12-4 Look at the open and closed circuits. What is the difference between the two circuit diagrams?

more current than the circuit can safely carry. A good example of this is when too many appliances are plugged into one outlet. A *short circuit* is a path of low resistance resulting in high current because the

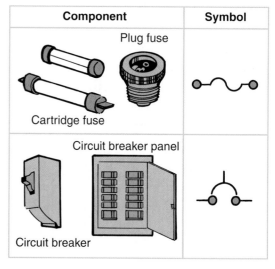

Figure 12-6 Overload devices protect the wiring and other devices in a circuit.

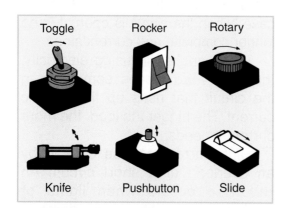

Figure 12-5 Where have you seen these different switches used?

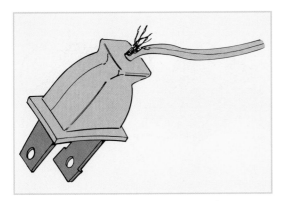

Figure 12-7 A short circuit is a fault in a circuit that allows current to return to its source without traveling through the entire circuit. Frayed wires can produce a short circuit.

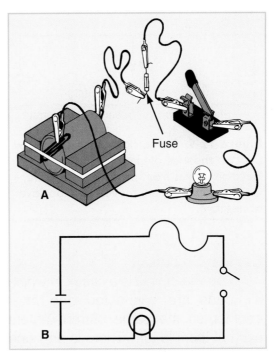

Figure 12-8 A fuse stops the flow of electricity when an overload occurs.
A—This is a simple circuit with a fuse.
B—Here is a schematic of the same circuit.

current avoided a section of a circuit, such as a lamp or a motor (load), **Figure 12-7**.

How Fuses and Circuit Breakers Work

When too much current attempts to pass through a fuse, **Figure 12-8**, a thin wire, called a filament, melts. Current stops before the wires can be damaged. When high current enters a circuit breaker, it heats a bimetal (two metals) strip. The strip bends as one metal expands more than the other. This opens contacts so no current can pass.

Once a fuse blows, it must be replaced. A circuit breaker's contacts can be reset.

Direction of Current

Earlier, you learned that current is the flow of electricity in a conductor. The direction of this flow in a circuit can be shown either as "electron flow" or "conventional current," **Figure 12-9**.

Electron flow is based on the electron theory. This theory states that current moves from negative to positive. Conventional current is based on an older theory of electricity. Early scientists assumed a current moved from positive to negative.

Both theories are acceptable. In this book, however, all explanations will be based on electron flow.

Types of Circuits

A circuit is a pathway along which electricity may travel. The simple circuit shown in **Figure 12-9** contains only one lamp. If two or more lamps are put into this circuit, they can be connected in one of two ways: in series or in parallel.

Useful Web site: Students can learn about electron flow by visiting www.qrg.northwestern.edu/projects/vss/docs/Power/2-whats-electron-flow.html.

Enrichment
Introduce students to a wide range of symbols used in architectural plans, including those for bells, buzzers, telephones, and television outlets.

Activity
Have students describe what happens when different switches in the circuit, shown in Figure 12-12, are closed or opened.

Activity
Connect a bulb and dry cell so that it can be turned off by one switch and turned on by the other. This simulates turning a light on at the bottom of the stairs, and turning it off at the top of the stairs.

Examples
Connect examples of series and parallel circuits. Ask students to explain why bulbs in series circuits are dimmer than those in parallel circuits.

Safety
Review the hazards and risks involved when working with series and parallel circuits. Examine how working carefully and using the correct procedures can manage the risks.

Resource
Transparency 12-1, *Types of Circuits*

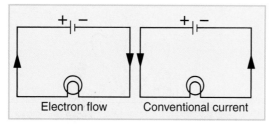

Electron flow   Conventional current

Figure 12-9 These circuit diagrams show the two theories of electric current. Electron flow is from negative to positive. Conventional current moves from positive to negative.

Series Circuits

A *series circuit* is a circuit in which all loads are connected one after another so the same current enters each of them in turn. There is only one path for electron flow, **Figure 12-10**.

When lamps are connected in series, each gets part of the voltage. For example, three lamps connected to a 1.5 V cell each receive 0.5 V. Therefore, each bulb will be dimmer than if only one lamp is in the circuit. However, current remains the same across each lamp.

In this series circuit, if one lamp burns out, all of the lamps go off! This is because the circuit would be open, and current would stop flowing. For this reason, very few series circuits are used in our homes. Sometimes, however, Christmas tree lights are wired in series.

Parallel Circuits

A *parallel circuit* is a circuit having more than one path for electron flow, **Figure 12-11**. You can see that when lamps are in parallel, the current splits. It goes through each of the lamps without passing through any others first. In parallel circuits, the current

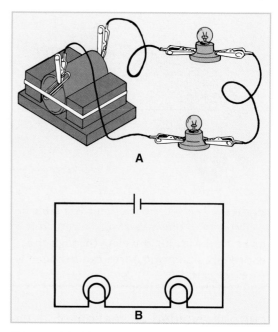

Figure 12-10 Since these lamps are connected in series, electricity flows through each lamp in turn.

varies in each path. However, voltage is always the same. If one bulb burns out, the circuit is not broken, and the other bulbs continue to burn. That is why most circuits in the home are parallel circuits.

Series-Parallel Circuits

The third kind of circuit is the *series-parallel circuit*. These circuits exhibit characteristics of both series and parallel circuits because they are wired in such a manner, see **Figure 12-12**. The different branches react to current and voltage in different ways. If a particular branch is a series branch, it behaves like a series circuit. Current is the same throughout that branch of the circuit, with the voltage dividing between the components. This matter is vice versa with parallel circuits. Voltage remains

Useful Web site: Have students work through the interactive tasks at www.schoolscience.co.uk/content/3/physics/circuits/circh2pg1.html to learn more about series and parallel circuits.

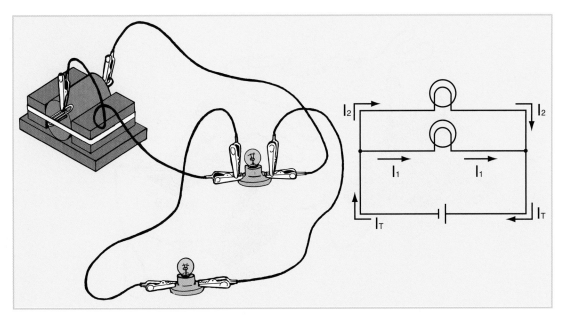

Figure 12-11   Since these lamps are connected in parallel, there are two paths for current. At the left intersection, the current splits into the two branches. At the right intersection, the separate currents add back together and equal the original current.

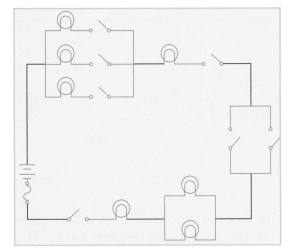

Figure 12-12   A series-parallel circuit has separate branches that are either series or parallel and, thus, exhibit those characteristics. Notice the circuit wiring and figure out which branches are series and which are parallel.

Discussion
Do Figures 12-14 and 12-16 show AND or OR circuits?

Safety
When is an AND gate important for safety reasons?

the same throughout the parallel branches, and current is divided between the parallel components.

AND and OR Gates

In the circuits just described, lamps were connected in series, parallel, or series-parallel. Switches may also be connected in these ways. Look at **Figure 12-13**. For the bulb to light, both switch A and switch B have to be closed. In **Figure 12-14**, how many switches must be closed to close the circuit?

In electronics, switches connected in series are called an AND gate. Can you think of a way in which we use AND gates? Look carefully at how an elevator works. Notice that there are two sets of doors. It is only when both

Useful Web site: Students can learn how AND and OR gates are based on Boolean Logic by visiting http://computer.howstuffworks.com/boolean.htm/printable.

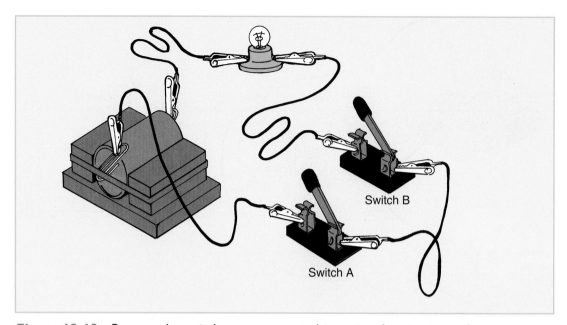

Figure 12-13 Because the switches are connected in series, this circuit conducts current only when both switches are closed.

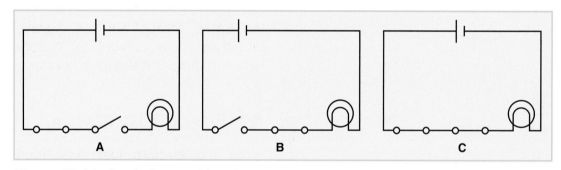

Figure 12-14 In which circuit(s) is the bulb lit?

the outside doors *and* the inside doors are closed that the electrical circuits are complete. Only then will the elevator move.

What happens when switches are connected in parallel, **Figure 12-15**? Switch A *or* switch B completes the circuit and turns on the light. Switches in parallel are called OR gates.

Many homes contain an OR circuit. One example is a doorbell. Push-button switches are located at both the front and back doors. Pressing one *or* the other will ring the bell. Look at the diagrams in **Figure 12-16**. The concepts you just learned of the AND and OR gates are just part of the basics of digital electronics.

Conductors, Insulators, and Semiconductors

Materials that allow electric current to flow easily are called *conductors*. Copper, aluminum, silver, and most

Links to other subjects: Mathematics—Students can learn more about the rules of Boolean algebra by visiting www.play-hookey.com/digital/boolean_algebra.html.

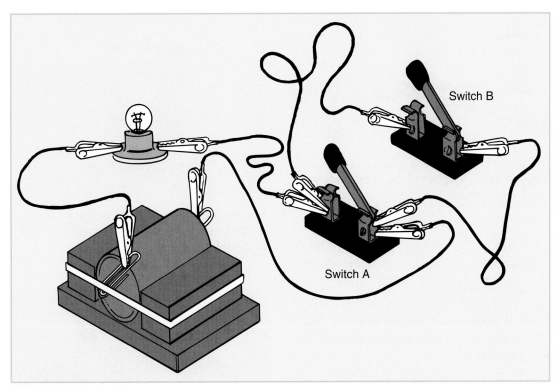

Figure 12-15 This parallel circuit conducts current when either switch A or switch B is closed.

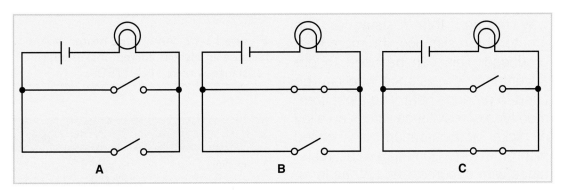

Figure 12-16 In which circuit(s) is the bulb lit?

Discussion
Discuss what is special about the molecular structure of materials that conduct electricity.

Enrichment
Aluminum will conduct electricity, but is not used in house wiring. Aluminum forms an oxide on the surface when exposed to air, which occasionally results in a spark when trying to complete the circuit.

Activity
Bring samples of different materials to class. Ask students to predict which items will be good conductors and which will be poor conductors. Verify their answers by connecting each in a simple circuit. What do the materials in both categories have in common?

other metals are examples of good conductors. Copper is used most often for house wires. Strength, low cost, and low resistance to current make it a good choice, **Figure 12-17.**

Materials that do not allow current to pass are called *insulators*. Glass, rubber, plastic, porcelain, and paper are good insulators. Insulators play an important part in controlling electricity. They are wrapped around a conductor to prevent it from passing current to another conductor. This keeps the current in the correct path.

Useful Web site: Students can learn more about why materials conduct or insulate by visiting www.pbs.org/transistor/science/info/conductors.html.

Discussion
Materials used to conduct electricity may also be used to resist the flow of electric current. When is this resistance beneficial and can be of use in a circuit and when is it not desired?

Enrichment
Have students browse the Internet to identify materials that are semiconductors.

Vocabulary
Define the term *semiconductor.*

Activity
Look for examples of where insulators are used. For example, identify the part of an electrical appliance that provides insulation for the user.

Activity
Have students dismantle a small appliance, such as a toaster or an iron, and identify where insulators are used.

Safety
Ask students to recall times when part or all of their bodies were wet. What types of electrical appliances were nearby that may have resulted in an electrical shock?

Resource
Activity 12-3, *Conductors and Insulators of Electricity*

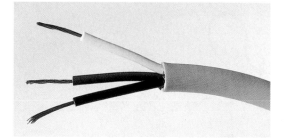

Figure 12-17 Here is a three conductor copper cable. Copper is the most widely used conductor.

Figure 12-18 Insulators protect against electric shock. How is insulation being used in this photograph?

Figure 12-19 An electric heater works because an electric current makes a high-resistance wire red hot. (TEC)

One of the most important uses of insulators is to protect us from electric current, **Figure 12-18.** Our bodies will conduct electricity especially when wet. If you touch a live wire by accident, you get a dangerous shock. The shock may kill you.

Some materials are better conductors than others, and almost all materials have some resistance to the flow of electricity. Therefore, electricity produces heat as it forces its way through this resistance. The greater the resistance, the more heat produced. This can be used to our advantage. A resistance wire can be used to produce heat. The most common type of resistance wire is an alloy of nickel and chromium. It is called nichrome wire. Sometimes the resistance wire becomes red hot, as in an electric stove, toaster, and other heating appliances, **Figure 12-19.** Sometimes the conductor becomes white hot. This is the case with an electric lightbulb.

Semiconductors are materials that do not conduct as well as copper or silver. Nor do they insulate as well as rubber or glass, **Figure 12-20.** They have some characteristics of each.

Class	Materials	
Conductors	• Copper • Silver • Tungsten • Nichrome	
Semiconductors	• Silicon • Germanium	
Insulators	• Rubber • Glass • Porcelain	• Nylon • PVC • Mica

Figure 12-20 Materials commonly used in electronics/electrical applications and their classifications.

Useful Web site: Students can learn more about how semiconductors work by visiting http://science.howstuffworks.com/diode.htm/printable.

These traits are useful in electronic circuits. They allow electron flow only under certain conditions. The most common semiconductor materials are silicon and germanium. Transistors and other electronic components are made from them.

Superconductivity

When electricity flows through transmission lines it meets resistance and loses some of its energy in the form of heat. For many years, technologists have dreamed of producing a material that will conduct electricity without resistance. Such a material is called a superconductor. Until recently, superconductivity was possible only at low temperatures close to absolute zero, –459° F (–273° C). Absolute zero is 0° Kelvin. Very recently, materials have been discovered that conduct at higher temperatures. Researchers believe that, within a short time, superconductor materials will be found to operate at room temperatures.

When these superconductors become widely available, they will revolutionize the electronics industry. The absence of electrical resistance reduces the amount of heat produced. This allows components to be packed more closely together. This reduces the size of components and products. Superconductors will also enable computers to operate at much greater speeds.

Measuring Electrical Energy

When measuring electrical energy, three terms are important: volts, ohms, and amperes. One way to understand their meaning is to compare electricity to water. Water flows through a hose under pressure. Electricity inside a wire is also under pressure. With electricity, the pressure is called voltage and is measured in *volts*. If pressure is increased, more water flows than before. The same is true with electricity, **Figure 12-21**.

As the water flows through the pipe, it meets resistance. The smaller and longer the pipe, the greater the resistance. So it is with electricity. Flow is affected by diameter and length of the wire. Also, electricity flows more easily through some materials than through others, **Figure 12-22**. Electrical resistance is measured in *ohms*.

The amount of water that flows out of the end of the pipe in a given period depends on pressure and resistance, **Figure 12-23**. The higher the pressure and the weaker the resistance is, the greater the amount of water leaving the hose.

Ohm's Law

The amount of electricity that passes a point in a conductor in a given period also depends on pressure and resistance. Electrical current is mea-sured in *amperes* (A). The higher the pressure (volts–V or E) and the weaker the resistance (ohms–R), the higher the current (amps–I). This relationship between voltage, resistance, and current is described by a formula known as *Ohm's Law*, **Figure 12-24**. Ohm's Law is written as:

Voltage = Current × Resistance

(or)

Volts = Amps × Ohms

(or)

E = I × R

Useful Web site: Students can learn more about the history of superconductors and how they work by visiting http://superconductors.org/Index.htm.

Examples
Use an ammeter to show the measurement of a flow of electrons or current. Use an Ohmmeter and a resistor of known quantity to demonstrate resistance.

Resource
Activity 12-6, *Practicing Ohm's Law*

Enrichment
The amount of current needed in a house varies. Typically, 100 amps are needed for general use, 150 amps for electrical heating, and 200 amps for air-conditioning. Resistance to the flow of an electrical current in a conductor primarily occurs for two reasons: each atom tries to keep an electron attached to its positive nucleus and there are collisions of countless electrons and atoms. Together, these resistances cause heat in the conductor.

Examples
Demonstrate voltage drop by connecting a simple circuit, such as a lamp, battery, and knife switch. Attach one voltmeter (VM 1) to the battery terminals and a second (VM 2) to the lamp terminals. VM 1 reads the applied voltage and VM 2 shows the drop across the lamp.

Resource
Transparency 12-2, *Ohm's Law*

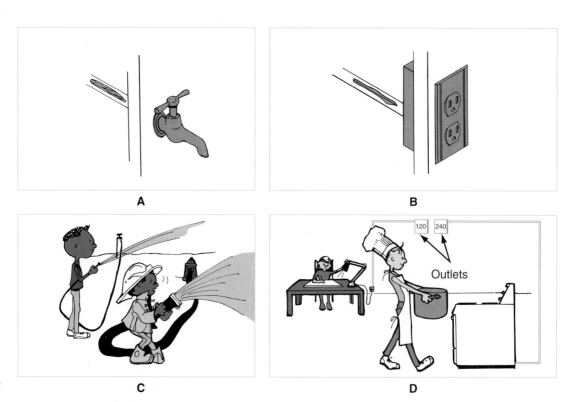

Figure 12-21 Electricity pressure principles are easily compared to those of water. A—Water behind a faucet is always under pressure. B—Electricity behind an outlet is also under pressure. C—To use water, you must attach a hose and turn on the faucet. D—To use electricity, you must connect a wire to an outlet and usually toggle a switch.

Figure 12-22 Left—The greater the resistance, the smaller the flow of water. Right—Why does an air conditioner require a larger diameter wire than a clock?

Useful Web site: Have students calculate the values in series circuits using Ohm's Law at http://webhome.idirect.com/.

Figure 12-23 Let's compare the high pressure and the low resistance principles of both water and electricity. A—High pressure and a large hose equal heavy water flow. B—High voltage and a large conductor equal high electron flow.

Watt's Law

Another unit of electrical measurement is the watt. A *watt* is the unit used to measure the work performed by an electric current. To calculate the

Figure 12-24 To remember Ohm's Law, just think of listening to your instructor with your EIR (pronounced like 'ear'). Volts are often represented using the letter E for electromotive force.

power (P) in watts, use *Watt's Law* by multiplying the voltage by the current, **Figure 12-25**.

Power = Current × Voltage

(or)

Watts = Amperes × Volts

(or)

P = I × E

The monthly electricity bill for your home is based upon the number of watts used, **Figure 12-26**. The utility company provides a meter for each home. The meter measures how many watts are used. The watt is a small unit, so the basic unit used by power companies is a kilowatt. (This is equal to 1000 watts.)

Most appliances are being constantly switched on and off, so the electrical usage of the home varies. The electricity used is measured over periods of one hour. The unit is therefore called one kilowatt-hour. A kilowatt-hour means 1000 watts used for a period of one hour.

Useful Web site: Have students investigate how to conserve electricity and reduce costs by visiting the Web site of your local electric company or the Hydro Quebec Web site at www.hydroquebec.com/residential.

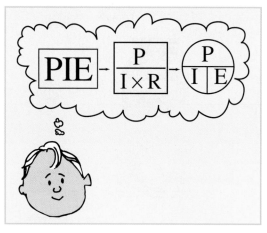

Figure 12-25 Remembering Watt's Law is as easy as PIE.

Figure 12-26 This meter measures the amount of electricity used in kilowatt-hours (kW-h). (TEC)

Figure 12-27 shows the dials of a typical electrical meter. When reading the dials, be careful because some of them revolve clockwise and some revolve counterclockwise. To read a dial, write down the number the pointer has passed. In **Figure 12-27**, the correct reading is 23,642.

Electronics

In this chapter, you have learned about the flow of electrons in a circuit. *Electronics* is the use of electrically controlled parts to automatically control or change current in a circuit. It is the technology of controlling electron flow. Electrons can be used to control, detect, indicate, measure, and provide power. A variety of electronic components carry out these functions.

Resistors

Resistors are electrical devices that control how much current flows through a circuit. Resistors make it more difficult for current to flow. In **Figure 12-28**, bulb A will be brighter than bulb B. The current in bulb B is smaller because there is a resistor in that loop of the circuit.

Resistors are made in many sizes and shapes, **Figure 12-29**. All do the same thing; they limit current. In a typical carbon composition resistor, powdered carbon is mixed with a glue-like binder. Changing the ratio of carbon particles to binder changes the resistance. The greater the amount of carbon used, the less the resistance, **Figure 12-30**.

Resistors are made in a wide range of values. The value corresponds to the degree to which they limit the flow of electrons, their *resistance*. Resistors are often quite small, which would make it difficult to write their values on them. To overcome this problem, resistors are usually marked with four colored bands, **Figure 12-31**.

Useful Web site: Have students visit www.electronic-circuits-diagrams.com to investigate electronic circuits and circuit diagrams related to the home, garden, computer, radio, and robotics.

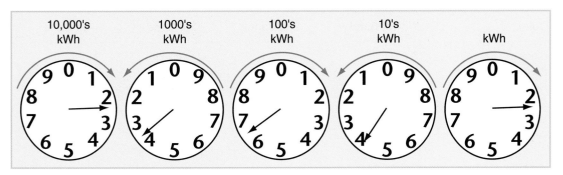

Figure 12-27 What is the reading of the meter at your home?

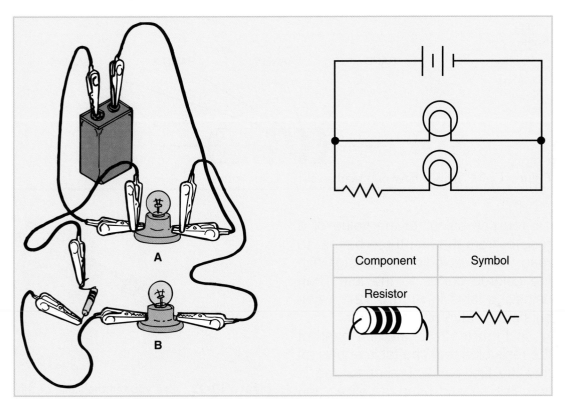

Component	Symbol
Resistor	

Figure 12-28 This parallel circuit has a resistor in one loop.

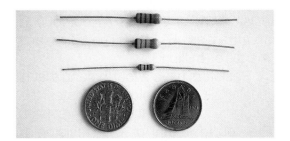

Figure 12-29 Resistors come in different shapes and sizes. They are often quite small.

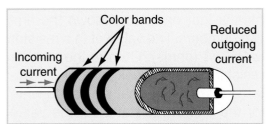

Color bands

Incoming current

Reduced outgoing current

Figure 12-30 Look at the inside of this resistor. Note how current is having difficulty moving through the resistor.

Discussion
Telling a lie will make you sweat. How does a lie detector that measures skin resistance work?

Enrichment
Students can learn more about resistors at www .Williamson-labs .com/resistors.htm.

Activity
Have students learn more about the uses of fixed value resistors in electronic circuits by visiting www.doctronics .co.uk/resistor.htm.

Useful Web site: Have students use the interactive Web site www.the12volt.com/resistors/resistors.asp to calculate a variety of resistor values.

Color	1st Band	2nd Band	3rd Band	4th Band
Black	0	0	Ω	
Brown	1	1	1 zero	
Red	2	2	2 zeros	
Orange	3	3	3 zeros	
Yellow	4	4	4 zeros	
Green	5	5	5 zeros	
Blue	6	6	6 zeros	
Violet	7	7	7 zeros	
Gray	8	8	8 zeros	
White	9	9	9 zeros	
Gold				5
Silver				10
None				20
	BAND 1	BAND 2	BAND 3	BAND 4

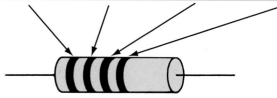

Figure 12-31 This is how one reads resistor color bands.

You can calculate the value of a resistor from the first three bands. To read the value, hold the resistor with the colored bands to the left. Then using the table in **Figure 12-31**, calculate the value of the resistor.

In **Figure 12-32**, the first band of the resistor is red. The table shows us that the first band number for red is two. The second band is yellow, and the second band number for yellow is four. The third band is brown. Since the number for brown is one, this means that one zero follows the first two numbers. The value of this resistor is 240 Ω (ohms). Ohm is the unit for resistance. The Greek symbol Ω (written as *omega*) is what we use to write for the unit of resistance. If the value were larger, for example, 47,000 Ω, its value would be written as 47 kΩ (The

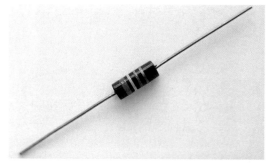

Figure 12-32 Can you tell the value of this resistor? Use the chart in Figure 12-31. (TEC)

k stands for "kilo" or thousand.). A resistor whose value is 47,000,000 Ω is written as 47 MΩ. (The *M* means "mega" or million.)

Resistors have a fourth band that is usually silver or gold. It indicates the accuracy or tolerance of the resistor. The fourth band of the resistor in **Figure 12-32** is silver, indicating

Useful Web site: Have students access www.circuit-innovations.co.uk/resistors.html to learn more about the use of resistors in circuits.

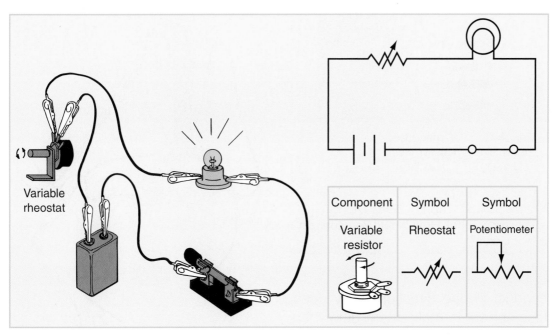

Component	Symbol	Symbol
Variable resistor	Rheostat	Potentiometer

Figure 12-33 A rheostat is connected in series.

that the resistor has a tolerance of ±10%. Ten percent of 240 is 24. Therefore, the actual value is between 216 and 264 Ω.

Two types of variable resistors are rheostats and potentiometers. Rheostats lower or raise the current in a circuit, **Figure 12-33**. The dimmer switch used for car dashboard lights is one example. Potentiometers lower or raise the voltage, **Figure 12-34**. They are used as volume controls in stereos.

Diodes

Diodes are devices that allow current to flow in one direction only. There are two ends to a diode: positive (anode) and negative (cathode). A dark band indicates the negative end, **Figure 12-35**.

Figure 12-36 shows two bulbs in parallel. Each has a diode in its loop

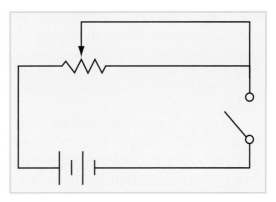

Figure 12-34 A potentiometer is connected in parallel.

of the circuit. Only bulb A will light because the diode next to it is positioned correctly to allow electrons to flow. Diodes are most commonly used in rectifiers, which are used to change alternating current to direct current. Some different types of diodes can be seen in **Figure 12-37**.

Another type of diode is a light emitting diode (LED). LEDs also only

Discussion
What can be "stored" in a sponge, suitcase, compact disc, and capacitor?

Enrichment
Have students learn more about how capacitors work by visiting http://science .howstuffworks.com/ capacitor.htm.

Activity
To demonstrate how a capacitor works connect it to a battery, remove the battery, and connect the charged capacitor to a lamp. The lamp will light for a moment as the charge from the capacitor flows through it.

Safety
Review the hazards and risks involved when working with capacitors.

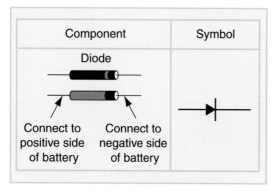

Figure 12-35 This is a diode and its schematic symbol. How would you connect it in a circuit?

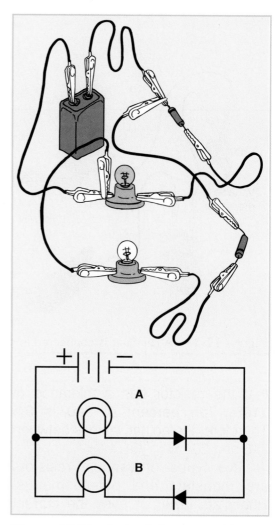

Figure 12-36 A circuit is set up with two lamps in parallel. Each lamp has a diode connected in series, but only one lamp will light. Decide which lamp will light and explain why.

conduct in one direction. They need less current to make them glow than do bulbs, but they are not as bright. LEDs, therefore, are used where the brightness of the bulb is not important. They show that electrical equipment is turned on and working, **Figure 12-38**.

Capacitors

Capacitors are designed to store an electrical charge. The simplest capacitor is made of two metal plates (conductors) separated by an insulator (called a dielectric), **Figure 12-39** and **Figure 12-40**. The insulator may be air, but it is often a thin sheet of plastic.

A capacitor smoothes the pulsating current from a rectifier into a steady direct current. A capacitor connected in a direct current circuit can store a charge for a considerable time after the voltage to the circuit has been switched off. NEVER touch the leads of a capacitor before it has been discharged.

Transistors

A *transistor*, **Figure 12-41**, is an electronic device with three terminals and many applications, such as a signal amplifier or an electrically-controlled switch that can turn an electric current on and off. It is like an electric light switch. The switching, however, is done by voltage, instead

Useful Web site: Students can learn about the development of the capacitor at www.meridianelectronics.ca/gadgets/caps/caps.html.

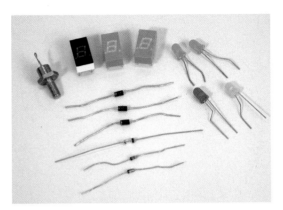

Figure 12-37 Here are some of the different types of diodes. (Tom Severson)

Figure 12-38 How many light emitting diodes do you see in this sound system? (Tom Severson)

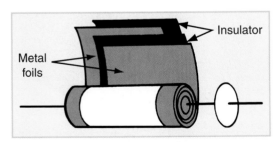

Figure 12-39 A capacitor is a "sandwich" made up of conductors and insulators.

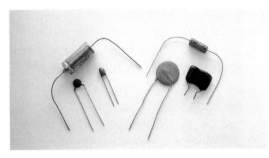

Symbol

Figure 12-40 These are some different types of capacitors and a capacitor schematic symbol.

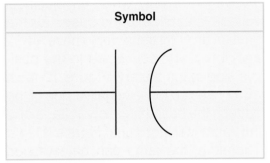

Component	Symbol
Transistor	NPN PNP

Figure 12-41 There are two types of transistor, PNP and NPN. Negative voltage controls the output of a PNP transistor. Positive voltage controls the NPN transistor. (Tom Severson)

Discussion
In a transistor, the base current is much smaller than the collector current. However, the small base current controls the larger collector current. What other small devices control a larger object? For example, how is a boat steered and how does an airplane move up and down?

Enrichment
There are many types of transistors, but they all have three working leads. They are the *collector* (entrance), the *base* (entrance), and the *emitter* (exit). Current flows through two entrances and out through one exit.

Activity
Have students learn more about how transistors work by visiting www.intel.com/ education/transworks.

Safety
Review the hazards and risks involved when working with transistors.

Links to other subjects: History—Have students investigate the development of the transistor. A useful starting point is www.pbs.org/transistor.

Useful Web site: Students can learn more about electrical and information technology at www.ieee-virtual-museum.org.

Designing and making
Designers frequently create products that operate automatically, rather than manually. What items around you could be altered to be controlled with little or no human input?

Discussion
Integrated circuits are very small. What are some of the applications where small size is a benefit?

Enrichment
Silicon is an insulator. To become a semiconductor, arsenic is added and it becomes negatively charged. When aluminum is added, it is positively charged.

Vocabulary
What is *nanotechnology*?

Activity
Have students browse the Web to learn about nanocircuits.

Resource
Transparency 12-4, *Electronic Components*

of by hand. A transistor can be used to switch the current to an electric lamp on and off. The lamp will not light unless electrons can flow through the circuit by way of the transistor.

All transistors have three terminals: a collector (c), a base (b), and an emitter (e). An electric current will flow through the transistor only when an electrical voltage is applied to the base. For the base of an NPN transistor to have a voltage, it must be connected to the positive side of the battery, **Figure 12-42**. If it is connected to the negative side of the battery, the base has a low voltage and the lamp will not light, **Figure 12-43**. Therefore, the lamp can be switched on and off by changing the voltage on the base of the transistor.

Many transistors are fast switches. They are used in timing, counting, and computer circuits. In these circuits, the signal is either on or off. Amplifying transistors are used in radios and stereos where a weak signal must be amplified in order to be heard over a speaker.

Integrated Circuits

The electronic circuits described in this chapter so far have been built using separate components. These

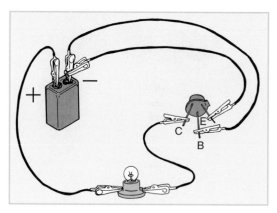

Figure 12-43 With the base terminal of the NPN transistor connected to the negative terminal of the battery, the lamp will not light.

components include resistors, capacitors, diodes, and transistors. It is possible to build an electronic circuit, called an *integrated circuit*, on a single chip (IC chip). These replace whole groups of separate components. Today, one integrated circuit may contain well over one billion components.

Most integrated circuits are built on a tiny piece of silicon called a silicon chip, **Figure 12-44**. Resistors, capacitors, diodes, and transistors, with their connections, are formed in miniature on the surface of the chip. We owe our advances in computer technology to the development of IC chips.

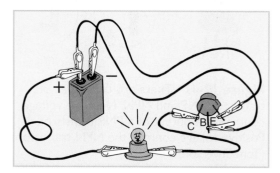

Figure 12-42 The base terminal of the NPN transistor is connected to the positive terminal of the battery. The lamp will light.

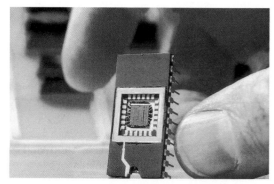

Figure 12-44 The term "integrated circuit" is usually shortened to "IC." (Telecom)

Useful Web site: Have students investigate nanocomputing, the use of molecule-based electronics to increase computational power in computers. A useful starting point is http://pubs.acs.org/cen/topstory/7946/7946notw4.html.

Chapter 12 Review
Using Electricity and Electronics

Summary

In order for any electrical device to operate, it must contain one or more circuits. An electric circuit is a continuous path from a source, through a load, and back to the source. Schematics use symbols to show components and their connections. There are three basic types of circuits: series, parallel, and series-parallel. Direction of current flow in a circuit may be shown in terms of electron flow or conventional current flow.

Materials that allow current to flow are called conductors. Those that do not allow current flow are called insulators. Between these two are materials known as semiconductors. They allow current under certain conditions.

Electrical energy is measured in terms of volts, ohms, and amperes. Work performed by an electrical current is measured in watts.

Electronics is the study of the way in which electron flow may be controlled. Control is carried out using electronic components. These include resistors, diodes, capacitors, and transistors. Integrated circuits replace a large number of these separate components.

Modular Connections

The information in this chapter provides the required foundation for the following types of modular activities:

◆ Electronic Systems

◆ Computer Technology

◆ Digital Design

Answers to Test Your Knowledge Questions
1. A continuous path from the source through a load and back to the source.
2. Refer to Figures 12-2 through 12-6 on pages 311–312.
3. Refer to Figure 12-8B on page 313.
4. Overloaded.
5. Short circuit.
6. When a circuit becomes overloaded, the thin filament of wire in the fuse becomes hot and melts. The circuit is then broken and the flow of current is stopped.
7. Circuit breaker.
8. See Figure 12-9, page 314.
9. See Figure 12-10, page 314.
10. See Figure 12-11, page 315.
11. In parallel so that when one bulb burns out, the others will stay lit.
12. Series is AND gate. Parallel is OR gate.
13. A conductor allows an electric current to flow through it easily. An insulator does not allow a current to flow through it. A semiconductor allows an electric current to flow through it only under certain conditions.
(continued)

◆ Digital Music
◆ Digital Electronics
◆ Music and Sound
◆ Electrical Systems
◆ Control Logic
◆ Technological Systems
◆ Digital Transportation
◆ Radio-Controlled Transportation

Test Your Knowledge

Write your answers to these review questions on a separate sheet of paper.

1. An electric circuit can be defined as _____.

2. Draw the symbol for (a) a lamp, (b) a dry cell, (c) a fuse, and (d) a switch.

3. Draw a circuit diagram of a simple circuit, which includes a lamp, dry cell, fuse, and switch.

4. If too many appliances are plugged into the same receptacle, the circuit may become _____.

5. If two bare wires carrying a current touch each other a(n) _____ _____ occurs.

6. Describe how a fuse protects a circuit from overload.

7. An alternative to a fuse for protecting the circuits in a home is a(n) _____ _____.

8. Using diagrams and notes, explain the difference between electron flow and conventional current flow.

9. Draw a circuit diagram to show two lamps and a dry cell that are connected in series.

10. Draw a circuit diagram to show two lamps and a dry cell connected in parallel.

11. Is it better to connect your Christmas tree lights in series or parallel? Explain your answer.

12. A circuit in which two switches are connected in series is called a(n) _____ gate. A circuit in which two switches are connected in parallel is called a(n) _____ gate.

13. A conductor is a material that _____. An insulator is a material that _____. A semiconductor is a material that _____.

14. Name four materials that are conductors of electric current and four insulators.

15. Electric pressure is measured in _____.
    A. joules
    B. watts
    C. ohms
    D. volts

16. Electric resistance is measured in _____.
    A. ohms
    B. amperes
    C. watts
    D. volts

17. Electric current is measured in _____.
    A. joules
    B. volts
    C. amperes
    D. watts

18. A portable electric heater with a resistance of 15 ohms is connected to a 120 volt AC outlet. The current flow in the circuit will be _____.

19. What is an integrated circuit?

20. Give the value of each of the following resistors.

	1st band	2nd band	3rd band	4th band	Value
(A)	red	green	yellow	silver	
(B)	orange	blue	brown	gold	
(C)	white	brown	red	none	
(D)	violet	green	orange	silver	

Answers to Test Your Knowledge Questions *(continued)*
14. Common conductors are copper, aluminum, silver, gold, and brass. Common insulators are glass, rubber, plastic, ceramic, and porcelain.
15. D. Volts.
16. A. Ohms.
17. C. Amperes.
18. 8 amperes.
19. A single component usually built of silicon. It contains miniaturized resistors, capacitors, diodes, and transistors.
20. A. 250,000 ± 10%
B. 360 ± 5%
C. 9100 ± 20%
D. 75,000 ± 10%

Apply Your Knowledge

1. State the advantages of a parallel circuit over a series circuit.

2. Collect samples of different materials. Design and build a method to test each sample to determine if it is an insulator or a conductor.

3. Sketch three different electronic components. Use colors where appropriate. State the function of each component.

4. Design and build a model of an electromagnetic crane to pick up scrap iron.

5. Design and build a land vehicle using a 6–9 V DC motor. The objective is to minimize the time taken to travel a distance of 30′ (9 m).

6. Research one career related to the information you have studied in this chapter and state the following:
   A. The occupation you selected.
   B. The education requirements to enter this occupation.
   C. The possibilities of promotion to a higher level at a later date.
   D. What someone with this career does on a day-to-day basis. You might find this information on the Internet or in your library. If possible, interview a person who already works in this field to answer the four points. Finally, state why you might or might not be interested in pursuing this occupation when you finish school.

The future of technology holds unlimited possibilities.

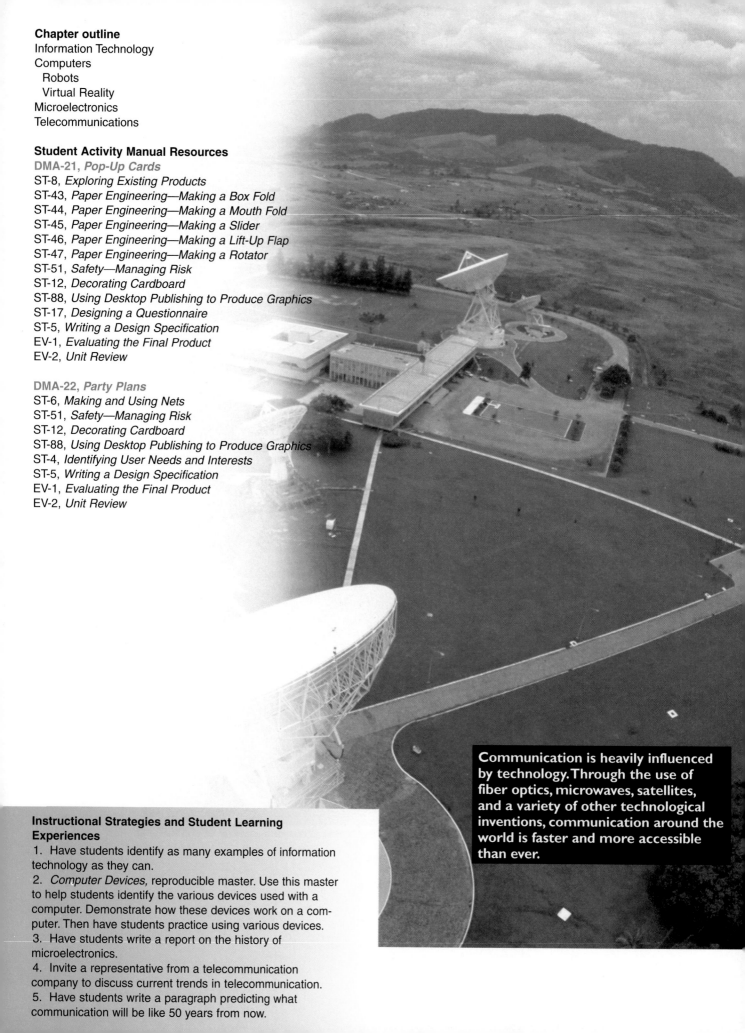

Chapter outline
Information Technology
Computers
  Robots
  Virtual Reality
Microelectronics
Telecommunications

Communication is heavily influenced by technology. Through the use of fiber optics, microwaves, satellites, and a variety of other technological inventions, communication around the world is faster and more accessible than ever.

Instructional Strategies and Student Learning Experiences
1. Have students identify as many examples of information technology as they can.
2. *Computer Devices,* reproducible master. Use this master to help students identify the various devices used with a computer. Demonstrate how these devices work on a computer. Then have students practice using various devices.
3. Have students write a report on the history of microelectronics.
4. Invite a representative from a telecommunication company to discuss current trends in telecommunication.
5. Have students write a paragraph predicting what communication will be like 50 years from now.

Information and Communication Technology

Key Terms

binary
bit
byte
chip
computer
data
fiber optics
information

information technology
laser light
microelectronics
microprocessor
robot
telecommunications
virtual reality

Objectives

After reading this chapter you will be able to:

- ◆ Understand the benefits of the information society.

- ◆ Compare old and new information and communication technologies.

- ◆ Describe the role and importance of information and communications technology in society.

- ◆ Describe the increasing role of robots in our daily lives and in manufacturing.

- ◆ Recognize how microelectronics is being used to progress information processing.

- ◆ Identify the major components of a telecommunications system.

Discussion
Discuss the importance of communication in the daily life of each student and the various means used to communicate.

Career connection
Have students identify careers that focus on information and communication technology. Have students browse the Web to find examples of work completed by people in these careers.

Electronic appliances have become smaller, thinner, and better since they were introduced decades ago. Early radios and televisions were huge, due to the use of vacuum tubes. In 1938, a typical TV screen was only 7″ (18 cm), but the cabinet was a large, imposing piece of furniture. Only when the transistor was invented in 1947 did radios and TVs become truly portable. The transistor also led to the home computer and subsequently to major changes in the activities of the tertiary (service) sector. A larger percentage of service workers are now employed in information and communication technology. Using computer-based technologies like robotics and artificial intelligence to solve complex problems, they create, process, and distribute information. *Information* is a collection of words or figures that have meaning or that can be combined to have meaning.

Information Technology

In our lives, we deal with huge amounts of information. For instance, every day you use telephone numbers to call friends. You refer to a bus schedule and class schedules. You can remember some of this information, but much of it must be stored, processed, and communicated by machines. The technologies used in storing, processing, and communicating information are referred to as *information technology*.

The old information technology relied on telephones, the postal service, printed materials, and film. Much of the equipment used was mechanical. Today, we rely more and more on electronics to handle information, **Figure 13-1**. Machines with moving parts have almost entirely disappeared. Replacing them is an electrical current.

A good example of this change is how print materials, such as the book you are reading, are printed. China was the first country in which people printed by using carved wooden blocks and ink to transfer words to paper. The oldest surviving book printed with block printing dates from 868 ad. It is a Buddhist scripture called the Diamond Sutra. It is probable that Chinese printing technology was spread to Europe through trade links to Johannes Gutenberg, who lived in Germany. In 1440, he improved on this earlier method by using a system of raised and movable type and a machine similar to a wine-press to press the inked image onto paper.

Figure 13-1 Telephones are now built using solid-state devices and IC chips. Many also come with special features, such as redial, speed dial, hands-free speakerphone, and video screen.

Useful Web site: Students can use the dictionaries at www.techtionary.com and www.hyperdictionary.com to find the definition of any new technical and technological terms used in this chapter.

Technology and society: Ask students to imagine they are office managers or representatives of an important company. What kinds of inquiries would they expect to receive and how would they respond?

Movable type and hot type (molten lead that created words) were commonly used until the advent of the computer in the 1960s. The newer printing methods use computerized, or electronic, printing. This process uses computers, typesetting and page-design software, optical scanners, and computer printers. A variety of materials, such as newsletters, reports, and books, can be produced. An operator can input text through a computer keyboard. Typesetting software is used to select type style and create page designs. Optical scanners, which feed information to the computer in the form of electronic signals, can read illustrations.

In addition to the computer, the new information technology uses a variety of equipment, including robots, satellites, microprocessors, and television, which are frequently linked. For instance, the combined technologies of the computer, telephone, and television may be merged into a single communications system, **Figure 13-2**. This system can transmit data as well as provide instantaneous interaction between people and computers.

Such a communication system is used in modern offices. Secretaries type and correct documents on a computer using a word processing program. A disk or CD stores input. A simple computer command causes a number of copies to be automatically printed. Meanwhile, the secretary may be speaking directly to someone on the other side of the world on a telephone. If the number is busy, the phone will automatically redial until the line is free.

Figure 13-2 A—Using cameras, monitors, microphones, speakers, high-speed communication lines, and other specialized equipment, instructors in one city can teach students located in another city. Businesses can also conduct these "teleconferences" with two or more parties in separate locations. B—Telephones are still widely used in the business world as a means of communication. (Jack Klasey)

Standards for Technological Literacy

3 17

Designing and making
What must designers consider to ensure that automatic tellers are well designed?

Reflection
Who do you know that uses a computer in their daily work? What functions are performed on these computers?

Enrichment
Electronic game playing has changed from being a solitary pursuit to one where human players go head-to-head with one another over the Internet.

Vocabulary
Discuss how mailing, online banking, and automatic teller machines (ATMs) relate to paying bills.

Activity
List new online games available from companies such as Epic Games, Looking Glass, Activision, and Sega. Choose one that you find exciting and share it with the class.

Useful Web site: Have students look at electronic versions of more traditional games, including backgammon and checkers, at www.zone.com.

Technology and society: If you were able to communicate with a star in the entertainment business or an important person who could change the world, who would you telephone and what would you discuss?

Phototelegraphy permits the sending of telex and facsimile pictures directly to clients. While the secretary is out of the office, an answering machine can receive telephone calls. Messages may be printed on a telex. Is overnight delivery by regular mail too slow? A letter can be fed into a document scanner for transmission over telephone lines. At the receiving end, a printer will print copies for immediate delivery.

Today's department store is another example of how machines are linked in a system. A computer terminal is combined with a cash register. The system adds the consumer's bill. At the same time, it sends information on the sale to the store's headquarters where another computer uses the information to monitor inventory. Payment for these purchases may be electronically processed. Another system that is known as an automatic teller offers round-the-clock banking services, **Figure 13-3**.

Figure 13-3 What impact does round-the-clock banking have on shopping habits?

This rapid exchange of information between people and machines and from machine to machine also applies to robots. Robots can be programmed with information that enables them to perform tasks in industry, in entertainment, and in a variety of dangerous situations.

Information technology relies on three complex technologies that have recently converged. They are the following: computers, microelectronics, and telecommunications. These three are combined into systems that:

- Create, collect, select, and transform information.
- Send, receive, and store information.
- Retrieve and display information.
- Perform routine tasks.

The following are some examples of how this takes place:

- The head office of a chain of stores can receive, collect, and store sales information transmitted from the electronic cash registers in each store, **Figure 13-4**.

- A night watchman responsible for the security of a large building can use a closed-circuit television system. It receives, displays, and stores information about the condition of different parts of the building.

- Hospitals use electronic machines for various purposes. Machines can measure the pulse, heartbeat, and other vital signs of a patient. This

Figure 13-4 How does the computer "read" the cost of each item being purchased?

information can be displayed on a screen. A nurse can read the screen of an EKG or other machine to keep track of a patient's condition, **Figure 13-5**.

◆ Suppose a rock group makes a recording of its music. The recording equipment collects information. Sounds from individual instruments or vocalists are placed on separate recording tracks. Afterwards, individual sounds are selected and mixed. This produces the final recording, **Figure 13-6**.

◆ An engineer or drafter feeds data into a CAD system. The machine turns the information into three-dimensional drawings, **Figure 13-7**. The engineer can then check the design of machine parts.

◆ An automower uses sensing technology to stay on a lawn and avoid obstacles like trees, plants, and even your dog. When its battery runs down, it finds its docking station and recharges itself.

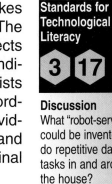

Standards for Technological Literacy

3 **17**

Discussion
What "robot-servants" could be invented to do repetitive daily tasks in and around the house?

Reflection
Stonehenge (UK) may have been a kind of prehistoric computer, allowing people to create an early version of a calendar based on the position of shadows made by the sun (stones = computer, sunlight = input, and calendar = output).

Vocabulary
Have students define the terms *computers*, *microelectronics*, and *telecommunications*. Discuss the difference between a thinking machine and an information processor.

Resource
Activity 13-1, *The Information Society*

Vocabulary
Define the term *ubiquitous*.

Activity
Investigate and describe the characteristics and advantages of a "smart" classroom.

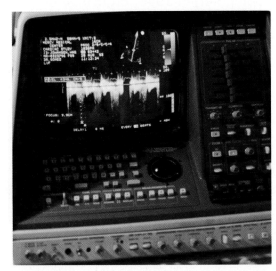

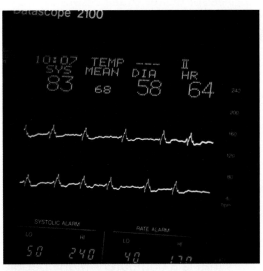

Figure 13-5 An EKG (electrocardiograph) measures and produces a record of the electrical signals on a person's body. They are used to monitor heart activity and detect heartbeat abnormalities.

Links to other subjects: How is transmitting large amounts of data used in subjects other than technology education?

Useful Web site: Have students complete one of the tutorials on telecommunications and Internet technologies at www.techtionary.com.

Links to other subjects: History—Investigate the history of papermaking or printing.

Figure 13-6 The equipment in a recording studio collects, stores, and retrieves the music for the final recording.

Figure 13-7 A CAD system allows an engineer or drafter to see designs in two and three dimensions.

Let us now look at the three technologies that make all of these systems possible:

◆ Computers.

◆ Microelectronics.

◆ Telecommunications.

Computers

What is a computer? Many people think of a computer as almost human. It seems to have a "brain" that allows it to think. Computers do not have brains, and they cannot think for themselves. *Computers* are primarily machines that can store, retrieve, and process data at high speeds using arithmetic. However, the humans who feed computers data do the really important thinking.

Until 1940s, computers were largely mechanical devices. World War II presented a new challenge: how to crack secret enemy codes that changed three times a day. Using the ideas of mathematician Alan Turing, a computer called Colossus was made in 1943. It used 2000 vacuum tubes. Intercepted messages were fed into Colossus as symbols on paper tape. The computer processed them at the rate of 25,000 characters per second. Colossus became the world's first programmable computer, but it was limited to breaking codes. The first electronic computer capable of tackling many different jobs was ENIAC (Electronic Numerical Integrator and Computer) in 1945, **Figure 13-8**.

Figure 13-8 ENIAC was the world's first electronic digital computer. It was designed to calculate ballistic firing tables. (John W. Rauchly Papers, Rare Book & Manuscript Library, University of Pennsylvania)

The modern computer is primarily a calculating machine. However, it can also store a vast amount of data. It can be programmed to carry out logical operations. For instance, it can transfer data from one part of the machine to another. On command, it will sort this data and compare it with other data. From the comparison, it is able to provide new information.

The computing process involves three main stages: the input stage, the central processing stage, and the output stage. **Figure 13-9** shows these three stages and a variety of devices used for each.

Discussion
Why would some people consider the Internet to be the greatest invention since Gutenberg invented the printing press in 1454? In what ways are the two inventions similar?

Vocabulary
What is the meaning of the phrase *knowledge economy*?

Activity
Divide the class into two groups. Have each group prepare to debate opposite opinions about the statement: "Knowledge equals power. You are valued by an employer for what you know".

Input	Central processing unit	Output

Digitizer
Disk
Graphics tablet
Joystick
Keyboard
Speech
Touch sensitive
  screen
Modem
	Uses instructions
	  (programs), and data
	  stored in memory,
	  to carry out
	  calculations
		Disk
		Machines
		Monitor
		Other computers
		Plotter
		Printer
		Robot
		Voice
		  synthesizer
		Modem

Monitor

Central processing unit (including modem)

Disk drive

Joystick

Printer

Mouse

Mobile robot

Graphics tablet

Keyboard

Robot arm

Figure 13-9 A computer with its three stages and various devices.

Community resources and services: Which of the following resources are available in your community: training courses, financial transactions, different kinds of products to buy, access to museums or art galleries, and discussions with people who have the same interests as you? What additional resources are available through the Internet?

A computer may receive and process many types of information:

◆ Sales information from electronic cash registers.

◆ Pulse rate, heartbeat, and vital signs for a patient in a hospital.

◆ Sounds from musical instruments and vocalists.

◆ Three-dimensional engineering drawings.

Information processed by a computer must be input in a form that the computer can handle. Any set of symbols that represents information is correctly called *data*. For example, the letters A, B, and C are data.

Computers use a simple code made from electrical signals. There are only two signals in this code, ON and OFF. These are written as 1s and 0s. This is called a *binary* digital code. Binary means two, and digital means number. Inside the computer, an ON condition is used to represent a 1. An OFF condition represents a 0.

To provide for a code with more than two elements, signals are combined to produce patterns. Look at **Figure 13-10**.

Imagine a set of four lightbulbs. Each one can be turned on or off. Each of these bulbs can be assigned a value. Since, in the binary system, each digit has a value twice as large as the one on its right. The bulbs must be assigned values of 8, 4, 2, and 1. When a bulb is off, it represents 0. When it is switched on, a bulb represents the value assigned to that position. These values can be added together. In **Figure 13-10**, the first bulb and the third bulb are on. This represents the numbers 5 (1+4). If the first bulb is off and the second, third, and fourth are on, the number represented is 14 (2+4+8). This group of four bulbs can represent any number from 0 (all off) to 15 (all on). Each on or off signal is known as a *bit*, which is short for *bi*nary dig*it*. The example above uses a four-bit word. Since computers need to work with numbers larger than 15, a system of eight-bit words is used, **Figure 13-11**. Each eight-bit word is called a *byte*.

Robots

A *robot* is a computer that can interact with its environment. Without computer technology, we could not have robots. The future of both is interrelated: as one improves, so does the other. A true robot is one that can perceive its environment and respond automatically without any intervention from humans.

There are two main categories of robots:

◆ Mobile ones that can move, **Figure 13-12**.

◆ Stationary ones that cannot move from where they are set, **Figure 13-13**.

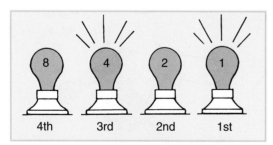

Figure 13-10 The first and third bulbs switched on would represent the number five.

Technology and society: How has the increased use of computers and robots impacted the nature of work and the quality of the workplace?

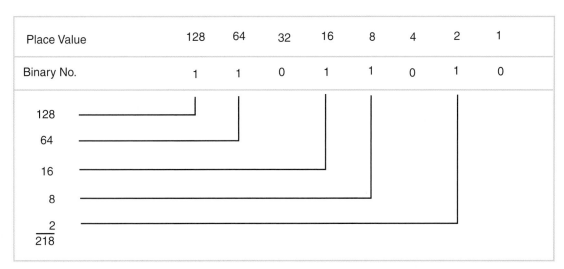

Place Value		128	64	32	16	8	4	2	1
Binary No.		1	1	0	1	1	0	1	0

```
128  ────────────────┐
 64  ──────────────────┐
 16  ──────────────────────┐
  8  ──────────────────────────┐
  2  ──────────────────────────────────┐
─────
218
```

Figure 13-11 The eight-bit word, 11011010, represents the number 218.

Designing and making
How can robots make the lives of handicapped people easier, safer, and more enjoyable?

Discussion
Discuss what robots can do well, such as repetitive tasks, and what they are incapable of doing, such as feeling emotions.

Reflection
Would you like to have a robot that can express when it is anxious or experiencing pleasure?

Enrichment
Have students investigate the life and work of Carl Capek.

Activity
What practical uses can you think of for the robots in Figure 13-12?

Resource
Reproducible Master, *Computer Devices*

Figure 13-12 A—The AQUA robot cruises effortlessly over coral reefs during successful sea trials. The operator sees the robot on a computer screen and directs it using a joystick, much like a video game. (CIM McGill University) B—Some mobile robots move around on legs. (Centre for Intelligent Machines, McGill University)

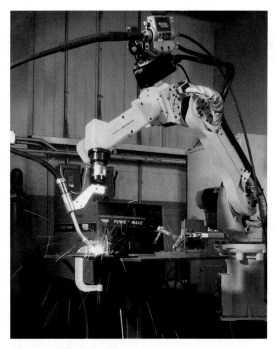

Figure 13-13 Pivoting on an axis, a stationary robotic welder performs repetitive welding procedures quickly and efficiently. (The Lincoln Electric Company)

Useful Web site: Have students browse www.androidworld.com to review descriptions of robots that look like humans. Students should print information they find interesting and share it with the class.

Both types have three main parts:

◆ Mechanical parts that move.

◆ Computer electronics hardware that controls movement.

◆ Computer software that provides instruction.

As you read this chapter, think of the thousands of robots that are right now doing the type of work humans find boring or dangerous. They work in:

◆ Material handling, transferring parts between areas of a factory, and stacking and storing.

◆ "Pick and place" situations, picking up parts and putting them in place ready for machining.

◆ Assembling parts together to make a product.

◆ Welding car bodies using arc and spot welding techniques.

◆ Spray painting finishes onto all kinds of manufactured goods.

◆ Inspecting finished products in quality control departments.

Robots used in nuclear power plants must be rugged, capable of doing inspection, detection, and decontamination. They must not break down, because on-the-job humans might be at great risk if they were to try to rescue them. Robots used for bomb disposal must be very strong, capable of climbing stairs, and maneuvering around objects.

Robots can vacuum your house or search a collapsed building for possible life. They have gone on missions to Mars and explored the insides of active volcanoes. Robots have searched Antarctica for meteorites,

and they have searched war zones for land mines. Their use is increasing in surgical operations. Elderly and sick people, as well as people with disabilities, are increasingly using robots in situations in which they need frequent help. Robots can sing and dance, run, play instruments, search for people, tell jokes, ride a bicycle, and wrestle.

If robots are ever to become equal to humans, they must be able to feel and to experience physical sensations like joy and pain. Just as our emotions play a large part in the decisions we make, robots will be built with emotions. Of course, machines are no closer to being able to feel than they are to thinking. If they ever reach this stage, will they have all the rights of humans and would it be wrong to shut them off?

Virtual Reality

Virtual reality blurs the line between reality and fantasy. *Virtual reality* is an artificial environment provided by a computer that creates sights, sounds, and, occasionally, feelings through the use of a mask, speakers, and gloves. It is a relatively new technology with great potential. Some simple systems already exist. For example, a user can enter a virtual world by putting on a headset equipped with a small image screen. He or she is immersed in the sights and sounds of a computer-generated world. With a special glove, there is the added experience of touching and moving objects in virtual space. Virtual reality systems exist for architects, pilots, surgeons, and for special types

Technology and society: Would it be wrong to permanently switch-off a "thinking and feeling robot"?

of video games. Pilots take training in flight simulators while wearing helmet systems that immerse them in computer-generated airspace. Workers in some manufacturing plants are trained in the operation and maintenance of robotic systems using virtual reality systems. Athletes can be immersed in an environment that simulates the course they will be steering or jumping. There is considerable development ahead, but developers of this technology have a dream: to make it possible for anyone to recreate a day in history, experience outer space, or walk through a museum simply by switching on a computer and using a special headset.

Microelectronics

The second of the technologies that led to the growth of information technology is microelectronics. *Microelectronics* is the result of miniaturization, in which the switches and circuits of processors and their accessories are made incredibly small. This miniaturization has resulted from the invention of new manufacturing processes and the use of new materials.

Of all the electronic elements resulting from these new processes and materials, the most important is the *microprocessor* or "computer chip." This is commonly referred to simply as a chip, **Figure 13-14**. A *chip* is a tiny flake of a substance called silicon. It is covered with microscopic electronic circuits. These chips may be mass-produced in the tens of thousands. Thus, a single chip costs

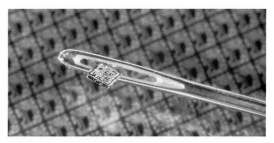

Figure 13-14 A microprocessor, also known as a "chip." Notice how it is "dwarfed" by the eye of a needle. (Bell)

less than a textbook. However, it still contains most of the switches and circuits needed by a computer. Chips are becoming cheaper and more powerful each year and are reaching into every area of modern life.

Over 90% of all microchips are not in desktop computers. They are in cars, homes, and industrial machines. They are around us to such an extent that we don't think about devices like the microchip that controls antilock brakes on cars. Microprocessors provide a machine with decision-making ability, memory for instructions, and self-adjusting controls. In pacemakers, the miniature components on a chip time heartbeats. Microprocessors also set thermostats, switch VCRs on and off, pump gas, and control car engines. Robots and on-board computers in satellites rely on them.

In the future, microprocessors that are embedded in machines will become even smaller and communicate wirelessly. To do this, future chips will likely be made of materials other than silicon. Experiments are being conducted to make optical microchips, instead of electronic ones. Particles of light, rather than electrons, could be used to control

Technology and society: Should we find a way to ensure that everyone has access to the Internet? If this universal access were achieved, what would it mean for a global society?

and power circuits. Switches would conduct light rather than electricity.

Telecommunications

Personal computers started out as machines for word processing. The fact that they double in speed approximately every 18 months has made them much more powerful and increased the number of tasks they can do. Whether they are in homes, schools, or on-the-job, computers are now very versatile because they can be linked to virtually any source of information via the Internet.

Telecommunications can be described as the sending and receiving of information over a distance, making use of a variety of technological devices and systems. The Internet is the way of connecting to all kinds of data sources on the World Wide Web, **Figure 13-15**. Think of the Internet as being like a telephone system that connects callers, and the information they know, around the world. Giant computers, called servers, are linked together

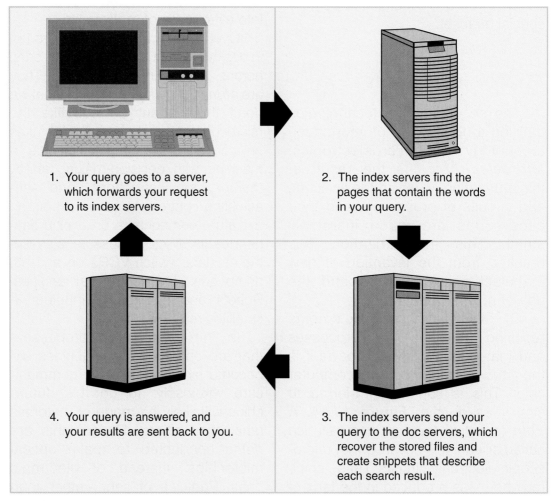

1. Your query goes to a server, which forwards your request to its index servers.

2. The index servers find the pages that contain the words in your query.

4. Your query is answered, and your results are sent back to you.

3. The index servers send your query to the doc servers, which recover the stored files and create snippets that describe each search result.

Figure 13-15 This diagram explains the process of what happens when you use a search engine on the Internet.

Technology and society: How does access to the Internet change personal relationships?

A   **B**

Figure 13-16   A—Fiber optic cables are replacing copper cables in communications. A thin glass fibers cable can handle the same number of telephone messages as the thicker copper cable. (Bell) B—In telecommunications, fiber optic cables and their protective conduit are usually identified by the color orange. (Jack Klasey)

via copper telephone lines and fiber optic cables. *Fiber optic* cables can carry light-coded messages over long distances. A fiber optic strand four-thousandths of an inch (0.1 mm) in diameter is capable of carrying 2000 two-way telephone conversations at once, see **Figure 13-16**.

At the transmitter end of a fiber optic-cable network, telephone signals are converted into pulses of light usually using an infrared LED. These pulses travel through the glass fibers to the receiver end. There the pulses are converted back into electrical signals that are carried to a telephone or computer.

As the Internet gives us access to such a great amount of information, some people believe it is the greatest invention since Gutenberg's printing press made books more available in the Middle Ages. This is because they know that information equals power. In this information age, your value as an employee increases according to what you know.

The greatest use of the Internet is to send messages via e-mail. It is also used to research companies and

their products, download software, visit cultural sites, and access chat groups. When searching the World Wide Web (often called the Web), most people use a search engine, such as Alta Vista™, Google™, Yahoo!™, or Mamma™, **Figure 13-17**. A typical search engine will contain over a billion Web pages being used by 10 million different people a month and process 100 million searches a day. There are also web sites to help you get exactly the information you need. One site is www.searchenginewatch.com that

Figure 13-17   How are search engines changing the way in which people work, study, and communicate with others? (TEC)

Useful Web site: To learn more about using a search engine efficiently, visit www.searchenginewatch.com

Community resources and services: Investigate the special resources available on the Internet for the hearing impaired.

gives you tips on how to search better, how major search engines work, and lists all the major search engines.

The Internet is also helping persons who are hearing impaired. Instant messaging (IM) is a convenient way to carry out online conversations. Chat rooms are becoming one of the main ways that members of the deaf community communicate among themselves or with the hearing population. IM is available for the laptop, pager, or cell phone.

Bar codes are also being used to access information. The bar code owes its beginning to the Morse code invented by Samuel Morse in 1838. Messages could be sent and received as dots and dashes on a paper strip. Today's bar codes are dots and dashes extended downward to make thin and thick lines. Widespread use of bar codes was made possible by scanners that use lasers to sweep across the bar code, **Figure 13-18**. *Laser light* is a form of radiation that has been boosted to a high level of energy. It produces a strong, narrow beam of light. Lasers have made possible such popular conveniences as automatic supermarket checkouts, fiber optic communication, and a new generation of printing devices. Lasers are even behind the compact disc used in entertainment. Lasers have changed the aspects of medicine. Laser surgery is now commonplace.

A more recent method of identification than bar codes is called radio frequency identification (RFID). RFID tags are essentially tiny microchips equipped with radio transmitters. They have been implanted in dogs, cats, and even humans. The chips

A

B

C

Figure 13-18 A—This bar code contains encoded information about a specific product by the varying widths of the lines.
B—Warehouse workers often use these types of wearable scanning systems to help track orders and inventory. (Symbol Technologies)
C—This model has a keypad, viewing screen, and a finger switch for activating the bar code reader. (Symbol Technologies)

can store information ranging from product color and expiration date to medical records and ownership details. RFID tags might someday be used for protection in firearms. A smart gun stolen from a police officer

Technology and society: What are the advantages and disadvantages of allowing customers to use bar code readers to check out their own products instead of having a clerk do so?

would not be able to be fired, as it would not sense the corresponding chip in the owner's hand.

Some computers are equipped with bar code readers that give access to Web sites. The next generation of bar codes will use wireless microchip technology. These chips will communicate their information without being scanned. Information could be sent directly to a checkout register at a supermarket or to a machine in your home, such as a refrigerator. It will be like one machine talking to another. For example, if your supply of milk, ice cream, or any product gets low, you could be notified by a display on the front of your refrigerator.

Wireless technology will be the key to the future when most devices will contact one another at a distance. This will give us access to information wherever we are. It is the ultimate in remote control technology.

Satellites in space can relay information. To do this, messages must be converted to electromagnetic waves or pulses. After they have been sent through the atmosphere, they are beamed into space and bounced off orbiting satellites. Communication satellites can handle more information than cables. They can also transmit it faster.

Information is beamed to and from satellites as microwave signals. These signals must be picked up by special dish-shaped antenna, **Figure 13-19**. The antenna concentrates the weak signals. After the microwaves bounce off the inside of the dish to a focal point, the signals are then converted to electricity. The current travels to a receiver such as a TV set. Satellites are powered with electricity generated by sunlight received by the solar panels. As satellites spend some time in the earth's shadow, they also rely on electricity stored in batteries.

Standards for Technological Literacy

Discussion
Discuss how modern information and communication technologies increase the speed of communication and travel. When is speed a good thing and when is it best to do things "nice and easy"? What is the effect of an ever-increasing pace on peoples' lives?

Enrichment
The first satellite launched was Sputnik 1 on October 4, 1957.

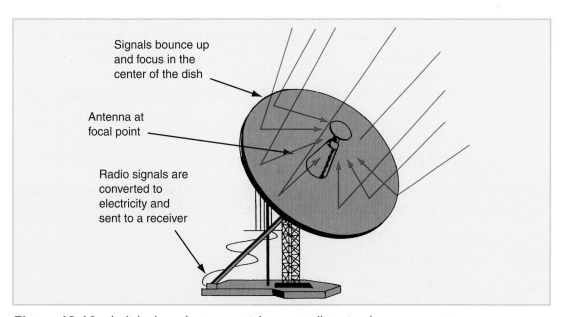

Signals bounce up and focus in the center of the dish

Antenna at focal point

Radio signals are converted to electricity and sent to a receiver

Figure 13-19 A dish-shaped antenna picks up satellite signals.

Technology and society: Is constant change a good thing? When is change not beneficial? Should consumer products be built to last a lifetime?

Chapter 13 Review
Information and Communication Technology

Summary

Most workers in the tertiary sector are now employed in the creation, processing, and distribution of information. Most of this information must be stored, processed, and communicated by machines. The old information technology relied on telephones, the postal service, telegraph, printed materials, and film. The new information technology uses computers, microprocessors, and telecommunications.

Computers are primarily machines for doing arithmetic at high speeds. They can also store a vast amount of data and carry out logical operations. Information to be handled by a computer must be changed into the binary digital code.

A robot is a computer that can interact with its environment. There are mobile ones on wheels or tracks and stationary ones that can grab and hold items.

Microelectronics makes use of tiny switches and circuits. The most important component is the microprocessor or the microchip. Microprocessors can be used to control almost any device.

Telecommunications involves the moving of data between points. The three systems that are used depend on electrical waves or pulses transmitted through wires, the transmission of light pulses along fiber optic cables, or electromagnetic waves broadcast through the atmosphere. Wireless technology holds the key to significantly extending our capability in remote control.

Modular Connections

The information in this chapter provides the required foundation for the following types of modular activities:

◆ Computer Technology

◆ Computer Hardware

◆ Computer Construction

◆ Computer Repair

- Software Operations
- Purchasing a Personal Computer
- Home Computer Systems
- Computer Control
- Information Technology
- Networking
- Introduction to the Internet
- E-mail Technology
- Web Site Design
- Light Technology
- Laser Applications
- Fiber Optics Technology
- Computer Applications
- Control Logic
- Robots
- Robotics
- Radio-Controlled Transportation
- BioRobotics
- System Control Technology
- Communication Technology
- Audio/Video Technology
- Audio Broadcasting
- Digital Video
- Non-Linear Video Production
- Video Production
- Digital Photography
- Computer Special Effects
- Film Technology
- Digital Imaging
- Multimedia Design
- Multimedia Production
- Virtual Reality
- Portfolio Development
- Desktop Publishing
- Digital Transportation

Test Your Knowledge

Write your answers to these review questions on a separate sheet of paper.

1. A collection of words or figures that have meaning is defined as _____.
2. The technologies used in storing, processing, and communicating information are, together, referred to as _____.
3. What are the differences between the old and the new information technologies?
4. Copy and complete the chart below describing the equipment a secretary in a modern office would use to complete the tasks listed.

TASK	EQUIPMENT
Prepare a letter	
Store information	
Make a telephone connection to an engaged number	
Send a copy of a picture to Australia as rapidly as possible	

5. Give three examples of computerized systems that affect our daily lives.
6. What are the three major technologies that make up information technology?
7. The common term for a microprocessor is _____.
8. What is the major task of a telecommunications system?
9. Telecommunications technology uses three different systems to transmit data. Describe the devices used by each system.

Apply Your Knowledge

1. From your own community, give three examples of old information technology and three examples of new information technology.

2. Which of the communication systems described in this chapter do you or your parents use each week?

3. List the input and output devices used with computers. Give an example of the use of each.

4. Look through one issue of a newspaper. Cut out all the references to microelectronics. Try to find references from each section of the newspaper, including the advertisements.

5. Draw a block diagram to show the major components of a telecommunications system.

6. Selecting the right television for your home is not an easy task, as there are many different types. Assume that a friend or relative has asked you to give her advice on the purchase of a new TV set. Make a summary of the advantages and disadvantages of as many different types of TVs as are available in your local stores, including conventional television sets, rear-projection televisions, high-definition televisions (HDTVs), and plasma flat-panel displays.

7. Home theaters achieve great sound and visual effects, when they are well designed. Describe the conditions that make an ideal room for a home theater, including the floor plan, furnishings, natural and artificial light, methods of installing electrical and coax wiring, and ventilation requirements.

8. Information relating to almost any technological topic is available through an Internet search. Decide on a topic, obtain your teacher's approval of the topic, and then use a search engine to access the world's digitized information. Remember to look carefully at the source of the information to decide whether or not you should rely fully on its truthfulness.

9. Bloggers are cyberchroniclers who self-publish all sorts of information on the Internet using a Web log (blog). Blogs are updated frequently. The topics can vary from news sources to food and from what is happening in local entertainment to poetry. Check out blogs created by people who live in your

area. If you were to start a blog, what would your topic be? What witty name would you give it? Which pictures might you publish? Remember that you would not say or show anything that you did not want the world to see, as anyone could read your blog. In fact, some bloggers have been fired from their jobs for things they said about their employers!

10. Humanlike robots that can communicate with their owners or guard their owners' homes are often shown in newspapers and magazines. For example, Wakamaru has a vocabulary for 10,000 words. Identify other robots designed for use in homes and describe their capabilities.

11. Research one career related to the information you have studied in this chapter and state the following:
   A. The occupation you selected.
   B. The education requirements to enter this occupation.
   C. The possibilities of promotion to a higher level at a later date.
   D. What someone with this career does on a day-to-day basis.
   You might find this information on the Internet or in your library. If possible, interview a person who already works in this field to answer the four points. Finally, state why you might or might not be interested in pursuing this occupation when you finish school.

Wherever technology is taking us, it will always be important to strike a sustainable balance between the made world and the natural world. (Vision Quest Windelectric Inc.)

Chapter outline

Agriculture
Biotechnology
Environmental Technology
  Keeping Our Air Clean
  Regenerating Soil
  Purifying Water
  Using Appropriate Types of Energy
  Taking Action
    Reduce
    Reuse
    Recycle
Medical Technology
  Eating Right
  Maintaining a Healthy Body
  Repairing the Damage
  Surgical Techniques
  Prosthetic Devices

Student Activity Manual Resources

DMA-23, *Healthy Eating*
ST-89, *Using MyPyramid*
ST-90, *Analyzing Food Labels*
ST-2, *Brainstorming*
ST-88, *Using Desktop Publishing to Produce Graphics*
ST-38, *Cutting and Shaping Acrylic*
ST-36, *Cutting and Shaping Thin Plywood*
ST-51, *Safety—Managing Risk*
ST-6, *Making and Using Nets*
ST-13, *Writing a Design Brief*
ST-5, *Writing a Design Specification*
EV-1, *Evaluating the Final Product*
EV-2, *Unit Review*

DMA-24, *Prepacked Lunches*
ST-4, *Identifying User Needs and Interests*
ST-89, *Using MyPyramid*
ST-90, *Analyzing Food Labels*
ST-51, *Safety—Managing Risk*
ST-6, *Making and Using Nets*
ST-88, *Using Desktop Publishing to Produce Graphics*
ST-13, *Writing a Design Brief*
ST-5, *Writing a Design Specification*
EV-1, *Evaluating the Final Product*
EV-2, *Unit Review*

Instructional Strategies and Student Learning Experiences

1. Help students understand the significance of biotechnology by asking them to describe a current news item related to biotechnology, such as gene therapy, DNA testing, genetically modified/hybrid plants and crops, or break-through medication.

2. *MyPyramid*, reproducible master. Use this master to help students summarize their daily consumption in each of the food groups.

3. Ask students to create and report on a unique and beneficial product of genetic engineering.

4. Have students investigate the efforts of a local (school or community) environmental group.

Biotechnology combines the knowledge and theories of science and the instrumentation and advancements of technology.

Agriculture, Biotechnology, Environmental Technology, and Medical Technology

Key Terms

acid rain
bioprocessing
biotechnology
carbohydrate
cell
dairy
DNA
fats, oils, and sweets

fruit
genetic engineering
human genome
methane
protein
recycling
vegetable

Objectives

After reading this chapter you will be able to:

◆ Describe how food is grown, harvested, and processed.

◆ Discuss the use of biotechnology in foods, medicines, and other products.

◆ Debate the implications of biotechnology.

◆ Assess the environmental impact of various technologies.

◆ Describe the nutritional requirements to maintain a healthy body.

◆ List medical technology devices that can improve quality of life.

◆ Design and make products to promote healthy eating habits.

Discussion
Discuss the following statement: "The nineteenth century was based on coal, the twentieth century on oil and natural gas, and the twenty-first century could be based on carbon dioxide and water captured by plants and turned into products by using biotechnology."

Career connection
Have students identify careers associated with agriculture, biotechnology, environmental technology, and medical technology. Students should browse the Web to find examples of work completed by people in these careers.

Standards for Technological Literacy

Designing and making
What types of foods are grown and made available to consumers in their natural form? What types are processed? What particular problems must designers resolve when packaging these products?

Discussion
Are foods grown on high-yield farms as tasty and nutritious as foods grown in small market gardens?

Reflection
Do you enjoy looking at plants and flowers? How can you increase your contact with the "natural" parts of your environment?

Enrichment
Should farmers be encouraged to grow more organic fruit and vegetables? If they were, what would be the advantages and disadvantages to consumers?

Vocabulary
What do the terms *agriculture* and *subsistence farming* mean?

Activity
Have students grow flowers or vegetables from seeds in simple clay pots. Students can choose whether to grow a flower, green plant, vegetable, or herb.

Agriculture

Farmers own and operate farms that grow grain, cotton and other fibers, fruit, and vegetables. They are responsible for tilling the land, planting, fertilizing, cultivating, spraying, and harvesting. Some also package, store, and market the crops. Livestock, dairy, and poultry farmers feed and care for their animals in barns and other farm buildings, which they must keep clean and in good repair. Some farmers raise cattle on the plains and sheep on the hillsides. These animals are usually rounded up and taken to feedlots when they mature, so they can be fed to increase the meat on their bodies. At certain times of the year, farmers plan and oversee animal breeding.

Horticultural specialists are farmers who grow flowers, bulbs, shrubs, fruits, and vegetables in greenhouses. See **Figure 14-1**. A greenhouse is a building made mostly of glass or transparent plastic, where conditions, such as humidity and temperature, can be controlled. Aquaculture farmers raise fish and shellfish in seawater or freshwater. They usually raise the animals in ponds, floating net pens, or other systems in which they can feed and protect the animals.

Many farmers carefully plan to plant a combination of crops so, if the price of one crop drops, they will have income from another to make up for the loss. Farming is hard work, however, and crops are unpredictable, due to weather conditions. Weeds sometimes grow faster than the

Figure 14-1 Some people cultivate plants year round in greenhouses.

crops. The farmer also has to fight insects and plant diseases. An early frost in autumn or a late frost in spring can destroy an entire crop. Having too much rain can cause as much damage as having too little.

Farmers frequently work from sunrise to sunset, especially during planting and harvesting seasons. Specialized machinery helps to reduce some of the work, as it is used to till the soil, spray fertilizers and insecticides, irrigate fields, and harvest crops. See **Figure 14-2**. Farmwork, however, can be dangerous. Tractors and other machinery can cause serious injury, if proper operational procedures are not followed. Also, machinery cannot harvest all crops. Migrant workers often pick fruits by hand, as mechanical harvesters cannot pick grapes and other fruits from trees and bushes.

Links to other subjects: Geography—Have students investigate companies and regions that engage in fair trading practices.

Community resources and services: Visit your local supermarket and determine if it sells organically grown vegetables and fruit. Compare the prices between these products and similar products that are not organically grown.

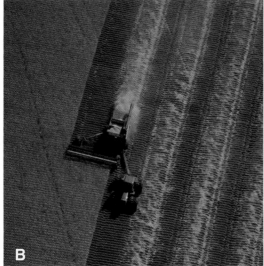

Figure 14-2 A—Computer-controlled irrigation systems water plants, lawns, and golf courses and keep them at their desired moisture. (Adcon Telemetry AG) B—Combines are used during harvest to cut, clean, and thresh various crops.

The foods we eat can be obtained and processed by various means. Natural methods include planting crops, such as the corn, tomatoes, and peppers you see in the supermarket. See **Figure 14-3**. Other foods are produced synthetically. Margarine, candies, pies, and many drinks are processed using both chemicals and natural ingredients.

When food arrives at the market, it is processed. Wastes, such as skins, bones, cobs, and shells, are removed first. The food might be cut, crushed, or ground. It is then sorted, and any substandard products are removed. Next, it is mixed and blended with additives, such as colorings, spices, and preservatives. For example, sausage can be made from a combination of pork, fat, spices, applesauce, corn syrup, and water. Further preservation is sometimes done by heating, pickling, dehydrating, refrigerating, or freezing. These processes, along with irradiation, help provide

Figure 14-3 Technology aids in the harvesting, processing, and preservation of our favorite foods. (Ecritek)

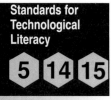

Standards for Technological Literacy

5 14 15

Vocabulary
Have students investigate the meaning of the term *fair trade*.

Safety
What types of safety precautions must be taken by people working in food production industries?

Designing and making
What recipes have students used to make their own meals? Select one to share with the class. Can it be made in your classroom? Discuss the design decisions involved in preparing and presenting a meal to guests. Include decisions about furniture, table settings, menus, and food preparation in the discussion.

Discussion
Discuss the pros and cons of increasing the number of genetically modified foods available to the consumer.

food with safe long-term storage capability and minimize the risks associated with eating spoiled food. Finally, foods might be formed into a certain shape, such as forming chocolate into bars and pastry dough into pie shells. See **Figure 14-4**. Foods are then packed to both protect them and keep them clean.

In the twenty-first century, farmers are using computers. For example, by using a database system, they can learn about weather forecasts or connect with the board of trade to see at what price corn or other commodities are trading. This information helps them decide the best time to sell their crops. Precision farming is another technological development in agriculture. It uses data from satellites and high-flying aircraft to pinpoint problems in drainage, insects, and weeds. The farmer can then spray only in the affected area. Satellite signals will soon control machinery. A Stanford research team has equipped a John Deere tractor with a satellite-based

Figure 14-4   Pies are a good example of a formed and processed food.

automatic control system that can guide a 20,000-pound farm vehicle more precisely than human drivers can. The ultimate application of this technology would be a tractor a farmer could control and monitor from his office.

Biotechnology

Improvements have been made in animals and plants over hundreds of years.

For example, a farmer wants to increase the yield of soybeans. The farmer would crossbreed plants with the highest yield with another variety of the same species that had a resistance to a particular problem, such as drought. This method, however, takes generations to produce results, and there is no guarantee of success.

Today, hardier plants and animals are possible in a single generation through genetic engineering. *Genetic engineering* is a technology used to alter the genetic material of living cells. *Cells* are the basic units of all living organisms. Inside a cell is a nucleus, which contains the hereditary information, or genetic material, of a cell. This genetic material may determine whether a person is tall or short, whether a cow can produce more or less milk, and whether a plant can withstand drought.

Steel and fossil fuels were the resources of the last century. Genes are the resource material of the twenty-first century. Every living thing, from the smallest beetle to the tallest oak tree, has a set of genes, which we call a genetic code. This code determines precisely what traits it will have.

Community resources and services: Local farmers sometimes use cooperatives to send their food directly to the consumers. Have students investigate how cooperatives work, the benefits to consumers, and the locations of cooperatives in their area.

Useful Web site: Have students visit the Biotechnology Institute Web site at www.biotechinstitute.org to search for articles on transgenic animals.

Biotechnology aims to find the most beneficial traits in terms of nutrition, flavor, and disease resistance, and move traits from one organism to another. In the past, it took centuries of evolution to change a life form. Now we can transfer genes from one species to a totally different species almost instantly. Human genes can be introduced into animals, and animal genes can be introduced into plants. "Antifreeze" protein genes from the flounder fish have been inserted into the genetic code of tomatoes to protect the fruit from frost damage.

Biotechnology is also changing the way animals are bred. We have been breeding animals and plants for thousands of years. However, in this long history, our practices have been restrained by the natural boundaries of nature. Two completely different animals cannot mate and have offspring. By recombining DNA, two unrelated organisms, which could not mate in nature, are combined by inserting a piece of DNA from each to form a new genetic material, **Figure 14-5**.

Simply stated, *biotechnology* is the use of living organisms to develop and improve foods, medicines, and other products. The modern age of biotechnology started in 1953 when James Watson and Francis Crick

Standards for Technological Literacy

5 **14** **15**

Enrichment
One hundred years ago, 80% of the population lived in rural areas; the figure is now about 20%. Consequently, most people are not in touch with how food is produced. The general population has even less information about genetically modified (GM) foods, especially because they have only been in use for 7 years.

Vocabulary
What do the terms *biotechnology, DNA, genetic modification, GM foods, genes,* and *transgenic* mean?

Enrichment
A large percentage of canola, soybeans, and corn are genetically modified.

Vocabulary
Critics of GM are concerned about damage to *biodiversity*. What does this term mean?

Activity
Have students investigate the work of James Watson and Francis Crick. In 1953, they proposed the double helix structure of DNA; the winding staircase of interwoven chemicals that contain the heredity of all living things. Make a model of the double helix structure representing DNA.

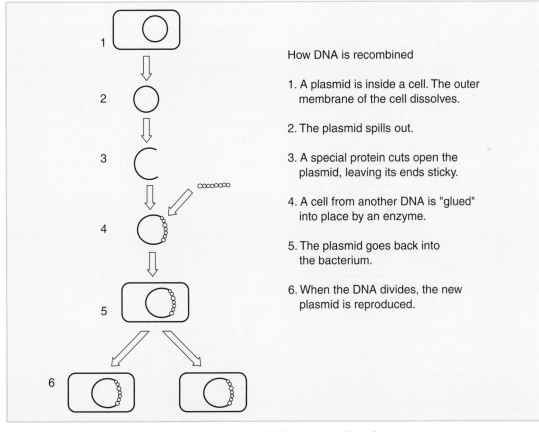

How DNA is recombined

1. A plasmid is inside a cell. The outer membrane of the cell dissolves.

2. The plasmid spills out.

3. A special protein cuts open the plasmid, leaving its ends sticky.

4. A cell from another DNA is "glued" into place by an enzyme.

5. The plasmid goes back into the bacterium.

6. When the DNA divides, the new plasmid is reproduced.

Figure 14-5 This diagram illustrates how DNA is recombined.

Useful Web site: Have students visit the following Web sites to gather information about the advantages and disadvantages of using GM foods:
Institute of Science in Society (a leading critic of GM) - www.i-sis.org
Site of global biotech giant Monsanto - www.monsanto.com
Council for Biotechnology Information - www.whybiotech.com
Economic, legal, and environmental impacts of GM crops - www.soilassociation.org/gm

Standards for Technological Literacy

5 14 15

Safety
Most of the GM plants sold today contain a single trait, such as the ability to resist herbicides. People are currently working on more complex genetic modifications that will alter the nutrient content. What impact will they have on human nutrition? How can consumers be sure that genetically modified foods will not cause short-term or long-term risks to their health?

Discussion
What are the most persuasive arguments for and against genetic modification?

Vocabulary
What is the meaning of the terms *toxins*, *allergens*, and *biodiversity*?

announced the discovery of the structure of *DNA*, which stands for deoxyribonucleic acid, **Figure 14-6**. Strands of DNA are the basic building blocks of life. This discovery opened the way for a more precise understanding of how living things work.

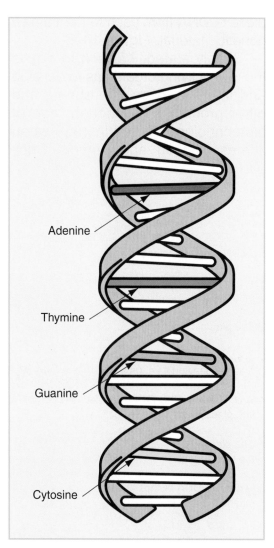

Figure 14-6 DNA is composed of the following three parts: phosphate, a sugar called deoxyribose, and four compounds called bases. The bases are adenine, guanine, thymine, and cytosine.

Adenine

Thymine

Guanine

Cytosine

Scientists and technologists started to take the basic blocks of living creatures apart and to put them back together in new and, possibly, improved ways. In this way, thousands of these transgenic-plants, animals, bacteria, and viruses could become part of the earth's ecosystem.

The Enviropig™ is an example of transgenic engineering, **Figure 14-7**. Manure from normal pigs can pollute rivers and lakes because of its high phosphorous content. Phosphorous increases the amount of algae in water. Algae rob fish and other organisms of oxygen. That is why reducing pig pollution is very important. Using genes from other organisms, the Enviropig™ processes feed more efficiently. This reduces phosphorous in its manure, and water supplies are protected.

A second example is genetically enhanced "super rice" being used in China. The project involves speeding up the growth of existing hybrid rice by inserting a gene from the corn plant. A certain gene from the wing-bean is also added to the rice through transgenic engineering in order to improve resistance to stress resulting from severe weather conditions. Transgenic crops are also known as genetically modified crops (GMC). They are increasingly popular in Argentina, Canada, China, Mexico, and the United States, where over half of crops such as soybean, corn, and canola are GMC. Herbicide-resistant crops and insect-resistant, transgenic crops are also popular worldwide. Within the near future, we may see vitamin-enhanced fruits

Technology and society: Europeans have a protocol (an established method) for testing and labeling the GM content of foods. What should a protocol include? What are the advantages and disadvantages of having this type of protocol?

Useful Web site: The following Web sites provide useful information about biotechnology:
Biotechnology and Biological Sciences Research Council - www.bbsrc.ac.uk/life

British nutrition foundation - www.nutrition.org.uk

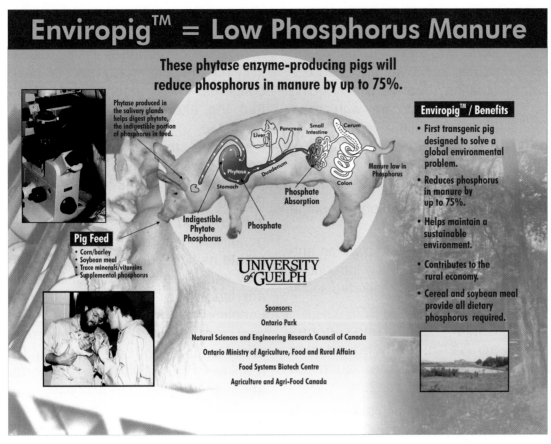

Figure 14-7 This chart explains how the Enviropig™ functions and lists its benefits. (University of Guelph)

and vegetables, allergen-free nuts, low-calorie bread, genetically decaffeinated coffee, and vaccine-enriched bananas.

These products will be introduced with some controversy. However, one of the best arguments for the development and use of biotech (Bt) crops is that they reduce the amount of insecticides used. Fewer insecticides mean less wasted time and energy in spraying, less risk of exposure to people, and less fuel used. Planting corn with a Bt gene that kills the corn borer insect can greatly reduce the amount of insecticides applied. It is also claimed

that less plowing and other conservation practices, such as leaving more crop residue on the surface of the soil, have reduced sediment runoff. The result is that rivers and streams stay cleaner for fish and wildlife.

In spite of these advantages, many people question whether we understand all the implications of the new technology:

◆ What effect will the genetically modified organisms have on wild animals and plants? Are the risks greater than the benefits?

Technology and society: What are the ethical considerations of GM foods? Is it acceptable to move genes between plants or animals that do not normally interbreed? What religious, health, or other objections may people have against GM foods?

Technology and society: The World Health Organization (WHO) pointed out that the lives of 10 million children in developing countries who die each year from infectious diseases could be saved with vaccines. How could GM foods help fight against infectious diseases? What are the implications of saving the lives of these children?

Vocabulary
What is *bovine spongiform encephalopathy* (mad cow disease)? How does it affect animals?

Activity
Have students browse the Internet to investigate how poultry or pigs are raised and sent to market. Which countries have outlawed battery cages, farrowing crates, and veal crates?

Activity
Have students investigate the work of each of the following people and state why they agree or disagree with their views:
(a) American law professors Tom Regan and Steven Wise who want to protect animals by defining their basic rights.
(b) Princeton philosopher Peter Singer who, in his book *Animal Liberation,* speaks of basic moral principles.
(c) Matthew Scully, who wants us to treat animals with common human decency and kindness.

◆ Are we sure that it is safe to mix genetic material from different life forms? Have the tests been carried out over an adequate duration? Are we blindly going into this new era with great hopes but little idea of the unpredictable outcomes?

◆ If we transfer genes from one plant to another, will it trigger allergic reactions in some people?

◆ Will vegetarians know if an animal gene has been introduced into a plant and marketed solely as a vegetable?

◆ Could transferring genes across natural boundaries cause epidemics, such as mad cow disease?

◆ Could there be pressure to change less desirable genetic traits in humans?

◆ Might employers use genetic information to make decisions about hiring?

◆ Will the smaller number of varieties be a source of increased risk for farmers, as genetically homogenous fields might be more vulnerable to disease and pest attacks?

There is also concern about who controls the industries responsible for the new biotechnologies. For example, the ten top agrochemical companies control over 80% of the global market. This has an effect on costs. Genetically modified seeds are more expensive than standard seeds. Often the places that need the genetically altered seeds most are the ones that cannot afford them.

With such changes taking place, it is natural that many people have concerns. These people understand that the altered life forms may be restricted to one area. But at the same time, they know that Bt crops could migrate to another area where there is little control over them. People are also questioning whether it is a good idea to reprogram the genetic codes of life and artificially create new life forms. As a minimum, they would like to know that Bt foods are labeled so they have a choice. Labeling of foods would help to prevent allergic reactions, respect personal values and religious dietary requirements, and address concerns about environmental practices.

Biotechnology also affects humans directly. The ***human genome***, which is all the genetic material of the human species, is being mapped out right now, **Figure 14-8**. In the near future,

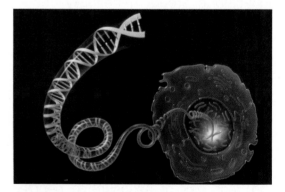

Figure 14-8 The 3 billion base pairs in DNA, known as the human genome, reside in each of our cells and contain the genetic information necessary to create a human being. (National Human Genome Research Institute)

Useful Web site: The Biotechnology Institute Web site at www.biotechinstitute.org provides extensive information about biotechnology.

Technology and society: What does the statement "Air, soil, water, and all living creatures should be given the same respect as we give our bodies and ourselves" mean?

we will know which of our 30,000 genes are responsible for certain traits and disorders. This means that it will be possible to manipulate genes. Certain defects could be removed and attractive features added. However, as far as we know, no gene gives us the important characteristics that make us human. No particular genes make a human being kind hearted, law-abiding, or capable of loving.

Figure 14-9 People of all races, religions, and ethnic groups are encouraged to accept with one another and to celebrate their diversity.

Environmental Technology

Each generation inherits a world shaped by their ancestors. Our daily actions will affect generations to come. We must remember that we do not own the world—we hold it in trust. For the sake of those who will follow us, we should not do anything that will harm the earth. Although technology provides many benefits and can solve problems, there is more air pollution, environmental destruction, and soil erosion happening now than ever before industrialization.

Not very long ago, women, various minorities, and people of several different ethnic backgrounds were treated as lesser human beings. We now know this is unacceptable. Our fundamental beliefs have changed so that all human beings are treated equally, **Figure 14-9**. We need a similar change of attitude regarding our environment. Air, soil, water, and all living creatures should be given the same respect as we give our bodies and ourselves. This is not easy to do when we live in an age where most

people want more and more. This type of consumer demand means continual economic growth. For this growth to occur, we are constantly taking from the environment, often without replenishing it. Living in harmony with all species on earth should be our goal, **Figure 14-10**. We do not have the right to dominate other species. It is easy to think only of ourselves. Buying products makes us feel good, but let us remember that this feeling only lasts for a while. Then we start thinking of the next purchase and the cycle continues.

It is easy to forget the needs of other creatures. Most humans live either in cities or suburbs surrounding the cities. We are rarely in direct contact with nature. Only 1 out of 100 people collects food from fields or milks a cow. We think of food coming to us in packages. Technology provides food to us neatly arranged in brightly lit, clean stores. But technology also bears responsibility for some major accidents, **Figure 14-11**. Some of the best-known technological

Standards for Technological Literacy

5 **14**

Discussion
Have students listen to a radio or television station and record the air quality index for their region. When are the peak times of pollution? What could be done to improve the situation?

Vocabulary
What does the term *organic farming* mean?

Activity
Choose one car in each of the following categories: compact, mid-size, full-size, and SUV. Determine which vehicle in each category uses the least fuel.

Useful Web site: Have students browse the Internet to find information on how they can contribute to keeping the air clean. Students can print information they find interesting and share it with the class. Helpful places to start are www.pembina.org and the David Suzuki Web site at www.davidsuzuki.org.

Technology and society: What lessons can be learned from the Three Mile Island, Bhopal, and Exxon-Valdez disasters?

Standards for Technological Literacy

5 **14**

Designing and making
Design and make a product to increase public awareness of the causes and problems of global warming.

Discussion
Discuss the steps your class can take to reduce the amount of carbon dioxide (CO_2) being pumped into the atmosphere.

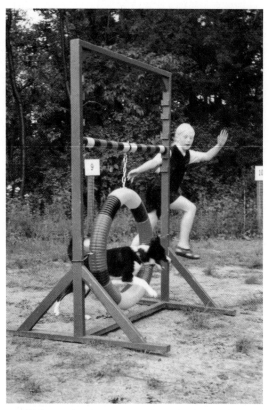

Figure 14-10 Owning and caring for a pet often breeds love and sympathy for other animals.

Figure 14-11 On March 28, 1979, ten miles from Harrisburg, Pennsylvania, a nuclear reactor core at the Three Mile Island power plant overheated and released radioactive gases. (Michael Rennhack)

tragedies were the chemical leak at Bhopal, India, the nuclear meltdown at Chernobyl, and the oil spill of the ship *Exxon-Valdez.*

The Intergovernmental Panel on Climate Change (IPCC) concluded that, since the Industrial Revolution, activities such as burning fossil fuels and clearing the land have increased the concentration of carbon dioxide (CO_2) in the atmosphere by 30%. While it is true that oceans and land vegetation release and absorb 200 billion metric tons of carbon into and out of the atmosphere annually, human activities have added an extra 7 billion tons each year. This is too much to be absorbed, and the result is a gradual warming of our atmosphere. The polar ice caps are melting, and sea levels are rising.

Keeping Our Air Clean

Clean air is our most vital need. If you try to stop breathing, before long your body will scream for air. We need air every minute of our lives. If we stop breathing, we will likely suffer brain damage after three minutes and will be dead within five minutes. Industrial companies and governments are working to reduce harmful emissions. The majority of air pollution comes from burning fossil fuels. Automobiles contribute about one half of our air pollution. In order to reduce the amount of automobile pollution, the Environmental Protection Agency requires cars to be tested at testing facilities. This ensures that vehicles meet certain standards.

Links to other subjects: History—Investigate how the Industrial Revolution began increasing impacts of people on the environment.

Useful Web site: Have students access the Web site of the Intergovernmental Panel on Climate Change at www.ipcc.ch to investigate climate change and its potential impacts.

Community resources and services: Investigate what your community is doing to improve air quality where you live.

We can start purifying our air at home. Indoor plants clean the air of pollutants. The pollutants are sucked into the leaves. They then go to the soil where microorganisms break them down turning them into plant food. One plant purifies 100 ft.^{2} (9 m^2). We should also limit use of aerosol sprays, use white toilet paper (not colored), recycle (glass bottles, tin cans, and plastic containers), and avoid using non-biodegradable containers.

Regenerating Soil

We need to eat to survive. The plants or animals that give us our food have absorbed whatever was around them in the air and soil. Whatever we do to the air and soil, we do to ourselves. Everything we eat comes directly or indirectly from the soil. Soil is only present as a thin outer layer on the earth. You could say that it is much like the skin on an apple. Modern farming practices use up soil by decreasing organic matter and causing soil erosion. Soil needs time to regenerate, otherwise erosion takes place. Chemical fertilizers only serve to increase productivity, not to renew the soil.

Destroying forests is a worldwide problem that affects not only the soil but also the entire ecosystem. For example, in Brazil, slash-and-burn agriculture destroys millions of acres of tropical rain forest every year. Tropical forests form a green belt around the equator. Almost one half of all growing wood and other plant species can be found there.

It is not just that trees are destroyed. Plants and animals, even humans, are dependent on one another. In forested areas on our Pacific Coast, bears play a vital role. They fish for salmon and take them into the forest so they can eat them. What they leave feeds other animals and insects and even provides nitrogen fertilizer for the forest soil. The forest itself recycles oxygen, nitrogen, and carbon. Forests act as enormous sponges collecting and distributing water. They protect the soil from water and wind erosion.

What can we do? Some forested areas should be left to provide places for birds and animals. Fish-bearing streams should be left undisturbed. Frequently, it is what you do not see that is important. Mountain lions, giraffes, polar bears, and penguins may inspire us with awe, but they are not the creatures that help us most. We should have more respect for the small creatures and microorganisms that form most of our living world. They keep the soil fertile so it acts as a filter for cleaning water as it goes through its recycling process.

Purifying Water

Compare your surroundings with the Sahara Desert where the sun beats down every day and little is growing. The main difference is water. Water is vital to life. It enables hundreds of organisms to grow. Without water we would die. Over half of our body by weight is water. We lose it from our skin, lungs, every opening in the body, and when we breathe.

Standards for Technological Literacy

5 14

Discussion
Discuss how deforestation, especially in developing countries, leads to floods, drought, and sometimes to the formation of deserts.

Enrichment
The Northern Great Plains region, stretching from Alberta and Saskatchewan to Wyoming and Nebraska, has been plowed into cropland. As a result, endangered species like the bison and swift fox have lost their habitat and are struggling for survival.

Links to other subjects: Biology—How do plants purify air? Have students investigate how carbon dioxide is used in photosynthesis and oxygen is released.

Useful Web site: Have students investigate and report on techniques for soil conservation. Three useful Web sites are: www.cav.pworld.net, www.agroforester.com, and www.panda.org.

Community resources and services: What waste materials are recycled in your neighborhood? Where are they taken and how are they sorted?

Standards for Technological Literacy

5 14

Designing and making
What types of products could designers create to help safeguard fresh water supplies?

Enrichment
The five Great Lakes hold 20% of the world's fresh water. Canals dug to make these waters navigable have brought "foreign invaders," such as zebra mussels. New invaders are being discovered, on average, every 8 months.

Activity
Have students build a model of the global greenhouse. Discuss with students what their model must show.

Water from our bodies evaporates into the atmosphere, enters plants, soil, or other water sources. It becomes part of an endless cycle between all living things.

There is a huge amount of water on earth, but what we need is freshwater. About 97% of the water is saltwater. Most of the rest is either in glaciers or underground. Less than 1% of all water on our planet is available, but often a large amount of that ends up as floodwater. Think how we use this amount of available water—toilet (45%), washing ourselves (30%), washing our clothes and dishes (20%), for cooking or drinking (5%).

A recent pollution problem on land occurs when cattle and pig farms become very large. Manure from factory farms often contains a variety of deadly pathogens and viruses, such as E. coli, in addition to smells, flies, dust, and noise. A farm with 18,000 pigs creates as much waste as 60,000 people. The difference is that waste-treatment plants process human waste, whereas animal waste is often allowed to pollute streams, lakes, and our drinking water.

When rivers and seawater become polluted, fish die. Fish breathe through gills. If their gills are clogged with oil, they can't breathe. Oil can also get into sea plants. Small fish and animals eat the plants including some oil. People who eat oil-polluted foods become sick. The oil can also cause cancer in some animals. We might think that pollutants somehow disappear in water, but traces of the pesticide DDT have been found in remote parts of Antarctica.

In some cases, bioprocessing can use microorganisms to help clean up the environment. Microorganisms are living cells that must eat to survive. Fortunately, some microorganisms will eat the plant and animal matter in garbage. *Bioprocessing* is the use of microorganisms to break down and purify plant and animal matter. When this occurs, a gas called *methane* is produced. This gas can be used for cooking and heating.

In a similar way, sewage, containing human and industrial wastes, can be purified. The microorganisms eat some of the solid matter and help to decompose the wastes. Using genetically engineered microorganisms can increase the capacity and efficiency of waste treatment plants.

Using Appropriate Types of Energy

You may have heard the expressions "global warming" or "greenhouse effect" to describe how the earth may be overheating. If a "greenhouse" did not exist on earth the temperature would be about 14° F (−10° C). The oceans would be frozen, and carbon dioxide and water vapor in the atmosphere would have made Earth unlivable. The greenhouse doesn't threaten our lives, but overheating the earth does.

Burning fossil fuels produces massive quantities of greenhouse gas emissions that are increasing the temperature of our planet and changing our climate. Global warming is causing droughts, melting

Useful Web site: Have students access the following Web site and click on "Drinking Water and Health: Telling a Story" to learn about the connection between water and health: www.waterandhealth.org/drinkingwater.

Technology and society: What has happened to the cod fishing industry? What steps are being taken to remedy the problems encountered by commercial fishers?

Figure 14-12 If we continue our present practice of releasing heat-trapping gases into the atmosphere (principally carbon dioxide), we may melt all the natural ice on Earth. (David Suzuki)

Arctic ice, and contributes to lower water levels and more forest fires. The main fossil fuels are oil, gas, and coal, **Figure 14-12**.

By burning fossil fuels, billions of tons of carbon dioxide are sent into the atmosphere. The Scripps Institute of Oceanography that monitors CO_2 levels in Hawaii reports that CO_2 levels have increased 25% since the 1800s. These amounts are impossible for the atmosphere to absorb. They form a shield, rather like a pane of glass. This shield prevents the sun's heat from being radiated back into space, **Figure 14-13**. The result is a steady

Discussion
World oil consumption in 1940 was 2000 million barrels. By the year 2000, it had increased to 20,000 million barrels. A barrel is equal to 35 US gallons (132 L). Which of our daily practices have changed and caused this increase?

Reflection
What steps do you take to minimize your consumption of fossil fuels?

Enrichment
Polar bears fast all summer and in the fall, the hungry bears gather at the edge of Hudson Bay. They wait for the water to freeze, so they can go out onto the sea ice and hunt for seals.

Vocabulary
What does the acronym *OPEC* mean?

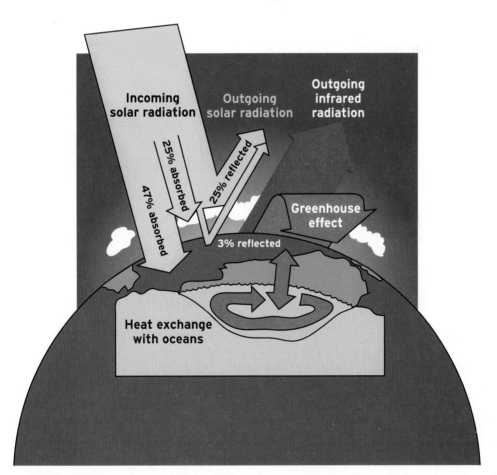

Figure 14-13 The greenhouse effect allows solar heat to pass through the atmosphere but does not allow it to escape from the atmosphere. (David Suzuki)

Useful Web site: Have students check predictions concerning future world oil demand. A helpful place to start searching is www.savvymotoring.com.

Technology and society: What could society be doing to ensure adequate supplies of oil and gas for future generations?

Designing and making
What actions can designers take to help conserve finite reserves of fossil fuels?

Discussion
Discuss options for reducing our continued dependence on burning fossil fuels for heating and transportation. Scientists claim this source of energy is a major contributor to the overheating of the earth's atmosphere.

Enrichment
NASA was a pioneer of using plants to purify air. In 1973, NASA found that the air inside Skylab 3 was contaminated with more than 100 toxic chemicals.

Vocabulary
What is meant by the term *acid rain*?

Activity
An analysis of air trapped in ice, preserved since the eighteenth century, shows that carbon dioxide concentrations began to rise early in the 1900s and have continued to do so ever since. Have students browse the Internet to investigate why this change occurred.

rise in temperature. More and more plants and animals will have to permanently migrate to find a suitable habitat. Polar bears who depend on a shrinking icepack for hunting are underweight and birth rates are falling fast. Many caribou herds are becoming smaller due to food shortages brought on by earlier springs and warmer summers.

Some of the chemicals from fossil fuels unite with moisture in the atmosphere to form *acid rain*. When acid rain falls on forests and lakes it raises the pH level of the water and kills aquatic life and vegetation. The damage caused by this type of pollution may take place at a great distance from the original cause. The areas in North America that are particularly sensitive to the impact of acid rain are shown in red while the sources of sulfur dioxide emissions are indicated by black dots in **Figure 14-14**.

Another concern for our atmosphere is ozone depletion. While there are some factors common to both, the greenhouse effect and

ozone depletion are two largely different issues. The greenhouse effect is related to global warming. Ozone depletion is concerned with shielding the earth against ultraviolet radiation.

Our atmosphere consists of the troposphere (we live in the lower portion), and the stratosphere, which begins above 7.5 miles (12 km). This includes the ozone layer, a layer of gas that starts about 15.5 miles (25 km) from the earth's surface, **Figure 14-15**. It shields us from damaging short wavelength radiation coming from the sun before it can strike the earth. It is

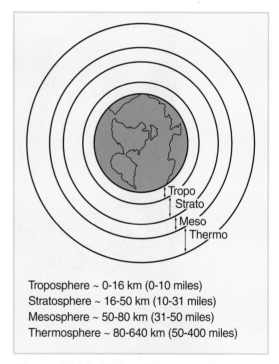

Troposphere ~ 0-16 km (0-10 miles)
Stratosphere ~ 16-50 km (10-31 miles)
Mesosphere ~ 50-80 km (31-50 miles)
Thermosphere ~ 80-640 km (50-400 miles)

Figure 14-15 The earth's atmosphere is composed of four layers. They are the following: troposphere (0–10 miles; 0–16 km), stratosphere (10–31 miles; 16–50 km), mesosphere (31–50 miles, 50–80 km), thermosphere (50–400 miles; 80–640 km). The ozone layer is the lower part of the stratosphere.

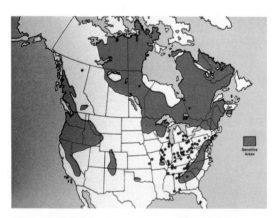

Figure 14-14 Is acid rain affecting the area where you live? (Science Dimension)

Links to other subjects: Science—Students should investigate the composition of a molecule of ozone, compared to a molecule of oxygen, to determine how ozone screens out harmful radiation from the sun.

Useful Web site: Have students visit the Center for Atmospheric Science (University of Cambridge) at www.atm.ch.cam.ac.uk/tour and take "The Ozone Hole Tour."

More reference information about the ozone layer can be found at www.nas.nasa.gov/Services/Education and www.science.org.au/nova/004/004key.htm.

continually being formed. However, synthetic chemicals and other pollutants can easily destroy it. If radiation gets through it can cause skin cancers, eye cataracts, and other problems. It could even damage DNA, which is the basic building block of all life on Earth.

In Chapter 10, you studied alternative sources of energy that could replace fossil fuels or at least reduce the amount used. Several of these were renewable including hydroelectric power, tidal and wave power, and energy from plants and garbage. Alternative sources often cost more; however, government subsidies can help to reduce costs. In the US, a federal tax credit is provided to allow wind energy to compete with other sources. Wind and solar power are ideal sources of renewable energy that have no destructive effect on the environment. Even hydropower is not entirely environmentally friendly. Damming rivers displaces humans and harms animals and fish habitats. The reservoirs themselves emit greenhouse gases due to the breakdown of plants that were submerged when the reservoir was built.

Taking Action

What can each of us do to reduce the negative effects of technology while enjoying its benefits? First we must recognize when a problem exists. This book and your technology course have made you aware of some of the issues, but you should be constantly evaluating the benefits of any new technology and comparing it with what has been displaced. Second, you should realize that technology is neither good nor bad. Its success depends on how it is used. For example, lasers are being used to perform very delicate surgery. They can also be made into highly destructive weapons. Third, we must act when there is a problem or a better alternative exists. For example, heating and cooling our homes has a big impact on the environment. Making our homes energy efficient can mean improving insulation, stopping drafts, and installing better windows. For new homes, it means building to the R-2000 standard. One feature of the R-2000 home is that as little as half as much energy is used for heat, light, and hot water. Rapid advances in technology lead to a shorter life span for electronic products. Roughly 120 million used cell phones are discarded each year in North America, and it is estimated that more than 500 million personal computers will become obsolete within 10 years in the United States alone. These devices contain lead, mercury, cadmium, and polyvinyl chloride (PVC) plastics that cause significant environmental and health risks. There are currently few recycling facilities for them.

On a day-by-day basis, we should think of the three Rs: Reduce, Reuse, and Recycle. See **Figure 14-16**.

Designing and making
What new type of container could be designed to separate various materials for recycling and composting?

Discussion
Discuss the ways in which students can become "green" consumers.

Enrichment
In addition to the three Rs (reduce, reuse, and recycle), manufacturers add two other Rs: Refine (materials should be of a high quality so less is used) and Recover to energy (waste materials can often be used as a source of energy).

Activity
Have students list the ways in which they, and their family, can reduce the total amount of energy consumed in their home. They should include information related to insulation, vapor barriers, thermostats, the energy rating of appliances, lighting, and weather stripping.

Resource
Reproducible Master, *The 3 Rs*

Useful Web site: The National Science Digital Library Web site contains useful materials on all aspects of reducing the impact of human consumption on the environment.

Technology and society: Scientists fear that as more ultraviolet light reaches the earth, due to a hole in the ozone layer, there will be more problems on earth. What steps should governments and individuals take to minimize the depletion of the ozone layer? What are the costs to society as a result of holes in the ozone layer?

Designing and making
How can designers help with the recycling of waste materials?

Discussion
Have students debate the benefits and disadvantages of buying quality, long-lasting products that cost a little more versus buying cheaper products that wear out more quickly.

Reflection
Buying in bulk reduces the amount of packaging, but you prefer juice, drinks, and yogurt in smaller refillable containers. What could you do to solve this dilemma?

Activity
Have students investigate which natural products, such as vinegar and baking soda, could replace commercial cleaners. Their research should include cost and effect on the environment, compared to commercial products.

Activity
Have the class collect waste products that could be put to further use with slight modifications.

Figure 14-16 Remember the three R's: Reduce, Reuse, and Recycle.

Figure 14-17 This woman is reducing nonbiodegradable waste by drinking from her own mug. Can you think of any ways to reduce your waste? (TEC)

Reduce

The United States, Australia, and Canada create the most garbage per capita. To reduce the amount, it is preferable to use natural packages made of paper and cardboard. Although some plastics can be recycled, it is far more common to recycle paper. Fast-food is popular and can be eaten in minutes, but foam containers can be very damaging to the earth's environment. This is why you should buy it in paper bags. Use your own mug when buying a cup of your favorite beverage, **Figure 14-17**.

Reuse

We seem to live in a throwaway society, but we can reuse many items. Some products may continue to be used by replacing the broken parts and keeping the product in use. A car engine can be rebuilt. Cloth diapers can be used instead of disposables. Why not hold a garage sale to sell your unwanted things to someone who needs them? Or you could sell them on the Internet. Think of new ways to use old items. Your favorite mug can be reused thousands of times. Even when the handle breaks, it could hold pens and pencils. Wood from an old wooden bench might be refinished and made into a coffee table.

Recycle

Many towns and cities have programs to recycle glass, plastics, paper, and cans, **Figure 14-18**. *Recycling* is the process of treating certain materials so that they can be made usable again. By using the same material a number of times, less energy is used to shape it into new products. Every ton of paper recycled saves up to 19 trees. One

Useful Web site: Have students take the tour of "Recycle City" at www.epa.gov/recyclecity.

Have students investigate how plastic coffee cups can be recycled to make a pencil at www.designcouncil.info/educationresources/studies/pencil/index.html.

Have students visit the Recycled Promotional Products Web site at www.recycled.ca to see the range of useful products that can be made from recycled coffee cups, plastic bread tags, and foam trays.

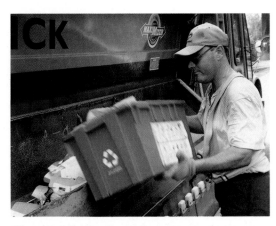

Figure 14-18 Recycling paper, plastic, glass, and aluminum has become a common practice in recent years. Trucks collect the contents of recycling bins or bags. (Ecritek)

ton of aluminum from recycled cans uses two-thirds less energy than extracting the metal from ores. Recycled plastic is formed into planks to make playground equipment, decks, and outdoor furniture. Recycled tires can be used to make artificial turf for sport stadiums. T-shirts can even be made from recycled soft drink bottles. To make a T-shirt, the plastic from about 14 bottles is chopped into small flakes, cleaned, melted, and extruded into fibers that are spun to create yarn. Persons with yards may start a compost pile and produce organic fertilizer from kitchen and garden waste.

The principles of reduce, reuse, and recycle should apply to everyone, whether they live in developed societies or third world countries. We all need to protect our environment for future generations. At the same time, it is important to be aware that individual actions will only get us part way. If government regulations promote coal-fired electricity generation or vehicles with poor fuel-efficiency, then consumers have less choice and the environment suffers. When we know the environment is not being protected, we have to apply pressure to governments to act.

Medical Technology

What do we need to enjoy good health? It helps to have comfortable surroundings, quiet moments for ourselves, and people who love and support us. In addition, fresh air, water, sunshine, rest and sleep, and the right foods are vital for a healthy body.

Eating Right

If we want to eat right, we need to know the tools that are available to us to prepare delicious and nutritious meals. Anyone who likes to prepare meals has a certain set of tools. There are tools for cutting, including graters, peelers, cutters, and knives. Other tools are used to mix or combine the ingredients, including measuring spoons and cups, blenders, mixers, mixing bowls, pastry blenders, rolling pins, spatulas, spoons, and wire whisks. Finally, there are tools to bake or cook, including baking pans, cookie sheets, cake pans, frying pans, saucepans, tube pans, and wire racks to cook in the oven and to cool the food on a counter after cooking, **Figure 14-19**.

Community resources and services: Ask a person in the community who is involved in the design and planning of kitchens to speak with the class. The ergonomic requirements of space are a particularly interesting consideration.

Technology and society: Three engineers at the University of Guelph have developed edible packaging made from soybeans. This packaging is designed to replace food trays made of polystyrene. What are other ways to reduce the volume of throwaway containers and packaging?

Standards for Technological Literacy

14

Resource
Transparency 14-1, *The 3 Rs*

Resource
Reproducible Master, *MyPyramid*

Discussion
Divide the class into small groups. Have each group choose one food item and list the different ways that it can be prepared.

Discussion
Discuss the importance of including foods with a low glycemic index into your daily diet.

Vocabulary
What does the term *glycemic index* mean?

Vocabulary
What does the term *hydrogenated* mean?

Activity
Develop an eating plan for yourself for one week, based on the Dietary Approaches to Stop Hypertension (DASH) Eating Plan, which the National Heart, Lung, and Blood Institute (www.nhlbi.nih.gov) developed.

Figure 14-19 These are some common tools used in cooking and preparing different foods. (Ecritek)

It is easy to remember what each of us needs for a balanced diet using MyPyramid, **Figure 14-20**. The US Department of Agriculture developed MyPyramid, which is an individualized food guidance system. This system divides foods into groups, according to the nutrients they provide. By going to www.MyPyramid.gov and entering your age, gender, and activity level, you can learn the amount of each food group you should eat daily.

The largest part of the pyramid is the grains section. This section includes *carbohydrates*, which provide much of the energy we need. A slice of bread, a bowl of cereal, and a half cup of rice are all servings of grains. The next two sections of the pyramid are the vegetables and fruits sections. *Vegetables* are the group of foods including broccoli, spinach, and carrots. A half cup of raw or cooked vegetables and a small bowl of lettuce are servings of vegetables. *Fruits* are the food group containing

Figure 14-20 MyPyramid is a great tool for planning healthy, nutritious meals. (US Department of Agriculture)

Links to other subjects: Social Studies—Divide the class into small groups. Using a World map, have each group select one country and describe a typical meal eaten by its inhabitants.

Useful Web site: Have students investigate the guidelines for healthy eating that the American Diabetes Association (www.diabetes.org) provides.

apples, bananas, and oranges. The other two major sections of the pyramid are the milk section and the meat and beans section. The milk section includes all *dairy* products, such as milk, yogurt, and cheese. The meat and beans section includes *protein* sources, such as meat, beans, poultry, eggs, fish, and nuts. The smallest section of the pyramid is the oils section. The oils section includes *fats, oils, and sweets*, such as butter, honey, and candy. These types of food supply calories but few nutrients, so they should be eaten in moderation. A varied diet, coupled with regular activity, will help you maintain your health. MyPyramid offers a low-fat, well-balanced diet.

Maintaining a Healthy Body

With millions of dollars spent annually on health care, there are still thousands of people who die every day from heart disease and cancer. Fat and cholesterol are part of the food we eat, but too much can make us fat and can clog up our arteries so that blood cannot get through. The right food and lots of exercise slows down or stops this happening, **Figure 14-21**. You should:

◆ Eat plenty of fruit and vegetables.

◆ Have whole-grain bread, cereal, pasta, and rice often.

◆ Drink skim and low-fat milk.

◆ Snack on fresh fruit, frozen fruit bars, raw vegetables, and nuts.

You have more than 200 bones in your body, **Figure 14-22**. Each is stiff

Figure 14-21 Fruit makes a healthy, low-calorie snack instead of chocolate bars or candy.

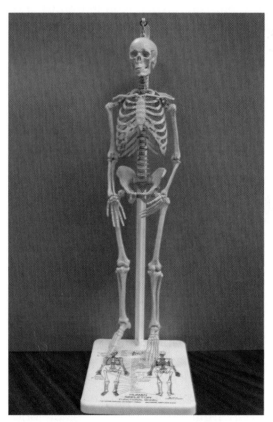

Figure 14-22 Eating plenty of calcium-rich foods is the best way to maintain strong bones. (TEC)

Useful Web site: Have students investigate the guidelines the National Heart, Lung, and Blood Institute (www.nhlbi.nih.gov) produced for reducing the risk of developing hypertension.

Technology and society: Which diseases are people developing as a result of consuming high-fat diets? What are the long-term consequences of these diseases for individuals and society?

and will not bend. Movement is at the joints, such as those in your elbow or knee. They are actually continuing to grow until you are about 30 years old, and they need calcium to grow. Calcium-rich foods include milk, cheese, yogurt, and tofu. Also certain greens including broccoli, collard greens, okra, kale, and oranges are sources of calcium.

Almost one-half of your body weight is made of muscles. Skeletal muscles pull on bones to make them move; smooth muscles are found on the walls of blood vessels, stomach, and organs. They work automatically to help these organs perform their work. The cardiac muscle is in the heart and contracts every second or so to keep blood moving. Protein is used to repair and build muscle. Beans, peas, meat, poultry, fish, eggs, milk, and dairy products all supply protein.

Repairing the Damage

Modern medicine can help when there are problems, but keep in mind that there are people who regularly take prescription drugs every day and do not improve their health. Why is this? They don't understand that good health comes from healthy living. If you overextend yourself during the day, you need to pay your body back later. Overeating, overwork, or too many hours of study are all excesses that will catch up with you. Eating nuts, seeds, fresh fruits and vegetables, or food that has not been altered in any way will give you far better health than

eating processed food that is filled with chemicals.

Balance is not only important for the food we eat but for our other activities, **Figure 14-23**. By having the right amount of rest and sleep, we are recharging for the next day. This includes regenerating our nerve energy, repairing tissue, and healing and replacing old cells. Through exercising we send more blood, nutrients, and energy to that area to strengthen and invigorate it. Being exposed to sunshine (not sunburn) with the full spectrum of natural sunlight also provides a vital nutrient for the body. Finally, having the love and attention of those you care for keeps us psychologically healthy. But remember that loving starts with loving oneself and having an optimistic outlook on life.

Surgical Techniques

For many procedures, patients spend several hours in surgery, a week in the hospital, and a month recuperating at home. With the latest

Figure 14-23 Exercise increases muscle strength, endurance, energy, and even life expectancy.

technologies, experienced surgeons can do procedures in a fraction of the time. New procedures are less invasive. This means either no incisions are made in the body or they are very small. Surgeons may make small cuts to permit tiny instruments and a miniature camera to enter. The camera sends views of the internal organs to a video monitor. The surgeon then manipulates the instruments according to the images shown on the video monitor, **Figure 14-24**. Using techniques, such as laser surgery, the surgeon may complete operations such as removing tumors or scar tissue or draining cysts. Patients have little pain and spend less time recuperating.

Technologies of this type have even been used for transatlantic operations. On one side of the ocean, a robot, equipped with a tiny fiber optic camera, scalpel, and tweezers, enters the body through a small incision. On the other side of the Atlantic, the delicate movements of the surgeon's hands are transmitted to the operation site. This type of "telemedicine" uses a high-speed fiber optics network so that the time delay between the surgeon's movements and those of the robot is only in milliseconds.

Robotic surgery is starting to become routine, as it can be more precise than traditional surgery. There are three robotic arms at the operating table, a computer controller, and a

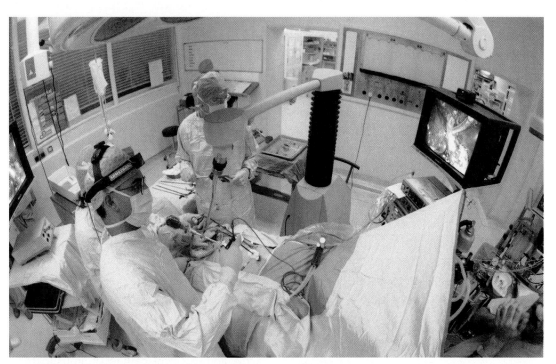

Figure 14-24 This surgeon is using a foot switch and special headband to control a robot, which holds and directs the camera. (Armstrong Healthcare Limited)

Technology and society: Research suggests that as many as 40% of American teenagers are overweight. Have students browse the Internet to investigate what will be the cost to society as these people become older.

Standards for Technological Literacy

14

Activity
Have students compare the US Department of Agriculture MyPyramid at www.MyPyramid.gov with the pyramid for a traditional healthy Latin American diet at www.oldwayspt .org/.

Designing and making
What issues must designers consider when designing new equipment that encourages people to engage in exercise?

Discussion
Discuss the advantages of engaging in vigorous activities such as fast walking, cycling, swimming, dancing, and playing tennis.

Reflection
How much exercise do you get on an average day? How many cups of coffee or cans of soft drinks do you drink in a day?

Reflection
Do you engage in hazard identification, risk assessment, and risk management before you engage in any potentially dangerous activity?

Vocabulary
What is an *endoscope*?

Vocabulary
What does the term *noninvasive surgery* mean?

control console for the surgeon. One robotic arm positions the endoscope (a device for minimal-access surgery), and the other two manipulate the surgical instruments. The surgeon can view the operation in two and three dimensions and operates this device continuously with voice commands. The technique is now used for minor and major operations, including gall bladder removals and pacemaker operations. It is estimated that most surgeries will soon be performed in this way. Patients are exposed to less risk, as the operations are minimally invasive. Minimal invasion (making small, if any, cuts into the body) is a goal of modern surgical procedures. For example, it takes only 10 minutes for a microwave therapy to kill a tumor and any precancerous cells without damaging healthy neighboring tissue. Being noninvasive, the procedure eliminates the spread of cancer within the body, which can happen when a scalpel dislodges cancer cells during surgery.

More medical devices are now available for consumers to monitor their own health or assist in an emergency. The U.S. Food and Drug Administration decided it is now safe for consumers to buy and use an automatic external defibrillator to revive a heart attack victim. These devices are similar to those found in ambulances and fire trucks.

Prosthetic Devices

Biomedical technologies can assist people whose limbs have been damaged. Many types of artificial components have been designed to improve mobility. At age 18, Victor was diagnosed with cancer in his left arm, which had to be removed. He loved sports and still wanted to compete. He tried mountain biking using a prosthetic arm. The arm was made from modified Teflon prosthetic leg components, **Figure 14-25**. It is fitted with a clip that connects to an adapter on the handlebar grip. Together, with help from a special hydraulic braking system he is able to compete with the best!

Through medical technology, artificial parts can be made to replace natural ones, but they are not better

Figure 14-25 There are many different kinds of prosthetics. Prosthetic hands, arms, and legs replace human limbs lost due to war, disease, and industrial or recreational accidents.

Useful Web site: Have students investigate the design of prosthetic devices. Three useful Web sites are: www.waramps.ca/nac/limbs.html, www.animatedprosthetics.com, and www.alatheia.com.

Technology and society: Should there be a system of state health insurance to ensure that everyone has access to medical services when necessary?

than natural ones. Even though we have seen fictional bionic characters that seem to have superhuman powers, this is not reality. However, most replacement parts are successful, if only for a period of time. They can be mechanical, electronic, or organic. Mechanical transplants such as metal hip joints, plastic valves, or electronic pacemakers are called implants. Living organs that replace ones that are giving problems are transplants and are usually from another human donor.

You might think that doctors and surgeons would be the ones who design and build these implants. This is not so. A doctor may recognize the need and have a basic idea of how to solve the problem, but it will be material and design engineers who must consider the loads placed on the implants. Their calculations will ensure that the devices can be designed for sufficient structural strength.

Material choices must take into account biocompatibility with surrounding tissues, the environment and corrosion issues, and friction and wear of the articulating surfaces. For example, the amount of wear of the polymer cup against the metal or ceramic ball in a knee or hip joint is a major problem. Mechanical parts depend a lot on the material used and its ability to trick the body into not rejecting it. Titanium is used for artificial hip joints, acrylic plastic can make strong lenses for eyes, and silicones can be molded into shapes for reconstructing a face or replacing a finger.

Some body parts, such as the heart and kidneys, are far too complex for artificial parts to be made at a reasonable cost. Others are now routinely transplanted or implanted including:

- Pacemakers keep the heart beating at the required rhythm. An electrocardiograph machine can trace the electrical activity of the heart to determine if one is needed.

- Heart valves work like a snorkel. A ball made of Teflon moves in a titanium cage and makes seal.

- Hearing aids can help to magnify sound. Cochlear implants are used by hearing-impaired people who cannot be helped by hearing aids. They use a tiny microprocessor and a transmitter to send and receive signals from the inner ear.

- Speech devices, fitted in the mouth, can be used to replace a person's larynx. Computers are also used especially for those who suffer from Lou Gehrig's disease where much of the body is paralyzed. The user may touch keys to form a combination of symbols and letters.

- Intraocular lenses of plastic correct eye cataracts. When a person's vision becomes clouded the cataracts must be removed. An artificial lense made of an acrylic-like plastic is implanted behind the iris.

- Knee and hip joints repair athletic injuries. Playing fooball or any game that requires sudden starts and stops is hard on the knees. Twisting

Standards for Technological Literacy

14

Vocabulary
What is the meaning of the term *regeneration*?

Activity
Have students investigate the ingredients of soft drinks to determine the percentage and types of stimulants included.

Enrichment
Thirty minutes of moderate physical activity a day is ideal for most people. However, the exact amount depends upon age, condition, health, type of work, and genetics. The right amount of sleep per night is important, but the answer is not to take sleeping pills. They become ineffective after a few weeks and are addictive. Physical exercise is a much better solution. To help in getting a good night's sleep, stay away from stimulants like coffee and colas in the evening.

Activity
Have students discuss the physical activity guidelines for children and youths described by Health Canada at www.hc-sc.gc.ca/ hppb/paguide/guides/ en/index.html. Students can compare the exercise they do on a daily/weekly basis with the recommendations.

Useful Web site: Have students investigate the exercise guidelines for teenagers at http://Bodyteen.com (click "Exercise" on the home page and then "Benefits").

Community resources and services: List the types of emergency services that are available in your community. For each, provide the necessary contact information for a person needing the service, in the event of a crisis.

Standards for Technological Literacy

14

Discussion
Ask students if they know anyone who has been fitted with artificial body parts, such as a replacement hip, implants to replace teeth, or an artificial limb. Discuss the advantages and limitations of each device, as well as its life expectancy.

Vocabulary
Define the terms *bionic* and *prosthetic*.

Vocabulary
What is meant by the term *quality of life* as compared to the *quantity of life*?

Activity
The materials used for medical purposes are similar to those used in manufacturing appliances, machinery, and sports equipment. Refer to Chapter 4 and compare those materials with the ones mentioned in this section.

Safety
Invite a doctor or nurse from the emergency unit of a hospital to describe the most common types of accidents to the class. Ask students to describe the precautions they take to avoid sustaining injury in their daily lives.

motions are made worse by cleats or running shoes that keep feet firmly fixed to the ground when the upper body twists. Artificial joints may be made of stainless steel, titanium alloys or ceramics.

◆ False teeth have been made for centuries. Nowadays, titanium screws can be implanted into the jawbone. The screws fuse with the bone to become permanent. They form a support for a single artificial tooth or for a complete mouthful.

◆ Artificial limbs used to replace natural ones have limitations. A human arm is capable of 27 different motions, but an artificial arm may only have six motions.

If the body itself cannot be repaired, the injured person may be helped by external devices. If a spinal cord were to be damaged in a severe motor accident, it is probable that the person would never walk again. In this case, they might become mobile or even compete in a Special Olympics, using an athlete's wheelchair.

Community resources and services: Invite a doctor or nurse from the emergency unit of a hospital to visit the class for a discussion on the latest surgical procedures and the types of body parts that are commonly replaced.

Chapter 14 Review
Agriculture, Biotechnology, Environmental Technology, and Medical Technology

Summary

Food can be obtained by the natural methods (planting and harvesting) or the synthetic method (processing foods with chemicals and natural ingredients). Special machinery is often used to water, maintain, and harvest many natural foods. However, some natural foods must be dug or picked. Processing foods includes removing skins, bones, cobs, and shells and crushing or grinding. The food is then mixed with spices and preservatives.

Genetic engineering is the altering of the genetic material of living cells. It has allowed plants to be bred hardier, stronger, more pest-resistant, and with many other traits. Genetic engineering allows cells from one organism (plant or animal) to be transplanted into the genes of another organism.

Remember that many people oppose genetic engineering of crops and other foods because of their religious beliefs or personal convictions. With the increase of greenhouse gases in the atmosphere, is the possible disaster of tomorrow worth the convenience of today? Global warming, melting ice, increased forest fires, and lower water levels are the consequences of increased greenhouse gases in our atmosphere.

One way of reducing the negative effects of technology is to remember the three R's: reduce, reuse, and recycle. Recycling is the process of treating certain materials so they can be made usable again.

In order to maintain a healthy body, one must eat right, exercise regularly, and sleep amply. Eight hours is a good recommended amount of sleep each day. A healthy diet includes carbohydrates; vegetables; fruit; dairy; protein; and a low fat, oil, and sweets intake.

Medical technology makes living possible and easier on people who suffer from certain ailments. Medical robots, cameras, and less invasive surgeries let surgeons work more quickly and allow patients to heal faster. Prosthetic devices allow people with missing limbs to do things they would normally be unable to do, such as walking, picking up objects, and even competing in sports. Pacemakers help people with heart problems by keeping their heart beating at the proper rhythm.

Modular Connections

The information in this chapter provides the required foundation for the following types of modular activities:

◆ Technology and the Environment
◆ Water Management
◆ Soil Reclamation
◆ Ecology
◆ Dynamic Earth
◆ Environmental Issues
◆ Environment and Ecology
◆ Environmental Technology
◆ Aquaponics
◆ Hydroponics
◆ Plants and Pollination
◆ AgriScience and Society
◆ Recycling
◆ Home Recycling Station
◆ Landfills
◆ BioEngineering
◆ Biotechnology
◆ Body Systems
◆ Cell Structure
◆ Microbiology
◆ Genetics
◆ Animals
◆ Immune System
◆ BioRobotics

Test Your Knowledge

1. How does carbon dioxide affect temperatures on Earth?

2. Why has the amount of CO_2 in the atmosphere increased?

3. What alternative sources of energy are available in your area?

4. What example of an artificial environment do florists and horticulturists often use to grow flowers, shrubs, and herbs year round?

5. What discovery in 1953 by James Watson and Francis Crick started the modern age of biotechnology?

6. The majority of pollution comes from burning _____.
   A. garbage
   B. biomass
   C. fossil fuels
   D. methane

7. Bioprocessing is the use of _____ to break down and purify plants and animal matter.
   A. microorganisms
   B. plants
   C. animals
   D. nanobots

8. What is produced from bioprocessing?
   A. fossil fuel
   B. nitrogen
   C. chlorine
   D. methane

9. What are the three main fossil fuels?
   A. biomass, nanobots, and microorganisms
   B. methane, nitrogen, and oxygen
   C. greenhouse gas, methane, and acid rain
   D. oil, gas, and coal

10. The ozone layer protects us from _____.
    A. nanobots
    B. ultraviolet radiation
    C. greenhouse gases
    D. nuclear energy

Answers to Test Your Knowledge Questions
1. Carbon dioxide creates a shield around the earth that allows heat from the sun into our atmosphere, but it does not allow it to radiate back into space. The heat is then trapped in the earth's atmosphere causing a steady rise in temperature.
2. The CO_2 levels have risen due to the burning fossil fuels.
3. Possible answers include hydroelectric, solar, wind, tidal and wave, and plant and garbage power.
4. A greenhouse.
5. The discovery of the structure of DNA.
6. C. fossil fuels.
7. A. microorganisms.
8. D. methane.
9. D. oil, gas, and coal.
10. B. ultraviolet radiation.

Apply Your Knowledge

1. It takes about a year for garden waste to change into good compost on a compost heap. Design and make a composter that speeds up this process. It should be esthetically pleasing and environmentally sound.

2. Schools and businesses often produce a lot of wastepaper. Design and make a system that will convert this waste into a yeast-based animal food.

3. Design and make a hydroponics unit for a location that needs attractive plants without continuous maintenance. Technologies will be different depending on where you live.

4. Suppose that two small towns wanted to develop a new source of power. One town was in Central Africa and the other in North America. Describe the type of power that each would choose if they were to use renewable resources. Comment on whether it would be low in capital cost, whether local materials could be used, and whether local skills would be employed. How might it be possible for the community to seek a collective decision on the type of power?

5. Find out which agencies in your region are responsible for environmental protection. Describe the main issues they are dealing with such as the clean up of rivers, wildlife preservation, and the fight against toxic substances.

6. Form two debating teams in your class to debate the following topic. "New technologies allow human beings to exploit nature. It is always done for short-term gain at the expense of polluting and depleting a portion of our biosphere in the process."

7. Form two teams to debate the following topic. "The new methods of genetic engineering are, by definition, eugenic instruments. Whenever DNA is recombined, decisions are made as to what should be altered, inserted, or deleted from the hereditary code. In other words, decisions are made about what genes to preserve and which bad ones should be eliminated. Is something really defective or simply an alternative?"

8. Almost all North Americans eat a fast-food meal sometime in our lives, perhaps several times a week. We may also have heard someone say that fast-food is bad for your health. To find out whether this is true for you, answer yes or no to the following quiz questions concerning the food you ate yesterday.
   A. Did you eat three servings of vegetables and two of fruit?
   B. Did you eat three servings of low fat dairy products?
   C. When you ate grain was it whole-grain (brown) rather than refined (white)?
   D. Did you restrict you consumption of sugar and fats?
   E. Did you exercise for at least 30 minutes?
   If you answered "yes" to these questions, then you can eat a fast-food meal occasionally and not put on excess weight.
   Check out which of the following meals are offered by your favorite fast-food restaurant.
   - Regular hamburger with lettuce and tomatoes (not double hamburger with cheese and special sauce)
   - Regular chicken, roasted or baked, or chicken fajitas (not deep-fried or breaded)
   - Pizza with thin crust, mushroom, eggplant, and peppers (not deep-dish with sausage, pepperoni, and extra cheese)
   - Baked potatoes (instead of french fries)
   - Salads that have lots of raw vegetables (not ones that come with dressings or lots of macaroni and potato salads)
   - Low-fat desserts such as low-fat yogurt with chocolate sprinkles (not hot fudge sauces and whipped toppings)

   Which of these low-fat alternatives are offered and which will you choose next time you eat at a fast-food restaurant?

9. MyPyramid has six categories. Organize a page with seven columns. The first column will list the foods that you eat (toast, eggs, etc.), and the other six are for the six categories in MyPyramid. Each time you eat something, estimate whether it is a full serving or a fraction of a serving, and enter the amount in the appropriate column. At the end of the day, total up the number of servings you ate from each food group. What changes will you make for tomorrow?

10. Three important needs of your body are a healthy heart, building strong bones, and strengthening muscles. Prepare food at home that is good for these three needs:
    A. For a healthy heart, select a recipe for food that is low in fat.
    B. To build bones, select a recipe that is calcium-rich.
    C. To strengthen muscles, select a recipe that will give you high-quality protein.

11. What methods are available in your community to recycle electronic equipment? Which nonprofit charity groups accept donations? Check out the following Web sites to investigate recycling: www.freecycle.com, www.think-food.com, www.dell.com/recycling, and www.rogers.com.

12. Use the Internet to investigate ways in which the following products are recycled: aluminum foil and plates, antifreeze, appliances, asphalt, batteries, bottles, books, boxes, cans, cars, CDs and DVDs, cellular phones, clothes, concrete, furniture, medicines and pills, metals, oil, paint, paper, plastics, and reading glasses.

13. Use the Internet to investigate new techniques and devices being used to prevent or cure illnesses.

14. Research one career related to the information you have studied in this chapter and state the following:
    A. The occupation you selected.
    B. The education requirements to enter this occupation.
    C. The possibilities of promotion to a higher level at a later date.
    D. What someone with this career does on a day-to-day basis. You might find this information on the Internet or in your library. If possible, interview a person who already works in this field to answer the four points. Finally, state why you might or might not be interested in pursuing this occupation when you finish school.

Constructing large buildings, such as high-rises, and smaller buildings, such as single-family homes, requires the skills of different trade workers. These workers range from carpenters to operators of large, heavy machines. A bulldozer is needed to move earth (top), and a crane can be used to lift building materials to the higher floors or roof of a home (bottom).

Chapter outline

Student Activity Manual Resources

DMA-25, *Fundraising*
ST-3, *Using a Collage to Generate Ideas*
ST-36, *Cutting and Shaping Thin Plywood*
ST-37, *Cutting and Shaping Thin Aluminum*
ST-38, *Cutting and Shaping Acrylic*
ST-42, *Tools and Materials*
ST-51, *Safety—Managing Risk*
ST-4, *Identifying User Needs and Interests*
ST-5, *Writing a Design Specification*
ST-7, *Using a Gantt Chart to Plan*
EV-1, *Evaluating the Final Product*
EV-2, *Unit Review*

DMA-26, *Getting a Job*
ST-91, *Choosing a Career*
ST-92, *Investigating Résumés*
ST-93, *Using the World Wide Web to Investigate Career Opportunities*
ST-94, *Writing a Cover Letter*
EV-1, *Evaluating the Final Product*
EV-2, *Unit Review*

Technology offers a wide range of careers in a large number of fields in the world of work.

Instructional Strategies and Student Learning Experiences

1. Have students choose an item, such as a book, bicycle, jacket, radio, etc. Then have them describe the steps in producing the item as it moved through the primary, secondary, and tertiary sectors.

2. To help students review careers in the primary, secondary, and tertiary sectors, write the names of various careers on index cards. Have students pick a card and identify the sector and occupational cluster from which it is taken.

3. Have students read articles on employment trends. Have students write a paragraph about how they feel these trends might affect their own career choices.

4. Arrange for students to participate in a "shadowing program" where they spend a day at work with people who have careers in which students are interested. Have students write reports about their experiences describing their impressions of the world of work.

5. *Career Ladder,* reproducible master. Have students select a career. Have them investigate the career by talking with people in that field, visiting the library, talking with a guidance counselor, or exploring the Internet. On the ladder, have students fill in jobs that could be obtained at each educational level. Have them share their career ladders with the class.

The World of Work

Discussion
Discuss the variety of work that is encompassed by the primary, secondary, and tertiary sectors of business and industry.

Career connection
Have students identify specific careers in each of the primary, secondary, and tertiary sector industries. Have students use the Internet to find out the training and licensing requirements to enter a career in one of the sectors.

Key Terms

artisan
assembly line
automation
career
Computer Numerical Control
(CNC)
degree of freedom
entrepreneur
factory
Flexible Manufacturing System
(FMS)

Industrial Revolution
mass production
primary sector
production system
raw material
robot
secondary sector
tertiary sector

Objectives

After reading this chapter you will be able to:

◆ Differentiate between primary and secondary materials processing.

◆ Compare traditional and modern manufacturing processes.

◆ State the purpose of mass production and identify its advantages and disadvantages.

◆ Describe the elements of a production system.

◆ Describe how new technologies have replaced, outmoded, or created jobs.

◆ Explain the recent growth of the tertiary sector.

◆ Discuss the range of careers in the primary, secondary, and tertiary sectors.

◆ Design and make a product that will raise public awareness of an environmental issue.

The wood used to build the chairs you have in your home came from a tree growing in a forest. What is involved in changing a tree into a chair? From **Figure 15-1**, you can see that producing a chair from a tree involves many steps. The steps in producing any product are organized into three sectors:

◆ *Primary* (first) *sector:* obtaining and processing raw materials.

◆ *Secondary* (second) *sector:* changing raw and processed materials into a product each of us can use.

◆ *Tertiary* (third) *sector:* delivering and servicing the product.

1 Trees are felled in the forest, transported to a sawmill, and converted into boards.

2 Designers submit alternative chair designs. The best is developed in detail. Working drawings are produced.

3 *FOAM covered with wool cloth* / *maple dowel joint* / *urethane finish*
Models and prototypes are made. Style, materials, and construction techniques are reviewed.

4 Prototypes are tested to determine the chair's strength and durability. Weaknesses in the design are corrected.

5 A mass production system is planned. Skilled workers operate and maintain machines to mass-produce the chair.

6 Office staff keeps track of materials and supplies. Salespersons receive orders and notify warehouse of addresses for delivery.

7 During the life of the chair, it may become dirty or broken. Service personnel may visit a home to clean or repair the chair.

Figure 15-1   Changing a tree into a chair takes much planning and work.

Useful Web site: Students can use the dictionaries at www.techtionary.com and www.hyperdictionary.com to find the definition of any new technical and technological terms used in this chapter.

The Primary Sector: Processing Raw Materials

The primary sector is concerned with obtaining and processing *raw materials*. These materials come from nature in one form or another. Some materials are renewable and can be reproduced continually. Others are nonrenewable; once used, they cannot be replaced.

Renewable Materials

Renewable raw materials come from plants or animals. Some are found in a wild state, while others are produced on farms. For instance, forests and tree farms provide wood for the lumber industry, **Figure 15-2**. The fishing industry harvests fish and other marine life from oceans, lakes, and waterways, **Figure 15-3**. Wild animals are hunted and trapped for their furs, hides, and meat.

Nonrenewable Raw Materials

Nonrenewable raw materials are of three types: fossil fuels, nonmetallic minerals, and metallic minerals. Fossil fuels include coal, peat, petroleum, and natural gas. These were once living organisms that decayed and were trapped in layers of sediment in the oceans and on land. Over millions of years, the heat and pressure from the sediment created mineral fuels. In addition to providing fuel, petroleum is the raw material used in making many products. These include plastics, synthetic fibers, and drugs.

Nonmetallic minerals include such construction materials as sand, gravel, and building stone. They also include abrasive materials, such as corundum. Among the metallic minerals are those from which iron, copper, and aluminum are extracted. Metallic minerals, also called ores, are removed from the earth by underground mining or surface mining, **Figure 15-4**. Drilling is

Figure 15-2 Lumberjacks harvest trees, a renewable raw material. Left—Logs are floated down rivers to be processed into usable form. Small logs may go to a mill to be processed into pulp for paper. Large logs will go to a sawmill to be sawed into lumber. Right—Log booms are towed by tugboats. (NFB)

Useful Web site: Have students investigate the number of employees, in 2002, involved in natural resource-related industries. One source for this information is the US Census Bureau Web site at www.census.gov, "Section 18 - natural resources."

Discussion
Ask students if they know someone who works in the primary sector. Answers will vary according to whether the community is largely industrial or agricultural. However, the response should be used as a springboard for discussion. For example, the responses will probably illustrate that various communities differ in terms of available employment. People who are interested in a particular career may need to relocate.

Vocabulary
What does the term *career* mean?

Activity
Have students select one of the careers listed in Figure 15-6 and identify the education required, employment opportunities, and average salary.

Safety
Discuss the hazards and risks faced by people working in the primary sector and how they manage the risks.

another method of extracting nonrenewable materials, **Figure 15-5**.

Careers in the Primary Sector

People with careers in the primary sector frequently work outside, **Figure 15-6**. A *career* is an occupation, a way of making a living. Farmers plant, fertilize, cultivate, and harvest crops. Ranchers breed, feed, and care for animals. Horticulturalists grow and maintain plants, shrubs, and trees.

Forestry workers cut, transport, and process trees for papermaking, construction, and furniture manufacturing. Geologists locate minerals and fossil fuels. Miners and oil riggers operate the machinery to extract raw materials. Oceanographers study the plant and animal life of the oceans. Environmentalists try to find solutions to problems relating to land use, pollution, conservation of natural resources, and the preservation of wildlife.

Figure 15-3 Colombian fishermen are hauling in their nets. (Ecritek)

Figure 15-5 Offshore drilling rigs drill wells in the ocean's floor. (Seeds)

Figure 15-4 Coal lying near the earth's surface can be strip-mined. (Seeds)

The Secondary Sector: Manufacturing Products

The secondary sector changes raw and processed materials into useful products. It is concerned not only with the manufacture of products but also with the construction of structures. Today's manufactured products include computers, jet planes, glues, lasers,

Community resources and services: Invite a person working in the primary sector to describe his/her career to the class.

Primary Sector Careers		
Occupational Cluster	**Professionals**	**Skilled Workers**
Agriculture	agriculture economist agronomist animal physiologist botonist soil conservationist soil scientist veterinarian	animal breeder animal inspector beekeeper dairy farmer farm equipment operator field crop farmer fruit farmer livestock rancher poultry farmer ranch worker
Horticulture	landscape architect	greenhouse manager groundskeeper landscape contractor landscape gardener lawn service worker nursery worker tree surgeon
Forestry	botanist	chainsaw operator logger sawyer
Marine science	oceanographer marine biologist marine geologist	fisher fish farmer
Natural resource extraction	geologist petroleum engineer	miner oil rigger
Environmental control	conservationist ecologist meteorologist urban planner wastewater treatment engineer	

Figure 15-6 Which of these primary sector careers would you choose?

plastics, medications, photocopying machines, and bubble gum. Construction involves the building of structures that people use for living, working, traveling, and playing. Among these structures are houses, office towers, and sports stadiums. Also included is the construction of road tunnels, bridges, towers, and dams.

The secondary sector has changed in the last 200 years. It has evolved through three stages:

◆ The individual artisan.

◆ Mechanization and mass production.

◆ Automation.

The Individual Artisan

Before the eighteenth century, the entire production of articles was in the hands of individuals. One person, an *artisan*, made products. Artisans used hand methods. Each product evolved by trial and error. One generation passed on acquired skills to the next through an apprenticeship system. Each artisan was normally responsible for every step in the process. He or she did everything from obtaining the raw materials to completing the finished product. For instance, **Figure 15-7** illustrates the work of a chair maker. This example of an artisan, working alone, produced Windsor chairs.

Useful Web site: Students can learn about the work of a living artisan and how a chair is made from green wood by visiting the Web site of Steve Vickers at www.workingwoodlands.info/bodgingintro.htm.

Community resources and services: Arrange for students to visit a local museum to view exhibits related to how things were made a century or more ago.

Standards for Technological Literacy

Designing and making
Discuss the effects of mass production on the design of products.

Vocabulary
What is meant by the statement "mass production reduces the unit cost of each item"?

Activity
Have students make a list of all the manufactured items in one room of their home. Discuss the range and types of mass-produced products they own and use.

Safety
Discuss the hazards and risks faced by people working in factories and how they manage the risks.

Figure 15-7 Steps in making the legs and spindles for a Windsor chair. A single artisan working alone made the entire chair.

Mechanization and Mass Production

The second stage in the evolution of the secondary sector divided the production process into specialized steps. Machinery replaced handwork. This was done to reduce the unit cost of the product. Through mass production, products could be made in large quantities. This change, occurring first in England, was known as the *Industrial Revolution*.

Early machines, powered by steam, replaced the muscle power of workers and animals. Burning coal produced this steam. The English had known the uses of coal for several centuries. However, Watt's steam engine was the first machine to convert the chemical energy of coal into steam and then into mechanical energy. Mechanical energy then powered the machinery.

Factories

As a result of the increased use of steam-powered machinery, production had to be located in larger buildings called *factories*. Towns and cities developed rapidly as people moved to live near these factories.

Links to other subjects: History—Have students browse the Internet to find information about the Industrial Revolution that began in England. What were the conditions that led to this revolution?

Technology and society: What were some of the social conditions that existed during the late eighteenth and early nineteenth century in towns that industrialized?

At first, factories had to locate near the coalfields. Moving the coal long distances was too costly. After the mid-nineteenth century, however, transportation became cheaper. Coal could be more readily transported. Factories could be built almost anywhere.

The factory system expanded because it was thought to be more efficient. Efficiency means that good use is made of energy, time, and materials. This efficiency was achieved through the use of new methods of production. These included:

◆ Division of labor.

◆ Use of machines to build parts.

◆ Use of interchangeable parts.

◆ Introduction of mass production and assembly lines.

The division of labor means each person is assigned one specific task in the making of a product. Through constant repetition, the worker becomes skilled in that task. The task can be performed at a more rapid rate. With specialized machinery, the same part can be made again and again with no variation in size or shape.

Because more products could be produced in the same amount of time, the cost for producing each item dropped. This meant the item could be sold for less, too. Having parts that were all alike was important. It meant any new part could be substituted for one worn or broken. Such parts are said to be interchangeable. Each new part is identical to the old one. As we shall see later in this chapter, assembly lines further increased the efficiency of a factory.

Factory Working Conditions

When efficiency was the only concern of factory owners, working conditions were often poor. Machinery improved, but little was done to improve life for the workers. The factories were built very close to one another. Their interiors were dark and dismal. No provisions were made for proper lighting. Children as young as four or five cleaned the floors. Older children worked twelve hours a day cleaning and maintaining machinery. They were poorly fed, ill-treated, and completely at the mercy of the factory owners.

Assembly Lines

Use of assembly lines made factories more efficient. A continuous assembly line could quickly produce a large number of identical items. An *assembly line* allows assembly of parts in a planned sequence. This is possible only when parts are manufactured to uniform standards.

Assembly usually begins with one major part. Other parts are added as the product moves to other stations along the line.

Making a large number of products on an assembly line is called *mass production*. Henry Ford first used it in 1914. Ford's assemblers worked side by side in long lines. The parts were brought to them. Each worker had only one assigned task as the Model T cars moved slowly by on the assembly line. His line cut production time per car from 13 hours to one, **Figure 15-8.**

Standards for Technological Literacy

19

Discussion
As many manufactured goods are being made offshore, local companies are changing what they do from being labor intensive to working smarter. What is meant by "working smarter" and what does this mean to your future education?

Discussion
Why was Henry Ford's system of manufacturing cars so revolutionary in the early part of the twentieth century?

Enrichment
Many mass produced goods are now made offshore, in places like China, Mexico, and India, where they can be produced at lower cost. People in these countries may be working for as little as $1.00/hour. Most North Americans are happy to buy more goods for less cost. However, others would like to use protectionist methods to keep manufacturing jobs at home.

Activity
Research the life of Henry Ford. What were his strong points and weaknesses?

Useful Web site: Have students learn about the design, development, mass production, and marketing of a new product by reviewing the Mono Bug Clamp case study in the Design Council's list of Millennium Products at www.designcouncil.info/educationresources/studies/index.html.

Technology and society: What are the advantages and disadvantages of increasing the amount and availability of cheaper, imported, mass-produced products, rather than encouraging more expensive home-based production?

Figure 15-8 Henry Ford perfected the first assembly line to produce his Model T car. (Ford Motor Company)

Today, most automobile plants are huge assembly points. Each plant contains a number of small assembly lines. Some lines produce subassemblies. These are components having a number of parts. Engines, gearboxes, and suspensions are examples. Subassemblies are later assembled into complete automobiles.

While the assembly-line techniques pioneered by Ford increased the rate at which products could be produced, they were still limited. Humans could work only so fast. To produce products faster, one must use an automated assembly line.

Automation

In this stage, the worker builds, monitors, and maintains machines that make products. *Automation* is when computers or other machines control machine operations. Machines have been developed that:

◆ Perform many different shaping operations.

◆ Can receive a number of parts and assemble them at high speed in the correct sequence.

Typical is the machinery used to package macaroni, **Figure 15-9**. The display packages have to be opened, filled, sealed, weighed, and boxed. Almost all of the operations are automated. Even on final inspection, machines check each package for the correct amount of macaroni.

Robots

The 1980s brought further development in automated manufacturing. Industrial robots revolutionized production lines. *Robots* are computer-controlled that can be programmed to perform various functions, such as the following:

◆ Handling — loading and unloading components onto machines.

◆ Processing — machining, drilling, painting, and coating.

◆ Assembling — placing and locating a part in another compartment.

◆ Dismantling — breaking down an object into its component parts.

◆ Fixing — assembling objects permanently by welding or soldering.

◆ Performing dangerous tasks and operations.

◆ Transporting materials and parts or delivering mail.

To understand the operation of a simple robot, imagine that you have

Useful Web site: Students can learn more about robots and view a "robot timeline" at http://inventors.about.com/library/inventors/blrobots.htm.

Automated Packaging

A — Step 1. Boxes arrive flat. One end of the box is opened and one end is sealed.

B — Step 2. Boxes are filled, and then open end is glued and sealed.

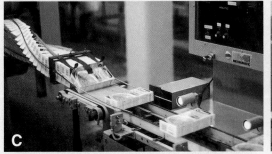

C — Step 3. A machine weighs each package. Packages that are too light or too heavy are sorted automatically.

D — Step 4. Packages are placed in cartons. They are moved to storage and then shipped.

Figure 15-9 An automated packaging line. (Ecritek)

Resource
Activity 15-2, *Individually Made Versus Mass-Produced Products*

Designing and making
The device at the end of a robot arm should be designed specifically for the tasks it is going to perform, such as picking up a part and moving it to where it will be altered in some way. What items can be moved by each of the following end-of-arm toolings for robots: grippers, magnets, and suction cups?

Discussion
What are the similarities and differences between the mechanized system used by Henry Ford and a robotic system?

Enrichment
Robots aim to increase productivity by reducing manufacturing and assembly time, the number of defects, the space needed for production, and the inventory of parts.

been blindfolded and tied to a chair. You are able only to move one arm and to rotate at the waist. Your one arm has joints at the shoulder, elbow, and wrist. Robots also have joints much like a waist, shoulder, elbow, and a wrist that can move in two or three directions. Each joint, or direction of movement, in a robot arm is called a *degree of freedom*. Most robots have five or six degrees of freedom, **Figure 15-10**.

Robots can even be found on farms. A milking robot can be located in a barn next to a cow pasture. If a cow is feeling heavy with milk, it heads for the robotic milking stall, **Figure 15-11**. A chip embedded in

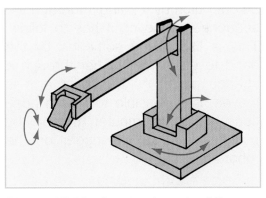

Figure 15-10 Arrows show the different movements of a robot arm. Each movement is known as a "degree of freedom."

Links to other subjects: Literature—Have students read William Gibson's famous science fiction book, *Neuromancer*. In 1986, his book anticipated what is known today as virtual reality. Have students read one or more of Isaac Asimov's robot short stories in *The Complete Robot*.

Standards for Technological Literacy

19

Figure 15-11 This picture illustrates the automatic milking machine. (Lely Astronaut robotic milking system)

the cow's collar sends a signal to the stall's computer, which recalls how the udder is shaped on a particular cow. The robotic machinery scrubs the udder clean. An arm equipped with four suction cups positions itself under the udder. At the same time, the cow is fed. If a cow enters the stall prematurely, the computer opens the exit gate immediately.

A robot can be "taught" to perform an operation by leading it through the sequence of moves it has to follow. This is like taking someone by the hand to guide him or her through a strange place. As the robot arm is moved, it is possible to record positions into the computer memory by pushing a button or trigger on the robot arm.

Advantages of Automation

Robots have several advantages over humans. They work better in hot, noisy, or dangerous situations. Robots do not take coffee breaks, go home ill, or sleep. They continue working 24 hours a day and will operate for thousands of hours before they require maintenance, **Figure 15-12**. Automation,

Figure 15-12 Robots are programmed to assemble automobiles on this automated assembly line. (Ford Motor Company)

including robotics, has certain advantages for industry:

◆ It has improved the quality of work. A machine set up to produce one product to a high standard will continue to produce parts to the same standard. The aim is to have zero defects. All parts are perfect.

◆ It has increased production. Thus, the cost of each item is reduced. Each worker can produce more products in the same amount of time.

◆ It has decreased the amount of waste material and the

Useful Web site: Have students browse www.androidworld.com to learn about the many different types of robots being developed. Students can print one picture, or information they find interesting, and share it with the class.

number of parts that need to be scrapped. Each machine produces parts accurately.

◆ It has reduced the amount of time needed to teach a worker a new task when a new product is being produced. Each worker is responsible for only one small task.

Disadvantages of Automation

Automation brings increases in production and savings in time and money. However, it still has some disadvantages:

◆ Now and then, a machine will malfunction. It may have been set up incorrectly. Products will be inaccurate or substandard. Such mistakes, if not detected, can be very costly. Thousands of defective items may have been produced.

◆ There is a loss of jobs as machines and robots take over tasks in the workplace.

Certain industries are more likely to be affected by automation. Those concerned with the assembly of finished goods, such as the car industry, rely more and more on automation. Without it, they cannot compete for world markets. They must modernize to keep costs down and improve quality.

How Automation Affects Jobs

The prospect of being replaced by a machine frightens or angers many people. Yet, in the long run, machines are likely to create more jobs than they destroy. Anyone doubting this should think about times when similar revolutions have taken place. In the late nineteenth century, farm mechanization displaced more than two thirds of the farmhands. Until 1880, more than half of all workers in advanced nations worked on farms. At that time, tractors and other machines were developed. These enabled one person to produce what ten or more had done. In the last 100 years, about 90% of the farm jobs have disappeared.

Workers leaving the farms entered factories. They soon found that there, too, machines were replacing them.

Ford's mass production was another laborsaving technology. It was again feared that the assembly line would cause unemployment. This was not the case. Mass production created mass markets. Most people could now buy goods once affordable only to the rich. Products began to sell in the millions.

CNC and FMS

In many factories computers control machinery. The computer performs only a simple task. However, once its program of instructions has been written, it can do the same job again and again. Provided they do not break down, automatic machines can work day and night. Computer-controlled machines are often called *CNC* machine tools. The initials stand for *Computer Numerical Control*.

The most modern factories group their machine tools together into *FMS* cells. The letters FMS stand for *Flexible Manufacturing System*. An FMS cell consists of a number of CNC machine tools. Each machine is

Standards for Technological Literacy

Designing and making
How has the introduction of CNC machines changed the design of products?

Discussion
Have students discuss the advantages and disadvantages of automation.

Vocabulary
What is meant by the phrase "mass production creates mass markets"?

Activity
Have students do a library search or browse the Internet to gather information about the Luddites and their leader Ned Ludd.

Useful Web site: Have students investigate the process of rapid prototyping, the fabrication of physical objects directly from CAD data sources. A useful starting point is http://home.att.net/~castleisland.

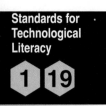

Standards for Technological Literacy

1 19

Designing and making
What is the role of the designer in a production system?

Vocabulary
Robots operating in CNC and FMS systems are *reprogrammable.* What does this mean and why is it important?

supplied with metal pieces, called blanks. These are carried either on moving conveyor belts, or by special robots called Automated Guided Vehicles (AGVs).

As the metal blanks arrive at the FMS cell, robots lift them from the conveyor or AGV onto the machine tools. Each CNC machine tool completes its task. A robot transfers the machined blank to the next machine in the FMS cell. In this way, each metal blank is machined into a finished product. Then, a robot lifts the completed product onto another AGV or conveyor. From there, it is either shipped or stored. A major advantage of an FMS cell is that it can be easily reprogrammed. It can then manufacture a totally different part or product.

A factory with a number of FMS cells may have as few as three people operating it during the daytime. At night, only one or two people need be present. They check the computers and supervise the loading of the blanks. They also supervise removal of the finished parts.

Factories vary greatly in the extent to which they have been automated. There is also great variety in the type of product they make. However, all factories are similar in two respects. They must have a management system and a production system.

The Management System

Operating a factory requires the combined effort of a team of people. The team may include:

- A president who has the responsibility for all decisions.
- A number of vice presidents who are in charge of major divisions of the company.
- Managers who direct particular operations.
- Department heads that supervise one major activity.
- Supervisors who direct the workers.
- Workers who complete tasks assigned to them.

Everyone needs to know who is responsible for each job. A chart is drawn to show lines of authority, **Figure 15-13.**

The Production System

The *production system* involves five basic operations.

- Designing — making original plans and drawings of products that satisfy consumer demands.
- Planning — organizing a system in which personnel, materials, and equipment can work together.
- Tooling up — acquiring and setting up tools and machines for production.
- Controlling — production using machines to make the product.
- Packaging and distribution — packaging, storing, and transporting the products to wholesalers and retailers.

Community resources and services: Invite one or more workers or professionals employed in manufacturing industries to visit your class and discuss their role in a production system.

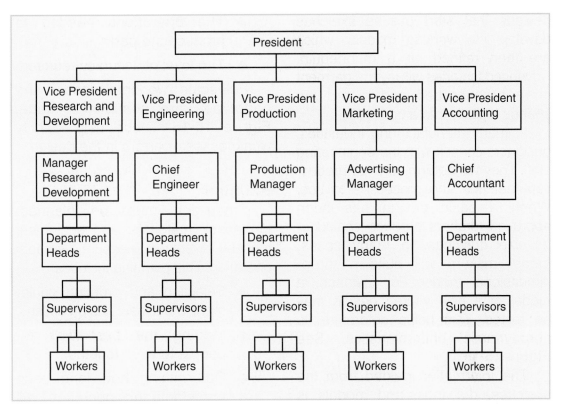

Figure 15-13 This organizational chart shows how many companies show the lines of authority and responsibility.

Standards for Technological Literacy

Designing and making
What general questions must a designer ask when developing a new product?

Discussion
Have you ever had an idea for a different kind of item that you thought other people might buy? How could you find out if you were right?

Vocabulary
What is the meaning of the phrase *market research?* Which people in a company would be responsible for conducting this research?

Designing

Customers want or need a variety of products. Manufacturers satisfy these needs. Before starting to design a product, a manufacturer will determine what buyers want. This will involve market research. The product will not be made if the number of potential sales is not large enough to make a profit after costs are paid.

Designing is the responsibility of engineers, drafters, and, sometimes, industrial designers. Designers first learn the needs of potential customers. Relying on research, they make preliminary sketches of the product. Next, they rework and refine these sketches, **Figure 15-14**. In the fashion industry,

Figure 15-14 Many designers make initial sketches on paper. They then rework and refine the sketches using computer-aided design (CAD). (Cadkey)

Technology and society: What problems occur when a company designs and manufactures a product that does not sell?

Standards for Technological Literacy

2 **19**

Designing and making
In what ways must a designer coordinate his/her work with a planning department?

Enrichment
Automation is increasingly used by the fashion industry. Pieces of cloth may be laid out on a scanner bed and recorded digitally; in the same way a FAX machine scans an image. Later, a computer operated cutter slices through many layers of cloth at a time. In other departments tests are made on various materials and using a variety of laundry detergents. This process helps to determine the wear qualities and how various procedures, such as stone washing, affect the material.

Vocabulary
What is the difference between a *model* and a *prototype*?

Activity
Assume you were in charge of planning operations to make a skateboard, a surfboard, or a kitchen cutting board. What would be some of your specific plans for each of the four steps listed under "Planning?"

designs that start out as sketches develop into working models, which are then refined on a mannequin. Advanced factories will lay the garment on a scanner bed and record the shapes digitally into a computer.

In many manufacturing industries, designers consult with the engineering staff. They will determine what production methods and processes to use. When a basic design has been approved, drafters produce scaled or full-size drawings. In many cases, model makers will build a three-dimensional model. For instance, a model of a new aircraft design may be tested. This helps determine its aerodynamic characteristics. See **Figure 15-15**.

The information gained from the sketches, drawings, and models is then converted into working drawings and specifications. These drawings show the exact size and shape of each part. Specifications include information about:

- The materials to be used.
- The number of parts to be made.

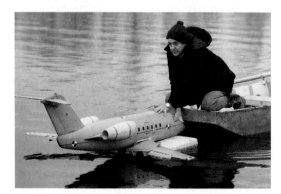

Figure 15-15 A model aircraft is tested to determine the effect of crash landing on water. (Bombardier)

- The operations needed to produce the part.
- The level of accuracy required.

Full-scale working models and prototypes are sometimes made, **Figure 15-16**. These help identify weaknesses or errors in the design.

Planning

Production must be planned. Personnel, materials, and equipment must be combined to ensure a smooth operation. This planning includes:

- Selecting and ordering equipment, machines, and processes.
- Finding the best way for people to work together.
- Determining how long each manufacturing operation will take.
- Gathering information on production costs.

An engineer will consider a variety of ways to complete each operation, **Figure 15-17**. He or she will select the most efficient one. The decision will be based on the time involved, the cost, and the quality and quantity of the product.

Tooling Up

Tools and machines may sometimes be purchased. At other times, tool and die makers must design and make them.

Tooling up involves four steps:

- Deciding which tools, equipment, and machines will be needed.

Useful Web site: Have students learn more about prototyping from the Prototypes Plus™ Web site at www.prototypesplus.com.

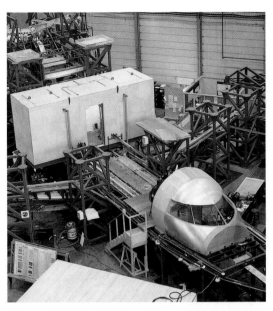

Figure 15-16 A full-scale working model of a plane is under construction. (Bombardier)

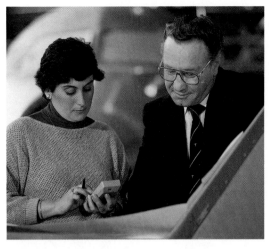

Figure 15-17 Engineers calculate costs of production before a product is made. (Bombardier)

◆ Selecting and ordering machines, tools, and equipment from manufacturers.

◆ Designing and making special equipment.

◆ Supervising the installation of machines, tools, and equipment and the organization of a trial production run, **Figure 15-18**.

Controlling Production

Production must be organized. Then materials and parts move efficiently from one operation to the next. This sequence begins in the receiving and storage area. This is where raw materials are kept. From storage, materials move through various stages of processing and assembling. A variety of procedures may be used. These may include machining, bending, shearing, folding, forming, and casting, **Figure 15-19**.

Each of the machines used in processing and assembly needs a power supply. Each also needs a system for controlling fumes, dust, and waste.

Figure 15-18 A machinist checks out the setup of a production machine to make certain parts are machined accurately. (Bombardier)

Standards for Technological Literacy

19

Designing and making
Why is quality control important to both the manufacturer and the consumer?

Discussion
Discuss the type of waste created, during processing or production, by the following industries: food industries, sawmills, and garment factories.

Vocabulary
What is the meaning of the phrase *quality control*?

Useful Web site: Students can learn more about controlling production by visiting www.2020software.com.

Discussion
Discuss what is happening in each operation shown in Figure 15-19. Include what waste, if any, resulted from the processing. How thick should material be to be processed in this way?

Vocabulary
Define each of the nine words describing the operations shown in Figure 15-19.

Activity
Which of the procedures shown in Figure 15-19 would be used to produce a: toolbox, heavy pendant, test tube holder, and saw bench table?

Activity
For each operation listed, name one product whose construction relies on the process: drilling a hole, milling a slot in a surface, shaping a flat surface, bending sheet metal at a right angle, shearing a shape out of sheet metal, forming a shape out of sheet metal, forging a shape using pressure, and casting hot metal into a mold.

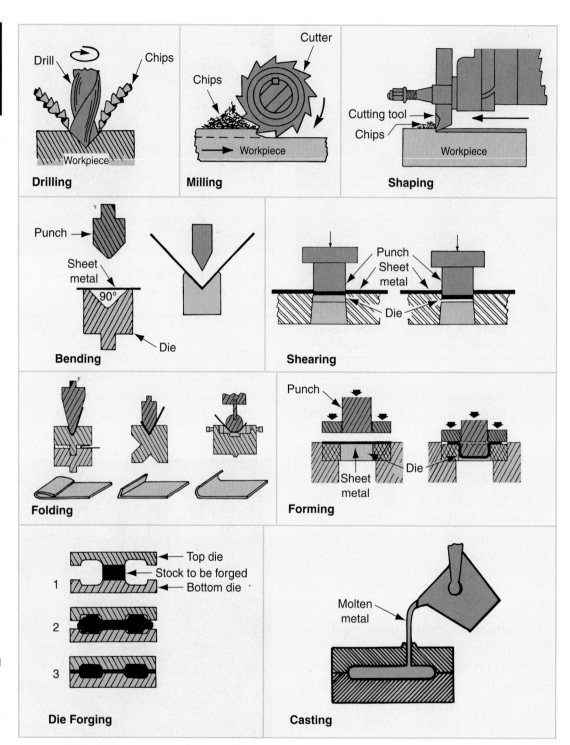

Figure 15-19 Production of a product may include one or several of these processes to change the shape or size of parts made of metal.

Useful Web site: Students can learn about the variety of machine tools by accessing http://globalspec.com/ProductFinder.

Production control finds ways to reduce waste. One type of waste is the space excess material or parts take up in storage. The ideal system is called "just-in-time delivery." Materials or parts arrive only a day, or even hours, before they are needed. One computer company runs their factory with only five or six hours of inventory on hand so warehouse space is drastically reduced. Another waste is time lost when workers or machines have nothing to do.

Production control is also in charge of:

◆ Ordering, routing, and scheduling materials.

◆ Dispatching job and work orders.

◆ Recording the performance of workers and machines.

◆ Taking corrective action when the production flow is interrupted.

Throughout the production process, a wide variety of inspection tools are used. Gauges check the sizes of parts. X rays check the internal structure of metal parts. The amount of inspection varies. In the manufacture of aircraft, it is important to check every part. However, for most consumer products, it is enough to check a small number of items from a large batch.

Packaging and Distribution

Products being shipped must be protected. Various forms of packaging are used. Packaged products must also be labeled. Bubble packaging, boxes, cartons, and crates are methods used to protect the product. They provide insulation and protection against moisture, weather, and rough handling. Labeling is needed so that the consumer can recognize the contents. Labels and other kinds of markings show the product, the name of the manufacturer, quantity, and directions for use and care. They also provide other special information.

Packaged and labeled products are usually stored in a warehouse to await shipment, **Figure 15-20**. They are organized in quantities that are convenient for handling, sorting, and counting. The machinery for handling bulk shipments includes conveyor belts, forklift trucks, and pallets. Pallets are wooden platforms on which the packaged products are placed. Then, a forklift can pick up the loaded pallet. Products may be loaded onto trucks, railroad cars, and ships. These vehicles transport products to wholesalers and retailers.

Figure 15-20 Parts and products must be warehoused until transported to wholesalers and retailers. (Bombardier)

Useful Web site: Students can learn how to design packaging by downloading *Guidebook #67: Designing Packaging* from the Small Business Town Web site at www.smbtn.com/books/gb67.pdf.

Standards for Technological Literacy

19

Safety
Discuss the hazards and risks involved with each of the processes shown in Figure 15-19, and how working carefully and using the correct procedures can manage the risks.

Designing and making
What general issues must the designer consider when developing the packaging for a product?

Discussion
Discuss with students why "just in time delivery" is important and how the use of computers and the Internet has helped.

Reflection
What kinds of protective packaging have you seen in products you have bought or packages you have received by mail? What are the features of packaging that encourage you to buy a product?

Activity
Have students make a collection of packaging, including examples that they like, dislike, think is effective, and think is ineffective. Discuss the criteria they used to make those decisions.

Standards for Technological Literacy

Safety
Discuss the hazards and risks faced by people working in a distribution warehouse and how they manage the risks.

Vocabulary
Define the term *apprenticeship*.

Activity
Have students select one of the secondary sector careers and research the job responsibilities, daily tasks, and the educational requirements. They should investigate whether the number of job openings in the chosen career is increasing or decreasing and why.

Resource
Reproducible Master, *Secondary Sector Careers*

Careers in the Secondary Sector

Most people employed in the secondary sector work in factories or on construction sites, **Figure 15-21**. Engineers carry out research to find new ways of changing raw materials into products. Market research analysts determine whether or not there is a market or need for the product. Factory workers install, operate, and maintain the machines that make the products. Management personnel oversee the production. They also ensure that the workplace is safe for workers.

People who work in the construction industry are responsible for the planning and building of homes, bridges, industrial plants, dams, hospitals, highways, pipelines, and shopping centers. Civil engineers and surveyors perform many tasks. They design and lay out structures, estimate costs, prepare material specifications. They survey building sites and organize work schedules. The work at a job site must be frequently supervised to ensure the structure follows the approved plans. Skilled workers specialize in a trade, for example, plumbing or carpentry. Laborers and hod carriers assist the skilled workers.

The Tertiary Sector: Providing Services

When was the last time you bought a hamburger, visited a library, or needed to have a checkup at the doctor or dentist's office? When did you last take clothes to be dry-cleaned or go to an amusement park? When you did any of these things, you made use of services in the tertiary (third) sector.

The tertiary sector is concerned with the servicing of products. It also

Secondary Sector Careers		
Occupational Cluster	**Professionals**	**Skilled Workers**
Manufacturing	drafter	assembler
	industrial designer	cutter
	industrial engineer	foundry worker
		instrument maker
	laboratory technician	jeweler
	market research analyst	machinist
	production manager	model maker
	quality controller	painter
	safety inspector	pattern maker
	supervisor	press operator
		sheet metal worker
		tool and die maker
		upholsterer
		welder
Construction	architect	bricklayer
	civil engineer	building inspector
	drafter	cabinetmaker
	electrical engineer	carpenter
	soil technologist	cement mason
	surveyor	contractor
		electrician
		floor covering installer
		glazier
		laborer
		painter
		paperhanger
		pipe fitter
		plasterer
		plumber
		roofer
		stone mason

Figure 15-21 How many people do you know who hold any of these jobs?

Useful Web site: Have students investigate the current number of employees in the secondary sector. One source for this information is the US Census Bureau Web site at www.census.gov.

Community resources and services: Invite a manufacturer or building contractor to speak with the class about the variety of jobs that exist in their sector.

provides services that add to the personal comfort, pleasure, and enjoyment of people.

Most people think of the service industry in a limited way. They see service as installing, maintaining, repairing, or altering products or structures. For instance, many of you use a computer at home. Once your computer is installed, it will require upgrades, troubleshooting, and regular maintenance. If you want to keep up-to-date with changing technology, you will need to upgrade your programs, operating systems, and hardware once every few years. If your computer crashes, you may need to hire a qualified computer technician to fix your computer. Other parts of the computer may fail. The fan may stop working, which would cause the computer to get too hot.

People who perform these services work for private companies. These companies aim to provide good service to their customers. They must compete with other companies who are providing similar services. They want to remain in business. To do so, they must also make a profit for their owners or shareholders.

Government agencies provide some services. These include education, health, and public works. This type of service is normally financed through local, state, or federal taxes.

Growth in Jobs

During the last 20 years, the number of service jobs has about doubled. There are reasons for this.

First, some workers cannot be more productive. A dentist is such a person. She or he can treat only a limited number of patients each hour. It is impossible to treat two patients at the same time. Thus, as population increases, the number of dentists needs to increase.

Second, many homes now have two wage earners. When people have more money, they consume more services but fewer goods. For instance, during his or her lifetime, a consumer may purchase three new television sets. The same person, however, may go to the bank once a week. She or he may eat at a restaurant twice weekly and use public transportation every day. Most products are purchased infrequently. However, many services are used daily.

Finally, the service sector faces little foreign competition. A television may have been manufactured abroad, but local people provide all services.

Looking to the Future

The growth in the service sector is likely to continue for three reasons. First, the aging population will need more health care.

Second, many businesses are now franchised. This means that one or more persons form a local company and enter an agreement with a parent company to use its name and provide its services. Many of the fast-food chains, like McDonalds, are franchised. There may be branches throughout the world. Franchised businesses generally grow quickly.

Discussion
Discuss the following statement and whether your local stores fit this description:
"In order to sell, it is necessary to know what people want and make it available to them in a pleasant and convenient way. The perfection of selling goods is to appeal to people individually and see the situation through their eyes. A top salesperson is very knowledgeable and can provide the right information to people who are price and value conscious."

Vocabulary
What is happening when a country is experiencing a *trade deficit*?

Activity
Make a list of services that you or your family members use over the period of one month. Why do you use the listed services? Where are they provided? Is it the type of services they offer or the people who work there that attracts you?

Activity
Have students use the Internet to find out which countries are experiencing a trade deficit and the size of the deficit. What are the long-term consequences of a country having a trade deficit?

Resource
Activity 15-3, *The Tertiary Sector*

Useful Web site: Have students investigate the current number of employees in the tertiary sector. One source for this information is the US Census Bureau Web site at www.census.gov.

Failure rates are also lower as a result of sound management practices.

Third, people like starting new small businesses providing services. This enthusiasm is likely to continue.

A person who organizes, manages, and assumes the risks of starting a business is called an *entrepreneur*. Small businesses are frequently operated by entrepreneurs who started their own enterprises. Entrepreneurs find unique ways to respond to the busy lifestyles of consumers. Most consumers are constantly looking for faster, simpler, and more convenient ways to obtain the goods and services they need. Entrepreneurs play a number of roles in the economy. They can:

◆ Create new products and services in response to consumer demand.

◆ Tailor the business to suit local needs and offer a quality of service, which might not be available from a large corporation.

◆ Help to maintain or lower prices through competition.

◆ Provide employment opportunities.

◆ Help contribute to the economic growth of a country and improve the country's place in international competition.

To become a successful entrepreneur, you need to keep in mind two points. First, you should make a link between the idea of a career and what you like to do. What passion do you have that you would like to develop? It should be something you are excited about, rather than just a money-making scheme. If you do not really love your career, you probably will not succeed. Next, you need to get good advice. No one knows everything, but we are all responsible for finding out what we do not know, either by searching for information or asking others.

There is no limit to the types of careers entrepreneurs can create for themselves. These careers vary from making pastries to creating holistic health centers, from removing junk to designing interiors, and from selling specialized items on e-Bay to conducting science demonstrations at children's parties. Entrepreneurs work in areas, however, in which they have become experts. They dedicate whatever time it takes to their work and are particularly good at it because they enjoy what they are doing.

Careers in the Tertiary Sector

People employed in the tertiary sector work in one of the following occupational clusters.

◆ Business and office.

◆ Communications.

◆ Health.

◆ Hospitality and recreation.

◆ Marketing and distribution.

◆ Personal services.

◆ Public and social services.

◆ Transportation.

Business and Office

Employees in this occupational cluster are involved in five areas. These include: administration, management, accounting, secretarial, and clerical tasks. See **Figure 15-22**. Professional employees solve problems, analyze data, and make major administrative decisions. They also prepare reports, design computer systems, and oversee matters of finance.

Office workers keep businesses and organizations running smoothly. Clerical workers maintain accurate records and files. They also operate office machines. Some ship and receive merchandise.

Communications

Some jobs in communications require creative skills. Among these are writing, editing, and producing information. Other jobs demand technical skills, **Figure 15-23**. People in these jobs operate, maintain, and repair equipment. **Figure 15-24** shows a person at work in a communication job.

Health

Employees in health occupations keep people healthy. They also help people recover from injuries or illness, **Figure 15-25**. Physicians and other medical practitioners, such as optometrists, diagnose illnesses and provide treatment. Nurses carry out doctors' orders. They also see to the day-to-day care of the ill and injured. Medical laboratory workers conduct tests to discover the cause of patients'

Designing and making
What careers for designers are related to communications and health occupations?

Discussion
Discuss how the careers in the health cluster are changing as a result of an aging population.

Reflection
What abilities do you have that make you suited for a career in a communications or health occupation?

Tertiary Sector Careers		
Occupational Cluster	**Professionals**	**Skilled Workers**
Business and office	accountant actuary lawyer personnel manager programmer systems analyst underwriter	bank teller bookkeeper buyer business machine mechanic business machine operator cashier clerk file clerk receptionist secretary stenographer switchboard operator typist

Figure 15-22 Every company in the service field requires a business and office workforce.

Tertiary Sector Careers		
Occupational Cluster	**Professionals**	**Skilled Workers**
Communications	announcer cartographer commercial illustrator director drafter editor newspaper reporter photographer technical illustrator	bindery worker broadcast technician camera operator compositor disc jockey film editor lithographer press operator photoengraver radio dispatcher sign painter telephone operator telephone repairer television programmer

Figure 15-23 Which of these careers in communications would you choose?

Community resources and services: Have students select one career in the communications or health clusters and interview someone in that career. Develop a series of questions to ask before attempting the interview.

Figure 15-24 Photography combines artistic and technical skill. This photographer uses a digital camera to capture images. (Jack Klasey)

Tertiary Sector Careers		
Occupational Cluster	Professionals	Skilled Workers
Health	cardiologist chiropractor dentist gynecologist music therapist neurologist obstetrician optometrist orthopedic surgeon pharmacist physical therapist physician psychiatrist radiologist speech pathologist surgeon veterinarian	dental assistant dental hygienist dental lab technician dispensing optician licensed practical nurse nursing aide nursing assistant orderly paramedic X-ray technician

Figure 15-25 Which of these occupations in the health field would you choose?

illness. Many other people work behind the scenes in hospitals and clinics. They provide information and support to doctors, nurses, patients, and visitors. Another important aspect of this field is the health care of animals. See **Figure 15-26**.

Hospitality and Recreation

People who work in hospitality and recreation occupations help others enjoy their leisure time. A travel agent helps with travel arrangements. People who work in hotels provide comfortable lodgings. Amusement and recreation employees provide fun activities for your enjoyment. **Figure 15-27** shows a variety of occupations in this field.

Many jobs in this cluster are related to food, **Figure 15-28**. Some people work directly with customers. Some of these people are waiters and bartenders. Others, such as cooks and chefs, work behind the scenes.

Figure 15-26 Veterinarians specialize in health care of animals. (NFB)

In the performing arts, actors, musicians, and dancers dedicate themselves to their work, **Figure 15-29**. They are assisted by stagehands, designers, electricians, and costume makers.

Community resources and services: Contact one of the major airlines and ask for maps showing the routes they travel.

Tertiary Sector Careers		
Occupational Cluster	**Professionals**	**Skilled Workers**
Hospitality and recreation	actor/actress choreographer conductor dancer dietitian director hotel manager music director musician pop singer producer sports teacher	baker bartender bell captain camp manager chef choral singer cook cruise director executive housekeeper greenskeeper hotel desk clerk lifeguard park caretaker recreation leader ticket seller tour guide travel agent waiter/waitress

Figure 15-27 Here is a list of hospitality and recreation careers. Which ones involve working directly with people? Which involve working behind the scenes?

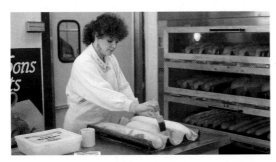

Figure 15-28 A baker at work preparing bread for the oven. (Ecritek)

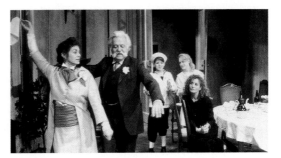

Figure 15-29 Actors and actresses spend many hours preparing for performances. (T. Diab)

Tertiary Sector Careers		
Occupational Cluster	**Professionals**	**Skilled Workers**
Marketing and distribution	ad copy writer advertising manager bank manager insurance agent insurance investigator loan officer market research analyst model real estate agent	buyer customer service representative fork lift operator loader packer purchasing agent salesclerk sales representative shipping clerk stock clerk store manager truck driver warehouse person window dresser

Figure 15-30 Can you think of other careers in marketing and distribution?

Marketing and Distribution

Employees in this cluster buy, promote, sell, and deliver goods and services, **Figure 15-30**. Manufacturers employ buyers. They are responsible for purchasing materials and supplies required to produce products. The manufacturer sells the products to wholesalers. Wholesalers sell to retailers who, in turn, sell to consumers.

Designing and making
What careers for designers are related to marketing and distribution and personal services occupations?

Discussion
Discuss cooperative education and how it can assist students in learning about future careers while still in school.

Reflection
What abilities do you have that make you suited for a career in a marketing and distribution or personal services occupation?

Vocabulary
What is meant by *quality management* and how does it apply to the Marketing and Distribution cluster of occupations?

Activity
Select a career in the Marketing and Distribution cluster and research its educational requirements, daily tasks, and job responsibilities.

Community resources and services: Have students select one career in the marketing and distribution or the personal services cluster and interview someone in that career. Develop a series of questions to ask before attempting the interview.

Most jobs in marketing involve meeting people, **Figure 15-31**. There are others, however, that involve the movement, storage, and inventory of products. Some employees must be able to organize large amounts of data. Others who work for insurance companies and banks are involved in finance.

Personal Services

Some people who work in this occupational cluster are concerned with the physical appearance of their customers. They help with personal grooming or physical conditioning. Others assist with tasks around the home. Still, others help keep places where people live, work, or play safe and clean.

Figure 15-32 lists some jobs in this cluster. **Figure 15-33** is an example of an occupation in this field.

Public and Social Services

People working in the public and social services cluster provide services to everyone in a community. See **Figure 15-34**. In some careers, helping people stay safe and comfortable is the

Tertiary Sector Careers		
Occupational Cluster	**Professionals**	**Skilled Workers**
Personal services	home nurse	animal trainer
		barber
		building superintendent
		butler
		chauffeur
		companion
		cosmetic demonstrator
		cosmetologist
		custodian
		exercise instructor
		exterminator
		funeral director
		hair stylist
		housekeeper
		kennel manager
		laundry worker
		manicurist
		nanny
		pedicurist
		seamstress
		shoe repairer
		tailor

Figure 15-32 Many people find satisfying careers providing personal services to others.

Figure 15-31 A salesclerk is working a register. People in sales occupations must enjoy meeting people. (Ecritek)

chief task. Other service workers help solve personal problems, while others teach, **Figure 15-35**.

Transportation

Transportation is meant to move people and materials. Employees in this cluster drive buses, taxis, trains, and trucks. See **Figure 15-36**. They fly aircraft or pilot ships. Other people provide services for customers.

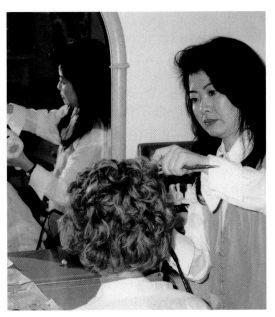

Figure 15-33 Hairdressers must be skilled and enjoy working with clients. (Drury)

Figure 15-35 Firefighters must respond to emergency situations. (Jack Klasey)

Tertiary Sector Careers		
Occupational Cluster	**Professionals**	**Skilled Workers**
Public and social services	coroner guidance counselor lawyer librarian minister notary police officer teacher	armed service personnel community youth worker firefighter health and safety inspector letter carrier mail clerk sanitation worker tax audit clerk

Figure 15-34 Every community has public and social programs.

Tertiary Sector Careers		
Occupational Cluster	**Professionals**	**Skilled Workers**
Transportation	air traffic controller captain co-pilot pilot ship's officer	able seaman aircraft mechanic automobile mechanic boatswain bus driver chief mate diesel mechanic dispatcher flight attendant locomotive engineer merchant marine passenger car conductor reservation agent service station attendant taxi driver ticket agent travel agent truck driver

Figure 15-36 Which of these transportation workers have provided you with service?

They make arrangements for their travel and for the movement of cargo. Other workers keep transportation equipment in good working order. See **Figure 15-37**.

Discussion
Do employers look for initiative in their employees? What does it mean to have workers who can think critically and act logically to evaluate situations, solve problems, and make decisions? Is the worker who identifies and suggests new ideas to get the job done valued?

Enrichment
Competitive pressures result in employees moving between jobs and responsibilities. They must retrain frequently and be able to absorb, process, and apply new information quickly and effectively.

Vocabulary
What is meant by the terms *blue collar* jobs and *white-collar* jobs? Why is the line blurring between these two types?

Activity
A person who learned to operate a hydraulic metal press in 1920 would probably be operating essentially the same machine at retirement in 1965. Have students investigate how this situation has changed today.

Resource
Reproducible Master, *Career Ladder*

Useful Web site: Have students access the United States Department of Labor Web site at www.dol.gov to find an article or fact that interests them and share it with the class.

Reflection
What can you do to prepare for living in a global village?

Vocabulary
What does the term *outsourcing* mean?

Figure 15-37 Helicopter pilots move people and material short distances by air. (NFB)

Outsourcing

Have you ever heard the expression "The world has become a global village"? We can travel around the globe in less time than it took our ancestors to travel 100 miles (160 km). Not only can goods be produced anywhere in the world and delivered within days, but faster ways of communication mean services can be outsourced. By outsourcing, corporations or even individuals subcontract parts of their operations to areas of the world where labor costs are much lower but labor is still reliable. For example, people in India are currently providing services for people in North America by doing such diverse jobs as reading computerized axial tomography (CAT) scans, preparing computer program presentations overnight, and operating call centers. People in China can manufacture practically any item for a fraction of our costs in any one of 160 cities. North America cannot compete in many areas of service or manufacturing. Instead, we must become more educated, more involved in research, and more creative, in order to develop new technologies, particularly in areas such as telecommunications, media, nanotechnology, energy, environmental protection, health care, transportation, and security.

Getting a Job

Regardless of whether you choose a career in the primary, secondary, or tertiary sectors, one thing is certain. What you know, as well as your ability to access information and what you are able to do with that information, will determine your value to society and your probability of acquiring a high-paying job. The longer you spend being educated, the less likely you will end up on the dole. This is particularly true when the economy slows down and people aren't buying as much. At this time, an even larger portion of undereducated people is unemployed. Also, the gap in salary between the well educated and the poorly educated is continuing to widen. One reason is that the well educated have jobs that provide them with more training, so they continue to move ahead faster.

Did you ever wonder why it is vital to continue your education when 25 years ago it wasn't as important? One reason is automation. As you have read, many routine or repetitive jobs are done faster by machine. A second reason is globalization. We now import products from around the world. We buy from

---

Useful Web site: Have students access the Web site www.answers.com/topic/global-village to investigate the work of Marshall McLuhan, who introduced the term *global village* in his book *The Gutenberg Galaxy*.

Technology and society: What could be the effect on people in your community of outsourcing to overseas producers?

whichever country can make quality products for the least cost. The next time you buy something, look at the label to see where it was made. You may be surprised to see how many things are made in China, Mexico, Korea, and India.

Many of the jobs that remain in North America are "smart jobs." They require more mental power than muscle power. You should have already read the chapter about Information and Communication Technology (ICT). Careers in this area have increased and will continue to increase. Engineers, scientists, and technologists are building machines that understand and obey voice commands. Managers and supervisors are inventing new ways to organize production. The products themselves are becoming increasingly sophisticated and the machinery more intelligent, so the workers must be smarter. Robotic kits, such as the one shown in **Figure 15-38**, can help you acquire the skills needed for ICT careers.

Employer Expectations

Many general qualities are needed to be successful in the workplace. Behavior required for professional success and advancement includes the following:

◆ **Cooperation.** An employee must cooperate with supervisors, other employees, and customers.

Figure 15-38 Experimenting with kits, such as this one, which simulates robotic operations, can help you prepare for a career in information and communications technology.

◆ **Dependability.** A dependable employee is timely, completes all assignments, and sets realistic goals for completing projects. A dependable employee is trusted by others.

◆ **Work ethic.** Good employees put an honest effort into their work.

◆ **Respectful.** In order to be respected, employees must show respect for others, the company, and themselves.

Today's workplace emphasizes *equality*—that is, the idea that all employees are to be treated alike. *Harassment* (an offensive and unwelcome action against another person) and *discrimination* (treating someone differently due to a personal characteristic such as age, sex, or race) are not tolerated. These negative behaviors often result in termination of employment.

Discussion
Discuss the statement "excellence is a journey, not a destination."

Reflection
Reflect on the career that interested you most in this chapter. Visit your local technical or community college to learn more about the qualifications needed to enter the career.

Enrichment
Regardless of the career students may have selected in this chapter, they should understand that both verbal and written communication skills are vital to entering and progressing on the job. Verbal skills involve not only asking the right questions, but also listening intently. Written skills include writing memos, notes, instructions, and presentations. A good command of the English language is needed to communicate information and concepts in a clear and concise way.

Useful Web site: Have students visit a Web site that helps with career choice, career planning, and job search. A useful starting point is www.careerkey.org/english.

Chapter 15 Review

The World of Work

Summary

The steps in producing any product are organized into three sectors: primary, secondary, and tertiary. The primary sector is concerned with obtaining and processing raw materials. Some of these materials are renewable, and other materials are nonrenewable. Careers in the primary sector include those in agriculture, horticulture, forestry, marine science, natural resource extraction, and environmental control.

The secondary sector is concerned with the manufacture of products and construction of structures. It has evolved through three stages: the individual artisan, mechanization and mass production, and automation.

Prior to the eighteenth century, the individual artisan was responsible for every step in producing a finished product. Mechanization and mass production occurred during the Industrial Revolution. At this stage, there were three important changes. Products were made with machinery, which were powered by engines. The products were made in factories. The efficiency of factories was increased by the use of assembly lines and later by automation. In an automated factory, the worker builds, monitors, and maintains machines that make the products. The latest development in automated production and assembly is the use of robots.

A production system involves five basic steps: designing, planning, tooling up, controlling production, and packaging and distribution. Careers in the secondary sector include those in manufacturing and construction.

The tertiary sector is concerned with the servicing of products and providing services. Private companies provide installation, maintenance, repair, or the alteration of products. Government agencies are responsible for education, health, and public works. Careers in the tertiary sector include those in: business and office, communications, health, hospitality and recreation, marketing and finance, personal services, public and social services, and transportation.

416

Modular Connections

The information in this chapter provides the required foundation for the following types of modular activities:

◆ Career Interviews

◆ Portfolio Development

◆ Marketing

◆ Business Presentations

◆ Career Exploration

◆ Entrepreneurship

◆ Leadership

Test Your Knowledge

Write your answers to these review questions on a separate sheet of paper.

1. Look at **Figure 15-1**, which describes the steps used to build a chair. Choose a simple object in your home. Describe the steps in its manufacture, from raw material to finished product.

2. When producing products, what is the main activity in each of the following sectors?

   Primary Sector.

   Secondary Sector.

   Tertiary Sector.

3. List three renewable and three nonrenewable raw materials.

4. Choose a career in the primary sector. Describe a typical day in the life of someone working in that career. Try to talk to someone working in your chosen career. Also, use the library resources.

5. List the three stages through which the secondary sector has evolved.

6. Until the Industrial Revolution, artisans worked alone or in small groups in villages. How did the Industrial Revolution change this?

7. How does the division of labor enable products to be made at a faster rate than by hand?

Answers to Test Your Knowledge Questions

1. Answers will vary depending upon the object selected. The steps will be similar to Figure 15-1 on page 390.

2. Primary sector—obtaining and processing raw materials. Secondary sector—changing raw materials into products. Tertiary sector—delivering and servicing products.

3. Renewable—such as lumber, fish, and furs. Nonrenewable—such as fossil fuels, nonmetallic minerals, metallic minerals, etc.

4. Answers will vary depending upon the career selected. The answers should be based on information found in Figure 15-6 on page 393.

5. First, individual artisan. Second, mechanization and mass production. Third, automation.

6. People were brought together in factories where products were made by machinery powered by steam.

7. Each worker becomes skilled at one small task, which can be performed repeatedly at a very rapid rate.

Answers to Test Your Knowledge Questions *(continued)*

8. Parts can be easily replaced when they are broken or worn out, since the new part is identical to the old one.

9. Factory buildings were dark, dismal, poorly lit, cold, and drafty. Workers labored for long hours and young children were part of the workforce.

10. Henry Ford, 1914

11. On a mass production assembly line, each worker adds one or more parts to the product. On an automated assembly line, workers monitor and maintain machines that make the product.

12. Handling—loading and unloading components onto machines. Processing—machining, drilling, painting, and coating. Assembling—placing and locating a part in another component. Dismantling—breaking down an object into its component parts. Fixing—assembling objects permanently by welding or soldering.

13. Robots work more efficiently in hot, noisy, or dangerous situations. Robots continue working 24 hours a day. Robots operate for thousands of hours without maintenance.

8. What is the advantage of products made with interchangeable parts?

9. What were some of the problems experienced by workers in early factories?

10. Mass production was first used by _____ _____ in the year _____.

11. Describe the difference between a mass production assembly line and an automated assembly line.

12. List five jobs that robots can perform in the manufacture of a product.

13. State three advantages of robots over human workers.

14. Describe the five basic steps in the production system.

15. Choose a career in the secondary sector. Describe a typical day in the life of someone working in that career. Interview someone working in your chosen career. Find out what this person did to become qualified or licensed to work in this field. What are the daily job duties, and what hobbies might provide some insight into the field? Also, use the library resources.

16. List five services provided by the tertiary sector that you have used.

17. State three reasons for the rapid growth in the number of tertiary sector jobs in the last 20 years.

18. Identify an entrepreneur in your neighborhood. State what new product or service is being offered. Describe how the product or service is helping you or your family. How many people are being employed?

19. Choose a career in the tertiary sector. Describe a typical day in the life of someone working in that career. Try to talk to someone working in your chosen career. Also, use the library resources.

20. List the names of 10 adults whom you know. Name their jobs and state the occupational cluster and sector for each.

14. Designing—designing and drawing products.
Planning—organizing personnel, materials, and equipment into a system.
Tooling up—acquiring and preparing tools and machines for production.
Controlling production—using machines to make the product.
Packing and distribution—packaging, storing, and transporting products to wholesalers and retailers.

15. Answers will vary depending on the career selected but will be based on the information found in Figure 15-21 on page 406.

16. Answers will vary according to the student's experiences.

17. Workers in some areas cannot increase their productivity; therefore, as demand grows, the number of workers will increase proportionately.
As families have become more affluent, they consume more services.
The service sector faces little foreign competition.

18. Answers will vary.

19. Answers will vary, but will be based on Figures 15-22, 15-23, 15-25, 15-27, 15-30, 15-32, 15-34, and 15-36 on pages 409–413.

20. Answers will vary.

Apply Your Knowledge

1. Select a raw material to research. Find out:
   A. Where is it found?
   B. In what form is it found in its natural state?
   C. How is it extracted, harvested, or farmed?
   D. How is it transported?
   E. How is it processed or refined?
   F. How have new manufactured goods increased production or changed its processing?

2. List the advantages of robots compared to human workers. List five tasks that a robot can perform better than a human worker.

3. Imagine that in your working life you started as an artisan. You then moved to a factory and worked on a mass production line. Your last job was in a fully automated factory where robots did the work. Describe the advantages and disadvantages of each of your three jobs.

4. Choose a small, common, mass-produced item. For each of the five steps in the production system, describe the activities that would occur to mass-produce the item you have chosen.

5. Refer to **Figure 15-19**. What process would be used to:
   A. Cut a 1″ (25 mm) diameter hole?
   B. Bend a piece of sheet metal into a 90° angle?
   C. Cast an irregular shape in metal?
   D. Produce a flat surface on a block of aluminum?
   E. Make a large number of 1″ (25 mm) diameter identity tag disks?

6. Research how new technologies have replaced, outmoded, or created new jobs in your hometown.

7. List the services in the tertiary sector that you have used in the past week.

8. Select one career from the primary, secondary, or tertiary sectors. Investigate the career to discover:
   A. The education or training required.
   B. Whether or not there will be a demand in the future for workers.
   C. The promotion possibilities.
   D. The salary, pension, vacation, and other benefits.
   E. The number of people currently employed in this type of work.

9. State the career you are thinking of selecting. What does it mean to act professionally in this career? How do people update their knowledge and skills in this career? How do they communicate effectively?

10. What is meant by "ethical behavior in business"? What does it mean when we say that technologists or engineers must act in an ethical manner?

11. Select one or more of the following Web sites and identify the careers described:

    www.aaas.org American Association for the Advancement of Science

    www.aiaa.org American Institute of Aeronautics and Astronautics

    www.asme.org American Society of Mechanical Engineers

    www.powerengineers.com National Association of Power Engineers

    www.sae.org Society of Automotive Engineers

    www.aaes.org American Association of Engineering Societies

    www.asabe.org American Society of Agricultural and Biological Engineers

    www.nsf.gov National Science Foundation

    www.ieee.org Institute of Electrical and Electronic Engineers

    www.sme.org Society of Manufacturing Engineers

    www.acs.org American Chemical Society

    www.asce.org American Society of Civil Engineers

    www.tms.org The Minerals, Metals, and Materials Society

    www.nspe.org National Society of Professional Engineers

    www.swe.org Society of Women Engineers

12. Bill Gates, Lee Iacocca, Guy Laliberté, Anita Roddick, and Oprah Winfrey have been immensely successful business leaders because they maximized opportunities in their chosen fields. What type of business has each one developed, and why are these people more successful than other people in similar areas of business?

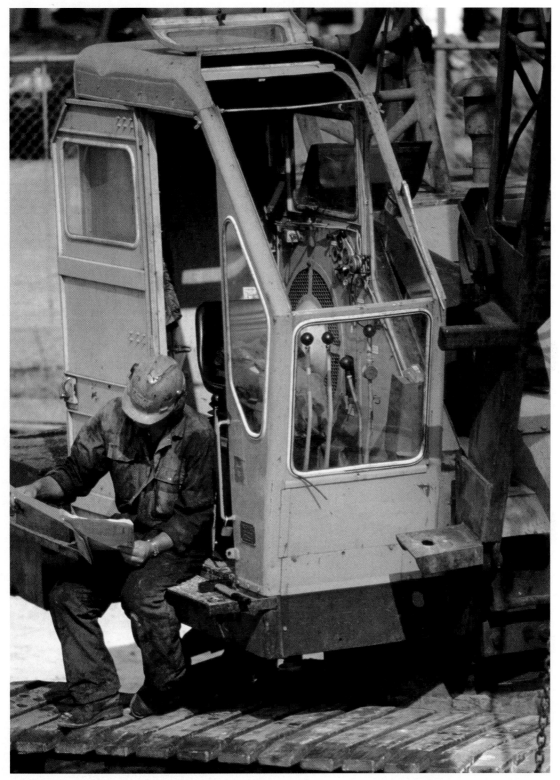

Heavy machine operators are just one of many people in the world of work. They must be able to read blueprints, precisely control large machines, and communicate well with other operators and construction workers.

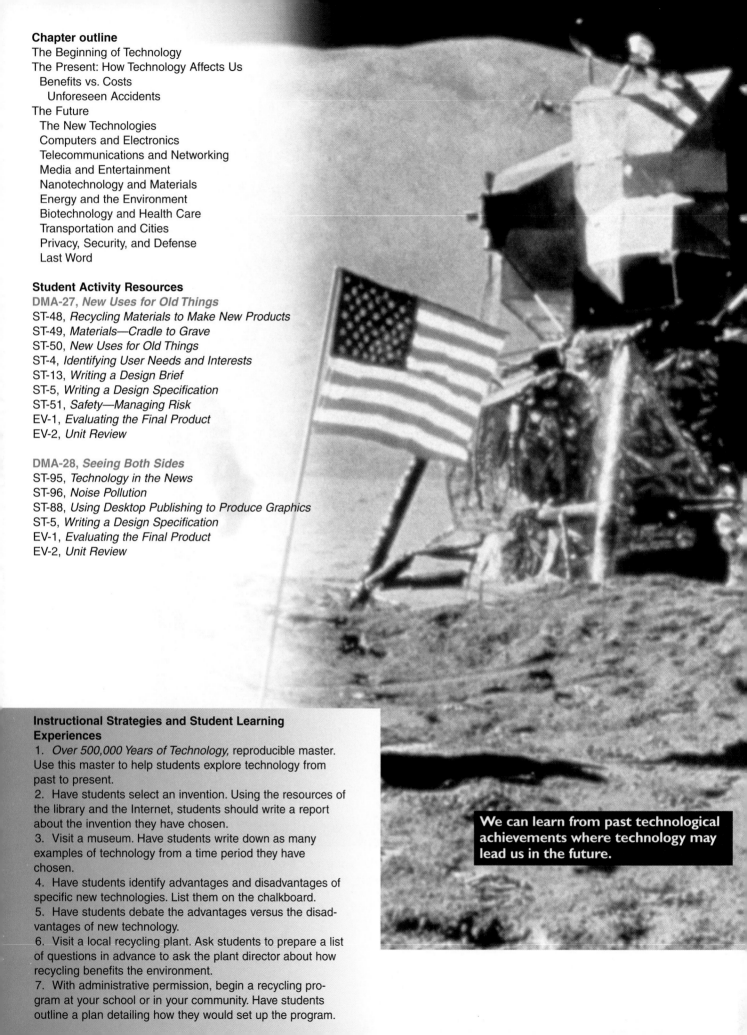

Chapter outline
The Beginning of Technology
The Present: How Technology Affects Us
  Benefits vs. Costs
    Unforeseen Accidents
The Future
  The New Technologies
  Computers and Electronics
  Telecommunications and Networking
  Media and Entertainment
  Nanotechnology and Materials
  Energy and the Environment
  Biotechnology and Health Care
  Transportation and Cities
  Privacy, Security, and Defense
  Last Word

Student Activity Resources
DMA-27, *New Uses for Old Things*
ST-48, *Recycling Materials to Make New Products*
ST-49, *Materials—Cradle to Grave*
ST-50, *New Uses for Old Things*
ST-4, *Identifying User Needs and Interests*
ST-13, *Writing a Design Brief*
ST-5, *Writing a Design Specification*
ST-51, *Safety—Managing Risk*
EV-1, *Evaluating the Final Product*
EV-2, *Unit Review*

DMA-28, *Seeing Both Sides*
ST-95, *Technology in the News*
ST-96, *Noise Pollution*
ST-88, *Using Desktop Publishing to Produce Graphics*
ST-5, *Writing a Design Specification*
EV-1, *Evaluating the Final Product*
EV-2, *Unit Review*

Instructional Strategies and Student Learning Experiences
1. *Over 500,000 Years of Technology,* reproducible master. Use this master to help students explore technology from past to present.
2. Have students select an invention. Using the resources of the library and the Internet, students should write a report about the invention they have chosen.
3. Visit a museum. Have students write down as many examples of technology from a time period they have chosen.
4. Have students identify advantages and disadvantages of specific new technologies. List them on the chalkboard.
5. Have students debate the advantages versus the disadvantages of new technology.
6. Visit a local recycling plant. Ask students to prepare a list of questions in advance to ask the plant director about how recycling benefits the environment.
7. With administrative permission, begin a recycling program at your school or in your community. Have students outline a plan detailing how they would set up the program.

We can learn from past technological achievements where technology may lead us in the future.

16

Learning from the Past, Predicting the Future

Key Terms

biometrics
cloning
downstream
genetic engineering
green chemistry
nanobot
pollution
upstream
wearable

Discussion
Discuss the importance of learning from the past in order to create a brighter future.

Career connection
Using information from the previous chapter, discuss how careers are changing and what the prospects for the future are likely to be.

Objectives

After reading this chapter you will be able to:

◆ Describe examples of early technology.

◆ List major inventions and innovations in different historical periods.

◆ Cite examples of the technological change that is occurring at an ever-increasing rate.

◆ Discuss the impact of evolving technologies on the individual and society.

◆ Design and make a product that reduces the negative impact of technology on the environment.

The Beginning of Technology

About two million years ago, prehistoric humans made the first tools. They discovered that when a large pebble is struck with great force against another stone, pieces flake off. Sharp cutting edges are formed, **Figure 16-1**.

The list of stone tools about 300 to 400 thousand years ago included hand axes, points, clubs, and scrapers. The points were made of roughly shaped flint, **Figure 16-2**.

The bow and arrow was invented about 10,000 years ago. At the same time, small pieces of flint were attached to bone and wood to make knives and spears.

Until 8000 years ago, humans had obtained their food by hunting and gathering. They were nomadic, moving around in search of food. In 6000 B.C., they discovered agriculture and domesticated animals. Stone tools became highly refined. About 2000 B.C., stone tools gradually gave way to tools cast in copper and bronze, **Figure 16-3**.

Later still, about 1000 B.C., iron began to replace copper and bronze. Iron was repeatedly heated and hammered to make ornaments, tools, and weapons.

People of Egypt and Mesopotamia were the first to use simple pulleys and levers. Later, Greek and Roman engineers developed these machines further.

The progress of technology was generally slow until about three centuries ago. In the 1700s, inventions occurred more rapidly. Many of these

Figure 16-1 For what purposes did prehistoric people use these stone tools? (Ecritek)

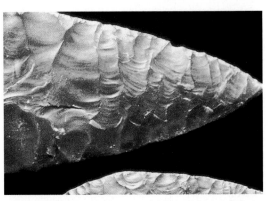

Figure 16-2 A spear point made from flint. Edges were shaped by chipping away the stone from both sides.

Figure 16-3 About 4000 years ago, humans learned how to cast tools from metal. These are examples of their craft.

inventions were machines to make products. They were run by a new power source, the steam engine. This mechanization led to the Industrial

Useful Web site: Students can use the dictionaries at www.techtionary.com and www.hyperdictionary.com to find the definition of any new technical and technological terms used in this chapter.

Technology and society: Have students type the phrase "traditional dress" into a search engine and browse one of the Web sites that describes clothing worn by indigenous people. Students should print information they find interesting and share it with the class.

Revolution. Applied to railways and ships, steam power completely changed transportation. The beginning of the twentieth century marked the first uses of automobiles, airplanes, telephones, radios, and electricity.

The number of inventions increased so that by the end of the twentieth century there were offshore oil rigs and nuclear reactors, combine harvesters and automated milking parlors, synthetic fibers and industrial robots, suspension bridges and hydroelectric dams, telephones with storage memory, compact discs, microwave ovens, digital sound systems, portable computers, DVD players, remote control television sets, and space shuttles, **Figure 16-4**.

From the Stone Age to the present day, invention has been one of the most important human activities. Some of the most important inventions are summarized in **Figure 16-5**. What conclusion can you draw about the rate at which inventions have occurred? Remember that human evolution has developed over about 4 million years or 50,000 generations, but television and computers have been around for only 2 generations. What might happen in the next 10 generations?

The Present: How Technology Affects Us

For the most part, technology has improved the quality of our lives. We live longer, healthier lives, and we have more goods and services. Machines have done away with drudgery and hard physical labor. We have more leisure time; however, we have not always taken the time to study fully the impact of each technology on our lives or our environment. No matter what the benefits of any new technology, history shows that there is a cost for each advance.

Usually we see the benefits immediately. We like new technology; it makes our lives easier or more fun. Once we have a new technology, it is hard to imagine life without it.

Benefits vs. Costs

The costs of using a new technology may only become known years after its introduction. For example, people have used chemicals such as DDT to increase food production. The chemicals killed unwanted insects that destroy crops. However, they also killed many birds, squirrels, cats, muskrats, rabbits, and pheasants. More recently, some of these chemicals have been found to cause cancer in humans.

Is there also a cost associated with television? The benefits deriving from this technology are easy to see. These include access to world news, live coverage of sports events, and famous celebrities to entertain us at home. What are the costs? What effect has television had on our family life? See **Figure 16-6**. Do people read in groups or as individuals anymore? Are viewers affected by watching so much violence?

Enrichment
Clothing was probably first worn about 30,000 years ago by Neanderthals living in caves in Central Europe. These early garments had no sleeves or fastenings. Around 6000 BC, the wool from sheep could be used because animals were domesticated. Cloth draped around the body was fastened at the waist. Metalsmiths in the Middle Ages made heavy suits of armor to protect knights in battle. In the colonies, yarn was spun on spinning wheels and woven into cloth on hand looms. Equipment developed during the Industrial Revolution sped up the process of making cloth and ready-to-wear garments.

Discussion
Have students debate the following issue: "Change is happening so fast we can't be sure it's for the best and we can't be sure that we are doing the right thing by letting it happen."

Activity
Have students make a list of the questions we should be asking about the benefits and costs of introducing new technologies.

Links to other subjects: Economics—Have students investigate the factors that must be included in an environmental cost/benefit analysis.

Technology and society: Divide the class into small groups and have each group choose a technological product they use every day. Have each group complete a cost/benefit analysis of the product, answering the question: "Do the social benefits of this technology outweigh its social costs?"

Discussion
Discuss the tension between economic growth and environmental protection. Economic growth requires the consumption of increasing amounts of nonrenewable resources. The technologies we employ are very energy intensive. They use synthetic chemicals instead of renewable natural resources and produce nondegradable products. Our economic system operates on perpetual growth and produces increasing levels of pollution.

Resource
Activity 16-2, *Life of the Pioneer*

Activity
Have each student select a manufactured item, such as a piece of jewelry, a refrigerator, or a lamp. Research the shape and form of the object as it existed 100 years ago, compared to its current design.

Resource
Activity 16-3, *Compare Your World*

Resource
Activity 16-4, *How Products Change over Time*

Figure 16-4   The technology of the last century has allowed us to build bigger, better structures and machines that help with harvest, repetitive tasks, and transportation of people and loads. A—Oil rigs drill deep into the ocean floor and pump crude oil into tanks. (Norsk Hydro) B—Combines cut, thresh, and gather crops. C—Robotic welders do fast, repetitive welding on cars and other products in manufacturing plants. (Motoman) D—The Hoover Dam functions as the following things: a bridge between Nevada and Arizona, a power plant supplying electricity to the American southwest, and a dam separating Lake Mead and the Colorado River. (U.S. Department of the Interior, Bureau of Reclamation) E—Space shuttles transport astronauts and equipment into outer space for research missions. (NASA)

Technology and society: What effect will the continually increasing number of available products have on the population's quality of life?

Period	Food	Shelter	Clothing	Defense	Transportation	Communication	Health
500,000–10,001 BC	-hoe -harpoon -nets for fishing	-cave -painting -hides over frame -oil lamp	-animal skins -needle	-fire -spear -bow and arrow	-dugout	-hieroglyphics	The six simple machines were developed during this period
	Technological characteristics: tools of ivory, bone, wood, antler, and a variety of stones						
10,000–1 BC	-Archimedes' water screw -waterwheel -quern to grind corn -trained animals to pull plows -pottery -spoon -fishhook -sickle	-sun-dried mud hut -lock and key -rope and pulley -gear -brick -arch -nail -glass -bath	-vertical loom -cosmetics	-monumental stone buildings -bronze and iron weapons -knives -swords -sling	-sled -wheel -sail -boat -skis -harness (oxen)	-ink and paper -cuneiform writing -Pheonician alphabet -astrolabe -Estruscan alphabet -first coinage -papyrus -maps	-false teeth
	Technological characteristics: a basic understanding of metallurgy						
AD 1–1399	-aqueduct -windmill -horse collar -horseshoe -porcelain drinking vessels	-stained glass -Roman central heating -clock -dome -chimney	-spinning wheel -trousers -felt hat -button -lace	-crossbow -bronze-cast cannon -gunpowder -gun	-magnetic compass -Roman roads -Viking longboat -horse stirrup and saddle -rudder -skates	-movable type -paper from bamboo (Chinese) -stencil -pen -printing	-hospital -spectacles
	Technological characteristics: small factories, foundries, forges, and mills by waterwheels						
AD 1400–1699	-ice cream -pressure cooker -bottle cork	-wallpaper -watch -theodolite -water closet (toilet) -thermostat -barometer -surveying instrument	-knitting machine -umbrella	-artillery shell -naval mine -hand grenade -rifle -submarine	-diving bell -dredger -telescope -wheelchair	-Gutenberg's press -arithmetic signs (+ - = × ÷) -newspaper -envelope -calculating machine	-toothbrush -artificial limbs -microscope -thermometer -inoculation
	Technological characteristics: blast furnaces to melt and cast iron; the development of many scientific instruments						
AD 1700–1849	-canned food -threshing machine -carbonated water -steam tractor -reaper -seed drill -sandwich -fertilizer	-electricity -street lighting (gas) -iron frame building -fire extinguishers -cement -matches -central heating	-spinning jenny -power loom -cotton gin -sewing machine -waterproof coat -dry cleaning	-Winchester rifle -standard parts for guns -machine gun -shrapnel	-pneumatic tire -bicycle -lifeboat -locomotive -hot-air balloon -sextant -roller skates	-metric system -photography -lithography -typewriter -Morse's telegraph -Braille -steel pen -eraser -postage stamp	-bifocals -anesthetics -sedatives -porcelain false teeth -ambulance -vaccination -plastic surgery -stethoscope -blood transfusion
	Technological characteristics: first mechanized factories and the start of mass production; the use of all machine tools						

Figure 16-5 Over 500,000 years of technology are represented in this chart. What will be invented in the next 100 years?

Useful Web site: Access www.geocities.com/Athens/Academy/7920 to read about life in the nineteenth century. Share information you find interesting with the class.

Period	Food	Shelter	Clothing	Defense	Transportation	Communication	Health
1850 – 1899	Technological characteristics: use of magnetism to produce electricity						
	-barbed wire -refrigeration -condensed milk -milking machine -margarine -cola -breakfast cereal	-high-rise building -plastics -linoleum -electric lighting	-jeans -man-made fibers -zipper -aniline dyes	-dynamite -submarine -automatic machine gun -torpedo	-hang glider -airship -glider -clipper ship -diesel engine -automobile -helicopter -modern bicycle -motorcycle	-telephone -typewriter -cinematography -wireless telegraph -radio -postcard -fountain pen	-hypodermic syringe -pasteurization -dental drill -antiseptics -incubator -X ray -aspirin
1900 – 1945	Technological characteristics: conquest of the skies						
	-tea bag -frozen food -insecticide (DDT) -combine harvester -supermarket	-prestressed concrete -air conditioning -fluorescent lighting -vacuum cleaner	-electric washing machine -nylon -artificial silk	-poison gas -tank -radar -gas mask -aerial bomb	-aircraft -tracked vehicles -safety glass -seaplane -traffic lights -helicopter -jet aircraft -subway system	-motion pictures -television -xerox -ballpoint pen	-electro-cardiograph -hearing aid -blood transfusion -chemotherapy -insulin -iron lung -kidney machine
1946 – present	Technological characteristics: use of atomic energy; of computers; and of microprocessors to control apparatus						
	-synthetic fertilizers -microwave oven -nonstick pan -domestic deep freezer -foods for use in space -new "miracle" strains of rice & wheat -ultrasonics to detect fish -cloning	-solar panel -geodesic dome -synthetic turf -fiberglass insulation -lightweight modular dwellings -space station	-synthetic fibers -automatic clothes dryer -permanent creases in clothes -metallized fabric	-atomic bomb -ejection seat in aircraft -portable atomic weapon -rocket -hydrogen bomb -ICBM	-aqualung -hovercraft -lunar vehicle -nuclear submarine -monorail train -space shuttle -ultra high-speed train -VTOL aircraft -space station -fuel injected engines -magnetic levitation trains -space orbiters	-photo typesetting -radar -transistor -satellite -instant camera -long-playing record -laser beam -fiber optics -computer network -hologram camera -silicon chip -videotape -microprocessor -pocket calculator -integrated circuits -digitized typesetting -desktop publishing -compact disc player -laptop computer -space telescope	-artificial voice box -heart-lung machine -artificial heart and other organs -equipment for organ transplants -high-speed dental drill

Figure 16-5   *Continued*

Useful Web site: Investigate which Web sites promote a counterculture. What ideas may become co-opted by large companies, in an effort to attract members of a younger generation? A useful starting point is: www.commondreams.org/views/092300-103.htm.

Figure 16-6 It is important for families to spend time together dining, talking, and doing other activities.

Unforeseen Accidents

Another cost of technology, often a very high one, is the unforeseen accident. Bhopal, capital of the Indian state of Madhya Pradesh, used to be a peaceful city. This peace ended December 23, 1984. A mysterious and deadly fog was discharged from a nearby pesticide factory. This fog began to settle into a section of Bhopal called Khazi Camp. Hundreds of families lived along a road bordering the factory. The fog was a gas, methyl isocyanate, used in pesticide production. This cloud of gas changed the lives of more than 250,000 people. Eighteen thousand of them have died. Others continue to have long-term health problems. Nobody knows exactly what effect the exposure will have on future generations.

Sometimes workers see the introduction of a new technology as a threat. Groups or individuals then resist change. In 1799, Ned Lud, an English mill worker, destroyed two textile machines belonging to his employer. By smashing these labor-saving devices, he had hoped to avoid unemployment. Was Ned Lud's fear of technology justified? In this case, it probably was not. History shows that advances in technology have usually produced more jobs than were lost. For example, when Henry Ford used an assembly line to mass-produce cars, the cost of each car was greatly reduced. More people could then afford to buy cars. This, in turn, led to an increase in the number of people required for service and repair, **Figure 16-7**.

The Future

We are all unique and different. That is what makes us special and interesting to other humans. It is also the reason why we will probably react in different ways to both new and old technology. Some things we can agree on. Almost all of us like the creature comforts it brings. Things like electricity to warm or cool homes according to the season, a phone to call our friends, and recordings of our favorite artist or shows for our leisure times.

Where we may start to disagree is whether we want to be surrounded by technology all the time. Perhaps you enjoy nature just as much. For example, there may be times when you would prefer to be splashing in waves on a beach or kicking leaves as you walk through an autumn forest rather

Useful Web site: Have students search the Internet to find more information about the Bhopal disaster. A useful Web site is: www.organicconsumers.org. Have students examine how disasters can occur in their homes by accessing the National Safety Council's Web site at www.ohsuhealth.com/ntrauma/stats.asp. Students should select a topic to research and share the information with the class.

Discussion
Which hobbies are based on old technologies? Why are these hobbies attractive to the people who enjoy them?

Activity
Multinational corporations market products that are sold in similar styles across a large part of the world. They are also involved in corporate sponsorship of major events. Is there a down-side to these new super-brands that market "cool" lifestyles? Do they rely on sweatshops in third world countries? Do they homogenize culture? Do they reduce our choices? Do they take away local jobs?

Resource
Activity 16-5, *Assessing the Impact of Technology*

Figure 16-7 A—The introduction of the assembly line by Henry Ford created job opportunities for people and gave them a chance to learn a skill. (Ford) B—Knowledgeable mechanics are needed to maintain and repair automobiles.

Figure 16-8 Some young people prefer to spend the day at a beach instead of playing video games at home all day.

than watching television in an air-conditioned house, **Figure 16-8**. There also might be times in which you would prefer to be in a nontechnology room—a meditation room—or a place to create your own private time to "switch off" the world. We may also disagree on whether the newest type of technology is always the best. Do you like to collect antiques? Perhaps you know someone who has a collection of vinyl records. Or maybe you would agree that flickering candlelight could be more beautiful than laser lights?

Another point on which we might all agree is that technological change is happening very fast. The speed of change may be one reason to wonder if all the changes can be of benefit to society. Can we be sure that we are doing the right thing unless we make a conscious decision about each change instead of just letting things happen? Certainly there are people with very impressive credentials telling us that change is inevitable. They say that we should simply sit back and enjoy the ride.

What each of us must recognize is that advances in technology are neither inherently good nor inherently bad. Every new technology has the potential to both solve problems and create problems:

◆ Burning fossil fuels can warm our homes, but it can also change the world's climate.

◆ Building large suburbs provides homes for families, but it also encroaches on wildlife habitat and endangers plant and animal species.

Technology and society: The Center for the Study of Technology and Society issued the following statement: "Advances in technology are neither inherently good nor inherently evil. Every new technology has the potential to cause problems, and the capacity to solve problems. New technologies can be brilliantly employed, or perniciously abused." Discuss this statement in class and find examples to illustrate its truth.

◆ Water resources piped in from a distant location may make our parks and lawns green, but could also cause a desert or climate change where the water has been taken from.

Each of these examples shows there is a tension between economic growth and environmental protection. For economic growth to take place, industries consume nonrenewable natural resources to make the goods that they hope we will buy. At the same time, they may pollute the air and water and spew out toxic wastes. Does this mean that we should simply stop making and buying new products? No, but we must ensure that we use technologies that are appropriate. They should use as few nonrenewable resources as possible. Also, they must have a minimum effect on the environment.

Some of the items we buy may have been produced overseas under very bad conditions. Some multinational companies manufacture products in Third World sweatshops where persons of school age are working for low wages in unsafe conditions. Although these young people are being paid, the amount is very small. The gulf between the poor and the very wealthy is growing. There are five times as many poor countries as there are rich countries. While it was hoped that technology would bring the developed and the developing nations closer together economically, this has not happened.

The New Technologies

Think about the buses, cars, and bicycles that we see in our streets or the stoves, refrigerators, and sewing machines that we have in our homes today, **Figure 16-9**. Have you seen antique ones that were built fifty years ago? Certainly the new ones are more streamlined, but they basically look similar and they work in much the same way. The products that were invented in the last century changed gradually over one hundred years.

Standards for Technological Literacy

Discussion
In what ways have cars remained the same over the past 100 years?

Discussion
Why do we tend to produce electricity using nonrenewable means that are bad for the environment, when renewable energies exist and will be less harmful?

Discussion
What is meant by the statement: "Often there is tension between economic growth and environmental protection"?

Enrichment
Technology itself is changing drastically. New technologies, such as genetic engineering, computers, and nanotechnology, are completely different than the technologies that preceded them. In contrast, the technologies developed during the twentieth century were mostly improved versions of earlier ideas or inventions. Examples of these include bicycles, buses, central heating, refrigerators, and the internal combustion engine.

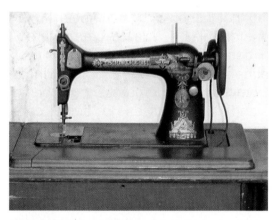
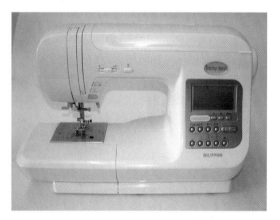

Figure 16-9 Compare the old and new sewing machines. What similarities and what differences do you see?

Technology and society: Have students access www.maketradefair.com to learn about trade in third world countries and the barriers faced by developing nations.

Vocabulary
What is the meaning of the terms *trade barriers* and *subsidies*?

Standards for Technological Literacy

Activity
Have students investigate how trade barriers and subsidies affect the ability of third world countries to reach the living standards of developed nations.

Activity
We may be able to predict new inventions by observing things that currently exist. Some inventions, however, bear no relation to anything created in the past. Have students give examples of both situations.

Discussion
List the ways we interact with computers today. How are these interactions likely to change in the future?

Vocabulary
What is the meaning of the terms *artificial intelligence, android, cyborg, nanomachine,* and *avatar*?

The emerging technologies are very different. These new technologies change much faster. Each generation is much more powerful than the previous one, and they can interact with each other. For example, computers can now be used to design more powerful computers. Information gained through biotechnology can be used to build nanobots that, in turn, will design and build improved nanobots. It is likely that in the next twenty years we will see more change than in the previous 100 years. There will be many incredible devices that could be built in the future. The important question will be: Just because it can be done, should it be done?

The technologies where we will experience the greatest change in the years to come are:

- ◆ Computers and electronics.
- ◆ Telecommunications and networking.
- ◆ Media and entertainment.
- ◆ Nanotechnology and materials.
- ◆ Energy and the environment.
- ◆ Biotechnology and healthcare.
- ◆ Transportation and cities.
- ◆ Privacy, security, and defense.

Computers and Electronics

One way to predict the future is to look at the past. Computer power has doubled every 18 to 24 months over the past 30 years. So it might seem reasonable to forecast that this doubling of power will continue.

However, there is a limit to how small silicon chips can be made. Future success may depend on finding a way to replace silicon. At present, various methods are being tried, including laser light beams, radio waves, and molecules of DNA. Much further into the future, computers will be able to reconfigure themselves using the atoms from which they were built. We might grow machines as if they are organic. If this happens, nature and technology will have fused completely.

We are approaching a time when computer power could enhance our brainpower. Our minds may seem to be large as they have 100 trillion neural connections, but they are limited as to the amount of operations that can be performed per second. In contrast, a computer can search a database with billions of records in fractions of a second. It can also remember billions or even trillions of facts perfectly while we may have difficulty remembering a few phone numbers.

Our interface with a PC will change. It will become much easier to connect using speech and handwriting, not just by using a keyboard. Your printer will be able to create three-dimensional objects, not simply two-dimensional texts and drawings. Computers will progress from being stand-alone devices to being connected together on a home network. With computer chips embedded into virtually every device, we will program any device so that it does its task automatically. Our cell phone will become a personal data access point to make contact with any machine in the home.

Useful Web site: Check the latest in computer innovation at www.ibm.com and www.apple.com.

Technology and society: What is the impact of computer gaming on young people's ability to engage in group activities?

Computers will become *wearables* for those of us who need access to lots of information on-the-job. Tiny cameras, display monitors, microphones, and cell phones could all be part of a wearable computer. Wearables have the advantage that they leave the wearers' hands free. Eventually your clothes will have all the electronics necessary to provide power and to receive and send signals embedded in them in a soft form.

There are already people who are trying out how they feel when they wear computers 24 hours a day. They call themselves "borgs," an adaptation of the word cyborg. Their computer is always on, as they are receiving information continuously from anywhere in the world. We could say that it is like having an artificial brain plugged into your own brain.

For other people, wearing a computer is just part of daily life on-the-job. Technicians on location can have easy reference to texts, manuals, and schematics. They can also have two-way communication with engineers and specialists who can provide them with technical information. In this way, they can be helped to solve complex problems quickly and efficiently, **Figure 16-10**. An army field sergeant can have constant contact with his base commander receiving maps and other visual data. Wearables will become commonplace when there are better batteries and when wiring and electronics can be woven into clothes that can be washed.

In past years, we could easily identify the difference between a

Figure 16-10 This wearable computer works with a see-through viewing screen, providing the technician with visual information. (Microvision, Inc.)

robot and a human—one was mechanical and electronic, and the other was a thinking, feeling, human being. Robots do many of the same tasks as humans, including working as assistants, caregivers, workers, and entertainers, although they have also been given dangerous tasks. These tasks can make them slaves to our needs and entirely dispensable. Ethical issues as to what we can require of a robot do not concern most of us, as robots are viewed as inanimate machinery.

The way we think about robots is likely to change, for several reasons. First, the hard plastic exterior might soon be modified by using electronically activated polymers that change stiffness and length so they shrink and grow, making robots more lovable. What would happen if we were to transplant a person's entire brain

Standards for Technological Literacy

3

Discussion
Think about a machine that has the ability to store the contents of your mind. This machine may not look like your physical body, but it could last forever. This machine could make backups of itself to ensure that your memories and skills continue on. Discuss whether this machine would really be "you."

Enrichment
Pilots in some countries can immerse themselves in computer-simulated worlds (virtual reality) using helmet-mounted visual display systems. The pilots simulate flying by manipulating two control sticks and respond to the entirely visual world they see through a helmet. In computer-generated airspace, they can experience day or night flying in any kind of weather and may encounter air-to-air attacks or air-to-ground combat.

Useful Web site: Browse some of the latest technological innovations at www.tecsoc.org/innovate/innovate.htm.

Discussion
Most personal computers (PCs) are "stand alone" devices that contain their own software programs. What would be the advantage of a terminal that is "bare bones," but would give you Internet access to its systems?

Activity
Check the Web sites of companies that make cell (mobile) phones, such as Ericsson, Kyocera, LG, Motorola, Nokia, and Samsung, to find out the features of new phones. Who would need the new features included on cell phones, other than people who enjoy owning high tech "toys"?

Resource
Transparency 16-2, *Computers, Electronics, and Telecommunications*

Examples
By 2020, it is predicted that a $1000 computer will match the power of a human brain, computers will read and understand documents, and we will port our mental processes into a computer to increase the capacity of our minds.

state and memories onto a hard drive in a robotic body when the old, arthritic, human body became unusable? Whether the robot's brain is silicone based or uses some other new material, we can be fairly sure it will not deteriorate like our own carbon-based brains. What ethical issues might arise as to how to treat the robot?

Telecommunications and Networking

The future is wireless. There will be universal wireless connectivity via hand-held devices and large numbers of low-cost, low-altitude satellites. Instead of phones and computers being separate, they will be hooked to the same Internet-based servers. Phones will be able to download full-color images, streaming video, and CD-quality audio. A cell phone could play your music collection and have a color screen and a mini-joystick for games.

When the Internet and the mobile phone finally merge, you will be able to do almost everything at a distance. Press a button and you can open your front door, turn on the oven, check bus schedules or movie listings, find the location of a family member, or even watch your younger sister at the day-care center. The device will be totally integrated with voice, data, and video. Any medium will be transferable to another medium.

The current challenge is getting all these devices to talk to each other. Until now, the solution has been

infrared beams; however, infrared only works when you can see the object. It can't go through walls or around corners. Once different types of devices can communicate with each other, the Internet will become a utility just like electricity. When connections become simple to use, we will not spend time thinking about wireless technology—it will just be there. It will be a mobile, anytime network of untethered devices. Internet users will literally surround themselves with a global network.

One goal of this interconnected world of the future is to personalize your purchase of any item from books to cars. Car companies will want you to use the Web to design customized built-to-order cars. E-books will become popular once there is copyright protection for the publishers, a large range of books, and attractive color screens available on hand-held devices. E-books will have special functions including highlighting, bookmarks, built-in dictionaries, the ability to change print size, and the ability to change from print to audio.

In North America, we are at the forefront of this change in media technology. Many other countries will also benefit from the information revolution. India has a growing class of high-tech workers and entrepreneurs. Russia and China will start to move beyond the urban areas to take the Internet to rural schools. Many South American countries have large telecommunications companies that can make relatively rapid progress in information and communications

Technology and society: Have students learn how wireless technology is changing the way people behave in public. A useful Web site to start their investigation is www.tecsoc.org.

technology. Most countries will want to promote this progress, but they will also be on guard against its harmful effects, such as replacing their own culture with one that is alien.

Many countries are concerned about the health risks of portable devices. It is feared that the microwaves from portable phones might damage human tissues by causing cell mutations or cancer. The long-term effects of prolonged use are not known. Other problems are well-known. The sound of phones in a restaurant or at the cinema is not what the patrons want to hear. Studies have shown that using a phone while driving quadruples the chances of an accident. That is why countries like Britain and Japan have already banned the use of phones by the driver of a moving vehicle, **Figure 16-11**.

Media and Entertainment

Can you imagine entering a museum and being met by a robot dinosaur that would show you around? This could happen some day. Troody is one technological advance leading to that. Troody is a robot dinosaur that is equipped with 16 joints and 36 sensors. It walks on two legs and has life-like movements. It is just one of the devices that show us how all facets of entertainment will become more and more like the real thing.

Games of the future, to be used on PCs and video consoles, will give access to alternative universes. These new worlds will mimic the real worlds in every detail. If you are not sure if you can water ski, you can try

Figure 16-11 What are the hazards of using a cell phone while driving? (Tom Severson)

it out in the virtual world first. Want to compare your driving skills with those of a racecar driver? You could compare by driving around a virtual Grand Prix racecourse. Many different Web sites will give you the opportunity to enter a virtual reality environment.

Virtual reality will not be the crude experience you find in today's arcade games or even in special rooms called Virtual Reality Caves. It will be realistic and detailed. So instead of just phoning a friend, you might meet in a virtual café in Paris and ride the elevator together up the Eiffel Tower. The difference between the virtual and the real world will become increasingly blurred over time until one is not really sure which is which!

Our leisure hours are greatly influenced by the type of clothing we wear. It may soon be possible to make shoes that act in the same manner as your foot so that they match the way your foot changes as you run and jump. There could be tiny motors embedded in them so as to give your step that extra bounce. Would you be interested in shoes or clothes that are impregnated with microorganisms that eat dirt, sweat, and body oils? You could wear the same gym suit several times without having any offensive smells.

The garments we wear will also be high-tech. Experiments are being conducted to change the usual round shape of fibers to make them oval, rectangular, or square. When this happens, it will be possible to make a garment contract or expand so that it becomes looser or tighter. It could also become warmer or cooler depending on the season. The color of a sweater could be changed to match the shirt you decided to wear that evening. Future shirts might be made that gradually discharge your daily dose of vitamins to be absorbed through the skin.

Other types of "smart clothing" will become a possibility when electronic components become even smaller. A "smart" vest, used as an undergarment, could have devices plugged into it. They would monitor heart rate, track body temperature and respiration, and count how many calories the wearer is burning. This information could be transmitted to your trainer or to your computer, which is located at home or even woven into the vest. It could be played back later to a cell phone, personal computer, or wrist monitor.

For people who need vision correction, there will be self-adjusting eyeglasses that measure the distance from the wearer to the object being observed using infrared rays. The lenses would then make the necessary adjustments so that what you see is in perfect focus. While on the

beach, the amount of sunlight allowed to penetrate the lenses will be varied, so that the sun will not blind you.

Other eyeglasses will have built-in display screens and surround-sound headphones. When plugged into a DVD player, game console, or camcorder, you will feel as if you are watching a huge screen up close. There will also be smarter TVs that will be able to recognize your program preferences and record your favorite shows automatically so that you can watch them when you want, not just when the TV station broadcasts them.

Starting soon, your favorite magazine may have tiny bar codes attached to articles and ads. With a special free scanner that one connects to a computer, you will be linked automatically to a Web site that will provide more information. This kind of tagging will become commonplace for any product. A radio tag could be linked to an Internet database to give consumers information, including whether the product is genetically modified, if it conforms to a religious diet, or where it is produced. Consumers equipped with personal digital devices could check which products meet their requirements.

Your computer can be the pathway to many other sources of information so that learning becomes a lifelong process. Learning can also become a personalized experience, as it will be available anytime, anywhere. We will need to be critical about the information we access over the Internet. In the past, library collections have been the most useful source of new learning materials. They are stocked with writings of men and women who have taken time and used critical judgment to shape their ideas. Thinking in this way is quite different from much of the spontaneous writing that takes place on the Internet. We must also remember that the information we access on the Internet usually belongs to someone. It is copyrighted material that we can use in some cases, but it does not belong to us.

With more and more technology surrounding us there will be more noise. Some devices, such as sound systems, are designed to make noise. Other noise just happens. Vehicles rumble and roar, tools whir and clatter, and gadgets beep and gong. We must be always conscious that the "noise" we find acceptable, such as that from a powerful motorbike or when we talk on a cell phone, may be annoying and stressful to someone else, **Figure 16-12**.

Nanotechnology and Materials

Developments in nanotechnology are likely to change the way almost everything, from vaccines to computers to objects not even imagined, is designed and made. Technologists are talking about creating useful materials and devices from molecules and atoms.

Nano comes from the Greek word for dwarf. When used as a prefix for a unit of measurement, it means one billionth of that unit. This technology

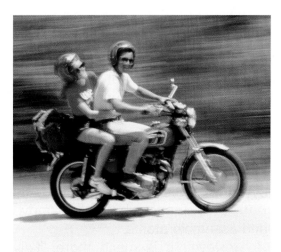

Figure 16-12 Motorcycle manufacturers produce different mufflers and baffles which can change the sound emitted by the motorcycle. Individuals often modify their motorcycles to produce a desired pitch or volume. Though this may please the rider, it may displease pedestrians.

only became possible with the development of the scanning tunnel microscope in the 1980s. With this microscope, researchers could view single atoms for the first time.

The size of a nanometer (nm) is difficult to visualize. One molecule of DNA is about 2.5 nm. One atom measures about one third of a nanometer. Just one red blood cell in your body is over one thousand nanometers and the dot at the end of this sentence has a diameter of 100,000 nm.

Imagine a machine so small it could be swallowed in a gel capsule. This machine could be programmed to seek out and destroy a particular virus. You might not be able to see this micro-machine, as it would be no larger than the diameter of a human hair. This is not science fiction. Already

Standards for Technological Literacy

Discussion
What types of devices can be identified by their sounds? Make an audiotape of sounds (household equipment, car, motor bike, hairdryer, and others) and ask students to identify each sound. Which of the sounds are annoying?

Enrichment
Nanotechnology did not emerge as an experimental science prior to the 1980s, because no tools existed to observe and manipulate individual atoms. In the 1980s, two new microscopy techniques were developed: atomic force microscopy and scanning tunnel microscopy.

Activity
Research new advancements in nanotechnology. Are there potential benefits to manufacturing, space, or transportation? What is meant by the phrase "direct assembly"?

Links to other subjects: Science—What developments, in the past 50 years, have resulted in the ability to manipulate (pick up and rearrange) atoms?

Enrichment
The gasoline used by a personal vehicle constitutes only part of the energy requirements. There are at least three components: the gasoline in the tank, the diesel in the truck delivering the gas to the gas station, and the energy used to refine the gas from crude oil.

Resource
Transparency 16-3, *Material Harvesting*

Resource
Transparency 16-4, *Manufacturing*

a nanocopter has been built. This microscopic helicopter has a single nickel blade that rotates eight times a second with rotors that are powered by a chemical in the human body.

Everything in the world is made up of molecules and atoms. A future nanomachine could dismantle anything and construct new items atom by atom. In the future, miniscule *nanobots* (nano-robots) will move and assemble atoms.

However, it is much more likely that they would use super-strong, nano-engineered materials. When used for aircraft wings and engine parts, these materials would be stronger and lighter, as well as more powerful and fuel-efficient. Paints created with nanotechnology might change color by altering the molecules on their surface. Fibers as strong as diamonds could be used to reinforce other materials or spun to make new super-strong, super-lightweight materials. It is expected that these new materials will make products that are smart, environmentally compatible, and customizable to our different lifestyles.

Future computers will also depend on nanotechnology. Today's more powerful computers chips contain roughly 40 million transistors. Through nanotechnology, that same chip would hold about one billion transistors. As these chips become more powerful, the size of components will shrink. It will then be possible to embed them in any device, even clothes that we wear. Connections may be made using nanowires made from silicon strands as small as ten nanometers in diameter.

Nanotechnology may be the most exciting and promising technology of the future, but there is always a potential that any new technology may be misused. The internal combustion engine gave us more freedom to move around, but it also increased global warming. Nuclear power produced cheap electricity and the nightmares of Chernobyl and Three Mile Island. There are already people who are warning us that self-replicating nanobots could get out of control.

What we can learn from these warnings is that it is a good idea to start public discussions early. Any new technology has the potential of bringing a whole sequence of problems and issues that we will have to deal with as a society.

Energy and the Environment

Fossil fuels, including coal, gasoline, and natural gas, are among the most widely used fuels. At the moment, they are relatively easy to extract from the ground and in abundant supply. Because they are nonrenewable resources, they will eventually be used up. In the immediate future, we need to find cheap, nonpolluting, renewable alternative sources of energy.

Both solar energy and wind energy have been around for a long time, but their use has been limited. This is partly because they depend on the right climate and advances in technology. In areas where the wind often

Useful Web site: Have students access www.roswellproof.homestead.com/debris9_nanotech.html to learn more about the properties of nanotechnology materials.

Community resources and services: Does a disposable lifestyle contribute to increasing waste in your community? Contact your local Waste Management Department to find out if the amount of waste is increasing.

blows, the number of wind farms, with dozens of modern wind generators, is increasing. Solar power has been limited in the past to projects such as powering satellites and communications towers. However, it also may provide more energy in the future. New plastic materials have recently been developed that could be used to produce inexpensive solar cells.

Several years ago it was thought that nuclear power would be the solution to providing large amounts of cheap, reliable electrical power. We now know that there are too many risks associated with nuclear power to rely on it as our principal source of power. Fuel cells may be the answer we have been seeking. They are silent, have zero emissions, and have proven their usefulness in NASA spacecraft, **Figure 16-13**. New lightweight fuel cells may soon power lawn mowers, cellular phones, and laptop computers, just to name a few devices. Whether they are used to run cars, buses, or just to make electricity, fuel cells operate by converting hydrogen to electricity without combustion. They create electricity with no environmental damage and only harmless water vapor as the by-product.

When we respect our environment and everything growing in it, we respect ourselves as we depend on each other. Unfortunately, as the population grows and more waste is produced per person, we need to find new solutions to manage our waste. In the past, landfill disposal has been the primary way of getting rid of waste.

Alternative solutions are now becoming more popular such as

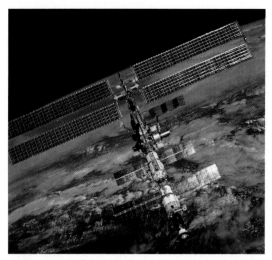

Figure 16-13 Satellites and other orbiting spacecrafts use solar cells for providing a renewable power source. Notice that the majority of the International Space Station is covered with solar cells. (NASA)

recycling, reusing, and composting. According to the Clear Air Council, in the United States, the average person creates 4.39 pounds of trash per day and up to 56 tons of trash per year. Each day, the United States throws away enough trash to fill 63,000 garbage trucks. The largest waste component is paper, followed by yard waste, food waste, plastics (including 2.5 million plastic bottles thrown away every hour), metals, and glass, in that order. If we are to avoid being surrounded by garbage, increasing the recycled portion will be the challenge and goal of the future.

Climate change might be one of the biggest challenges humanity faces this century. We need a long-term plan for meeting emission reduction targets—one including mandatory emission cuts for large factories and power plants and updated standards for more

Standards for Technological Literacy

Discussion
What does it mean to "respect our environment?"

Discussion
Does your school have a composting program? If not, how could your class start one?

Activity
What is the Kyoto Protocol? What are its emission-reduction targets? Research the arguments for and against ratifying the Kyoto Protocol. Decide for yourself whether this is an effective step toward reducing our dependence on fossil fuels.

Resource
Activity 16-6, *Design Your World— Biosphere*

Useful Web site: Have students visit http://auto.howstuffworks.com/fuel-cell.htm to learn more about how fuel cells work.

Visit the Web site of David Suzuki at www.davidsuzuki.org to learn more about how to protect our environment.

Visit the US Department of Energy Web site at www.em.doe.gov/em30/ to learn more about what is happening in waste management.

Standards for Technological Literacy

Designing and making
In what ways does protection of the environment impact the design decisions a designer makes?

Vocabulary
What does the term *green chemistry* mean?

Activity
Have students conduct a product life cycle analysis (PLCA) on a common object the class chooses.

energy-efficient buildings, homes, vehicles, and appliances. Measures to fight climate change will also involve using less energy, which in turn, will create new jobs and cut energy costs.

There is an increasing awareness that the designer of a chemical is responsible for considering what will happen to the world when the chemical is used. This is very different from the past, when little thought was given to the end result. In the past, someone else had to deal with the mess created later. The relatively new branch of industry considering the long-term impacts of new chemicals is called *green chemistry*. Green chemistry considers the total life cycle of a product, including how to reduce the amount of raw material used to make it, the amount of energy consumed in its manufacture, and any waste or harmful substances that might be emitted from the product.

What happens both upstream and downstream is important in green chemistry. Industry and individuals use chemicals in factories and homes, and this is where *upstream* effects appear. *Downstream* effects occur when the chemicals later wash into the environment and into organisms. Both the upstream and downstream effects impact many people. People work upstream to develop environmentally safe processes, and people work downstream to remove pollutants already in the environment.

An example of upstream technology was demonstrated at the 2000 Olympic Games in Sydney, Australia, which attracted 8 million visitors. Large quantities of food and mountains of plastic cutlery were served. The cutlery, however, was biodegradable, due to the invention of a starch-based, injection-molded thermoplastic. The costs of waste disposal were reduced, and the environment was not adversely affected.

A huge amount of electronic and electrical equipment is discarded each year, and the amount is growing. This presents a downstream problem, as scrap, consisting of printed circuit boards (PCBs) with microchips soldered to them, is dumped on landfill sites. Recently, processes have been developed to treat PCBs so their components, including harmful but valuable metals, can be recovered and recycled. These metals include copper, platinum, gold, tin, and lead.

Green chemistry is an emerging industry that will hopefully have long-term positive effects on our environment. The main idea of this movement is to take responsibility for the impacts of technology before technology causes any further negative impacts on our world. The following are the major principles of green chemistry:

◆ Prevent, rather than cure. The aim is to eliminate pollution by preventing it from happening in the first place. It is better not to produce waste than to have to clean it up later.

◆ Seek safer alternatives. Whenever possible, products and methods of manufacture should not be harmful to human health.

Useful Web site: Have students conduct a Web search to identify consumer products made from recycled materials. They can begin by looking at furniture made from recycled plastic bottles at www.kohestan.com/page/Kohestan/CTGY/Recycled_Plastic_Furniture and pencils that Remarkable Pencils Ltd. makes from recycled coffee cups.

They also should not create other kinds of dangers, such as explosions or fire hazards.

◆ Minimize energy needs. Processes should be designed and selected for the greatest energy efficiency with a minimal amount of energy wasted.

◆ Use renewable resources. Raw materials should be renewable ones derived from growing plants, rather than irreplaceable materials, such as petroleum.

◆ Consider the end result. Use materials that are biodegradable (easily digested by microorganisms in the environment) so, at the end of their use, they do not create mountains of solid waste.

Each individual can play a part in recycling and composting. We also have to ensure that our government is fully responsible for proper waste management of dangerous materials. These wastes can range from highly radioactive and toxic wastes. Radioactive waste may come from nuclear fuel, **Figure 16-14**. Other hazardous waste includes chemicals, explosives, solvents, and pesticides; all of which could damage the environment and enter our food supply.

Biotechnology and Health Care

During the next twenty years, major achievements are likely in the biotechnology revolution. Changes will be experienced in combating diseases

Figure 16-14 Specially trained workers suited in radiation protection suits deal with the handling, clean up, and disposal of toxic and radioactive materials. Nuclear power plants employ many people trained in radiation protection. (Michael Rennhack)

and increasing food production. Unfortunately, many of these developments will be very costly, especially in the medical field, so they will be primarily available only to citizens of wealthier nations. Also, some biotechnologies will continue to be controversial for moral and religious reasons.

As you learned in Chapter 14, every living thing has a set of genes. In the future, it will be possible for individuals to know their personal genetic codes (DNA profiles) and have them recorded on microchips. What if this chip could then be implanted into your brain and linked

Useful Web site: Learn about Gray Goo, Green Goo, Nanocopters, and the work at Cornell University by visiting www.lamontanita.com/docs/newsletterarticles/2003/Jun2003/nanotechnology.htm.

Enrichment
Although the US Patent Office has granted the right to patent genes, it is still controversial. Those who disagree with the decision point out that when the first anatomists identified the use of various organisms in the human body, they were not granted title to those organisms. So, why should genome scientists be able to claim ownership of genes just because they identified their functions? They note that it is possible to patent a process for making aluminum, but you can't patent the aluminum.

Vocabulary
What does the term *cloning* mean?

directly to a computer? This would allow the possibility of operating technological devices, such as lights, the stove, or the heater, simply through thoughts. If you think of your genetic make up as being like a loose-leaf book, then each individual page might be available to be moved or transferred to another book. *Genetic engineering* aims to select certain genes and move them, often from one species to another, in order to produce a specific result. Examples might include vitamin fortified vegetables, crops that are less affected by frost, or even more intelligent humans. The DNA sequence in each of us took billions of years to evolve. No one can guess exactly what effect these engineered changes will have in the future.

The United States Patent and Trademark Office has stated that genes can be patented. The argument that genes are part of nature and not an invention was rejected. This cleared the way for companies to *clone* genes in the laboratory and define their function. Companies have already been awarded thousands of gene patents for genetic sequences, and intense international opposition to these patent rights is mounting. In effect, they are patenting life!

Identifying this genetic information has only recently been possible due to advances in computing. The amount of gene data of all living things, from bacteria to humans, is incredibly large. Only computers are able to organize and store billions of data.

Not only will computers organize information but also with three-dimensional models, they can combine molecules and make predictions as to how they could interact.

Knowledge of the human genome is also making it easier to test for genetic diseases. Like many technologies, there are both good and bad sides to this possibility. It will be advantageous for doctors to be able to find out that an individual might be prone to diseases, such as cancer or heart disease. With this knowledge, people could make lifestyle plans. They could select the right diets or exercise levels to increase their chances of remaining healthy. This knowledge, however, could also result in genetic discrimination if an institution or individual treats a person, family, or group differently because tests show they have a reduced life span or tendency toward certain behavior. Some questions that might have to be answered in the future include the following: Should people be forced to undergo genetic testing? Should third parties have access to the genetic information of others?

Another controversial field of medical technology is cloning. Future debates will not be on cloning animals, which is largely accepted, but on cloning people. Since animals were first cloned, it took nearly eight years for the first human embryos to be cloned. In 2004, what many people feared and others hoped for happened.

Useful Web site: Have students access the Human Genome Project Information page at www.ornl.gov/sci/techresources/Human_genome/home.shtml and explore the site to find out about the project.

Technology and society: What ethical issues would have to be resolved if scientists and technologists were able to clone humans?

A further controversial development in biotechnology involves research on stem cells. These cells, taken from the hollow ball of cells that make up a human embryo, have the potential to grow and develop into new tissues or replace worn out or diseased parts of our bodies. This area of biotechnology is in a very early stage of development, but its potential for medical use is enormous. Due to the fact that human embryos are used, it is very controversial.

Huge changes will also be made in biomedical engineering. "Smart" materials will make possible new surgical procedures and systems including improved organic and artificial replacement parts for humans. We have already started down this path. There is an artificial leg, known as a smart leg, which calculates speed and position 50 times per second so that the wearer can walk with a normal gait. We often use transplants right now, but the wave of the future will be artificial heart valves and other body parts.

The Internet is starting to revolutionize home health care, as it is a means to transmit data from a home to a database containing a person's medical file. For example, imagine a toothbrush with a biosensing chip that checks your blood sugar and bacteria levels while you are brushing your teeth. This brush would come with a holder that could transmit data. Such a medical device would provide convenience to the user and reduce health care costs. It would give timely warnings of illness so early treatment can occur at a time when it will do the most good. This type of device would also make it possible for people to have a greater responsibility for their own health.

Technology is also helping millions of older people stay in touch with their doctors and nurses. Telemedicine connections can enable a nurse to ask a patient health questions and receive information. Devices used in telemedicine might include blood pressure cuffs, stethoscopes, and thermometers, as well as television monitors and cameras. They can even include heart monitors connected to personal computers that allow people to track their own heart rates and other vital information and transmit them to health care providers.

Presently, we have cochlear implants to give hearing to deaf people. A retina implant is being developed to give sight to those who are blind. It is predicted that within twenty years neural implants may be available to anyone who wants to improve perception, memory, and logical thinking. These implants will also plug us directly into the World Wide Web. They may enable us to have virtual reality experiences with real or simulated people without requiring any other equipment.

This kind of "embedded computing" could also be used to keep you healthy. Instead of taking two aspirins, a future patient might swallow microscopic processors that could release medications but also send wireless reports on a patient's response.

Useful Web site: Have students access the Human Genome Project Information page at www.ornl.gov/sci/techresources/Human_genome/home.shtml and explore the site to find out about the project.

Technology and society: What ethical issues would have to be resolved if scientists and technologists were able to clone humans?

Useful Web site: Have students access the two Web sites www.biohealthmatics.com and www.advcomms.co.uk/telemedicine/definition.htm to investigate the advantages and disadvantages of telemedicine.

These embedded computers could one day monitor blood sugar, cholesterol, or oxygen levels in the body.

Would you swallow a camera-in-a-pill? Italian researchers have designed a new camera-in-a-pill that, when swallowed, can move or stop, according to what a doctor wants to see. This camera-in-a-pill is radio controlled and could be used to replace a colonoscopy, a procedure that can be very uncomfortable.

Surgical procedures are being automated. Medical robots will be used to perform surgery. They will be steadier than a surgeon's hands and less invasive. The surgeon will sit in a control booth and direct the robot arms to insert needles, hold tissue, dissect, suture, and do everything with great precision through tiny access holes, **Figure 16-15**. When it comes time to recuperate, virtual nurses will be able to make home visits several times a day.

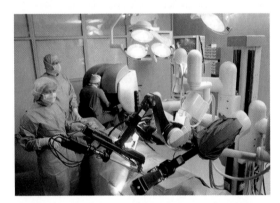

Figure 16-15 Medical robots are currently being developed by biotechnology medical companies to perform surgeries. Robots that exist today hold and direct cameras or position tools and keep them steady for special types of surgery. (Intuitive Surgical, Inc.)

Our health can also be improved through better nutrition. The debate will continue as to whether biotechnology foods are the best way to improve nutrition. Genetically modified (GM) foods will attempt to combine such foods as a banana that can deliver an effective vaccine for hepatitis and a type of rice that has added beta-carotene to fight blindness in people living in developing countries. People who are in favor of GM foods point to the fact that the world's population will increase. There will be less arable land available. Water and minerals will be reduced. This means that more food must be produced with fewer resources. GM foods can be grown that are more nutrient-dense on the same amount of land.

Biotechnology is probably the most controversial of the developing technologies. Never before in our history have we had the means to bypass species boundaries and create new creatures by combining genetic information from any plants or animals. We are at a point where living things are no longer seen as birds or bees but as bundles of genetic information.

There are many questions that will need answers. Will the world's gene pool become the patented intellectual property of a few multinational corporations? Will we experience increased genetic uniformity, a narrowing of the gene pool, and a loss of genetic diversity? New transgenic crops and animals are designed to grow faster, produce greater yield, and withstand more severe weather conditions. Will they

Useful Web site: Go to www.worldheart.com and www.heartsaver.com to learn about a heart pump for the left ventricle (the chamber that pumps re-oxygenated blood to the far reaches of the body).

Access www.nasaexplores.com/lessons/02-030/9-12 to learn more about prosthetic limbs.

cause other species to become extinct? Who will make the decisions as to what is a good gene that should be added to the gene pool or what is a bad gene that should be eliminated? Will third world countries be able to afford these technologies? Will the gap between the "haves" and the "have-nots" increase?

There are many ethical, moral, and religious questions to consider. Should couples be able to select the gender of their baby? Should they be able to select other features such as eye or hair color? We may be approaching the time when a child can be grown completely outside the mother's womb. Is this acceptable? Finally, how do our ethics and morals compare with those of other countries? Currently, there are approximately 100 deceased bodies frozen in American cryogenics facilities waiting to be brought back to life if the right technology should be developed. France has outlawed the practice of freezing corpses.

Transportation and Cities

Transportation is essential to our lives and is central to the development of our economy. In the twentieth century, road construction became the solution to most transportation problems. Many people are now realizing that when more roads are built, there are even more traffic jams, and the places where we live become more scattered over a larger area. The average commuter spends 500 hours a year in a car, much of this time in traffic jams surrounded by stress, noise, and discomfort.

Road construction provides only temporary relief to traffic congestion. New roads are filled to capacity within four or five years, and then people think of building even more roads. No matter how many are built, it is never enough! We cannot keep moving farther and farther out from the city, otherwise towns will join together right across the country.

Each time new highways are added, more destruction of the countryside and more water and air *pollution* occur, **Figure 16-16**. Nitrogen pollution from motor vehicle exhaust harms water quality in rivers and streams. Historical buildings, wetlands, and farms are destroyed. Simply building more roads can be bad for taxpayers—bad for our health and for the environment.

In the twenty-first century, we must learn to respect our environment more and everything growing in it, because we depend on each other. Sprawling suburbs adjacent to cities take away land from nature. Since homes are separated further from shopping centers, we drive everywhere. The land that we build our houses and shopping centers on were often used to grow food or provide homes for animals. The streams with smallmouth bass and the hickory trees with squirrels scampering up their branches are lost forever.

What can be done? One simple way is to provide more facilities for pedestrians, cyclists, and human transporters (HTs), **Figure 16-17**. Cycling and walking are cost-effective

Standards for Technological Literacy

3 **5** **14**

Discussion
Should society embrace genetic modification for medical reasons only and disallow "designer babies" with good looks and a high IQ?

Enrichment
The British Nutrition Foundation (BNF) and the Design and Technology Association (DATA) classify smart foods as foods with novel molecular structures (such as fat replacers), functional foods (such as cholesterol-lowering spreads), meat analogues (vegetable protein and tofu), encapsulation technology (such as flavors in confectionery), and modern biotechnology (such as modified soya bean).

Activity
Should biotech food be labeled? Research both sides of the issue and state your opinion.

Activity
Select one of the 10 questions listed on this page about biotechnology and research an answer.

Resource
Transparency 16-7, *Biotechnologies and Medical Technology*

Useful Web site: Investigate topics related to biotechnology at www.whybiotech.com, www.policynut.com, and www.agr.gc.ca.

Investigate smart foods at www.nutrition.org, www.foodfuture.org.uk, www.foodforum.org.uk, and www.foodingredientsonline.com.

Discussion
Are there too many cars on the road in major urban areas? If we were to establish a system of car sharing, how would it work?

Enrichment
Private transportation is extremely expensive. In Miami, for example, the costs of transportation exceed those of shelter. How can we reverse our love of car transportation? In Singapore, people pay a fee according to the area, the hour, and the level of pollution when entering certain areas. In London, drivers are assessed each time they enter the city.

Activity
Investigate the Commission on the Future of Transportation, Virginia. What problems does it identify and what solutions are proposed? A useful Web site to start the investigation is www.selcga.org.

Figure 16-16   A—How much time do your parents spend commuting by car? Can you think of better systems of transportation? (Bud Smith) B—The SkyTrain, in Vancouver, is an advanced rapid transit system providing driverless urban transit. (Bombardier)

ways to reduce demand for new roads and to provide alternative ways for people to move about. Over one quarter of all trips that we make are less than one mile and could easily be made on foot or by bicycle. Town planners may soon realize that when pedestrians can cross streets easily, sidewalks are continuous, and local streets are connected, there is a reduction in vehicular travel.

For people who must travel one hour or more to their work, better mass transit systems will be built. Special collection points will be designed that are convenient to use. Mass transit, however, will not become popular until it is cheaper, faster, and more convenient than car travel.

Personal cars may play a smaller role in our future lives, but they will not disappear. They may change in quite radical ways. One possible way is to have dualmode cars. These will travel in the normal way on local streets but will move automatically on "guideways." These automated highways will be separated from regular roads. A driver wishing to enter the guideway system will drive into an entry point and shut off the motor. Next, the exit number will be entered into the keypad. After computers check the vehicle's identification, the system will accelerate the car to guideway speed and merge it with the other traffic. Dualmode cars will be electric or fuel cell driven.

Revitalizing the inner cities leads to people moving back. After years of neglecting our cities, governments will start to recognize their importance

Useful Web site: Research ways to improve the cost and quality of housing by using advanced technologies at www.pathnet.org.

The Web site www.carsharing.net addresses car sharing as a viable alternative to car ownership.

NASA's Morphing Project aims to broaden the shape of aircraft wings, simulating the way birds spread their wings. More information available at www.nasa.gov.

Browse www.thinkmobility.com to see how cars of the future may be designed.

Figure 16-17 Walking, cycling, and battery-powdered vehicles are environmentally friendly forms of transportation. (Segway LLC)

once more. This will lead to neighborhood redevelopment, **Figure 16-18**. People will discover that it is more convenient to live closer to their jobs. They will find that it is fun to live in the glitter, bustle, and lights of the city.

Changing cities so that they are not only places to work but also to live is a

challenge. One way is to separate vehicles from pedestrians so that people can be more relaxed. Creating pedestrian malls where vehicles are banned from the inner core can do this. People can then stroll around the streets to see a favorite artist, visit a museum, sit at a sidewalk café, or have a choice of many theaters. Another way is to improve neighborhoods by rearranging streets so that there are fewer through streets, making local traffic slower. At the same time, speeds would be increased on other roads around the district so as to encourage traffic to take alternative routes.

Many cities have continually grown in concentric circles around existing cities. An alternative is to create satellite cities. These would be smaller cities separated from the bigger one and from each other by green space. Each satellite city should have employment for its citizens, leisure time activities, its own central core, and services. In this

Discussion
Discuss the advantages of living in older, established cities that incorporate ethnic diversity, cultural offerings, and places of historical interest.

Enrichment
Fuel cell cars are a reality, but most are currently in the testing phase. The Honda FCX has a 45% energy efficiency rating, compared to 18% for gasoline vehicles. The fuel cell is located in the floor of the vehicle and the compressed hydrogen tanks are under the rear seat. The trunk houses the chemical battery, leaving virtually no additional space. The engine compartment holds three radiators for cooling, the electric motor that drives the wheels, and the powertrain control module or computer brain.

Figure 16-18 Many older buildings in cities are being renovated and used for purposes other than their original purpose. This old house is now used as an art gallery. (Jack Klasey)

Activity
Many plans have been proposed to avoid urban sprawl, including protective green belts, the Lineal City, the Garden City, and satellite towns. Make sketches to show how each of these plans can be arranged.

Useful Web site: Find out what organisms may exist on Mars and what plans there are to study life on Mars at www.nasa.gov.

Check the specifications and capabilities of the Segway transporter at www.segway.com.

Community resources and services: Arrange for a person in charge of city planning to meet with the class and discuss his/her work.

way, we can avoid having a commuter society in which people work in central cities and live in dormitory communities.

Master plans should be created for all cities especially new ones. These plans will include optimum levels of development and restrict communities from going beyond these levels. In the last century, downtown cores became the place to work. In the future, we will see more industrial parks set up outside the cities in technological hubs located near shopping centers. In this way, the jobs will come to the workers rather than the other way.

Privacy, Security, and Defense

You have learned about how new technologies will provide access to data and the ability to control devices from a distance. Most people want this convenience, especially access to the Internet, but if you can do it there is also the possibility that someone else can. Personal messages to someone close to you, medical records, and bank statements would all be things we want to keep private. Song producers and authors of written texts also want to protect their data when it is transmitted.

All information transmitted by electronic means could potentially be accessed with the right software by spy agencies, hackers, or even your parents. Furthermore, data may be altered or deleted by those who spread a computer virus, whether that data belongs to you, your government,

or a space agency. We are entering the age of anywhere, anytime, anybody surveillance.

Personal surveillance devices are being used to protect private homes. A video camera might be hidden behind the face of a clock or the eye of a teddy bear. However, outsiders may access other cameras installed in your home. Using the right software, hackers could turn on small cameras attached to PCs. If your PC has an "always-on," high-speed connection, someone else could use it as an Internet server.

Security systems will become big business in this century. Security systems are increasingly in use to record your identity when you enter a building. An identification card can give access to offices, parking lots, and elevators, **Figure 16-19.** Video cameras capture your image as you enter. Governments and private industries have given themselves new rights to citizens. Police in every major country have asked for new powers to intercept Internet messages.

You will increasingly be watched when you use electronic devices. At the ATM, the bank machine might record time-and-date-stamped video images of you as you withdraw cash. When you use a cell phone, the phone company, your employer, or your parents will soon be able to tell which wireless cell you are in to a few yards. At home on your PC, your Internet service provider (ISP) can watch where you surf. Any member of your family could also check which Internet sites you were accessing by using

Useful Web site: Investigate the future of space travel at www.sciam.com.

Research methods proposed for ensuring that your personal information remains private and confidential at www.privcom.gc.ca, www.privacy.org, and www.efc.ca.

Technology and society: Contact a marketing research agency and ask for a copy of one of their surveys. The most common survey probes the habits of ordinary people (such as what radio stations they listen to, what they eat for breakfast, what type of deodorant they use, and much more).

Figure 16-19 Even small businesses are equipping themselves with security cameras and monitors used to record the faces of possible thieves and criminals. (Tom Severson)

special software downloaded from the Net. Your ISP also has all the information related to your e-mails, the contents and source of every Web site you visited, and the length of time you spent there. Police could get a warrant to seize and examine the ISP logs.

Techniques for security and defense can be very sophisticated or

very simple. Two examples of elaborate systems are Echelon and IT-ISAC. Echelon is an eavesdropping network operated by five English-speaking countries (Australia, Canada, New Zealand, the UK, and the US). It was originally developed to monitor foreign satellite communications. It has now evolved into a system capable of intercepting phone and e-mail messages across the globe.

IT-ISAC was formed by Microsoft Corporation in an alliance with eighteen other large companies dealing in software and hardware. The goal is to share information about cyber-attacks. Members who discover a new cyber-threat, such as a new strain of virus or a break-in method that foils existing electronic defenses, will be able to send detailed warnings to other group members. It is expected that other private alliances will be formed covering different sectors of industry.

Some smaller devices are equally important for our daily living. *Biometrics* is a fast-growing technology that uses a characteristic of your body to identify you. Biometric security systems work by storing a digitized record of some unique human feature. Systems use fingerprints, voices, retinas, faces, and even body odors. When a person wishes to enter a facility the system scans that person's characteristic and attempts to make a match with its records.

Police can use cameras to scan the faces of people and compare them to hundreds already stored in its database. These may be people in a

Technology and society: Access www.tecsoc.org/natsec/natsec.htm to see how the government is trying to protect citizens.

Standards for Technological Literacy

3

Discussion
The various tracking devices available can locate almost any object. Why would you want to locate fire trucks, hydro crews, your family car, or a cell phone user?

Reflection
Give examples from your readings in this book to illustrate the following statement: "Science studies what is; technology creates what has never been."

Enrichment
Security is just one component of privacy. The consequences of accessing information can range from the merely annoying, such as junk/spam email, to stealing someone's identity for financial gain.

Vocabulary
Define the term *biometrics* and describe how it is used for identification and security.

Vocabulary
Define the terms *cyberwarfare* and *cyberterrorism*.

Resource
Activity 16-7, *Futures Wheel*

crowd or someone who has just been arrested. Facial recognition technology is likely to be used more in the future. If certain individuals have been arrested for shoplifting in one store, they could be put in a database in another store in a different city. When trouble is anticipated at a sports event, fans entering the stadium might be scanned and matched with those who have previous records.

How much access to personal data do we want government and business to have? This is an important question because the trend toward collecting even greater amounts of information will continue. Advertising and marketing companies are searching to find people's spending habits. Governments are concerned about national security issues. The answer has to respect the right of an individual to privacy versus the legitimate needs of others to gather information.

Are you prepared to give up some of your privacy? What would you say if you were offered a free cell phone as long as you agreed to receive advertisements? These ads might be on the screen. Also, your cell phone might beep when you pass certain stores where there were items on sale that might interest you.

It may seem that we are being given something for free, but there is always a price to pay. In our future world, you will be called on to decide whether you want to pay the price of having your privacy invaded by so-called free products or whether you prefer to pay for alternatives.

In the future, you will be a citizen with the responsibility for making decisions about the use of new technologies. Having read this book, you should be able to make more informed decisions.

Chapter 16 Review
Learning from the Past, Predicting the Future

Summary

Emerging technologies will change a lot faster and be more powerful than previous ones. Our interface with computers will change as microchips become embedded in virtually all devices so that they can be programmed to operate automatically. Telecommunications will become wireless. By using a mobile phone and the Internet, most things could be done at a distance. In the world of entertainment, virtual reality will give access to alternative universes that will increasingly appear real.

Nanotechnology operates in the world of molecules and atoms. Future machines and components could be made in microscopic size. New applications will be seen in health care, materials, and computers. With regard to energy sources, it is unclear when petroleum will be replaced as our main fuel. It may depend on when a cheap replacement is found as well as advances in fuel cell technology. In the meantime, we must do our part in reducing our waste, reusing what we can, and recycling.

Major achievements are likely in biotechnology when individual genes are isolated. This will remain the most controversial of the new technologies. Transportation is an essential part of our lives; however, the insistence on driving high-powered automobiles may change as we run out of space to build roads in and around cities. Finally, privacy, security, and defense will remain top priorities both for our bodies and for personal data.

Modular Connections

The information in this chapter provides the required foundation for the following types of modular activities:

◆ Life Skills

◆ Practical Skills

◆ Introduction to Technology

◆ Development of Technology

◆ Explorations in Technology

◆ Technology Issues

◆ Technology and the Environment

◆ Technology Before I Was Born

◆ Technology Transfer

◆ Assessing Technology in News

◆ Inventions and Innovations

Test Your Knowledge

Write your answers to these review questions on a separate sheet of paper.

1. What do you consider to be the most important invention in each of the following time periods? Tell why each invention is important.
   A. 500,000 to 10,000 B.C.
   B. 10,000 B.C. to 1 B.C.
   C. AD 1 to 1400.
   D. 1401 to 1700.
   E. 1701 to 1850.
   F. 1851 to 1900.
   G. 1901 to 1945.
   H. 1946 to present.

2. What statement can be made about the number of objects invented during different periods throughout history?

3. Name 10 technical objects that have been invented in your lifetime.

4. Technical objects are designed to help humans, but they sometimes have negative side effects. One example is shown in the following chart. Copy and complete this chart, adding three more examples.

Technical Object	Intended Effect	Negative Side Effect
DDT	Destruction of unwanted insects	Kills birds, pets and wildlife and is poisonous to humans.

5. List items that you throw away that could be recycled.

6. Biotechnology is defined as _____.

7. Genetic engineering is defined as _____.

8. A superconductor is a material _____.

9. Describe how your life would change if computers could completely understand and reproduce the human voice.

10. Name five technical objects you believe will be invented 20 years from now. Describe (a) how each will work, (b) who will benefit from their use, and (c) what potential problems may occur as a result of using them.

Apply Your Knowledge

1. Make a model of a tool used prior to A.D.

2. Select what you consider to be the most important invention in each of the historical periods in Figure 16-5. State the reasons for your choice.

3. Make a list of five products that have been invented since you were born. Is the number of products invented each year increasing or decreasing? Explain your answer.

4. Choose one product and discuss its impact, both positive and negative, on (a) society and (b) the environment.

5. Give one example of "leading edge" technology in each of the following areas. If possible find examples that are NOT described in the textbook.
   A. Computers and electronics.
   B. Telecommunications and networking.
   C. Media and entertainment.
   D. Nanotechnology and materials.

E. Energy and the environment.

F. Biotechnology and healthcare.

G. Transportation and cities.

H. Privacy, security and defense.

6. From a science-fiction book, comic, TV show, or movie in which events occur in the future, identify three technical objects or systems that do not exist today. Describe how each one operates.

7. The following list shows how change took place from the beginning of the twentieth century to the end. Find or draw a picture that illustrates each of the changes. For example, in the first one you could have a picture of a steam train and another of an electric train.

◆ Steam to electric

◆ Rural to urban

◆ Cottage industries to factories

◆ Cart tracks to railways

◆ Personal contact to telephone

◆ Horse to car and plane

◆ Brick and stone to steel and aluminum

◆ General store to department store

◆ Gunpowder to atomic bomb

◆ Natural materials to synthetic materials

8. Visit Web sites where e-books can be downloaded. Which books would be interesting for you to read? To start this activity, you could access a computer search engine and type in a keyword search using the words "electronic books."

9. School composting programs have gained in popularity. Discuss what methods could be employed by your school to compost organic waste.

10. Locate an area in your town where new roads have been built. Find out what was demolished to make way for the roads. Ask people what they think of the destruction and construction. Do they miss any parts of the old buildings, farmland or other features that were once part of their environment? Was there a different way to solve "the problem"?

11. There have been many proposals for planning new towns. These include Garden City, Lineal City, Radial City, and Satellite Towns. Investigate at least one of these plans. Draw a diagram to show how the town would be arranged. Discuss its advantages and disadvantages.

12. Can someone use another person's information without that person's permission? Can organizations collect, use or disclose personal information about you without first telling you their intentions and obtaining your consent? How clear must their explanations be as to what they intend to do with the information? To find the answers to these questions access a computer search engine and type in a keyword search using the words "electronic privacy" and "computer privacy".

13. China is expanding industrially at a phenomenal rate and is, at the same time, experiencing huge environmental problems. These include polluted air and water and problems with garbage disposal. People in North America have had experience dealing with these same problems for several decades. As China's industries continue to expand and more cars are produced, the problems will only increase. Assume you are to give advice to the people in China, based on the information you have learned by reading this text. What advice would you give them to reduce pollution in each of the areas mentioned?

14. What advances can you predict for new materials that will reduce environmental damage?

15. Visit the Internet, and identify three smart materials and their uses.

16. Historically, some very successful societies seem to have disappeared. Among them are the Easter Islanders and the Norse Greenlanders. How did these ancient societies interact with their environments to cause the demise of their communities? What might they have done differently that would have helped them survive? Are there lessons to be learned by us in the twenty-first century?

17. Use the Internet to investigate new uses for hydrogen power.

18. Video games have been around since Atari invented the first one in 1972. What have been some of the most important developments since then?

19. Draw a picture of what you think a robot might look like 50 years from now.

20. Many people like to play the type of computer game that simulates the real world. Why do you think they like to simulate the real world, when they can experience what is real?

21. The following Web site lists the "Top 20 Greatest Engineering Achievements of the 20th Century": www.nspe.org/media/mr1-top20.asp. Place them in order from what you consider to be the most important to the least important.

Last Word

The technology of early humans was simple. Today's technology is often complex. A new technology may cause change throughout the world. We must learn to evaluate the effect that any new product or service has on our lives and our environment. While many technologies have beneficial results, we are increasingly aware that humans are capable of making mistakes. These mistakes may be costly for us and for future generations.

In the future, you will be a citizen with the responsibility for making decisions about the use of new technologies. Having read this book and completed the activities, you should be able to make more informed decisions.

Glossary

A

Abutments: the supports where a bridge arch meets the ground; they resist the outward thrust (push) and keep the bridge up.

Acid rain: an environmental problem that occurs when rain mixes in the air with pollutants from burning fossil fuels and other industrial emissions.

Acoustical properties: properties of a material that control how it reacts to sound waves.

Adobe: a mixture of clay and straw used as a structural material in the southwestern United States.

AGV: automated guided vehicle; robots that move by following a set path.

Alloy: a material that is a mixture of metals.

Alternating current: electron flow that reverses direction on a regular basis.

Alternative solutions: different ways of solving a problem.

Ampere: the unit that is used to measure the amount of current.

Antibody production: a branch of biotechnology in which researchers attempt to isolate and produce antibodies (substances the body's immune system produces to attack and kill bacteria) effective against specific diseases.

Arch bridge: a type of bridge in which the compressive stress created by the load is spread over the arch as a whole.

Artificial intelligence (AI): a technology that makes it possible for a computer to learn, solve problems, and reason.

Artisan: an individual responsible for every step in producing a finished product; also, someone skilled in a trade; a craftsman (for example, woodworking).

Assembly line: a continuous process that is used to quickly produce a large number of identical items by assembling parts in a planned sequence.

Atom: the smallest possible particle of matter that still retains the characteristics of that element.

Audio communication: messages that can be heard.

Audiovisual communication: ideas and messages that can be both seen and heard.

Automation: the use of computers or automatic machines to control machine operations and make a product.

B

Balance: in design, the arrangement of mass equally (or appearing to be equal) over the space used; the three types of balance are symmetrical, asymmetrical, and radial.

Batter boards: boards that support the lines set up to locate the building so excavating can begin for the foundation.

Battery: a single package containing several cells connected together to produce a DC current.

Beam: a horizontal structural member, usually used to support floor or roof joists.

Bending: a method of shaping sheet material by folding it like a sheet of paper.

Binary: a simple electrical signal code used by computers using only two signals, on and off (written as 1s and 0s).

Biocompatible: a characteristic of biomaterials, described as when the material is not rejected by the surrounding body tissue.

Biomass energy: energy from plants that can be burned or processed as fuel.

Biomaterial: special materials that are able to function in intimate contact with living tissue.

Biometrics: a security technology that uses a characteristic of your body to identify you, including fingerprints, voice, retina, face, and even body odors.

Bioprocessing: the use of microorganisms to break down and purify organic wastes.

Biotechnology: the use of living organisms to make goods or provide services.

Bit: an on or off signal, known as a binary digit, in the binary code used by computers.

C

CAD: a method of making drawings using a computer; CAD stands for "computer-aided design."

Cantilever bridge: a type of bridge in which a beam is capable of supporting a load at one end when the opposite end is anchored or fixed.

Capacitor: a device designed to store an electrical charge, consisting of two metal plates (conductors) separated by an insulator (dielectric).

Carbohydrates: a food group in MyPyramid that includes bread, cereal, and rice.

Career: an occupation or way of making a living.

Cartographer: a mapmaker.

Casting: a method of shaping parts or products by pouring liquid material into a mold.

Catalyst: an agent that enables and speeds the process of chemical reactions.

Cell: a single dc power source consisting of a positive electrode, a negative electrode, and an electrolyte.

Cells: the basic units of all living organisms.

Ceramic: a material used for making pottery, bricks, and other products; originated from the Greek word *Kermos*, meaning "burnt stuff."

Chemical energy: energy locked away in different kinds of substances; these are often released by burning.

Chemical joining: a method of fastening joints by using chemicals such as glues, adhesives, solvents, and cements.

Chemical properties: properties of a material that affect how it reacts to its surroundings.

Chip: a tiny flake of a substance called silicon, covered with microscopic electric circuits that is used in computers; also known as a microprocessor.

Chiseling: a technique used to shape material by cutting away the excess with a chisel and mallet.

CIM: the use of computers to carry on a whole range of manufacturing functions and processes; abbreviation for computer-integrated manufacturing.

Circuit: a closed path in which current can travel; three types of circuits: series, parallel, series-parallel.

Cloning: a process of growing an identical individual from a single somatic cell using asexual reproduction.

Closed-loop system: a system that includes a feedback device to provide control.

Coatings: materials applied to a surface.

Communication: the process of exchanging information or ideas between two or more living beings.

Communication system: a means of transmitting and receiving information between two or more points.

Communication technology: the transmitting and receiving of information using technical means, such as special devices for transmitting and receiving information and special mediums for transmission.

Composite: combinations of different materials.

Compression: a squeezing force.

Computer: a machine that can be programmed (given a set of instructions) and that can store a vast amount of data; machines that can process data rapidly using many mathematical operations.

Computer numerical control (CNC): a system in which machine movements and operations are controlled by a computer program.

Conduction: movement of heat energy by passing from molecule to molecule in a solid.

Conductors: materials that will allow an electric current to flow easily.

Construction line: thin, faint lines used to start a drawing.

Contrast: an obvious difference between two things.

Convection: movement that occurs when expanded warm liquid or gas rises above a cooler liquid or gas.

Corrosion: the result of a chemical reaction in which a material is changed by its environment.

Crossbreed: the process of breeding two plants or animals of different breed, species, or variety.

Cure: the setting of a molded object.

D

Dairy: the food group that consists of milk, yogurt, and cheese.

Data: any set of symbols that represents information.

Degree of freedom: the term used to describe each joint, or direction of movement, in a robot arm.

Design brief: a statement that clearly describes what problem a design must solve.

Design process: a careful and well thought-out means of solving a problem by working through a number of steps.

Designer: a person who creates and carries out plans for new products and structures.

Designing: generating and developing ideas for new and improved products and services that satisfy people's needs.

Development: in sheet metal working, it is a pattern.

Diesel engine: an engine in which air is squeezed to very high pressure and very high temperature inside the cylinder, diesel fuel is injected and spontaneous ignition occurs.

Diode: a device that allows current to flow in one direction only; most commonly used as rectifiers to change alternating current to direct current.

Direct current: current that does not change direction in an external circuit.

Distribution lines: another name for transmission lines.

Division of labor: a system where each person is assigned one specific task in the making of a product.

DNA: deoxyribonucleic acid, the basic building blocks of life.

Drafting: drawings of ideas to be communicated, often called the "language of industry."

Drilling: a process used to make holes in wood, plastic, metal, and other materials.

Ductility: a material's ability to be pulled out under tension.

Dynamic load: a load on a structure that is always changing.

E

Efficiency: good use being made of energy, time, and materials.

Elasticity: the ability to stretch or flex but return to an original size or shape.

Electric circuit: a closed path around which electrons can move.

Electric current: the flow of electrons in a circuit.

Electric motor: a machine that changes electrical energy into mechanical energy.

Electrical energy: the movement of electrons from one atom to another.

Electrical properties: properties of a material that determine whether it will or will not conduct electricity.

Electrical system: the circuits that carry electricity in a product or for light, heat, and appliances throughout a home.

Electromagnet: a magnet that is energized by an electric current.

Electron: a negatively charged particle that is part of an atom.

Electronics: the use of electrically controlled parts to automatically control or change current in a circuit.

Electroplated: coated with nickel, chromium, copper, silver, or gold.

Elements of design: the things you see when you look at an object, including line, shape and form, texture, and color.

Engine: a machine composed of many mechanisms that converts a form of energy into useful work.

Entrepreneur: a person who organizes, manages, and assumes the risks of starting a business.

Ergonomics: the study of how a person, the products used, and the environment (our surroundings) can be best fitted together.

Estuary: a meeting place between a sea and a river.

Expert system: a computer program that uses a large base of data to solve complex problems in a specific area of knowledge.

External combustion engine: an engine that burns fuel outside itself.

Exude: to slowly discharge or diffuse.

F

Factory: a building in which products are manufactured.

Fatigue: the ability to resist constant flexing or bending.

Fats, oils, and sweets: the food group that consists of butter, honey, and candy; this food group should be eaten in moderation.

Feedback: in manufacturing, it is information provided to a manufacturer by consumers after they try samples of a product; in communication, a receiver's response to a source's question or statement.

Ferrous: any metal or alloy that contains iron.

Fiber optics: the technology of sending light-encoded information through a glass fiber strand for uses in communication.

Filing: a process used to smooth a material (usually metal) by removing small amounts from the surface with a toothed tool called a file.

Finishing: a process that changes the surface of a product by treating it or placing a coating on it.

Flexible Manufacturing System: a grouping of machine tools, controlled by a computer program that can perform a series of operations on a single manufactured part. By reprogramming the computer, the tools can make a different part.

Floor plan: a drawing that shows the arrangement of rooms in a building.

Footing: the lowest portion of a building's foundation; the foundation wall rests on top of the footing.

Form: a three-dimensional representation of an object.

Forming: changing the shape of sheet material (often through the use of a mold).

Foundation: the footing and foundation wall that support a building and spread its load over a large ground area.

Friction: a force that acts like a brake on moving objects.

Fruits: the food group that consists of apples, bananas, and oranges.

Fuel cell: a type of vehicle that mixes hydrogen from an on-board tank with oxygen from the air to power electric motors for turning the wheels.

Function: what an object does or how it works; a functional object or product solves the problem described in a design brief.

Fuse: to join two things as if by heat or chemical reaction.

G

Galvanized: coated with zinc.

Gasoline engine: engine in which a mixture of air and gas is ignited by an electric spark, pushing down pistons to drive a crankshaft in a rotary motion.

Gear: a rotating wheel-like object with teeth around its rim; they are used to transmit force to other gears with matching teeth.

Generating station: a facility that uses an energy source to operate turbines and produce electricity.

Generators: machines that produce electricity when turned by an outside force, such as a turbine.

Genetic engineering: a technology used to alter the genetic material of living cells.

Genome: the genetic material of an organism.

Geothermal energy: energy produced by hot rocks changing underground water to steam.

Gravitational energy: the natural force of attraction created by an object that draws other objects toward itself.

H

Hardness: the ability of a material to resist cuts, scratches, and dents.

Hardwood: trees that have broad leaves, which they usually lose in the fall. They are also known as deciduous trees.

Harmony: a condition in which chosen colors or designs naturally go together.

Heat energy: energy that occurs as the atoms of a material become more active.

Heat joining: a process that melts the material itself or a bonding agent (such as solder or another filler metal) to secure a joint.

Heating system: the furnace and associated ducts or pipes used to distribute heated air or water to the different rooms of a home

HPV: human powered vehicle; any vehicle that is powered solely by one or more humans.

Human genome: the genetic material of humans.

Hydraulics: the study and technology of the actions and reaction of liquids at rest and in motion.

Hydroelectricity: electricity produced by using the energy of moving water.

Hydrogen: one of the two elements in ordinary water; when separated from water, it is a very combustible gas that can be used as a fuel.

I

Inclined plane: a simple machine in the form of a sloping surface or ramp, used to move a load from one level to another.

Industrial Revolution: the change from a single artisan performing all manufacturing operations to dividing the production process into specialized steps in which machinery replaced handwork.

Information: a collection of words or figures that have meaning or that can be combined to have meaning.

Information technology: the equipment and systems used in storing, processing, and extending human ability to communicate information.

Insulation: material used in the walls and ceiling of a building to help control heat loss (in winter) and heat gain (in summer).

Insulators: materials that will not carry an electric current.

Integrated circuit: a single electronic component that replaces a whole group of separate components (one integrated circuit may contain the equivalent of about 1,000,000 separate components).

Internal combustion engine: an engine that burns fuel inside itself.

Invention: a creation or design of something that did not exist before.

Investigation: the act of researching and gathering information; often used in the design process.

Isometric paper: paper that is used for sketching. It has a grid of vertical lines and other lines at 30° to the horizontal.

Isometric sketching: the simplest type of picturelike drawing; it is used to share an idea or record ideas for further discussion.

J

Jet engine: an engine that sucks in air at the front, squeezes it, mixes it with fuel, and ignites it; this creates a strong blast of hot gases that rush out of the back of the engine at great speed.

Jig: a tool useful when bending sheet metal.

Joist: horizontal member of a house framework that supports the floor.

K

Kinetic energy: the energy an object has because it is moving.

L

Laminating: a process that involves gluing together several veneers (thin sheets of wood) to form a strong part.

Landscaping: designing the exterior space that surrounds a home.

Larynx: the upper part of the trachea, which contains the vocal cords; also known as the voice box.

Laser light: a form of radiation that has been boosted to a high level of energy, producing a strong, narrow beam of light.

Lever: a simple machine, consisting of a bar and fulcrum (pivot point), that can be used to increase force or decrease effort needed to move a load.

Light energy: energy from the sun that travels as a wave motion. Also called *radiant energy*.

Line drawings: objects and ideas represented by lines and shapes.

Lines: design elements that describe the edges or contours (outlines) of shapes; they show how an object will look when it has been made.

Linkage: a system of levers used to transmit motion.

Load: the weight, mass, or force placed on a structure; in an electrical circuit, any current-using device.

M

Machine: a device that does some kind of work by changing or transmitting energy.

Magnetic: having magnetic properties; able to be attracted by a magnet.

Magnetic properties: properties of a material that determine whether it will or will not be attracted to a magnet.

Magnetism: the ability of a material to attract pieces of iron or steel.

Malfunction: to fail to operate as planned.

Malleability: a material's ability to be forced (compressed) into shape.

Marking out: measuring and marking material to the dimensions shown on a drawing.

Mass production: the process of making products in large quantities to reduce the unit cost.

Mechanical advantage: in a simple machine, the ability to move a large resistance by applying a small effort.

Mechanical energy: energy of motion, which is often associated with or caused by a machine.

Mechanical joining: the use of physical means, such as a bolt and nut, to assemble parts.

Mechanical properties: the ability of a material to withstand mechanical forces.

Mechanism: a way of changing one kind of effort into another kind of effort.

Medium: the agent over which information is sent.

Methane: a gas that is produced when microorganisms eat plant or animal matter in garbage.

Microelectronics: the process by which the switches and circuits of processors and their accessories are made incredibly small.

Microprocessor: a tiny silicon wafer, covered with microscopic electric circuits, sometimes called a "computer chip."

Model: a full-size or small-scale simulation (likeness) of an object, used for testing and evaluation.

Modular construction: a building system that involves basic room units of different sizes and shapes that can be combined on site.

Molding: a method of making shapes by forcing liquid material into a shaped cavity.

Molecule: a group of atoms.

Moment: the turning force acting on a lever, determined by the effort times the distance of the effort from the fulcrum.

Motor: an electrically powered device with a rotating shaft that provides the kinetic energy to operate machines and other devices.

N

Nanobot: a very tiny robot; part of a new developing technology.

Nonferrous: metals or alloys that do not have iron as their basic component.

Nonrenewable energy: energy from sources that will eventually be used up and cannot be replaced; examples are coal, oil, and natural gas.

Nuclear energy: energy produced from the nucleus of atoms.

Nuclear fission: energy produced by the splitting of atomic nuclei; this process gives off heat, which is used to produce steam to run turbines and generate electricity.

Nuclear fusion: the same energy source that powers our sun and the stars. It requires high temperatures. So far, technologists have not been able to produce it on earth.

O

Object line: line that is darker and thicker than a construction line; they are used to show the outline of an object.

Ohm: unit of measurement of electrical resistance.

Ohm's Law: the law of electricity stating that voltage equals the current times the resistance.

On-site transportation: types of vehicles that transport people and materials from one spot to another in a defined location.

Opaque: an optical characteristic of a material that allows no light to pass.

Open-loop system: a system that is not controlled through use of a feedback device.

Optical fibers: thin strands of glass that can carry laser-light-coded messages over long distances.

Optical properties: a material's reaction to light.

Orthographic projection: kind of drawing that shows each surface of the object "square on" (at right angles to the surface).

Overload: a condition that occurs when devices on an electrical circuit demand more current than the circuit can safely carry; a cause for a fuse or circuit breaker to interrupt current flow.

Ozone layer: a protective shield in the earth's atmosphere that absorbs much of the sun's ultraviolet radiation.

P

Parallel circuit: a circuit that provides more than one path for electron flow.

Pathogen: a bacteria or virus causing a disease.

Pattern: a design used to reproduce a shape many times.

Perspective sketching: sketching method that provides the most realistic picture of objects.

pH: a measure of acidity.

Photosynthesis: the process used by trees and plants to change water and carbon dioxide into carbohydrates using light as an energy source.

Photovoltaic: light induced electricity by dislodging electrons.

Physical properties: properties that give a material its size, density, porosity, and surface texture.

Pier: a structure used to support the center of a bridge.

Planing: a process for smoothing wood that uses a sharp blade to remove very thin shavings.

Plasticity: the ability to flow into a new shape under pressure and to remain in that shape when the force is removed.

Plasticizers: various chemicals that act as internal lubricants for thermoplastics.

Plumbing system: the means used to bring in clean water and dispose of wastewater in a home.

Pneumatics: the study and technology of the characteristics of gases.

Pollution: the presence of dangerous or unwanted materials in the air, water supply, or other areas of the environment.

Polymer: a chainlike molecule made up of smaller molecular units; the scientific name for plastic.

Polyurethane: a form of plastic.

Post and lintel: the simplest form of a framed structure, with horizontal framing members (beams) supported by vertical members (posts).

Potential energy: energy that is stored until it is released (used).

Power: the rate at which work is done or the rate at which energy is converted from one form to another or transferred from one place to another.

Prefabrication: a system of building in which components, such as wall framing or roof trusses, are built in a factory, rather than on the job site.

Pre-production series: a small number of samples made up by a manufacturer to obtain feedback through testing with consumers.

Pressure: the effort applied to a given area.

Primary cell: a device that chemically stores electricity; its electrode is gradually consumed (used up) during normal use, and it cannot be recharged.

Primary colors: the three most important colors of the spectrum – red, yellow, and blue.

Primary material: a natural material; a material that exists in nature.

Primary sector: the portion of the economy concerned with gathering and processing raw materials.

Problem: a situation or condition that can be solved or improved through the application of technology.

Production system: a system involving the five basic operations of making a product: designing, planning, tooling up, controlling production, and packaging and distributing the finished product.

Proportion: the relationship between the sizes of two things.

Protein: the food group that consists of meat, fish, and nuts.

Protocol: an agreement.

Prototype: the first working version of the designer's solution to a problem.

Pulley: a simple machine in the form of a wheel with a groove around its rim to accept a rope, chain, or belt; it is used to lift heavy objects.

R

Radiant energy: another name for light energy.

Radiation: the particles or rays thrown off by unstable atomic nuclei; also, one of the methods by which heat travels.

Raw materials: materials that come from nature in one form or another.

Receiver: in communication, something or someone who gets a message.

Recycling: the process of separating different types of wastes, and reclaiming those that can be reused.

Reinforced concrete: concrete in which steel rods have been embedded to increase the concrete's resistance to tension.

Renewable energy: energy from sources that will always be available, such as the sun, wind, and water.

Replenish: to give back.

Resistance: opposition to the flow of electricity.

Resistor: an electrical device that controls the amount of current flow through a circuit by making that flow more difficult.

Rhythm: a quality or feeling of movement, provided by repeating patterns.

Robot: a computer-controlled device that can be "taught" to perform various production or material handling operations.

Roof truss: a structure that forms a framework to support the roof and any loads applied to it.

S

Safety: the practice of working in a way that will avoid injury or damage.

Sawing: the process of cutting material with a tool that has a row of teeth on its edge.

Scale drawing: a drawing that is larger or smaller than the object by a fixed ratio.

Schematics: diagrams of electrical/electronic circuits using symbols.

Science: the field of study that is concerned with the laws of nature.

Scientists: persons whose field of study is the laws of nature.

Screw: a simple machine that is an inclined plane wrapped in the form of a cylinder.

Secondary cell: a device for storing electrical energy as a chemical; it can be charged, discharged, and recharged.

Secondary color: a color obtained by mixing equal parts of two primary colors.

Secondary sector: the sector of the economy that changes raw and processed materials into useful products.

Semiconductor: materials that allow electron flow only under certain conditions; they have some characteristics of conductors and some characteristics of insulators and are widely used in electronic equipment.

Series circuit: a circuit that provides only one path for electron flow.

Shank: the unthreaded part of the screw below the head.

Shape: a two-dimensional representation of an object.

Shear: a multidirectional sliding and separating force.

Shearing: a process used to cut thin material with a scissors-like tool, called shears or snips.

Short circuit: a condition that occurs when bare wires in an electrical circuit accidentally touch, causing more current to flow than the circuit can safely carry; a cause for a fuse or circuit breaker to interrupt current flow.

Sintered: formed by heat.

Site: the land on which a building is to be built.

Softwoods: coniferous trees that retain their needlelike leaves and are commonly called evergreen trees.

Solar energy: energy from the sun, the most important of the alternative sources of energy.

Sound energy: a form of kinetic energy that moves at about 1100 ft. (331 m) per second.

Source: in communication, the starting point of message.

Space shuttle: a reusable vehicle, developed by NASA, which is capable of operating in space as well as in the Earth's atmosphere.

Static load: a load that is unchanging or changes slowly.

Stays: cables that support a bridge deck from above.

Storage: information that is understood and stored for later use.

Strain energy: the energy of deformation, possessed by materials that tend to return to their original shape after being stretched or compressed.

Structure: something that encloses and defines a space; also, an assembly of separate parts that is capable of supporting a load.

Strut: a rigid structural member that is in compression.

Style: an individual, unique way of designing an object or solving a problem; also, an identifiable set of common elements that give a common appearance to related objects.

Subfloor: a covering over joists that supports other floor coverings.

Subsystem: a smaller system that operates as a part of a larger system.

Superconductivity: a quality of a material that will allow it to conduct electricity without resistance.

Supersonic: faster than the speed of sound. Used to describe an aircraft that flies at such speeds.

Suspension bridge: a bridge in which the deck is suspended (hung) from hangers attached to a continuous cable, which passes over towers and is anchored to the ground at each end.

Symbol: simple picture or shape used as a means of communicating without using words.

Synthetic: a manmade material; usually something which was originally natural that is treated and changed chemically (usually by heat or chemicals) to become a synthetic material.

System: a series of parts or objects connected together for a particular purpose.

T

Technologist: a person who uses the laws of nature to solve problems by designing and making products or structures.

Technology: the knowledge and process of using the laws of nature, to solve problems by designing and making products or structures.

Telecommunications: the sending and receiving of information over a distance, making use of a variety of technological devices and systems.

Tension: a pulling force.

Tertiary sector: the economic sector that is concerned with the servicing of products and with providing services that add to the personal comfort, pleasure, and enjoyment of people.

Texture: a design element that determines the way a surface feels or looks.

Thermal energy: another name for heat energy.

Thermal expansion: expansion of matter caused by heat.

Thermal properties: properties that control how a material reacts to heat or cold.

Thermoplastics: materials that can be repeatedly softened by heating and hardened by cooling.

Thermosets: a material that assumes a permanent shape once heated.

Thrust: pushing power, based on the principle that for every action there is an equal and opposite reaction.

Tidal energy: energy produced by the rise or fall of tidal water in the oceans.

Tie: a rigid structural member that is in tension.

Torsion: a twisting force.

Torque: a measure of turning effort.

Toughness: the ability to resist breaking.

Transformer: an electrical device that changes voltage and amperage and consists of a pair of coils and a core.

Transgenic: a term used to describe organisms whose genome has been changed by transferring genes.

Transgenic engineering: the process of changing the genome of an organism by rearranging the genes of an organism, often done by added genes.

Transistor: an electronic device with three terminals and many applications, such as a signal amplifier or an electrically controlled switch that can turn an electric current on and off.

Translucent: an optical characteristic of a material that allows some light to pass.

Transmission lines: lines that transport electricity to wherever it is needed.

Transparent: an optical characteristic of a material that allows all light to pass.

Transportation: the process of moving people or material from a point of origin to a point of destination.

Transportation system: an organized set of coordinated modes of travel within a set area that are usually run by the local government; an internationally coordinated collaboration of businesses utilizing jets, planes, helicopters, boats, barges, semi trailers, trucks, and trains to deliver passengers, foods, and other goods.

Truss: a structural element made up of a series of triangular frames.

Turbine: an energy converter that includes a large wheel with blades turned by water, a jet of steam, or hot gases.

Turbofan: a type of jet engine in which the gas stream drives a large fan located at the front of the engine. Thrust is as great as a simple jet but the engine is quieter.

Turboshaft: an engine that uses a stream of gases to drive turbine blades connected to a shaft which, in turn, is connected to rotors or propellers.

V

Vegetables: the group of foods including broccoli, spinach, and carrots.

View: in orthographic projection, a drawing of the front, top, or right side of an object as seen looking at right angles to the surface.

Virtual reality: an artificial environment provided by a computer that creates sights, sounds, and, occasionally, feelings through the use of a mask, speakers, and gloves or other sensory-enhancing devices.

Viscosity: a liquid's resistance to flow.

Visual communication: ideas in a form that we can see.

Volt: a unit of measurement of electrical pressure.

Voltage: a measure of electrical pressure.

Voltaic cell: a simple cell consisting of copper and zinc rods immersed in a weak sulfuric acid solution.

W

Wall studs: vertical framing members to which gypsum board, paneling, or other wall coverings are attached.

Watt: the unit used to measure the work performed by an electric current.

Watt's Law: the law of electricity stating that the power equals the voltage times the current or the current squared times the resistance.

Wave energy: energy produced by the movement of ocean waves.

Wearable: a computer that one wears that consists of cameras, monitors, and microphones.

Wedge: a simple machine that consists of two inclined planes placed back-to-back.

Wheel and axle: a simple machine that is a special kind of lever; effort applied to the outer edge of the wheel is transmitted through the axle.

Wind energy: one of the oldest sources of kinetic energy, used for centuries to grind grain and pump water; today, wind is being used to spin wind turbines and generate electricity.

Work: measurement of the amount of effort needed to change one kind of energy into another; the result of a force applied on an object and the object's movement in the direction of the force applied on the object.

Index